OVIAN MEDICINE

fluids - pg. 33

Visit our website at **www.mosby.com**

MANUAL OF 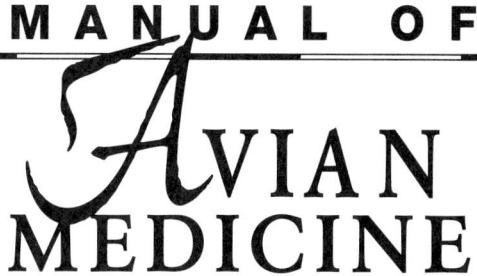 AVIAN MEDICINE

GLENN H. OLSEN, DVM, PhD
Patuxent Wildlife Research Center
National Biological Survey
Laurel, Maryland

SUSAN E. OROSZ, DVM, PhD
Department of Comparative Medicine
College of Veterinary Medicine
The University of Tennessee
Knoxville, Tennessee

with 96 illustrations

St. Louis   Baltimore   Boston   Carlsbad   Chicago   Minneapolis   New York   Philadelphia   Portland
London   Milan   Sydney   Tokyo   Toronto

Dedicated to Publishing Excellence

Publisher: John Schrefer
Executive Editor: Linda L. Duncan
Senior Developmental Editor: Teri Merchant
Project Manager: Linda McKinley
Senior Production Editor: Julie Eddy
Book Design Manager: Judi Lang

Copyright © 2000 by Mosby, Inc.

All rights reserved. No part of this publication may be reproduced or transmitted in any form or by any means, electronic or mechanical, including photocopy, recording, or any information storage and retrieval system, without permission in writing from the publisher.

Permission to photocopy or reproduce solely for internal or personal use is permitted for libraries or other users registered with the Copyright Clearance Center, provided that the base fee of $4.00 per chapter plus $.10 per page is paid directly to the Copyright Clearance Center, 222 Rosewood Drive, Danvers, Massachusetts 01923. This consent does not extend to other kinds of copying, such as copying for general distribution, for advertising or promotional purposes, for creating new collected works, or for resale.

Mosby, Inc.
A Harcourt Health Sciences Company
11830 Westline Industrial Drive
St. Louis, Missouri 63146

Printed in the United States of America

International Standard Book Number
0-8151-8466-2

Contributors

Louise Bauck, DVM, MVSc
Director of Veterinary Services
Hagen Avicultural Research Institute
Pointe Claire, Quebec, Canada

Patricia Bright, DVM, MS
Virginia-Maryland Regional College
  of Veterinary Medicine
University of Maryland
College Park, Maryland

Carol J. Canny, DVM, Dipl ABVP
Avian Resident
Department of Comparative Medicine
University of Tennessee
Knoxville, Tennessee

James W. Carpenter, MS, DVM
Professor, Exotic Animal, Wildlife,
  and Zoo Animal Medicine Service
Department of Clinical Sciences
College of Veterinary Medicine
Kansas State University
Manhattan, Kansas

Tracy L. Clippinger, DVM
Assistant Clinical Veterinarian
Wildlife Conservation Society
Bronx, New York

Victoria L. Clyde, DVM
Milwaukee County Zoo
Milwaukee, Wisconsin

Carolyn Cray, PhD
Assistant Professor and Associate
  Director
Division of Comparative Pathology
University of Miami
Miami, Florida

Michael D. Doolen, DVM
Oakhurst Veterinary Hospital
Oakhurst, New Jersey

Thomas M. Edling, DVM
Veterinary Services Manager
Kaytee Avian Research Center
Chilton, Wisconsin

Donita L. Frazier, DVM, PhD, Dipl ABVT
Associate Professor of Pharmacology
  and Toxicology
University of Tennessee,
College of Veterinary Medicine
Knoxville, Tennessee

Blake Hawley, DVM
Senior Marketing Manager
Hill's Pet Nutrition, Inc.
Lawrence, Kansas

Darryl J. Heard, BSc, BVMS, PhD, Dipl ACZM
Assistant Professor, Wildlife
  and Zoological Medicine
University of Florida
Gainesville, Florida

Michael P. Jones, DVM, Diplomate ABVP-Avian Specialty
Assistant Professor, Avian &
  Zoological Medicine
University of Tennessee, College of
  Veterinary Medicine
Knoxville, Tennessee

Victoria J. Joseph, DVM, Dipl ABVP-Avian Practice
Bird and Pet Clinic of Roseville
Roseville, California

Nancy E. Love, DVM, Dipl ACVR
Associate Professor, Radiology
College of Veterinary Medicine
North Carolina State University
Raleigh, North Carolina

Ted Y. Mashima, DVM
Projects Director
National Association of Physicians
 for the Environment
Bethesda, Maryland

Sharon Patton, MS, PhD
Professor of Parasitology
College of Veterinary Medicine
University of Tennessee
Knoxville, Tennessee

Simon R. Platt, BVM&S, MRCVS
Assistant Professor,
Neurology/Neurosurgery
College of Veterinary Medicine
Department of Small Animal Medicine
University of Georgia
Athens, Georgia

Christal G. Pollock, DVM
Resident in Zoo and Avian Medicine
University of Tennessee
College of Veterinary Medicine
Knoxville, Tennessee

Patrick T. Redig, DVM
The Raptor Center
University of Minnesota
St. Paul, Minnesota

Tracey Ritzman, DVM
Staff Veterinarian, Avian and Exotic
 Pet Medicine
Angell Memorial Animal Hospital
Boston, Massachusetts

**April Romagnano, PhD, DVM, Dipl ABVP
(Avian Practice)**
Staff Veterinarian
Avicultural Breeding and
 Research Center
Loxahatchee, Florida

David J. Rupiper, DVM
Capri Plaza Pet Clinic
Tarzana, California

Robert E. Schmidt, DVM, PhD, Dipl ACVP
Zoo/Exotic Pathology Services
College of Veterinary Medicine
University of California
Davis, California

Michael Taylor, DVM
Service Chief, Avian/Exotic Medicine
Ontario Veterinary College
University of Guelph
Guelph, Ontario, Canada

**Thomas N. Tully, Jr., DVM, MS, Dipl
ABPV (Avian Practice)**
Associate Professor, Section Chief
Bird, Zoo, and Exotic Animal
 Medicine
School of Veterinary Medicine
Louisiana State University
Baton Rouge, Louisiana

**Michelle Curtis Velasco, DVM, Dipl ABVP
(Avian Practice)**
Fleming Island Pet and Bird Clinic
Orange Park, Florida

Liz Wilson, CVT
Parrot Behavior Consultant
Levittown, Pennsylvania

Preface

With the rapid expansion of avian medicine and surgery over the past 2 decades, a number of excellent texts have been published to support the growing needs of practitioners. For example, the first edition of Petrak's *Diseases of Cage and Aviary Birds* proved a forerunner for later works. However, not since Steiner and Davis published their course material from the 1970s has there been a text targeted to the dual needs of avian clinicians in companion animal practice and students of avian medicine at veterinary colleges. Both audiences can benefit from a text emphasizing problems that clinicians and students are likely to encounter and that presents clear information about proper diagnosis and treatment.

In writing and editing this *Manual of Avian Medicine,* we have been conscious of this dual need and have developed this book to address these needs. We have chosen an outline format not only because of its familiarity to students at veterinary colleges but also for its power to quickly communicate the information needed to treat avian patients. The format is also familiar to both of the editors and many of our contributors who have taught as faculty at veterinary teaching institutions.

The avian clinician in private practice will find that we have chosen an unconventional format for many of the chapters, starting first with the symptoms presented by the avian patient. We hope this problem-oriented format will guide the busy clinician quickly to the information needed to properly diagnose and treat the patient. This feature will be especially important for clinicians in companion animal practice for which avian patients are a small but important part of practice.

We have emphasized psittacine and the small passerine species seen in companion animal practice when writing the chapters but have mentioned other species when and where appropriate, especially those species clinicians working in wildlife rehabilitation may encounter. In general the diagnostic and therapeutic recommendations given in the text for the companion bird species are also applicable to other avian species.

The *Manual of Avian Medicine* has been a group effort, and we the editors sincerely thank our friends and colleagues who have generously helped us by contributing the chapters and sharing their wealth of information. In addition, we want to thank the various editors at Mosby who have helped us bring this project to fruition: Paul W. Pratt, VMD, Linda Duncan, Teri Merchant, and Julie Eddy. Others who have helped with parts of this work and whom we wish to thank include Marie Maltese, Jennifer Kodak, Naya Brangengerg, Betsy Cagle, Shelia Hatcher, Eric Bergman, and those in the Avian and Exotic Animal Medicine Group at The University of Tennessee.

We dedicate this work to a real pioneer in avian medicine, Dr. Ted Lafeber, a man whom at one time or another taught the two editors and inspired us to continue striving as clinicians for our avian patients. Dr. Lafeber has done more to promote the health of avian patients than anyone we know. Through personal example and his active involvement in veterinary continuing education over 3 decades, he has been an advocate of the pet bird as a loved companion requiring quality veterinary care. We hope that this text will continue in his footsteps and provide students and practitioners with the knowledge to treat our avian friends.

Glenn H. Olsen
Susan E. Orosz

Contents

1 **Diagnostic Workup Plan** 1
Susan E. Orosz

2 **Supportive Care and Shock** 17
Michael P. Jones and Christal G. Pollock

3 **Dyspnea and Other Respiratory Signs** 47
Thomas N. Tully, Jr.

4 **Abnormal Droppings** 62
Louise Bauck

5 **Vomiting and Regurgitation** 70
Victoria J. Joseph

6 **Abdominal or Coelomic Distention** 85
Michelle Curtis Velasco

7 **Avian Dermatology** 95
April Romagnano and Darryl J. Heard

8 **Behavior Problems in Pet Parrots** 124
Liz Wilson

9 **Neurologic Signs** 148
Simon R. Platt and Tracy L. Clippinger

10 **Seizures** 170
Tracy L. Clippinger and Simon R. Platt

11 **Straining and Reproductive Disorders** 183
Michael D. Doolen

12 **Embryologic Considerations** 189
Glenn H. Olsen

13 **Problems of Neonates** 213
Glenn H. Olsen

14 **Avian Toxicology** 228
Donita L. Frazier

15 Ophthalmic Disorders 264
Patricia Bright

16 Avian Endocrinology 313
Carol J. Canny and Christal G. Pollock

17 Problems of the Bill and Oropharynx 359
Glenn H. Olsen

18 Avian Nutrition 369
Blake Hawley, Tracey Ritzman, and Thomas M. Edling

19 Imaging Interpretation 391
April Romagnano and Nancy E. Love

20 Parasitism of Caged Birds 424
Victoria L. Clyde and Sharon Patton

21 Endoscopic Diagnosis 449
Michael Taylor

22 Avian Anesthesia 464
Darryl J. Heard

23 Limb Dysfunction 493
Glenn H. Olsen, Patrick T. Redig, and Susan E. Orosz

24 Soft Tissue Surgery 527
Glenn H. Olsen

25 Necropsy 542
Robert E. Schmidt

26 Formulary 553
David J. Rupiper, James W. Carpenter, and Ted Y. Mashima

27 Blood and Chemistry Tables 590
Carolyn Cray

MANUAL OF
AVIAN MEDICINE

[Handwritten notes at top:]
History –
animal – type, origin, length ownership
environment – cage type, hygiene
cage mates, time in cage,
other animals, toxins, toys?

Diagnostic Workup Plan

Susan E. Orosz

I. Diagnostic Workup Plan

A. History

The history and physical examination provide most of the information the clinician needs to make a diagnosis. The history is one way that the avian veterinarian can learn more about the patient and the relationship between the patient and client.

1. Animal
   a. Species: Does the owner know the species? Is the bird a hybrid?
   b. Origin: Is it domestically raised? How was it raised—in a nursery with a large number of birds or in a home setting with just a few birds?
   c. Length of ownership: Does the current owner know anything about the previous owners and the patient's previous history?
2. Environment
   a. Cage description and size: Are any heavy metals or hazards present? Is the cage an adequate size?
   b. Time patient is caged: Is the bird kept in other locations? What are the conditions of these locations?
   c. Cage hygiene: How often are the cages, perches, and toys cleaned? How often are they disinfected? What types of cleaners or disinfectants are used?
   d. Cage mates: If cage mates are present, how do the birds interact and socialize?
   e. Animal contact: Are any other pets or animals present?
   f. Toxin exposure: Is the patient exposed to smoke or other possible sources of airborne toxins such as Teflon and cigarette smoke? Other types of toxins the bird is exposed to are important depending on physical examination findings, particularly with respiratory and neurologic signs.

g. **Environmental enrichment:** Are toys present? Are they the appropriate size? Are heavy metals or other possible toxins or foreign bodies present?
3. **Diet**
   a. **General health:** What does the bird eat compared with the food items that are provided?
   b. **Percentage of individual food items:** Does the owner give a commercial food pellet to the bird? The clinician may discover that the bird only eats sunflower seeds.
   c. **Food preference:** What kind of food does the bird prefer? What does the owner feed the bird, and of those items, what does the bird select?
   d. **Food storage:** How are the food items stored? Are pellets kept in the freezer or are they left out? How long is it between purchases of food items? The time between purchases may be critical with some vitamins in commercialized pellets or hand-feeding formulas if the food is stored for long periods.
   e. **Food preparation: How is the food prepared?** Preparation, timing, and temperature of the food are critical in neonates that are hand fed. The adult bird may become infected with *Salmonella* bacteria when fed scrambled eggs or other temperature-dependent foods that are left out too long.
   f. **Water:** Where is the water coming from? Bacterial or parasitic contamination can originate from the water source. Well water should be checked for pathogens.
   g. **Bowl hygiene:** How often are the bowls cleaned? What types of cleaners are used? What types of bowls are used? The bowl type is important when heavy metal exposure is possible.
4. **Problems**
   a. **Previous problems:** Toys are essential for environmental enrichment. Often cages have decorative bent metal pieces that contain lead, zinc, or both. Toys should be checked to ensure they are free of heavy metals and appropriate for the type of bird owned. The toys should be made of natural materials such as pinecones, nontoxic branches from unsprayed trees, seed heads, and cardboard paper rolls.
      (1) Hygiene: The clinician can evaluate cleanliness by requesting the owner to bring the bird and its cage to the hospital. The clinician should recommend easy ways for the owner to keep the cage clean. Putting newspaper on the bottom of the cage is an inexpensive option, and the owner can assess the health status of the bird by changing the newspaper daily. Food cups should be shallow and broad and cleaned daily. The cage, perches, and accessories should be cleaned as often as required to minimize fecal contamination. The cage, perches, and bowls should be rinsed thoroughly to prevent exposure to cleaning-product residue.
      (2) Client-animal bond: Is the bird bonding well or poorly to the owner? What is the relationship of the bird to the owner?

> Signs/symptoms
> – continuation of previous problem
> – progression
> – how long?
> – treatment
>   other vets, med. from store?

Commonly, owners are more observant with well-bonded birds.

(3) **Attitude:** Is the bird relaxed and exhibiting its normal behavioral characteristics, including vocalizations? Is the bird fluffed, ruffled, or sleepy with its head tucked over its back? These clues are important in the determination of its health.

(4) **Conformation:** To detect abnormalities, the clinician must know the normal conformation of various species. Clues to help the clinician diagnose disease include displacement of contour feathers, unusual body curves, abnormal posture, and a drooped wing or wings.

(5) **Movement:** The bird should be observed in its cage as part of the physical examination for attitude, conformation, and movement. The clinician should look for signs of disequilibrium and notice grip on the perch, stance, and use of legs and feet, including a shifting grip. A tail bob at rest suggests respiratory disease.

b. **Current problem:** What signs and symptoms does the owner notice currently? Is this a continuation of a previous problem?

c. **Progression of the problem:** How long has the current problem been observed? How observant is the owner? One owner may be very perceptive and able to detect slight changes, whereas another may have failed to observe previous symptoms until the bird is on the bottom of the cage. In the history, the latter owner will report the illness as an acute event, whereas it may be more chronic.

d. **Treatment for the problem:** What treatment has the bird undergone previously or currently? The clinician should tactfully inquire about medications being used that the owner may have obtained from another veterinarian or pet store and their frequency of use, concentration, and amount administered.

5. **Systems**

a. **Appetite and thirst:** Observant owners will notice small changes in appetite and thirst. Sometimes the owner who feeds the birdseed mistakes the hulls for the actual seed and assumes the bird is eating. An owner who provides food in large food bowls may not notice a decrease in consumption as readily as an owner who feeds small quantities more frequently. A bird's preferences for food items may change, which may reflect a change in breeding, weaning, taste patterns, or weather. Birds with health problems often consume large quantities of grit, which causes secondary effects.

b. **Behavior:** The clinician needs to determine normal behavior of the bird and the owner's perception of any changes. Some birds with systemic illnesses are quieter, fluffed or more ruffled, and sleep more than usual. Others display hyperexcitability, seizures, and increased activity. For example, seizures are common in lutino cockatiels, whereas pionus parrots make a sniffing sound when they are stressed.

The clinician should obtain a detailed history of the bird's feather picking and behavioral problems. Pubescent psittacine birds and weaning birds may also exhibit behavioral changes (see Chapter 8). Regurgitation, for example, may be a sign of systemic illness or a courtship behavior directed at an object or the owner.

    c. Coughing and sneezing: Birds rarely cough or sneeze, and these symptoms suggest respiratory tract involvement. Head shaking, which is more common in cockatiels, is another symptom and is associated with upper respiratory tract involvement.

    d. Oculonasal discharge: Discharge often accumulates on the feathers just rostral to the nares and may be serous or mucopurulent. Oculonasal discharge may also appear on the carpus. Birds may wipe the discharge with their carpi, and the mucus tends to stick to the feathers in that area.

    e. Voice: The clinician should inquire about the quality of the bird's voice and frequency of vocalization. Primary or secondary respiratory tract involvement is associated with changes in sound quality or the lack of vocalizations.

    f. Regurgitation: The expulsion of food from a bird's crop is called *regurgitation*. Regurgitation may secondarily cause aspiration pneumonia and is often a sign of a systemic illness, frequently of gastrointestinal origin (see Chapter 5). However, regurgitation can be a normal behavior. Some psittacine species, including budgerigars, cockatiels, and macaws, regurgitate their food, which is a common courtship behavior. Adult columbiformes regurgitate crop milk from the epithelial lining to feed their squabs.

    g. Droppings: Uric acid, urine, and fecal material are present in normal bird droppings.

       (1) White uric acid is produced in the liver and secreted by the reptilian nephrons of the avian kidney. The excretion of uric acid does not depend on tubular reabsorption of water; therefore it does not reflect the state of hydration.

       (2) The urine, or watery component, is produced by the mammalian nephrons of the avian kidney. Urine may be reabsorbed in the lower gastrointestinal tract by retroperistalsis from the urodeum up the coprodeum to include the ceca in galliforms.[1]

       (3) Fecal material is produced in the gastrointestinal tract. The clinician should note blood in droppings and color and frequency of droppings on the bird's history. If the color changes to lime green, the clinician should suspect liver failure (see Chapter 4).

6. Classifications

    a. New bird: A new bird[2] has been exposed either directly or indirectly to other birds within the past 2 years. Classifying a bird as new helps the clinician develop a diagnostic workup plan. Infectious diseases in new birds are the predominant concern. At risk of disease are newly acquired birds and any bird exposed to other birds during grooming at a local pet store or veterinary hospital, gatherings at bird fairs, and boarding with other birds

outside the home. Birds may also be indirectly exposed to disease; for example, the owner may use contaminated supplies from a local pet store where birds are sold.
  b. Old bird: An old bird[2] lives in a stable, uncontaminated environment. Old does not denote age but length of ownership without exposure to new birds or contaminated environments. Factors that affect differential diagnoses include the following:
     (1) Husbandry
     (2) Nutrition
     (3) Hygiene
     (4) Psychologic problems and stress
     (5) Chronic disease
     (6) Parasitism
     (7) Fungal disease
     (8) Chlamydiosis
     (9) Endocrine disturbances

B. **Physical Examination**
  1. Observations
     a. Cage setup: Toys provide environmental enrichment for the bird. Often cages have decorative metal pieces that contain lead and zinc, which can cause toxicosis. Toys should be checked to ensure they are free of heavy metals and appropriate for the type and size of companion bird owned. Natural materials may include pinecones and nontoxic branches from trees that have not been sprayed with chemicals or pesticides.
     b. Hygiene: The clinician can evaluate cleanliness by requesting the owner to bring the bird and its cage to the office. The clinician should recommend easy ways for the owner to keep the cage clean. Putting newspaper on the bottom of the cage is inexpensive, and by changing the newspaper daily, the owner can assess the health status of the bird. Food cups should be broad, shallow, and cleaned daily. The cage, perches, and accessories should be cleaned as often as necessary to minimize fecal contamination.
     c. Client-animal bond: Is the bird bonding well or poorly to the owner? What is the relationship of the bird to the owner? Commonly, owners are more observant when there is a strong bond.
     d. Attitude: Is the bird relaxed and exhibiting its normal behavioral, including vocalizations? Or, is the bird fluffed, ruffled, or sleepy with its head tucked over its back? These clues are important in the determination of its health status.
     e. Conformation: To detect any abnormalities, the clinician must know the normal conformation of various species. Clues to help the clinician diagnose disease include displacement of contour feathers, unusual body curves, abnormal posture, and a drooped wing or wings.
     f. Movement: The bird should be observed in its cage as part of the physical examination for attitude, conformation, and

movement. The clinician should look for signs of disequilibrium and notice grip on the perch, stance, and use of legs and feet, including a shifting grip.

2. Palpation and visual inspection
   a. Protocol: The clinician should follow the same protocol during each physical examination and should evaluate the following:
      (1) Cere: The normal cere should be dry, smooth, and firm. The presence of feathers on the bird is species dependent, and the clinician should know the characteristics of common companion species (e.g., budgies and cockatiels have no feathers and Amazons have small feathers). Rubbing or feather loss in the cere may suggest an upper respiratory tract infection. Color changes of the cere are associated with gonadal tumors, particularly in budgies, and the clinician should note any alterations when taking a history.
      (2) Nares: The nares should be bilaterally symmetrical. Shape changes in the nares are associated with a chronic upper respiratory tract infection. Feathers may be matted with mucus over the nares or carpus, which suggests an upper respiratory tract infection. Nasal plugs in cockatiels are common.
      (3) Beak: The beak's length, width, and shape and its appearance (e.g., shiny, powdered) are species dependent. The beak should be smooth and symmetrical, with some keratin along the side. A flaky or roughened beak suggests nutritional deficiencies. A fast-growing beak may be associated with hepatic disease.[3] Grooves in the beak are associated with an upper respiratory infection or may be secondary to trauma, nutritional deficiencies, or infections (see Chapter 3).
      (4) Oropharynx: The oropharynx should have only a mild odor, which should not be foul. Any pigmentation of the oropharynx should be uniform and shiny. The oropharynx is often dark, and the clinician can inspect it and the choanal slit with a penlight or an otoscope. Psittacines and galliformes should have caudally directed choanal papillae. When the papillae are blunted, the clinician should suspect hypovitaminosis A. White plaques on the hard palate may be a sign of hypovitaminosis A (resulting in squamous metaplasia of the small, multicellular salivary glands), candidiasis, trichomoniasis, avian pox, avian tuberculosis, or neoplasia. The clinician should also inspect the floor of the oropharynx and the laryngeal opening, or rima glottidis, and tongue for any abnormalities, including plaques.
      (5) Eyes: The eyes should be clear, the cornea should glisten, and no plaques or excoriations should be present. The nictitating membrane should move across each eye rapidly in a dorsomedial to ventrolateral direction. The margins of the lids should be symmetrical. Macaws and canaries have modified bristles as Amazons do. To diagnose an upper respiratory

infection and infraorbital sinusitis, the clinician should inspect and palpate the area immediately surrounding the eyes for periorbital swelling (see Chapter 3). The clinician should perform a retinal examination on any bird released into the wild and birds with head trauma. In addition, the clinician should note retinal detachments, hemorrhages, and lenticular or vitriol changes that lead to blindness.

(6) Ears: Otitis externus of the ears is uncommon in birds. A hemorrhage into the ear canal is associated with head trauma, and the clinician should carefully examine the ear. The infraorbital sinus has extensions rostral and caudal to the ear canal (see Chapter 3). An infraorbital sinusitis may cause swelling in the ear canal, which may secondarily lead to an ear infection.

(7) Trachea: A bird's trachea is longer and more flexible than a mammal's. The trachea is also ventrally located and has complete tracheal rings. In larger psittacines (Amazons and larger), the clinician should evaluate the trachea to the level of the syrinx at the tracheal bifurcation by using a rigid endoscope. The clinician can transilluminate the trachea when diagnosing air sac mites in passerines and other small birds.

(8) Pectoral muscle mass: The muscle mass on either side of the keel should be relatively firm without fat accumulations. In healthy birds the keel should not be prominent. However, the keel becomes prominent when the muscle mass is reduced. Muscle mass reduction occurs in birds who are emaciated and have chronic diseases. Companion birds such as budgies and Amazons who are caged and eat high-fat food commonly have more fat deposits over their ventral body walls.

(9) Crop: The clinician should palpate the crop in the right thoracic inlet, making sure that secondary aspiration does not occur. The crop may be doughy if infection is present. A doughy crop is common with candidiasis, particularly in hand-fed psittacine neonates. The clinician may also palpate foreign bodies in the crop and perform a crop flush using pH-balanced saline in a syringe attached to a red rubber catheter. The clinician can flush a small amount of saline into the crop and aspirate the fluid for cytologic screening and culturing. Results of pH should also be determined.

(10) Coelomic or abdominal space: The coelomic area is a small space between the sternum and pubic bones. The clinician can palpate the coelomic area with the little finger, or fifth digit. The slightly firm, rounded object between the pubic bones is the gizzard and can often be palpated centrally. On palpation the clinician may find eggs, tumors, or an enlarged liver. The liver is normally within the border of the sternum. Fluid accumulation and any enlargement of the coelomic space are considered abnormal findings.

(11) Vent: The vent is normally closed by tone from the external anal sphincter. Soiling, feather loss, or indication of inflammation in the area of the vent should not be present. The underlying epithelium of the cloaca should not be protruding or foul smelling. An acidic odor is often associated with cloacal papillomatosis, which is a differential diagnosis for a cloacitis.

(12) Thoracic limbs: The clinician should perform a musculoskeletal palpation of the thoracic limbs and girdle and should palpate the joints, the propatagialis tendon on the leading edge of the wing, and the propatagium, or the flap of skin and underlying structures also along the leading edge of the wing. The clinician should palpate the primary and secondary remiges, or flight feathers, which insert along the carpometacarpus and digits and the caudal border of the ulna.

(13) Pelvic limbs: The clinician should palpate the pelvic limbs, girdle, and the joints. The clinician should also inspect the ventral surface of the digits for any lesions and bumblefoot. A bird's nails should be an appropriate length, and birds kept in captivity should have their nails blunted to reduce injury to the foot pad area.

(14) Feathers: The contour feathers should be shiny and devoid of stress lines, color changes, and frays. The clinician should note signs of chewing or feather plucking (see Chapter 7).

3. Auscultation
    a. Auscultation of the respiratory system: The clinician should listen to the dorsal and ventral thorax and trachea during auscultation of the respiratory system. Abnormal sounds, such as clicks, wheezes, and squeaks, are associated with a narrowing of the parabronchi, which often occurs with lower respiratory tract involvement. These abnormal sounds may originate from the upper respiratory tract (see Chapter 3).
    b. Respiratory recovery time: Respiratory recovery time can reflect involvement of the lower respiratory tract. Prolonged respiratory recovery time is indicated when respiratory rate and rhythm do not return to normal within 2 to 3 minutes. Tail bobbing occurs with increased respiratory effort and may be a sign of upper or lower respiratory tract involvement.
    c. Auscultation of the heart: Because the bird's heart rate increases with stress, the clinician should listen to the heart early in the physical examination. A well-socialized psittacine often allows the clinician to auscultate the heart ventrally while the bird is perching. An abnormal rate or rhythm is associated with systemic illness or heart disease.

C. Database
1. Minimum examination and testing
    a. Physical examination: The physical examination should follow a consistent protocol.

b. **Weight:** All birds should be weighed on a gram scale at least once daily.
c. **Gram stains:** Clinicians perform Gram stains to analyze the flora and examine fecal or cloacal and choanal epithelial cells. Some controversy exists in avian medicine over the use of choanal Gram stains, but I use them to analyze epithelial cells to indicate hypovitaminosis A and discover the presence of spirochetes and other abnormal cellular elements.[4] A quick assessment of the flora can also be obtained. If the clinician finds abnormalities on a Gram stain, a culture should be performed.
d. **Positive and negative culture:** Some avian veterinarians prefer culturing the choanal, fecal, or cloacal cells instead of using Gram staining. Cultures allow the clinician to identify bacterial and fungal organisms that exist in the testing site. Some organisms such as spirochetes are difficult to culture; therefore the clinician should perform both a culture and a Gram stain whenever possible. Gram stains and culture results may differ because cultures sometimes result in the growth of abnormal bacteria and fungi that were not detected on Gram staining.
e. **Complete blood count** and differentials: The clinician can use several methods for obtaining complete blood counts and differentials.
f. **Chemistry profile:** The clinician should use blood serum when conducting a chemistry profile. The profile indices should measure the levels of total protein, glucose, Ca^{++}, PO_4, uric acid, blood urea nitrogen, creatine kinase, aspartate aminotransferase, lactate dehydrogenase, and bile acids. The clinician may also request measurement of glutamic dehydrogenase levels and cholesterol levels.

2. Additional tests
   a. Electrolytes
   b. Chlamydial testing (elementary body agglutination, immunofluorescent assay, and deoxyribonucleic acid–PCR probe)
   c. DNA acid probe (for PBFD and polyomavirus)
   d. Radiographs (an extension of the physical examination and a common test in practice)
   e. Ultrasound
   f. Fecal parasite evaluation (e.g., a fecal float and zinc turbidity test for *Giardia* bacteria) (see Chapter 20)
   g. Aspergillosis titer
   h. Fecal acid-fast stain and tuberculosis titer tests
   i. Urinalysis
   j. Cytologic examination of an aspirate or a biopsy
   k. Histopathologic examination of a biopsy through endoscopic evaluation
   l. Electrocardiogram
   m. Electromyogram

> Polychromasia – RBC ↑ w/ affinity for acid, basic, or neutral stains

3. **Venous collection sites**
   a. **Jugular vein**: The jugular vein is often largest on the right side of the neck and is considered the best site for the clinician to collect blood because it provides the most blood with the least interpretive error.[5]
   b. **Basilic vein**: The basilic vein is located over the ventral distal humerus.
   c. **Medial metatarsal vein**
   d. **Toenail clip**: The clinician should use this route only if other methods have failed.
4. Laboratories
   a. Laboratory: The clinician should use a high quality laboratory that routinely performs avian sample analyses.
   b. Hemolysis: The breakdown of red blood cells and release of hemoglobin occur in ethylenediaminetetraacetic acid (EDTA) with crows, curasows, hornbills, ostrich, and brush turkeys[5]; therefore the clinician should use heparin as an anticoagulant.
5. Complete blood count interpretations (Tables 1-1 and 1-2)
   a. Red blood cells: A packed cell volume <35% indicates anemia. The clinician should consider a blood transfusion for the patient when packed cell volume <20%; with low packed cell volumes, birds display adverse effects more rapidly than mammals. *Anemia with ≤5%, polychromasia* indicates a nonregenerative anemia or an inadequate response time. *Anemia with ≥10%, polychromasia indicates a regenerative anemia.* The clinician may also observe immature red blood cells on the smear with a regenerative anemia.
      (1) Nonregenerative anemias may be associated with the following:
         (a) Chronic disease (parasitism, chlamydiosis, aspergillosis, avian tuberculosis, and neoplasia)
         (b) Leukemias

Table 1-1
Common hematologic abnormalities: Changes in the PCV for common bird species

	Interpretation	Differential diagnosis
PCV ≤ 35% (<5% polychromasia)	Nonregenerative anemia Blood loss, inadequate response time	Chronic disease, including chlamydiosis, aspergillosis, parasitism, avian tuberculosis, neoplasia Leukemias, lymphosarcoma Toxicities including heavy metals, aflatoxins Deficiencies including iron, folic acid Hypothyroidism
PCV ≤ 35% (>10% polychromasia)	Regenerative anemia	Blood loss including trauma, parasitism, organ leakage, coagulopathic condition Bacterial septicemias Red blood cell parasites Toxicities including mustards, petroleum products
PCV > 55% (TP normal or ↑)		Dehydration Polycythemia

PCV, Packed cell volume.

Polycythemia ↑ normal RBC

   (c) Toxicities (such as exposure to heavy metals and long-standing aflatoxins)
   (d) Iron and folic acid deficiencies
   (e) Hypothyroidism
  (2) Regenerative anemias may be associated with the following:
   (a) Red blood cell parasites
   (b) Bacterial septicemias
   (c) Toxicities (from exposure to mustards and petroleum products)[5]
   (d) Blood loss (associated with the following):
      (i) Trauma
      (ii) Parasitism
      (iii) Leakage (from the gastrointestinal tract, kidney, or parenchymous organ)
      (iv) Coagulopathic condition
   A packed cell volume > 55% indicates dehydration and polycythemia; primary polycythemia is rare in birds.
 b. **White blood cells:** Normal counts vary between and within species of birds, making interpretation of results difficult. The clinician should use the laboratory's normal values for comparison when interpreting the patient's results.
  (1) Leukocytosis: For normal adult psittacines the average total leukocyte count is approximately 10,000/µl. The values may be elevated slightly because of the bird's response to handling. Additional considerations for leukocytosis include the following:
   (a) Stress (incurred through travel, handling, and breeding)

Table 1-2
Common hematologic abnormalities: Changes in WBC counts for psittacines

	Interpretation	Differential diagnosis
WBC ≥ 15,000/µl	Leukocytosis	Stress including handling, travel, breeding, hemorrhage Infectious causes, septicemia (see monocytosis) Trauma, inflammation Toxicities Immature birds
WBC ≤ 5000/µl	Leukopenia	Viral diseases Septicemia Early response to steroids Bone marrow suppression from infectious causes, noninfectious causes Toxins
Lymphocytes ≥4500/µl	Lymphocytosis	Antigenic stimulation Infectious causes Lymphocytic leukemia
Monocytes ≥5%	Monocytosis	Chlamydiosis Aspergillosis Mycotic infections Avian tuberculosis Bacterial infections, granulomas Massive tissue Zinc-deficient diet

WBC, White blood cell.

(b) Infection and septicemia
(c) Trauma and inflammation
(d) Toxicities
(e) Neoplasia (both solid and circulating)
(f) Hemorrhage
(g) Age (higher counts in immature birds)

The clinician may observe toxic heterophils on blood smears from birds with toxemias and septicemias. As the severity of the condition and prognosis worsen, the number of toxic heterophils increases.

(2) Components: Other cells may contribute to the leukocytosis, or the count may remain normal and the percentages of the following cells may increase:
   (a) Lymphocytosis can occur with the following:
      (i) Antigenic stimulation — infection/toxic
      (ii) Lymphocytic leukemia
   (b) Monocytosis can occur with the following:
      (i) Chronic infections such as chlamydiosis
      (ii) Aspergillosis and other mycotic infections
      (iii) Avian tuberculosis
      (iv) Bacterial infections and granulomas
      (v) Massive tissue necrosis
      (vi) Zinc-deficient diets
   (c) Eosinophils and basophils: The function of eosinophils and basophils remains unclear in birds. The clinician may have difficulty interpreting the reasons for an increase or a decrease in the numbers. However, eosinophilias have been associated in birds with parasitic infections and allergic reactions. Eosinophil counts may be normal with these disease processes. Basophilias may occur but not consistently with chlamydiosis and tissue necrosis.
   (d) Avian thrombocytes: Birds have thrombocytes that are involved in clotting similar to mammalian platelets. Additionally, avian thrombocytes phagocytose foreign substances in the blood stream. Thrombocytopenias can occur with severe septicemias and toxemias, and thrombocytosis can result from a marked regenerative response from anemia. A normal thrombocyte count is 20,000 to 30,000/µl of blood.

(3) Leukopenia: A white blood cell count $\leq 5000/\mu l$ can be interpreted as leukopenia or decreased leukopoiesis. Common causes of leukopenia and decreased leukopoiesis include the following:
   (a) Viral diseases can result in leukopenia
   (b) Septicemia can result in leukopenia and may also cause a decrease in leukopoiesis by depressing bone marrow production
   (c) Corticosteroid administration has resulted in leukopenia with lymphopenia

c. **Blood parasites:** In addition to changes in the numbers of cellular elements of the blood, the morphology of the cells may change. Blood parasites may commonly be found within blood cells or within the blood stream. Cells change with toxicity, presenting a toxic appearance. Basophilic stippling with lead poisoning, which can be observed in mammalian cells, rarely occurs in birds. Wild birds or birds housed outside are more likely to have blood parasites. Common blood parasites include *Haemoproteus, Plasmodium, Leucocytozoon,* and *Toxoplasma* organisms.

D. **Bone Marrow**
  1. Evaluation
      a. Cytologic evaluation: The clinician should consider performing a cytologic evaluation of bone marrow when thrombocytopenia, panleukopenia, heteropenia, or nonregenerative anemias are suspected. Evaluation is also needed when abnormal cells are found on the peripheral blood smear.
      b. Collection sites: Collection sites for bone marrow include the proximal tibiotarsus, ulna, and keel. The most common collection site is the proximal tibiotarsus.
      c. Surgery: With the patient under general anesthesia, the clinician surgically prepares the dorsal surface of the stifle and then moves a spinal or hypodermic needle across the tibial plateau in line with the cranial cnemial crest. Once the needle is in its proper position, the clinician inserts it into the marrow space using a twisting motion along the long axis of the bone. After removing the cannula, the clinician attaches a 12-ml syringe to the needle and aspirates the marrow into the needle. The clinician removes the needle and fills the syringe with air to gently expel the needle's contents onto a glass microscope slide. Placing a second slide on the first slide, the clinician moves the slides apart horizontally. The specimens are then sent to the laboratory.

II. Common Interpretations of the Chemistry Profile

A. ↑ **Aspartate Aminotransferase, → Creatine Kinase, ↑ Bile Acids, and ↑ Lactate Dehydrogenase Levels: These Findings Indicate Liver Disease.**
Additional diagnostic tests that the clinician should consider include the following:
1. Radiographs
2. Ultrasound (of the coelom)
3. Chlamydial testing
4. Blood tests for viral diseases such as polyoma and PBFD. Pacheco's disease also results in severe hepatic cellular necrosis.

5. Endoscopic evaluation and biopsies (The clinician should submit liver samples for histopathologic examination, aerobic, anaerobic, and fungal culture; immunofluorescence antibody testing for *Chlamydia* bacteria; and acid-fast staining. The clinician should submit larger biopsy samples for suspected organophosphate toxicities because an endoscopic biopsy is too small to obtain the sample needed. (See Chapter 21.)

B. ↑ Aspartate Aminotransferase, ↑ Creatine Kinase, → Bile Acids, ↑ Lactate Dehydrogenase Levels: These Findings are Associated with the Following:
   1. Muscle cell necrosis and wasting
   2. Intramuscular injections
   3. Vitamin E and selenium deficiency
   4. Convulsions
   5. Neuropathic conditions
   6. Chlamydiosis (Neurologic signs associated with chlamydiosis occur most commonly in cockatiels. The clinician should perform more diagnostic tests to rule out neurologic problems.)

C. ↑ Ca^{++}, → PO_4, → Albumen Levels: The Increased Level of Ca^{++} and Normal Level of PO_4 Can Occur with the Following:
   1. Egg laying
   2. Vitamin D increase and toxicosis
   3. Bone tumors
   4. Polyostotic hyperostosis
   5. Renal failure
   6. Dehydration

   Polyostotic - involv. more than 1 bone
   increase in bulk

   The clinician should request the following tests for the diagnostic workup: radiographs to look for eggs and osteopetrosis associated with physiologic lay or polyostotic hyperostosis (the clinician may also observe bone tumors); a urinalysis to look for casts and other abnormalities to rule out renal failure; a coelomic ultrasound to look for eggs, masses, and kidney detail; a test to determine the sex hormone levels such as estradiol, progesterone, and testosterone; and a history of vitamin D toxicosis (leading to dystrophic calcification of the kidney tubules).

D. ↓ Calcium, → Ionized Ca^{++}, and ↓ Albumen Levels: The Findings Indicate Hypocalcemia, Which Can Occur with the Following:
   1. Malnutrition
   2. Rickets (Rickets is a metabolic bone disease found in growing animals. The skeleton, including the beak and claws, can become pliable. Radiographic abnormalities include deformities along the joints of the sternal and costal ribs and bowing of the tibiotarsus. The ribs may be infolded as well, reducing the efficiency of oxygen capture from the respiratory system. The loss of oxygen results in tachypnea and secondary polycythemia leading to right ventricular failure.)

3. Osteomalacia or osteodystrophy (These disorders result from calcium deficiencies in mature birds. The bone thins as osteoclasts are induced by parathyroid hormone to resorb first medullary bone and eventually cortical bone to meet calcium demands.[6] Degeneration can result in pathologic fractures that often appear as greenstick fractures. Increased production of connective tissue coupled with bone resorption in osteodystrophy is not observed commonly in birds.)
4. Osteoporosis or cage layer fatigue (In poultry, these conditions occur as a normal physiologic consequence of the lay cycle. Osteoporosis or cage layer fatigue rarely occurs in most psittacines but occur in hens who continually lay eggs.)
5. Hypocalcemic syndrome (This syndrome occurs in African grey parrots and is characterized by periods of abnormal mentation and seizures. The cause of the syndrome remains unknown; however, the parathyroid gland and its hormone may be involved.[6] Parenteral calcium administration followed by increased dietary calcium often resolves the problem. The clinician should monitor total protein, albumen, Ca^{++}, and ionized Ca^{++} values.)

E. ↑ Uric Acid: Increased Uric Acid Levels Can Result from the Following:
1. Renal disease
2. Decreased glomerular filtration rate
3. Postprandial rise in uric acid levels
4. Tissue damage
5. Ovulation
6. Liver disease

Uric acid is produced in the liver and secreted by the kidneys. The clinician should consider the following diagnostic tests:
   a. Radiographs
   b. Ultrasound (of liver and kidneys)
   c. Urinalysis
   d. Heavy metal tests (for lead and zinc)
   e. Gram stains (Gram staining of the choanal cells can be used as a quick assessment for hypovitaminosis A. Squamous metaplasia of the kidney tubules with hypovitaminosis A can result in renal failure.)
   f. Polyomavirus testing
   g. Aspergillosis titer
   h. Chlamydial testing
   i. Endoscopic evaluation with or without biopsy for histologic examination, culture, acid-fast stain, and immunofluorescence antibody
   j. Parasite evaluation
   k. Toxoplasmosis titer

F. ↑ Cholesterol Level: Increased Cholesterol Can Occur with the Following:
1. High fat diet (Amazons may eat only seed and other fatty foods.)
2. Liver disease (see pp. 13-14)

3. Hypothyroidism
4. Bile duct obstruction (Amazons and other psittacines may have bile duct hyperplasia, which is often found concomitantly with hepatic fibrosis and hepatocellular lipidosis. Bile duct hyperplasia may lead to bile duct carcinoma.)
5. Starvation (Although mammals use glucose, the primary energy source in birds is fatty acids. Elevated fatty acid levels can occur during starvation and migration.)

REFERENCES

1. Duke GE: Alimentary canal: anatomy regulation of feeding and motility. In Sturkie PD, editor: *Avian physiology,* New York, 1976, Springer-Verlag, pp 269-288.
2. Harrison GJ, Harrison LR, Fudge AM: Preliminary evaluation of a case. In Harrison GJ, Harrison LR, editors: *Clinical avian medicine & surgery including aviculture,* Philadelphia, 1986, Saunders, pp 101-114.
3. Rupley AE: Dermatologic signs. In Rupley AE, editor: *Manual of avian practice,* Philadelphia, 1997, Saunders, pp 226-263.
4. Campbell TW: Cytology of the upper alimentary tract: oral cavity, esophagus and ingluvies. In Campbell TW, editor: *Avian hematology and cytology,* Ames, Iowa, 1995, Iowa State University Press, pp 47-59.
5. Campbell TW: Hematology. In Ritchie BW, Harrison GJ, Harrison RL, editors: *Avian medicine: principles and application,* Lake Worth, Fla, 1994, Wingers, pp 176-198.
6. Lumeij JT: Endocrinology. In Ritchie BW, Harrison GJ, Harrison RL, editors: *Avian medicine: principles and application,* Lake Worth, Fla, 1994, Wingers, pp 582-606.

2

Supportive Care and Shock

Michael P. Jones and Christal G. Pollock

Emergency medicine and supportive care of the avian patient is one of the most challenging aspects of avian medicine. A thorough knowledge not only of disease pathophysiology but also of therapeutic techniques and supportive care is required for successful case management. Many techniques can be extrapolated from companion animal medicine, with appropriate adjustments for differences in anatomy and physiology between avian and mammalian species.

Birds often are presented to the veterinarian in an extreme state of disability, which makes each clinical decision crucial to the overall outcome. Learning to recognize common emergencies and addressing problems in an appropriate manner greatly enhance the patient's chances for survival.

I. General Considerations

A. Clinical Examination
1. Prepare for patient's arrival.
    a. Prepare and warm fluids.
    b. Provide a warm environment.
    c. Have oxygen support available.
    d. Prepare other medications and supplies as initially indicated by conversation with owner or agent.
2. Perform triage.
    a. Obtain thorough history and observe patient.
    b. Rapidly assess and classify patient's condition, which determines the supportive and therapeutic approach.
    c. Treat patient according to the severity of illness and the degree of critical care required.
    d. Check ABC's of cardiopulmonary resuscitation—*a*irway, *b*reathing, and *c*irculation.[1]
        (1) Airway established (endotracheal tube or air sac cannula)
        (2) Positive pressure ventilation

(3) Cardiac massage
(4) Epinephrine (1:1000) (0.5 to 1.0 mg/kg given IT, intravenously, IC, or intraosseously)[2]
(5) Atropine (0.5 mg/kg given intramuscularly, intravenously, intraosseously, or IT[2]; 0.2 mg/kg every 3 to 4 hours as needed for organophosphate or carbamate toxicosis)[2]
(6) Doxapram hydrochloride (5 to 10 mg/kg given once intramuscularly, intravenously, or subcutaneously)[3]

e. Perform preliminary examination or observation of patient.
   (1) Posture
   (2) Degree of alertness
   (3) Cardiovascular system (Heart rate, rhythm, and pulse quality are assessed simultaneously with auscultation.)
   (4) Patient's environment
f. Treat life-threatening conditions of the respiratory or cardiovascular system.
   (1) Dyspnea
   (2) Circulatory collapse
   (3) Seizures
   (4) Severe trauma
g. Notify owners of severity and prognosis as soon as possible; *obtain informed consent*.

3. Conduct physical examination.
   a. Perform complete physical examination when patient's condition is stable or in stages as condition stabilizes.
   b. Remember to *do no harm*.
4. Formulate detailed plan to manage patient clinically.[4]
   a. Account for all diagnostic and therapeutic actions required by patient's condition.
   b. Assess patient's ability to tolerate handling.
5. Perform diagnostic procedures (perform only procedures necessary to evaluate patient).
   a. Minimum database
      (1) Packed cell volume (PCV)/TS
      (2) Glucose level
      (3) Uric acid level
      (4) Aspartate aminotransferase level
      (5) Creatinine level
      (6) Electrolytes
   b. Additional blood may be drawn for complete blood count (CBC) (Serum or plasma chemistry panel or other diagnostic tests may also be performed if the differential diagnosis requires it *and* if the patient can tolerate further diagnostics.)
   c. Ophthalmic examination
   d. Radiographs
   e. Endoscopy (laparoscopy or tracheoscopy)
   f. Cytologic or microbiologic studies
6. Allow patient's condition to stabilize before attempting any further diagnostic tests.

II. Emergency Medical Syndromes

A. Respiratory Emergencies
1. Dyspnea
   a. Clinical signs: Open-mouthed breathing, exaggerated respiratory effort, tail bobbing, change or loss of voice or song, wheezing
   b. Etiology includes the following:
      (1) Primary respiratory diseases: Bacterial, fungal, parasitic, chlamydial, and mycoplasmal infections; toxins, foreign bodies, trauma, neoplasia
      (2) Extrarespiratory diseases: Organomegaly (thyroid, hepatic, renal, gonadal), oral masses, coelomic fluid, cardiovascular disease, neoplasia
   c. Diagnosis includes the following:
      (1) Thorough history and observation of patient
      (2) Minimum database
         (a) CBC
         (b) Chemistry panel
         (c) Radiographs
         (d) Abdominocentesis (if fluid is present)
   d. Therapy: Therapy is based on the diagnosis; it may include oxygen (O_2) therapy (O_2 cage, intubation of air sacs) (Fig. 2-1) and fluid and nutritional support.

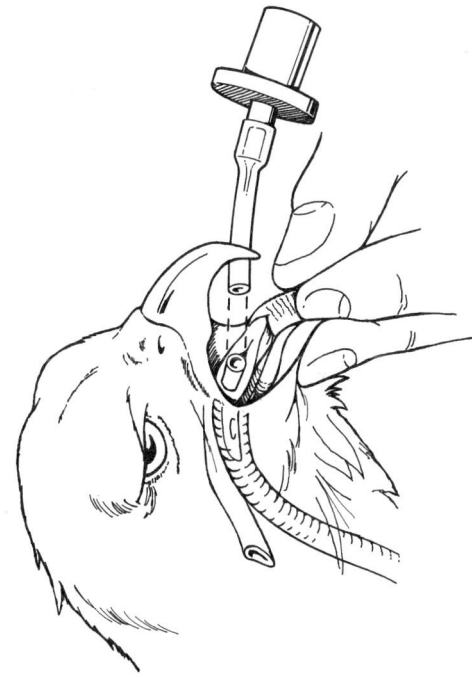

Fig. 2-1 Intubation.

B. Cardiovascular Emergencies

1. Shock has not been clearly defined in birds and does not occur in birds as it does in mammals.[5]
   a. Pathophysiology of shock[5-7]
      (1) Shock is not a single physiologic entity, nor is it a static condition; it is a multifactorial syndrome that results when the supply of oxygen to the tissues is inadequate or when the tissues are unable to use oxygen.
      (2) Circulatory collapse occurs as a result of insufficient capillary perfusion and cellular hypoxia.
      (3) Multiple organs are affected in different ways and at different times.
      (4) Pathogenesis at the tissue and cellular levels is different in birds and mammals.[5]
      (5) Birds do not develop lactic acidosis caused by cellular hypoxia that progresses to peripheral circulatory failure and pooling of blood in small vessels and terminal circulatory beds.[5]
      (6) General circulatory failure occurs as a consequence of hypovolemia (hypovolemic shock), heart failure (circulatory shock), and sepsis (septic shock).[5]
      (7) Hemorrhagic shock may not occur in birds[8]:
         (a) Birds tolerate severe blood loss better than mammals.
         (b) Birds have a higher rate of absorption of tissue fluid to replace lost blood volume, as well as baroreceptor reflexes; these processes help maintain normal blood pressure.[1]
   b. Classification of shock[6]
      (1) Cardiogenic shock
         (a) Results from impaired cardiac function and diminished cardiac output
         (b) Common causes include arrhythmias, pericardial effusion, and iatrogenic dysrhythmias
      (2) Hypovolemic shock
         (a) Has many etiologies but the common pathophysiologic mechanism is a decrease in circulating blood volume[6,7]
         (b) Common causes include hemorrhage, dehydration, vomiting, diarrhea, trauma, surgery, neoplasia, anaphylaxis, and burns
      (3) Vasogenic shock
         (a) Involves changes in peripheral resistance that impair circulatory distribution
         (b) Causes include sepsis, endotoxemia, toxicosis, anesthetic overdose, and anaphylaxis[6]
   c. Treatment of shock (The immediate goals are to restore blood volume and blood pressure to facilitate perfusion of tissues and organs.)[6]
      (1) Obtain baseline PCV, total protein level, and glucose and bicarbonate levels; postpone other diagnostic tests unless they are considered necessary.

(2) Place intravenous and intraosseous catheters (Figs. 2-2 and 2-3).
(3) Calculate dehydration and fluid requirements.
(4) Provide rapid volume expansion (Give lactated Ringer's solution at half of fluid deficit for the first 12 hours and remaining 50% of deficit over following 48 hours.)
(5) Give vitamin B complex (10 mg/kg thiamine), steroids, iron dextran, antibiotics if indicated, supplemental oxygen, and heat and nutritional support if necessary.
(6) Monitor PCV, total protein level, bicarbonate level, and urine output.
(7) Obtain a complete history and initiate diagnostic testing.
(8) Begin maintenance fluids and force feeding (Fig. 2-4).
(9) Monitor weight until bird is able to self-feed.
(10) Maintain as stress-free an environment as possible.

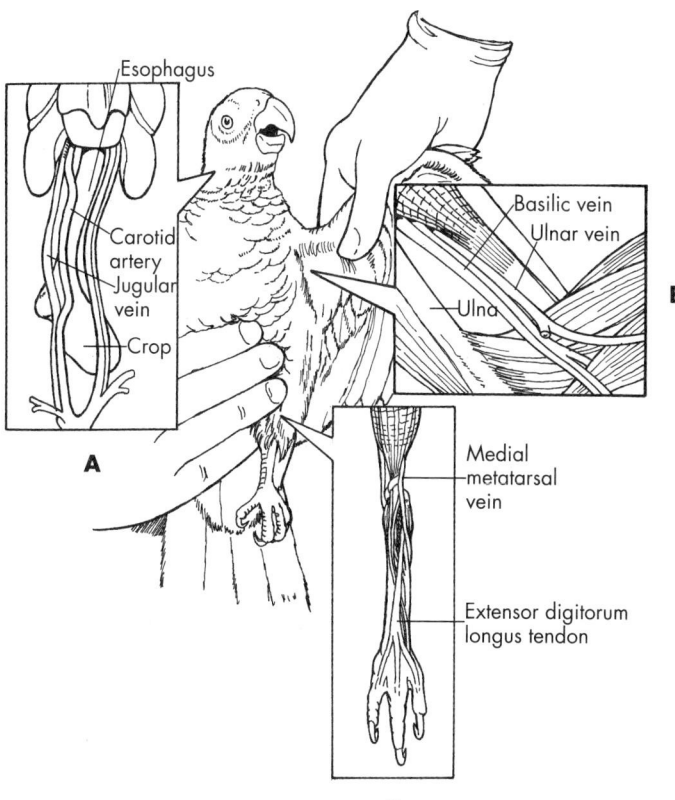

Fig. 2-2 Routes for intravenous lines, catheter placement, and withdrawal of blood. **A,** Right jugular vein. **B,** Basilic vein. **C,** Medial metatarsal vein.

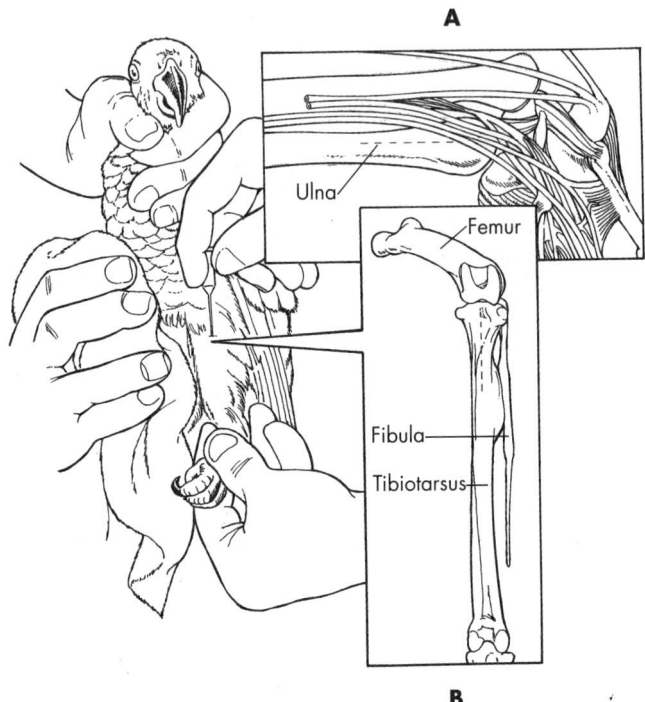

Fig. 2-3 A, Placement of intraosseous catheter in the dorsal ulna. **B,** Bone marrow tap into the tibiotarsus

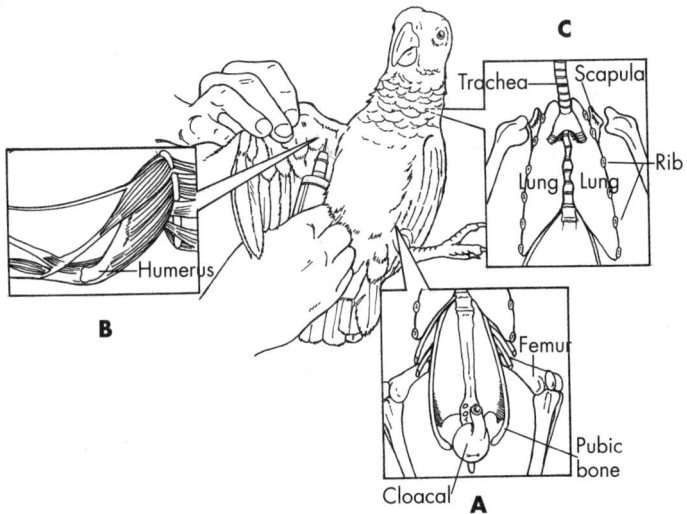

Fig. 2-4 Subcutaneous sites for fluid placement. **A,** Inguen. **B,** Propatagial or wing web site. **C,** Interscapular and intrascapular sites.

2. Blood loss or anemia and cardiac failure
   a. Clinical signs: Weakness, dyspnea, tachypnea, lethargy, depression, inactivity, pale mucous membranes (oral, conjunctival, and cloacal mucosa), and tachycardia
   b. Etiology
      (1) Decreased red blood cell production or increased red blood cell destruction; classified as regenerative or nonregenerative anemias
      (2) Trauma
      (3) Gastrointestinal bleeding
      (4) Genitourinary bleeding
      (5) Hemolysis and idiopathic hemorrhage
      (6) Coagulopathic conditions
      (7) Cloacal papillomas
      (8) Egg laying
      (9) Cloacal or uterine prolapse
      (10) Heavy metal toxicity
      (11) Infectious diseases (polyomavirus, psittacine beak and feather disease, and other infectious agents[1])
   c. Diagnosis
      (1) CBC and reticulocyte count
      (2) Radiographs
      (3) Specific diagnostic tests for infectious diseases
      (4) Tests for heavy metal toxicity
   d. Therapy
      (1) Control hemorrhage
      (2) Determine source of bleeding
      (3) Replace fluid volume
      (4) Give vitamin K, vitamin D_3, calcium, and antibiotics as necessary
      (5) Administer blood transfusions if PCV < 20%
      (6) Treatment for shock should include volume expansion, corticosteroids, bicarbonate therapy if metabolic acidosis is present, and parenteral antibiotics
3. Cardiopulmonary arrest[1]
   a. ABC's of cardiopulmonary resuscitation
      (1) Establish airway (endotracheal tube or air sac tube)
      (2) Positive pressure ventilation
      (3) Cardiac massage
      (4) Epinephrine and atropine as necessary
      (5) Doxapram

C. Gastrointestinal Emergencies
1. Thermal burns and fistulas
   a. Clinical signs: Commonly seen in hand-fed psittacine chicks; Initially, redness and swelling of the skin, which progresses to necrosis and drainage of food from the fistula

b. Etiology
   (1) Hand-feeding formula that is too hot
   (2) Puncture of crop by gavage tubes, with subsequent leakage of food subcutaneously and necrosis of overlying tissue (Fig. 2-5)
c. Diagnosis is usually based on clinical signs noted by owner
d. Therapy
   (1) Aimed at nutritional support of the patient while the fistula is corrected
   (2) If the injury is acute, closure should be postponed 5 to 7 days so that healthy tissue can be distinguished from necrotic tissue before closure; surgical closure is the treatment of choice

2. Gastrointestinal stasis
   a. Clinical signs: Changes in motility, delayed crop emptying, regurgitation, vomiting, weight loss, dehydration, and decreased fecal output[1,5]

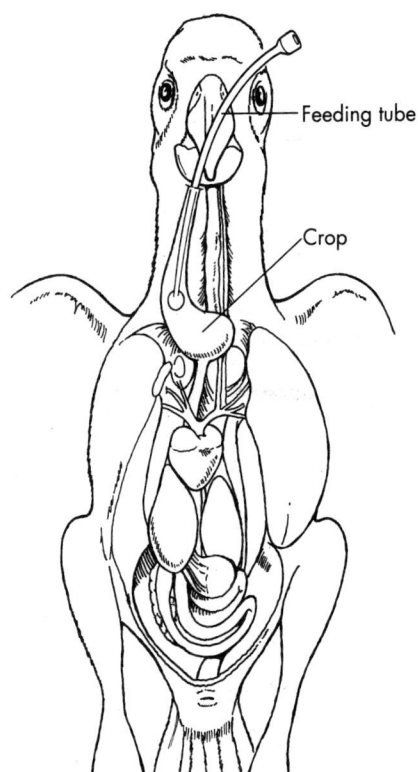

Fig. 2-5 Gavage feeding technique, showing crop and placement of feeding tube or needle.

b. Etiology
     (1) Poor sanitation
     (2) Inadequate environmental conditions
     (3) Insufficiently warmed formula
     (4) Gastrointestinal foreign body
     (5) Heavy metal toxicity
     (6) Neoplasia
     (7) Infectious disease[1,2]
  c. Diagnosis
     (1) History and physical examination
     (2) CBC, chemistry profile
     (3) Radiographs (plain films and contrast studies)
     (4) Crop flush or swab cytologic studies and culture
  d. Therapy
     (1) Supportive care
     (2) Fluid replacement
     (3) Antibiotic therapy (parenteral) (see Chapter 26)
     (4) Metoclopramide (see Chapter 26)
     (5) Parenteral nutrition or placement of duodenostomy tube
3. Regurgitation and vomiting
  a. Clinical signs: Food staining around beak and feathers on the head, weight loss, dehydration, head bobbing
  b. Etiology
     (1) Behavioral causes
     (2) Insufficiently warmed formula
     (3) Gastrointestinal foreign body
     (4) Obstruction of gastrointestinal tract
     (5) Infectious disease
     (6) Neoplasia
     (7) Heavy metal toxicity
     (8) Iatrogenic causes (from handling)
     (9) Toxins (plant)[1,5]
  c. Diagnosis
     (1) History and physical examination
     (2) CBC, chemistry profile
     (3) Cytologic study and culture of regurgitated material
     (4) Fecal examination
     (5) Radiographs (plain film and contrast studies)
     (6) Heavy metal analysis
  d. Therapy
     (1) Supportive care
     (2) Fluid replacement
     (3) Antibiotic therapy (parenteral) (see Chapter 26)
     (4) Metoclopramide (see Chapter 26)
     (5) Parenteral nutrition or placement of duodenostomy tube[1,5]
4. Diarrhea, hematochezia, and melena
  a. Clinical signs: Loose feces
  b. Etiology

(1) Infectious diseases
(2) Dietary changes
(3) Foods with high water content
(4) Foreign body
(5) Hepatic, renal, or pancreatic disease
(6) Toxins
(7) Heavy metal toxicity
  c. Diagnosis
   (1) History and physical examination
   (2) CBC, chemistry profile
   (3) Radiographs (survey and contrast studies)
   (4) Fecal examination for parasites
   (5) Cytologic studies
  d. Therapy
   (1) Parenteral fluids
   (2) Antibiotic therapy (parenteral) (see Chapter 26)
   (3) Nutritional support
5. Cloacal prolapse
  a. Clinical signs: Cloacal prolapse must be differentiated from cloacal papillomatosis. With the latter condition, the tissue's surface has a cauliflower-like appearance, whereas most prolapsed tissue has a smooth, glistening surface.[1] Papillomas often turn white when vinegar is dropped onto the mucosal surface. Both diseases may have clinical signs, including tenesmus, decreased fecal output, and bleeding from the cloaca.
  b. Etiology
   (1) Masses in the cloaca
   (2) Neurogenic disorders
   (3) Tenesmus caused by enteritis, cloacitis, or egg binding
   (4) Idiopathic causes[1]
  c. Diagnosis
   (1) Physical examination (a complete examination of the cloaca usually is performed under anesthesia)
   (2) Fecal examination for parasites
   (3) Gram stain, culture
   (4) Radiographs and biopsy if required for diagnosis
  d. Therapy
   (1) Prevent further trauma to the exposed tissue
   (2) Return the exposed tissue to its normal place by means of retention sutures or cloacopexy[1]
6. Hepatic disease
  a. Clinical signs: Lethargy, anorexia, weakness, diarrhea, regurgitation or vomiting, polyuria, polydipsia, dyspnea (caused by ascites if present), organomegaly, yellow-green discoloration of urates
  b. Etiology
   (1) Infectious or inflammatory condition
   (2) Trauma
   (3) Toxins

    c. Diagnosis
        (1) History and physical examination
        (2) CBC, chemistry profile
        (3) Fecal Gram stain and culture
        (4) Bile acid levels
        (5) Tests for *Chlamydia* organisms
        (6) Radiographs
    d. Therapy
        (1) Supportive care
        (2) Antibiotics (see Chapter 26)
        (3) Lactulose (see Chapter 26)
        (4) Vitamin K (see Chapter 26)

D. Urogenital Emergencies
  1. Renal disease or failure
     a. Clinical signs: Clinical signs commonly seen with renal disease include polyuria, polydipsia, lethargy, depression, seizures, dehydration
     b. Etiology
        (1) Infectious diseases (bacterial, viral, parasitic, fungal)
        (2) Toxicities
        (3) Ureteral obstruction
        (4) Trauma
        (5) Hypervitaminosis D (in macaws)
     c. Diagnosis
        (1) History and physical examination
        (2) CBC, chemistry panel
        (3) Urinalysis
        (4) Fecal Gram stain and culture[1]
     d. Therapy
        (1) Supportive care
        (2) Fluid and electrolyte replacement
        (3) Parenteral antibiotics (see Chapter 26)
        (4) Vitamin supplementation (contraindicated if hypervitaminosis D is suspected)[1] (see Chapter 26)
  2. Oviductal prolapse
     a. Clinical signs: Signs suggestive of uterine or oviductal prolapse include tenesmus, fecal staining of feathers around the vent, presence of uterine tissue outside the vent, decreased fecal output, and bleeding from the vent.[1]
     b. Etiology
        (1) Egg binding
        (2) Poor nutrition
     c. Diagnosis
        (1) History of egg laying
        (2) Physical examination
     d. Therapy
        (1) Supportive care

(2) Fluid replacement
(3) Parenteral calcium (see Chapter 26)
(4) Vitamins A and D_3 (see Chapter 26)
(5) Parenteral antibiotics (see Chapter 26)
(6) Corticosteroids (see Chapter 26)
(7) Replacement of prolapsed tissue after thorough cleansing, controlling bleeding, reducing swelling, and removing egg if present in prolapse[1]

3. Egg binding
   a. Clinical signs: Commonly seen in females that are first-time or chronic egg layers. Clinical signs include tachypnea, straining to pass the egg, bleeding from the vent, abdominal distension, leg paresis, lethargy, depression, weakness, sitting on the bottom of the cage, decreased fecal output, and death[1,9]
   b. Etiology
      (1) First-time or chronic egg layer
      (2) Poor nutrition
   c. Diagnosis
      (1) History and physical examination
      (2) Whole body radiographs
   d. Therapy
      (1) Conservative therapy
         (a) Warmth
         (b) Fluids
         (c) Vitamins A and D_3 (see Chapter 26)
         (d) Calcium (see Chapter 26)
         (e) Prostaglandins (see Chapter 26)
         (f) Nutritional support
      (2) Manual expression of egg (ovocentesis)
      (3) Transabdominal ovocentesis (surgical removal by means of laparotomy)

4. Egg peritonitis
   a. Clinical signs: Anorexia, lethargy, depression, weakness, preceding nesting or laying behavior, dyspnea
   b. Etiology
      (1) Traumatic rupture of egg or failure of ovum to enter infundibulum[1] (The peritonitis that develops usually is sterile but may be complicated by a secondary bacterial infection.[1])
   c. Diagnosis
      (1) History and physical examination
      (2) CBC, chemistry panel
      (3) Abdominocentesis with fluid analysis and cytologic studies
      (4) Whole body radiographs
   d. Therapy
      (1) Parenteral antibiotics (see Chapter 26)
      (2) Fluid therapy
      (3) Corticosteroids (antiinflammatory use) (see Chapter 26)

E. Neurologic Emergencies
  1. Head trauma
     a. Clinical signs: Bruising of skin, blood in ear canal, depression, lethargy, ataxia, paresis
     b. Etiology
        (1) Often caused when bird flies into a stationary object (e.g., window pane, wall, or fan)
        (2) When bird is stepped on or an object is dropped on the bird
        (3) Mate aggression
     c. Diagnosis
        (1) History and physical examination
        (2) Neurologic examination
     d. Therapy
        (1) Supportive care
        (2) Fluids
        (3) O_2 cage
        (4) Corticosteroids (see Chapter 26)
        (5) Lasix (see Chapter 26)
  2. Seizures (see Chapter 10)
     a. Clinical signs: Depending on phase of seizure, the signs include lethargy, depression, convulsions, ataxia, and inability to stand
     b. Etiology
        (1) Central nervous system disease resulting from trauma[1]
        (2) Cardiovascular disease
        (3) Hepatopathic conditions — damaging liver
        (4) Renal disease
        (5) Endocrinopathic conditions
        (6) Exposure to toxins
        (7) Poor nutrition
     c. Diagnosis
        (1) History and physical examination
        (2) Neurologic examination
        (3) CBC, chemistry panel
        (4) Bile acid levels
        (5) Toxicology tests
        (6) Radiographs[1]
     d. Therapy (control of seizures)
        (1) Diazepam (see Chapter 26)
        (2) Supportive care
            (a) Intravenous, IO, or subcutaneous fluids
            (b) Calcium ethylenediaminetetraacetic acid (see Chapter 26)
            (c) Vitamins A, D_3, and E (see Chapter 26)
            (d) Nutritional support
            (e) Parenteral calcium[1,9] (see Chapter 26)

F. Traumatic Injuries
1. Soft tissue and musculoskeletal injuries
   a. Clinical signs: Thermal burns, conspecific and interspecific aggression, weather extremes, flying into objects, owner-induced trauma; often associated with the musculoskeletal and neurologic systems
   b. Diagnosis
      (1) History and physical examination
      (2) Observation of patient
   c. Therapy
      (1) Supportive care
      (2) Correction of underlying problem

G. Toxins[10]
1. Primarily ingested toxins but may be skin contact or inhaled
   a. Clinical signs: Vary, depending on factors such as type and amount of toxin, duration of toxicosis, and patient's size
   b. Etiology (also see Chapter 14)
      (1) Alcohol
      (2) Ammonia
      (3) Chlorine
      (4) Chocolate
      (5) Detergents
      (6) Furniture polish
      (7) Heavy metals (e.g., lead, zinc)
      (8) Insecticides
      (9) Matches
      (10) Nicotine, cigarette smoke
      (11) Pine oils (phenols)
      (12) Plants
      (13) Polytetrafluoroethylene
      (14) Rodenticides
   c. Diagnosis
      (1) History of ingestion and physical examination
      (2) CBC, chemistry panel
      (3) Toxicologic analysis of blood, secretions, excretions, and vomitus
      (4) Radiographs
   d. Therapy
      (1) Stabilize patient's condition
      (2) Prevent further exposure
      (3) Delay further absorption
      (4) Institute physiologic antagonist
      (5) Facilitate removal of absorbed toxins
      (6) Provide supportive care

III. Dehydration and Fluid Therapy

A. Normal Distribution of Water in the Body
1. Total body water in an adult bird accounts for approximately 60% of the body weight, and the percentage is even higher in a young bird.[5,8,11]
2. Extracellular water constitutes approximately 18% to 24% of body weight, depending on the method used to determine its volume and on the bird's age, sex, and lean body weight.[5,8]
3. Blood volume (cells and plasma) constitutes approximately 4.4% to 8.3 % of body fluid volume in chickens.[5,8] In other avian species percentages as high as 14.3% have been reported.[8]

B. Loss of Water or Dehydration
1. Serious problem in avian patients.
2. Makes rapid, correct replacement of fluid deficits more important in them than in other companion animal species.[12]

C. Total Body Water
1. Higher in chicks than in adult birds.
2. Clinically recognizable dehydration is more severe in younger birds.[11]

D. Sources of Water
1. Include ingested water, water in foods, and water produced through metabolic processes.
2. Fluids are lost through urine, respiration, and feces.

E. Assessing Hydration Status
1. Patient's history often indicates the etiology and magnitude of dehydration.
2. Water intake, abnormal loss from the body, and duration of clinical signs should be considered.

F. Physical Examination
1. May provide an assessment of clinically recognized dehydration resulting from decreased intake (caused by systemic illness, anorexia, depression) or increased losses (caused by polyuria, diarrhea, vomiting, panting, hyperthermia, or trauma).
2. Parameters used to distinguish these conditions include the following:
   a. Skin fold elasticity—dorsal aspect of the metatarsus and other featherless areas
   b. Turgescence, filling time, and luminal volume of basilic artery and vein[1,11,13] (filling time > 1 to 2 seconds in the basilic vein indicates dehydration > 7 %[1]; dehydration < 5% is clinically difficult to recognize)
   c. Appearance of the eyes, corneal hydration, and ocular pressure[12]
   d. Physical assessment of dehydration[15]
      (1) < 5% (difficult to detect)

      (2) 5% to 6% (subtle loss of skin elasticity)
      (3) 7% to 10%
         (a) Definite loss of skin elasticity
         (b) Prolonged filling time of basilic artery and vein
         (c) Dry mucous membranes (possible)
         (d) Loss of brightness and roundness in the eyes or sunken appearance to the eyes
      (4) 10% to 12%
         (a) Tented skin stands in place
         (b) Possible signs of shock
         (c) Muddy color to scales of feet
         (d) Dry mucous membranes
         (e) Cool extremities
         (f) Increased heart rate
         (g) Poor pulse quality
      (5) 12% to 15%
         (a) Extreme depression
         (b) Signs of shock
         (c) Death imminent
  e. Laboratory evaluation
      (1) PCV and total protein level: Quick, inexpensive tests that support the clinical diagnosis of dehydration or reveal it if the physical examination does not indicate dehydration.
         (a) May be normal even in dehydrated patients if compensatory mechanisms are active
         (b) Normal PCV for most birds is 35% to 55%[16,17]
            (i) PCV < 35% indicates anemia
            (ii) PCV > 55% indicates dehydration
         (c) Plasma total protein of most birds is 3.5 to 5.5 g/dl[1,16]
            (i) Total protein level generally increases 20% to 40% with dehydration; elevated PCV and serum protein level suggest dehydration
      (2) Plasma urea nitrogen may be an accurate indicator of dehydration in some birds (e.g., pigeons)[18]; increases of 6.5 to 15.3 times normal may indicate dehydration.[17,18]
      (3) Urinalysis
      (4) In most cases that require supportive care (severe trauma or disease), a 5% to 10% level of dehydration may be assumed.

G. Calculation of Fluid Requirements and Planning Daily Fluid Therapy

1. The goal of fluid therapy is to correct fluid and electrolyte imbalances while meeting the patient's daily fluid maintenance requirements. When planning fluid therapy, the veterinarian should consider the status of the cardiovascular and renal systems.[11]
2. These questions should be considered[19]:
  a. Is fluid therapy indicated?

(1) Disease state and patient's physical condition
   (a) Decreased fluid intake (anorexia)
   (b) Increase fluid loss (vomiting, diarrhea, or polyuria)
   (c) Trauma or burns
(2) Surgical cases

b. What type of fluid should be given? (Table 2-1)
   (1) Isotonic crystalloids: Lactated Ringer's solution, Normosol-R, and normal saline (0.9%), 0.45% saline plus 2.5% dextrose, and Plasma-Lyte M
   (2) Nonisotonic crystalloid: 5% Dextrose in water and hypertonic saline
   (3) Colloids: Plasma, dextrans, and hetastarch

c. What route should be used for administration?
   (1) Intravenous route[6,11,19]
      (a) Preferred for critically ill patients, when blood or fluid loss is severe, or with acute fluid loss
      (b) Available sites include the jugular, basilic, or medial metatarsal vein
         (i) Jugular vein is most useful because it allows delivery of large volumes and administration of hypertonic or potentially irritating solutions; recommended for patients with long necks.
         (ii) Basilic vein is recommended for species weighing at least 300 to 400 g.

Table 2-1
Commonly used fluids[20]

Fluid type	Indications	Contraindications
Lactated Ringer's solution	Good choice for shock, rehydration, and diuresis	Contains Ca^{++}, which is not recommended for hypercalcemic patients May exacerbate hyperkalemia or hypernatremia
0.9% Saline	Shock therapy Hypercalcemia, hyperkalemia Hyponatremia Vomiting caused by GI obstruction	May exacerbate volume overload Liver failure Hypertension
0.45% Saline/2.5% dextrose	Hypernatremia, diuresis Liver failure Sepsis	Hypovolemic shock
Plasma-Lyte M	Maintenance fluid Heart failure, liver disease, and hypertension	Hypovolemic states Hyponatremia and hypokalemia
5% Dextrose	Short-term treatment of heart failure	Subcutaneous route Do not give large boluses IV
Hypertonic saline	Hypovolemic shock Head trauma without intracranial hemorrhage Sepsis, shock, burns	Dehydration Cardiogenic shock Intracranial hemorrhage

GI, gastrointestinal; *IV*, intravenous

hypernatremia ↑ plasma concentration sodium ions
hyperkalemia ↑ potassium ions

(iii) Medial metatarsal vein is especially useful in long-legged birds (cranes, storks, herons, egrets).
- (c) Catheters must be sutured or taped into position
- (d) Advantages of intravenous route
    - (i) Allows precise fluid administration
    - (ii) Allows administration of a large volume over several minutes
- (e) Disadvantages of intravenous route
    - (i) Requires maintenance of indwelling catheter if a butterfly catheter is not used.
    - (ii) Butterfly catheter frequently requires physical restraint of the patient.
    - (iii) Severe hemorrhage or hematoma formation is possible.
    - (iv) Patient may remove the catheter.
    - (v) Catheter must be replaced every 2 or 3 days.

(2) Intraosseous route
- (a) Useful in small patients or when peripheral veins are too small or constricted
- (b) Available sites include the distal ulna and proximal tibiotarsus
- (c) Advantages of intraosseous route
    - (i) With practice can be placed quickly
    - (ii) Provides excellent access to peripheral circulation, with absorption equivalent to intravenous route
- (d) Disadvantages of intraosseous route
    - (i) Placement and maintenance of catheter
    - (ii) Placement is painful unless patient has been anesthetized (however, the catheter may be placed without anesthesia in an emergency to achieve venous access)
    - (iii) Requires strict asepsis or the risk of osteomyelitis is greater
    - (iv) May require more than one attempt at placement
    - (v) Extravasation of fluid can occur (if more than one hole is present in the cortex)

(3) Subcutaneous route
- (a) Inappropriate for critically ill patients
- (b) Available sites include the inguen (inguinal region); propatagium or wing web (we prefer not to use this site, especially in patients undergoing rehabilitation for release); interscapular region
- (c) Advantages of subcutaneous route
    - (i) Can accommodate a significant amount of fluid without the risk of cardiac overload
    - (ii) Safe and reliable
- (d) Disadvantages of subcutaneous route
    - (i) Only a small volume (5 to 10 ml/kg) should be given per site

      (ii) May cause local vasodilation in birds that are in shock or hypothermic
      (iii) Possibility of poor absorption makes this route inappropriate for critically ill patients
    (4) Oral route
      (a) Preferred route if patient is conscious, standing, and normotensive
      (b) Advantages of oral route
        (i) Most physiologic route
        (ii) Ease of administration
        (iii) Allows administration of fluids with a wide variety of compositions
      (c) Disadvantages of oral route
        (i) Should not be used in critically ill, recumbent, or highly excitable patients because regurgitation, vomiting, and aspiration may occur
  d. How rapidly should fluid be given?
    (1) Birds can tolerate intravenous boluses of 10 ml/kg if they are given slowly (over 5 to 7 minutes)[11]
    (2) Monitor for signs of fluid overload[11]
  e. How much fluid should be given? (See no. 4 below.)
    (1) Suggested intravenous bolus volumes for some species[12]

    | Canary | 1 ml |
    | Budgerigar | 1 to 2 ml |
    | Cockatiel | 2 to 3 ml |
    | Conure | 4 to 6 ml |
    | Amazon parrot | 8 to 10 ml |
    | Macaw | 15 to 25 ml |

  f. When should fluid therapy be discontinued?
    (1) When hydration has been restored and the patient is able to maintain hydration without assistance
    (2) As the patient recovers, fluid therapy usually is tapered by decreasing the amount administered by 20% to 25% per day
3. Daily maintenance water requirement of healthy psittacine birds is approximately 50 ml/kg/day; the requirement for Passeriformes is much higher.
4. Calculation of fluid deficit

Body weight (g) × Percent dehydration (in decimal form) = Estimated fluid deficit (ml)

  a. Half the total fluid deficit should be replaced within the first 12 to 24 hours.[1,6]
  b. The remaining 50% is divided over the following 48 hours, to be administered with daily maintenance requirements.
  c. Example: Amazon parrot (weight is 500 g), 8% dehydration
    (1) Maintenance fluid requirement: 50 ml/kg × 0.5 kg = 25 ml
    (2) Fluid deficit: 500 g × 0.08 = 40 ml

- (3) Day 1: Provide maintenance = 25 ml
  - and half of fluid deficit = 20 ml
  - 45 ml
- (4) Day 2: Provide maintenance = 25 ml
  - and half of fluid deficit = 20 ml
  - 45 ml
   d. If force-feeding patient two to three times daily, 75% of maintenance fluid requirements may be met.[21]
   e. Warming fluids to 38° to 39°C (100.4° to 102.2°F) may prevent or correct hypothermia.[11]
5. Plasma and dextrans
   a. Very little information is available that fully assesses the use of colloids in avian supportive care.
   b. Plasma is an excellent source of a natural colloid or use albumin.
   c. Dextrans are high-molecular-weight polysaccharides with a particle size similar to that of albumin.
      (1) Effect is similar to that of hypertonic saline.
      (2) Duration is much longer than that of hypertonic saline (approximately 24 hours).[6]
      (3) A dramatic improvement in birds in shock has been reported with 6% dextran given at 10 to 20 ml/kg.[22]
      (4) Adverse effects reported in mammals include hypervolemia, hemorrhagic diathesis, and anaphylaxis.
6. Monitoring the efficacy of fluid therapy
   a. Patient should gain weight as hydration is reestablished.
   b. Monitor patient daily with PCV/TP and with physical examination if patient can tolerate it.
7. Complications of fluid therapy[11]
   a. Overhydration or fluid overload
      (1) Common complication of fluid therapy
      (2) Manifests as dilution anemia, hypoproteinemia, pulmonary edema, or excessive workload for the heart
      (3) Patients with underlying systemic disease (e.g., renal, cardiac, metabolic, or vascular disease) are especially at risk
   b. Electrolyte imbalances involving potassium, sodium, and chloride
   c. Bicarbonate administration (Administer carefully to prevent metabolic alkalosis, respiratory acidosis, or hyperosmolarity.)
   d. Phlebitis, osteomyelitis, hematomas, and skin sloughing from extravasation of irritating substances
   e. Aspiration pneumonia from orally administered fluids or gavage feeding
8. Supplementation of fluid[19]

H. Blood Transfusions
1. Blood transfusion appears to be useful in patients with a PCV ≤ 20%.
   a. Every attempt should be made to find homologous donors.
   b. Single heterologous transfusions have been successful.[20,22]

c. Transfusion should be 10% to 20% of recipient's calculated blood volume.[1]
d. Normal blood volume of a bird is approximately 10% of its body weight in grams.
2. Collection and storage of blood products
   a. Anticoagulant of choice is acid citrose dextrose (ACD, ACD solution, formulation A; Baxter Health Care Corp., Deerfield, Ill.)
   b. Amount used is 0.15 ml of ACD per milliliter of blood.[5]
3. Administration of iron dextran (10 mg/kg given intramuscularly [Fig. 2-6] and repeated in 7 to 10 days if needed) and vitamin B may benefit birds with severe hemorrhage and anemia.[23]
4. Complications
   a. Anaphylactic reaction to donor blood
   b. See Complications of Fluid Therapy, above

IV. Drug Therapy

A. Antimicrobial Agents
1. Antimicrobial therapy should be based on appropriate culture and sensitivity testing; however, this is not always possible.
2. Antibiotics often are indicated in critically ill patients in which an infectious etiology is suspected.
3. The most common pathogens recovered from avian emergency cases are gram-negative bacteria; therefore antibiotics should be effective against a host of gram-negative bacteria.

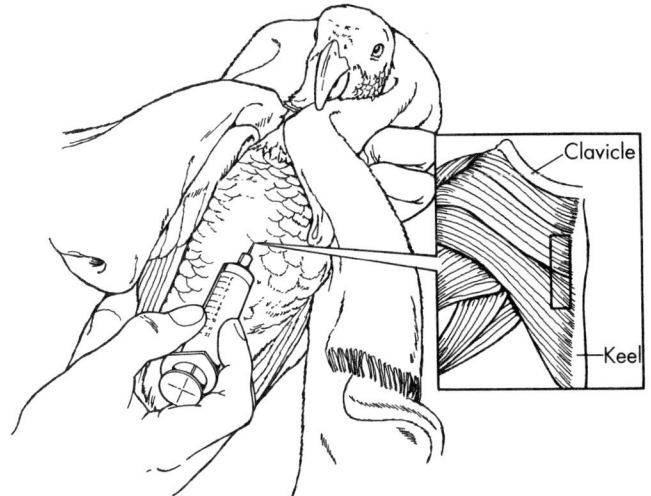

Fig. 2-6 Intramuscular injection.

4. Bactericidal antimicrobials are preferred in many cases because the immune system of debilitated patients often is compromised.
   a. Commonly used bactericidal antibiotics (see Chapter 26)
      (1) Penicillins (amoxicillin, ampicillin, carbenicillin, piperacillin, and ticarcillin)
      (2) Cephalosporins (cefotaxime)
      (3) Fluoroquinolones (enrofloxacin and ciprofloxacin)
      (4) Aminoglycosides (amikacin)
      (5) Sulfonamides (trimethoprim-sulfa)
5. Other commonly used bacteriostatic antibiotics are chloramphenicol and doxycycline.
6. Route of administration
   a. Parenteral routes (intramuscular, intravenous, intraosseous) are preferred.
   b. Subcutaneous and oral routes are reserved for maintenance therapy or for patients whose size dictates the use of these methods.
   c. Medicated water or food is useful in some situations; however, many ill birds do not eat or drink enough to achieve therapeutic levels.

B. Corticosteroids

1. Administration of corticosteroids for the treatment of shock remains controversial.
   a. Beneficial effects include improved capillary membrane integrity, stabilization of lysosomal membranes, improved tissue perfusion and microcirculation, and gluconeogenesis.
   b. Corticosteroid use for cases of shock that are not caused by sepsis is viewed favorably.[7]
2. Water-soluble products should be given intravenously or intramuscularly for the first 24 to 48 hours.
   a. Dexamethasone: 2 mg/kg
   b. Prednisolone sodium succinate: 2.4 mg/kg
3. Steroids that have a rapid effect are preferred in emergency situations.
4. Adverse effects[6,13]
   a. Immunosuppressive effects (Corticosteroids should not be used in patients with septicemia or an infectious disease.)
   b. Water and salt retention
   c. Delayed would healing
   d. Adrenal insufficiency
   e. Gastrointestinal ulceration
5. Most adverse effects are associated with chronic use.
6. Doses should be tapered when discontinuing steroid therapy.

C. Bicarbonate Therapy

1. Critically ill patients often are in a state of metabolic acidosis.
2. In severe acidotic states, fluid therapy should be augmented with bicarbonate therapy.
3. Bicarbonate deficit is estimated by subtracting the blood bicarbonate value from the normal avian bicarbonate value.[6]

Bicarbonate deficit (mEq/L) = 20 mEq/L–blood bicarbonate level (mEq/L)

Bicarbonate required (mEq/L) = Deficit × body weight (kg) × 0.4

- a. Generally, if no means are available for determining the patient's bicarbonate level, 1 mEq/L of bicarbonate can be administered at 15- to 30-minute intervals.
- b. Do not exceed 4 mEq/kg/day.

4. Adverse effects of bicarbonate administration
   - a. Hyperkalemia, especially in cases of severe tissue injury, catabolic state, or severe renal disease
     - (1) The addition of calcium gluconate (0.5 ml/kg), along with glucose administration, helps prevent hyperkalemia.[6,15]
   - b. Vomiting
   - c. Hypotension
   - d. Death

D. Vitamins

1. Potassium
   - a. Patients that are anorexic, vomiting, or hypokalemic from other metabolic problems may benefit from supplementation of lactated Ringer's solution with potassium (0.1 to 0.3 mEq/kg), not to exceed 11 mEq/day.[6,15]
2. Chronically ill birds or birds eating a poor diet benefit from multivitamin and mineral supplementation.[1,14]
   - a. Vitamins A, D_3, and E
     - (1) May be beneficial in patients that are debilitated, that have a poor diet, or that are suspected of having a calcium deficiency, neurologic disorder, or compromised immune system.
     - (2) Care must be taken in administering these vitamins to patients given formulated diets to prevent toxicities from oversupplementation.[1]
     - (3) 0.1 to 0.2 ml/300 g given once weekly.
   - b. Vitamin B complex
     - (1) Indicated for patients with anorexia and anemia; given on initial examination and daily
     - (2) 1 to 3 mg of thiamine per kilogram given daily or weekly
     - (3) May also add to fluids
   - c. Vitamin K_1
     - (1) Indicated in patients with extreme blood loss, anemia, or hemorrhagic or hepatic disease, or when warfarin toxicity is suspected
     - (2) 0.2 to 2.5 mg/kg as needed
   - d. Iron dextran
     - (1) Indicated in patients with extreme blood loss or anemia
     - (2) 10 mg/kg, repeated in 7 to 10 days as needed
     - (3) Used with extreme caution in birds subject to hemochromatosis
   - e. Calcium
     - (1) Indicated for patients that have a poor diet; in cases of egg laying, egg binding, or neurologic disorders; or in birds sus-

pected of having hypocalcemia (certain species, such as African grey parrots)
- (2) Calcium gluconate (50 to 100 mg/kg intravenously slowly, or intramuscularly diluted)[24]
- (3) Calcium glubionate (Neo-Calglucon), 25 mg/kg orally,[25] up to 150 mg/kg orally[12]; 23 mg/kg orally in psittacine neonates[26]

V. Nutritional Support

A. Normal Nutritional Requirements
1. The exact nutritional requirements of birds other than poultry have not yet been determined.
2. Even less is known about the nutritional requirements of birds that are critically ill because of disease or inadequate ingestion of nutrients.

B. Nutritional Requirements of Sick Patients
1. Birds have a high metabolic rate and are unable to function for extended periods on their energy reserves.[1,12]
2. Stress, sepsis, trauma, severe illness, and surgery increase the demand for nutrients (hypermetabolism) and cause the release of several hormones, including catecholamines, corticosteroids, and glucagon, that further increase the metabolic rate.[1,27]
3. When this increase in metabolism occurs with a decrease in food intake or inadequate absorption of nutrients, fat oxidation cannot keep up with energy needs, and body proteins subsequently are broken down for gluconeogenesis; glycogenolysis also occurs at a high rate.
4. Glucagon plays a more prominent role in energy metabolism in birds than in mammals.
   a. In mammals, insulin is a strong suppressor of lipolysis, whereas glucagon, epinephrine, and cortisol are strongly lipolytic.[1]
   b. Unlike mammals, birds do not exhibit a decrease in free fatty acid concentration with an increase in the blood insulin concentration.[1]

C. Assessment of Nutritional Status
1. Dietary and clinical history
2. Duration of clinical signs (e.g., anorexia, regurgitation, vomiting, diarrhea)

D. Nutritional Support of the Avian Patient
1. The goal is to provide the patient with a highly nutritious, easily digestible and absorbable product that halts protein catabolism and promotes protein synthesis.[12,27]
2. Indications
   a. Anorexia
   b. Loss of 5% to 10% of body weight
   c. Decrease in serum albumin or severe, continuing loss of protein (or both)[28]

3. Enteral feeding is the preferred route for administering nutritional products.

E. **Calculation of Energy Requirements**
   1. Basal metabolic rate (BMR) = K × (weight in kg)$^{0.75}$
      a. Passerine birds: K = 129
      b. Nonpasserine birds: K = 78
   2. Normal activities, including digestion, absorption, and formation and excretion of wastes, require additional energy.[1]
      a. Maintenance energy (ME) is the energy required for the above activities plus the BMR.[1]
      b. In passerine birds the ME varies from 1.3 to 7.2 times the BMR, depending on the energy required for a particular activity.
      c. Adjustments, as multiples of the ME, may be required, depending on the disease condition.

F. **Enteral Nutritional Support**
   1. Easiest and most physiologic method of providing nutritional support
      a. Gavage feeding
      b. Rubber French catheters
   2. Proper restraint techniques
      a. Maintain patient in an upright position next to body or lightly wrapped in a towel.
      b. Pass lubricated feeding instrument into the crop.
      c. Observe oral cavity for overflow of food from the crop.
      d. Release bird immediately into cage.
      e. Enteral feeding should be performed last, after diagnostics or other supportive measures, because handling the bird at this time may result in regurgitation and vomiting of food.
   3. Always warm enteral product before feeding.
   4. Frequency of feedings varies according to the amount to be fed.
   5. Suggested volumes to feed[12]

Canary	0.5 to 1 ml
Budgerigar	1 to 3 ml
Cockatiel	3 to 8 ml
Conure	10 to 12 ml
Amazon parrot	15 to 20 ml
Macaw	20 to 40 ml

   6. Caloric density of enteral feeding products varies, especially if homemade products are used.[1]
      a. Isocal: 1.0 kcal/ml
      b. Isocal HCN: 2.0 kcal/ml
      c. Traumacal: 1.5 kcal/ml
      d. Pulmocare: 1.5 kcal/ml
      e. Ensure Plus: 1.5 kcal/ml
      f. Clinicare Canine: Approximately 1.0 kcal/ml
      g. Clinicare Feline: Approximately 1.0 kcal/ml
      h. Hills Canine/Feline A/D: 1.2 kcal/ml

i. Eukanuba Canine/Feline Nutritional Recovery Formula: 1.9 kcal/ml
j. Exact Hand Feeding Formula: 3.98 kcal/g

7. Calculation of nutritional requirements
Example: 1.0 kg Macaw is brought in for treatment of anorexia and depression that have occurred secondary to severe trauma.

$$BMR = 78 \times (1.0)^{0.75} = 78 \text{ kcal}$$
$$ME = 1.5 \times 78 \text{ kcal/day} = 117 \text{ kcal}$$

Severe trauma adjustment:

1.1 to 2.0 ($\times$ 117/kcal) = 128.7 to 234 kcal/day as the total energy requirement for the macaw

G. Parenteral Nutritional Support

1. Indications
   a. Enteral feeding impossible
   b. Respiratory tract cannot be protected[28,29]
2. Parenteral products
   a. Aminosyn: 8.5% amino acid solution (0.085 g of protein per milliliter, equivalent to 0.34 kcal/ml)
   b. Liposyn: 20% lipid solution (2 kcal/ml)
   c. Dextrose: 50% solution (1.7 kcal/ml)
3. Use of parenteral products requires close monitoring of serum or plasma chemistry and electrolyte value; adjust nutritional formulas daily as needed.[1]
4. Energy requirements
   a. Calculate basal metabolic rate: $BMR = K \times (\text{weight in kg})^{0.75}$
   b. Calculate maintenance energy requirements: $ME = \text{Adjustment} \times BMR$
   c. Determine grams of protein required.
      (1) 6 g/kg has been recommended for most avian species.[5,29]
      (2) Calories provided by protein are calculated *in addition* to the ME requirements, or they are *subtracted from* the total calories needed if hepatic or renal disease is suspected.
   d. Determine nonprotein caloric requirement.
      (1) 60% of nonprotein calories are supplied as lipid unless hyperlipidemia or severe hepatic disease is present.
      (2) 40% of nonprotein calories are supplied as carbohydrates (dextrose).
         (a) Administer half the amount of dextrose on day 1 and increase to full amount on day 2 if no problems with glucosuria develop.
   e. Electrolyte requirements
      (1) Some amino acid solutions contain electrolytes.
      (2) For those that do not, add 20 ml total parenteral nutrition electrolytes and 5 ml KPO_4 to each *liter* of solution if electrolyte levels are normal. The resulting solution contains: Na^+

(35 mEq/L), CL^- (35 mEq/L), K^+ (42 mEq/L), PO_4 (15 mmol/L), Ca^{++} (4 mEq/L), and Mg^{++} (5 mEq/L).
      (3) Further supplementation with KCL may be required if the patient is hypokalemic.
    f. Vitamin requirements
      (1) Vitamin K: 0.5 mg/kg given subcutaneously on day 1, then once weekly
      (2) Multivitamin for infusion: 3 ml/10 kg/day in dogs and 3 ml/day in cats
    g. Fluid requirements
      (1) If additional fluid therapy is necessary, crystalloids may be added to the total parenteral nutrition solution or given separately via another catheter.
      (2) Indications include continuing losses (e.g., vomiting, diarrhea, exudation, or diuresis)[30]
    h. Solutions should be mixed aseptically.
  5. Vascular access: Studies using intraosseous catheters for administration of fluid, parenteral nutritional products, and drugs in birds have been performed; however, the full effect of intraosseous administration in birds has not been fully evaluated.[1,31,32]
    a. Catheters used for total parenteral nutrition should not be used for other purposes, including blood sampling or administration of other medications.
    b. Catheters may be intravenous, intraosseous, or attached to a vascular access device.[29,33,34]
  6. Administration
    a. Solutions should be mixed aseptically.
    b. Vascular access: Studies using intraosseous catheters for administration of fluid, parenteral nutritional products, and drugs in birds have been performed; however, the full effect of intraosseous administration in birds has not been fully evaluated.[1,31,32]
    c. Constant-rate infusion is most commonly used.
    d. Multiple boluses over a 24-hour period also have been given.
    e. Multiple short-duration infusion can be used.[29,33,34]
    f. Some clinicians change the administration set and bandage regularly (every 24 to 96 hours); other clinicians believe this practice increases the risk of line contamination.
    g. Catheter is changed only when necessary.
  7. Complications
    a. Metabolic: A variety of metabolic complications have been reported in human and small animal patients.
      (1) Azotemia, as a result of dehydration (a comparable rise in uric acid may be expected in avian patients; however, an increase in uric acid was not seen in pigeons receiving total parenteral nutrition)[29]
      (2) Hyperglycemia

(3) Hypophosphatemia (severe if supplementation is not provided); clinical signs include weakness and tremors
(4) Moderate lipemia that may persist for several days
(5) Retention of fluids or edema
(6) Hyperammonemia, allergic reactions, and hypermagnesemia have been reported in human patients

b. Catheter complications: Most commonly thrombophlebitis and catheter occlusion; however, the following also are seen:
(1) Edema
(2) Cellulitis
(3) Thromboembolism
(4) Sepsis
(5) Disconnection or breakage

c. Patient monitoring while administering total parenteral nutrition include daily physical examinations, weighing, and determination of serum and plasma chemistry and electrolyte values as frequently as possible.

VI. Oxygen Therapy

A. Oxygen Cages
Oxygen cages should be standard equipment for avian practice.
1. O_2 therapy is beneficial in most critically ill patients.
2. O_2 therapy may not be beneficial in severely anemic patients or those in circulatory shock; these patients may benefit more from adequate volume expansion and transfusions to improve tissue oxygenation.

B. Respiratory System
Birds have a unique respiratory system that allows them to extract oxygen efficiently from the environment; therefore they are far better able to withstand hypoxic episodes than are mammals.[1]
1. Dyspneic birds placed in O_2 cages tend to stabilize sooner; hyperoxygenation counteracts tachypnea and increases respiratory efficiency.[4,6]
2. 40% oxygen saturation is recommended when using an O_2 cage.
3. Administering oxygen at 50 ml/kg/minute with a face mask provides 40% O_2 saturation.

VII. Nebulization

A. Benefits
Nebulization, with or without oxygen, may be beneficial in an avian patient with respiratory disease.
1. Antibiotics and other medications may be administered.
2. Particles 3 to 7 μm in diameter generally are deposited in the trachea and on mucosal surfaces. (Particles must be < 3 μm in diameter to reach the air sacs and lungs.)

3. Nebulization sessions should last 10 to 30 minutes.
4. Many intravenous antibiotics can be mixed with saline for nebulization (see Chapter 26).
5. Mucolytic agents are not used.

VIII. Supplemental Heat[5,35,36]

A. **Regulation of Body Temperature**
Regulation of body temperature depends on feather condition, fat and muscle content, hydration status, food intake, and respirations.
1. Poor body condition greatly affects heat retention.
2. Many debilitated patients are hypothermic, making heat regulation crucial.
3. Core body temperature in birds is 38° to 42.5° C (100° to 108° F).
4. Heat is lost in the following ways:
    a. Radiation: From the surface of the patient's body.
    b. Convection: Air surrounding the patient is warmed and heat is lost from the bird.
    c. Conduction: Heat is lost through patient contact with conductive surfaces (e.g., metal).
5. Environment (incubator, intensive care cage, heating pad, heat lamp) should keep the temperature equal to or just below the patient's normal body temperature to eliminate heat loss.
6. Heat should be supplied by conduction, convection, and radiation to prevent heat loss by these routes in the patient.
7. Humidifying the air seems to reduce heat loss from evaporation.

REFERENCES

1. Quesenberry KE, Hillyer EV: Supportive care and emergency therapy. In Ritchie BW, Harrison GJ, Harrison LR, editors: *Avian medicine: principles and application,* Lake Worth, Fla, 1994, Wingers, pp 382-416.
2. Rupley AE: *Manual of avian practice,* Philadelphia, 1997, Saunders pp 502-517.
3. Clubb S: Therapeutics. In Harrison GJ, Harrison LR, editors: *Clinical avian medicine and surgery,* Philadelphia, 1986, Saunders, pp 327-355.
4. Harris DJ: Care of the critically ill patient, *Seminars in Avian and Exotic Pet Practice* 3(4):175-179, 1994.
5. Jenkins JR: Avian critical care and emergency medicine. In Altman RB et al, editors: *Avian medicine and surgery,* Philadelphia, 1997, Saunders, pp 839-863.
6. Hernandez M, Aguilar RF: Steroids and fluid therapy for treatment of shock in the critical avian patient, *Seminars in Avian and Exotic Pet Practice* 3(4):190-199, 1994.
7. Tobias TA, Shertel ER: Shock: concepts and management. In DiBartola SP, editor: *Fluid therapy in small animal practice,* Philadelphia, 1992, Saunders, pp 436-470.
8. Sturkie PD, Griminger P: Body fluids: blood. In Sturkie PD, editor: *Avian physiology,* New York, 1986, Springer-Verlag, pp 102-109.
9. Worell AB: Therapy of noninfectious avian disorders, *Semin Avian Exotic Pet Practice* 2(1):42-47, 1993.
10. LaBonde J: Toxicity in pet avian patients, *Semin Avian Exotic Pet Med* 4(1):23-31, 1995.

11. Abou-Madi N, Kollias GV: Avian fluid therapy. In Kirk RW, Bonagura JD, editors: *Current veterinary therapy*, XI, Small animal practice, Philadelphia, 1992, Saunders, pp 1154-1159.
12. Huff DG: Avian fluid therapy and nutritional therapeutics, *Seminars in Avian and Exotic Pet Practice* 2(1): 13-16, 1993.
13. Jones MP: Emergency care and nutritional support of the avian patient, *Proc Annu Conf Assoc Avian Vet* :325-331, 1994.
14. Tully TN: Formulary. In Altman RB et al, editors: *Avian medicine and surgery*, Philadelphia, 1997, Saunders, pp 671-688.
15. Redig PT: Medical management of birds of prey: a collection of notes on selected topics, St Paul, Minn, 1993, The Raptor Center, University of Minnesota.
16. Campbell TW: *Avian hematology and cytology*, Ames, Iowa, 1995, Iowa State University Press.
17. Martin HD, Kollias GV: Evaluation of water deprivation and fluid therapy in pigeons, *J Zoo Wildlife Med* 20(2):173-179, 1989.
18. Lumeji JT: Plasma urea, creatinine, and uric acid concentrations in response to dehydration in racing pigeons, *Avian Pathol* 16:377-382, 1987.
19. DiBartola SP: Introduction to fluid therapy. In DiBartola SP, editor: *Fluid therapy in small animal practice*, Philadelphia, 1992, Saunders, pp 321-340.
20. Trepanier LA: Fluid therapy in the critical patient (unpublished).
21. Rosskopf WJ et al: Pet avian emergency care, *Proc Annu Conf Assoc Avian Vet* :341-356, 1991.
22. Redig PT: Fluid therapy and acid-base balance in the critically ill avian patient, *Proc Annu Conf Assoc Avian Vet* :59-74, 1984.
23. Rupley AE: Emergency procedures: recovering from disaster, *Proc Annu Conf Assoc Avian Vet* :249-256, 1997.
24. Ritchie BW, Harrison GJ: Formulary. In Ritchie BW, Harrison GJ, Harrison LR, editors: *Avian medicine: principles and application*, Lake Worth, Fla, 1994, Wingers, pp 457-478.
25. Bauk L: *A practitioner's guide to avian medicine*, Lakewood, Colo, 1993, American Animal Hospital Association, pp 5, 36.
26. Joyner KL: Pediatric therapeutics, *Proc Annu Conf Assoc Avian Vet* :188-199, 1991.
27. Pollock CG: Practical total parenteral nutrition, *Proc Annu Conf Assoc Avian Vet* :263-278, 1997.
28. Labota MA: Nutritional management of the critical care patient. In Kirk RW, Bonagura JD, editors: *Current veterinary therapy*, XI, Small animal practice, Philadelphia, 1992, pp 117-125.
29. Degernes LA et al: Administration of total parenteral nutrition in pigeons, *Am J Vet Res* 55(5):660-665, 1994.
30. Grant JP: *Handbook of total parenteral nutrition*, Philadelphia, 1992, Saunders.
31. Lamberski N, Daniel GB: Fluid dynamics of intraosseous fluid administration in birds, *J Zoo Wildlife Med* 23(1):47-54, 1992.
32. Ritchie BW et al: A technique of intraosseous cannulation for intravenous therapy in birds, *Comp Cont Ed Pract Vet* 12(1):55-59, 1990.
33. Harvey-Clark C: Clinical and research use of implantable vascular access ports in avian medicine, *Proc Annu Conf Assoc Avian Vet* :191-208, 1990.
34. Degernes LA et al: Preliminary report on the use of total parenteral nutrition in birds, *Proc Annu Conf Assoc Avian Vet* :19-20, 1992.
35. Loudis BG, Sutherland-Smith M: Methods used in the critical care of avian patients, *Seminars in Avian and Exotic Pet Practice* 3(4):180-189, 1994.
36. Whittow GC: Regulation of body temperature. In Sturkie PD, editor: *Avian physiology*, New York, 1986, Springer-Verlag, pp 222-246.

3

Dyspnea and Other Respiratory Signs

Thomas N. Tully, Jr.

I. Diagnostic Workup Plan

The extensive avian respiratory system is susceptible to many environmental insults, nutritional deficiencies, and infectious and noninfectious disease processes. To develop an appropriate treatment plan, the veterinarian must have a thorough understanding of avian respiratory anatomy and of the individual patient's history. Owners need to understand the circumstances of the disorder so that they can prevent or control recurrence. An appropriate diagnostic plan follows the parameters established in this chapter (Box 3-1). For many avian respiratory patients, deviation from a correct diagnostic plan can lead to treatment failure.

A. Respiratory Diagnostic Workup Plan
1. Animal (species of bird)
2. Environment
3. Diet (see Chapter 18)
4. Problems (How long has the bird had the presenting problem? Does it have a history of chronic problems?)
5. Review of body systems with owner
6. Ownership (Had the owner recently acquired the patient? If not, what is the length of ownership? Any recent avian additions to the household?)

The basic history and initiation of a respiratory diagnostic protocol provide a foundation for more extensive diagnostic tests and an understanding of the problem. However, further investigation requires a basic understanding of the avian respiratory system.

Box 3-1

Diagnostic Plan for Upper and Lower Respiratory Tract Disease

I. Upper Respiratory Tract Disease
 1. History
 2. External physical examination
    a. Cere
    b. Nares
    c. Oropharynx (choanal slit)
    d. Eyes
 3. Auscultation
 4. Minimum database
    a. Weight
    b. Culture, Gram stain: Choana
    c. Complete blood count
    d. Plasma chemistry panel
 5. Additional tests
    a. Chlamydial testing
    b. Mycoplasma testing
    c. Radiographs
    d. Ultrasound scan of the sinus
    e. Endoscopic examination
    f. Infraorbital sinus aspiration and culture
    g. Aspergillosis testing
    h. Avian tuberculosis testing
    i. Nasal flush and culture

II. Lower Respiratory Tract Disease
 1. History
 2. External physical examination
    a. Trachea
    b. Coelomic palpation
    c. Same as for upper respiratory examination
 3. Auscultation
 4. Minimum database
    a. Weight
    b. Culture, Gram stain: trachea, air sacs
    c. Complete blood count
    d. Plasma chemistry panel
 5. Additional tests
    a. Chlamydial testing
    b. Mycoplasma testing
    c. Radiographs
    d. Endoscopic examination
    e. Endoscopic biopsy
    f. Fecal acid-fast stain
    g. Aspergillosis testing
    h. Tracheal wash for culture and cytologic studies
    i. Air sac wash for culture and cytologic studies

II. Basic Avian Anatomy[1]

A. Nares
1. In most psittacine birds, the nares are rounded and symmetric within the area of the cere.

B. Nasal Cavity
1. A complete nasal septum often divides the nasal cavity into two parts.
2. The rostral, middle, and caudal nasal conchae project from the lateral wall medially into the nasal cavity.

C. Infraorbital Sinus
1. The infraorbital sinus opens into the middle and caudal nasal conchae dorsally. It has numerous diverticula including the following:
    a. Rostral diverticulum: With its maxillary chamber, projects into the maxillary bill
    b. Preorbital diverticulum: Located cranial to the eye
    c. Suborbital diverticulum: Extends ventral to the eye
    d. Infraorbital
    e. Postorbital diverticulum: Located caudal to the eye
    f. Preauditory diverticulum: Located cranial to the external ear canal

g. Mandibular diverticulum: Projects into the mandible
h. Communicates with cervicocephalic air sac or cervicocephalic diverticulum

D. Paired Choanae
1. The anatomical terminus of the nasal cavity is at the level of the choanae.
2. Air exits the choanae (internal nares) at the level of the hard palate through the choanal slit.

E. Rima Glottis
1. The rima glottis is the laryngeal opening into the trachea.
2. The structure is not covered by an epiglottic cartilage.
3. The rima glottis can be easily visualized for intubation by pulling the tongue cranially or by moving a finger dorsally and cranially into the gular fold between each mandibular ramus.

F. Trachea
1. The trachea is composed of complete, signet-shaped rings. When the clinician is intubating the bird, care must be taken with the cuffed tubes to prevent pressure necrosis or inflammation.
2. The syrinx, or the true voice box of birds, is found at the tracheal bifurcation. The reduction in diameter and the presence of internal membranes at this point make it a common location for foreign bodies or fungal granulomas.
3. Emus have a tracheal cleft in the ventral aspect of the midcervical area; this area should be wrapped while administering anesthesia to prevent inflation.

G. Lungs
1. The lungs have a dorsal position in the thoracic cavity. They are fixed and firmly attached to the vertebral column and ribs.
2. The lungs undergo only limited changes in size and position during the respiratory process. The important pulmonary anatomical structures include the following:
   a. Primary bronchi: The primary bronchi have both extrapulmonary and intrapulmonary portions; they are formed at the bifurcation of the trachea at the syrinx.
   b. Secondary bronchi: The secondary bronchi are commonly divided into four groups, which are named by the area of the lungs they supply. Mediodorsal and medioventral secondary bronchi are part of the paleopulmonic system, a unidirectional system that promotes efficient movement of air through the lungs.
   c. Tertiary bronchi (parabronchi): The parabronchi are a large number of pathways that branch off the secondary bronchi. Their primary structures include the following:
      (1) Atria: Atria are channels that allow air to move from the parabronchi into the infundibula.
      (2) Infundibula: Each infundibulum leads to air capillaries, the primary site where gas exchange occurs.

(3) Air capillaries: Air capillaries, or air labyrinth, can anastomose with each other; the blood-gas barrier is much thinner in birds than in mammals, allowing greater diffusion of oxygen (O_2).
3. Birds have no functional diaphragm.
4. Ratites (ostriches, emus, and kiwis) are regarded as primitive species.
   a. Emus have only a paleopulmonic respiratory system, a structural arrangement that is more efficient for O_2 exchange than the combined paleopulmonic and neopulmonic system of the relatively advanced avian species.

H. Air Sacs
1. Cervicocephalic air sacs are connected to the infraorbital sinus and are considered another diverticulum of the sinus. They are not connected to the lungs or to the pulmonary air sac system.
2. Pulmonary air sacs
   a. Clavicular air sac: This air sac is located in the cranial mediastinum. Depending on the species, the air sac extends into the bones of the thoracic girdle, humerus, sternum, and sternal ribs. The blood vessels, nerves, esophagus, and trachea are found within pleural folds of the clavicular air sac.
   b. Cervical air sac: This sac surrounds the cervical vertebrae.
   c. Cranial thoracic air sac: This sac often is small in psittacines and gallinaceous birds.
   d. Caudal thoracic air sac: This sac surrounds the area of the liver and is commonly used for endoscopic examinations.
   e. Abdominal air sac: This sac surrounds the intestinal loops and, depending on the species, extends into the pelvic bones and femur. The abdominal air sac also is a site for endoscopic procedures.

I. Pneumatized Bones
1. The type of pneumatized bones varies, depending on the species.
2. Pneumatized bones are connected to the respiratory system through diverticula.
3. The humerus, clavicles, coracoid, cervical vertebrae, sternum, and sternal ribs are connected to the abdominal air sac.

III. Physical Examination with Emphasis on the Respiratory System[2,3]

A complete external physical examination protocol should be followed for each avian patient regardless of the presenting complaint.

A. Cere
1. The cere should be dry, smooth, symmetric, and normal in color. (The veterinarian should be familiar with the normal appearance for each species.)

2. No feather loss or matted feathers should be seen (matted feathers suggest an upper respiratory tract infection).

B. **Nares**
   1. The veterinarian should observe for symmetry, matted feathers, and nasal discharge or plugs.

C. **Oropharynx (Choanal Slit) (Fig. 3-1)**
   1. The opening should be smooth and thin with papillary projections extending toward the medial aspect of the choana. Blunting of the papillae suggests hypovitaminosis A.
   2. The veterinarian should observe for swelling, hyperemia, and papillary sloughing or blunting, which denote pathologic changes (observation can be improved by using a rigid endoscope or otoscope).
   3. White plaques can develop in the oropharynx as a result of fungal infections, bacterial granulomas, mycobacterial granulomas, trichomoniasis, capillary infections, pox lesions, and neoplasia.

D. **Eyes**
   1. Infraorbital or periorbital swelling or asymmetry associated with the position of the globe may indicate disease of the infraorbital sinus.

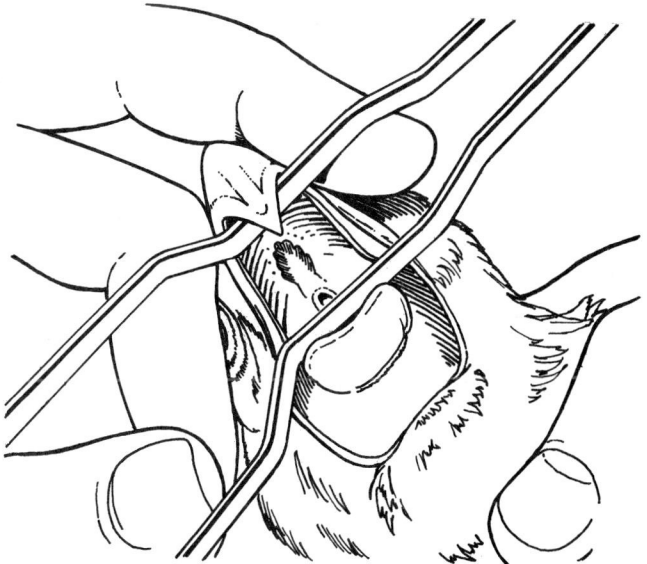

Fig. 3-1 Normal structure of a psittacine choana. Thin, sharp borders of the choana and pointed epithelial papillae project to the center of the cleft. Culture specimens are obtained from the rostral aspect of the choana.

E. Trachea
   1. The flexible trachea may be observed through the apterylae (the unfeathered areas of the skin of the neck). This inspection is particularly useful for detecting air sac mites in passerines by transillumination or, in larger birds, by using a flexible or rigid endoscope intraluminally.

F. Auscultation
   1. The stethoscope should be placed dorsally and ventrally over the thorax and the trachea.
   2. Abnormal sounds include clicks, wheezes, and squeaks.
   3. Tail feather movement up and down, called *tail bobbing*, usually indicates diminished respiratory capacity.
   4. With a normal ambient temperature and only brief handling, the respiratory recovery time should be < 3 minutes. A prolonged recovery time suggests primary or secondary respiratory involvement or cardiovascular disease, or both.

IV. Minimum Database

A. Required Procedures
   1. Physical examination
   2. Weight
   3. Gram stain: To determine if the normal flora and epithelial cells are present. Normal epithelial cells stain eosinophilic; basophilic staining cells are associated with hypervitaminosis A.
   4. Culture: Aerobic and fungal cultures of the choana and cloaca.
   5. Complete blood count: Total protein, glucose, calcium, PO_4, uric acid, creatine kinase, aspartate aminotransferase, lactate dehydrogenase, and bile acid levels. Glutamic dehydrogenase and cholesterol testing may be included.

B. Additional Tests
   1. Electrolytes
   2. Chlamydial testing
   3. Mycoplasma testing
   4. Psittacine beak and feather disease and polyomavirus testing
   5. Radiographs (survey and contrast)
   6. Ultrasound scan
   7. Tuberculosis testing (acid-fast stains or serologic testing, or both)
   8. Aspergillosis testing
   9. Endoscopic examination (coelomic observation, tissue or lung biopsy)
   10. Urinalysis
   11. Cytologic study and culture of aspirate (tracheal, infraorbital, sinus, air sacs) or biopsy[1]
   12. Electrocardiogram

13. Computed tomography scan
14. Examination of fecal droppings and material from choanal slit and trachea for parasites

C. Specific Respiratory Points
1. Auscultation
   a. A pediatric stethoscope is the best instrument for auscultation of the sinuses, trachea, and lungs, as well as the thoracic and abdominal air sacs.
   b. Increased inspiratory sounds often are associated with upper respiratory tract involvement.
   c. Increased expiratory sounds are associated with lower respiratory tract disease.
   d. Normally no alveolar sounds are noted in avian species because birds have nonexpandable air capillaries rather than expandable alveoli.
   e. Auscultation through a thin towel may help in detecting abnormal respiratory sounds.
   f. Auscultation of the heart is difficult because of the high heart rate in birds that are captured and held. Auscultating hand-fed or tamed birds on perches can enhance the ability to pick up changes in rhythms or abnormal sounds.
   g. An abnormal rate or rhythm is associated with systemic illness or heart disease.
2. Imaging (radiography and endoscopy with biopsy and culture)
   a. Radiographs and endoscopic examinations are effective diagnostic techniques for respiratory conditions.
   b. With generalized air sacculitis, air sac lines can be seen on lateral radiographs.
   c. The ventrodorsal view is the best radiographic view for detecting disorders of the lungs and pulmonary air sacs.
   d. Rhinography and sinography may aid in the diagnosis of upper respiratory tract problems.
3. Sinus aspiration
   a. Aspiration of the right and left infraorbital sinuses for diagnostic cultures and cytologic studies can be performed at a number of sites. Caution must be used if aspiration is done over the swelling[4] (Fig. 3-2).
4. Tracheal lavage
   a. General anesthesia must be used for tracheal lavage.
   b. The normal wash of a trachea has low cellularity, with few pulmonary macrophages or inflammatory cells.
   c. A sterile soft tube is placed into the trachea.
   d. Sterile saline (0.5 to 1.0 ml/kg body weight) is infused into the trachea and aspirated with a sterile syringe that is attached to the tube.
   e. Tracheal lavage (Place an 18- to 22-gauge Teflon catheter through the skin into the trachea.)

Fig. 3-2 Infraorbital sinus (preorbital diverticulum), may be used for sinus aspiration of cytologic and culture specimens.

D. Biopsy: Air Sacs and Lungs[5]
1. A small-diameter, rigid endoscope with a biopsy sleeve is used for the procedure; the patient is under general anesthesic.
2. Culture specimens also can be taken with the endoscope.
3. A 2.7 or 5 French biopsy forceps can be used for air sac or lung biopsies.
4. Lung biopsies are risky procedures and should be performed only when absolutely necessary.

V. Common Etiologic Agents of Avian Respiratory Disease[1]

A. Bacterial Organisms
1. *Chlamydia psittaci*, *Escherichia coli*, *Mycoplasma* spp., *Pseudomonas* spp., *Klebsiella* spp., *Salmonella* spp., *Mycobacterium avium*, *Proteus*

spp., *Haemophilus* sp., *Bordetella avium*, *Pasteurella* spp., *Streptococcus* spp., *Staphylococcus* spp.

B. Fungal Organisms
 1. *Aspergillus* spp., *Candida albicans*, *Mucor* sp., *Cryptococcus* sp.

C. Viral Organisms
 1. Adenovirus, paramyxovirus, laryngotracheitis virus, influenza virus, infectious bronchitis virus, avian pox virus

D. Parasitic Organisms
 1. *Sternostoma* spp. (tracheal mites), *Cytodites* spp. (tracheal mites), *Cyathostoma* sp., *Syngamus* sp., *Seratospiculum* sp. (nematode), *Cryptosporidia* spp., *Trichomonas* spp., *Coccidia* spp. (systemic), hematozoa, *Knemidokoptes* spp. (scaly face mites), *Sarcocystis* spp.

E. Nutritional Causes
 1. Vitamin A deficiency, iodine deficiency, obesity, general malnutrition

F. Toxic Causes
 1. Polytetrafluorethylene gas, formaldehyde, quaternary ammonium, creosote, chlorinated biphenyls, carbon monoxide, cigarette smoke, naphthalene, ammonia, environmental particulate matter

VI. Upper Respiratory Tract Disease[1]

A. Clinical Presentation
 1. Dyspnea
 2. Rhinorrhea
 3. Purulent or serous nasal discharge
 4. Periocular swelling
 5. Voice change
 6. Open-mouthed breathing
 7. Coughing
 8. Sneezing
 9. Yawning
 10. Neck stretching

B. Diagnosis
 1. History
 2. Examination of nares, choana, oropharynx, and larynx
 3. Palpation of neck, head, and thoracic inlet, with particular attention to periorbital area
 4. Gram stain and culture (choana)
 5. Radiographs (survey and contrast)
 6. Endoscopic examination
 7. Sinus aspiration or nasal flush

Normal sounds should not be confused with respiratory pathologic conditions. Many psittacine species commonly mimic noises and sounds, and the possibility of mimicking should be addressed during the history and physical examination.

C. Normal Sounds Not to be Confused with Pathology
1. African grey parrot: Growling
2. Cockatoos and cockatiels: Hissing
3. Pionus parrots: Moist respiratory sounds (sniffling sounds) when bird is excited

D. Differential Diagnoses for Rhinitis
1. Bacterial organisms (commonly *C. psittaci, Mycoplasma* spp., and avian tuberculosis; also gram-negative bacteria)
2. Fungal organisms (commonly *Aspergillus* spp.)
3. Reovirus
4. Parasites
5. Nutritional disorders (hypovitaminosis A)
6. Toxins (commonly cigarette smoke or some other type of smoke)
7. Respiratory irritants
8. Neoplasia
9. Choanal atresia
10. Allergies

E. Differential Diagnoses for Sinusitis
1. Bacteria
2. Fungal organisms
3. Hypovitaminosis A
4. Papillomas
5. Toxins
6. Respiratory irritants
7. Neoplasia
8. Choanal atresia
9. Allergies

F. Differential Diagnoses for Periorbital Swelling
1. Sinusitis
   a. Bacterial organisms and *Mycobacterium* sp.
   b. Fungal organisms
   c. Parasitic organisms
   d. Viral diseases
2. Subcutaneous emphysema
3. Trauma
4. Hemorrhage
5. Neoplasia
6. Noninfectious granuloma resulting from hypovitaminosis A
7. Abscess

VII. Lower Respiratory Tract Disease[6] (Fig. 3-3)

A. **Clinical Presentation**
   1. Coughing
   2. Dyspnea
   3. Open-mouthed breathing
   4. Tail bobbing
   5. Prolonged respiratory recovery time
   6. Depression

B. **Diagnosis**
   1. History
   2. Auscultation of lungs, heart, and air sacs
   3. Radiographs
   4. Transillumination of trachea
   5. Laparoscopic examination
      a. Culture, flush, or biopsy
      b. Visualization of lungs and air sacs

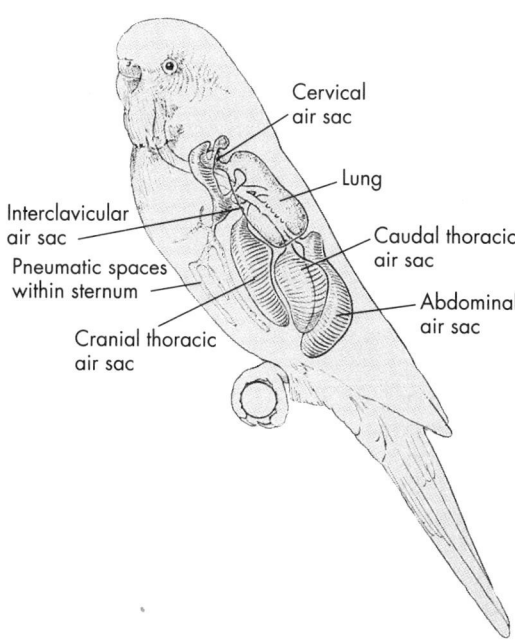

Fig. 3-3 Lower respiratory system of companion avian species.

C. Differential Diagnoses for Nonrespiratory Diseases with Respiratory Signs

1. Ascites secondary to the following:
   a. Liver disease
   b. Renal disease
   c. Neoplasia
   d. Low plasma protein level
   e. Reproductive problems
2. Hemocoelom from the following:
   a. Trauma
   b. Vitamin K deficiency
   c. Liver disease
   d. Polyomavirus
3. Malnutrition
4. Obesity (with associated lipid accumulation in the liver or coelom, or both)
5. Goiter (particularly "clicks" as enlarged thyroid glands compress trachea)
6. Cardiomyopathy
   a. Heart disease
   b. Congenital defects
   c. Heart failure
7. Viruses
   a. Paramyxovirus
   b. Herpesvirus
   c. Reovirus
8. Hemochromatosis (more commonly associated with toucans, aracaris, mynahs, and starlings)
9. Egg-related peritonitis
10. Parasites, allergies, and aspiration of a foreign body (including hand-feeding formula and crop contents) also may cause presentations of lower respiratory tract disease.

D. Differential Diagnoses for Presentation of Tracheitis

1. Amazon tracheitis virus
2. Avian pox
   a. Agapornis pox
   b. Psittacine pox
   c. Amazon pox
   d. Budgerigar pox
3. Parasites (commonly *Syngamus* spp., *Cyathostoma* sp., and *Sternostoma* spp.)
4. Malnutrition
5. Toxins

E. Differential Diagnoses for Laryngitis

1. Herpesvirus, avian viral serositis
2. Poxviruses

3. *Haemophilus*-like organisms
4. Hypovitaminosis A

F. Differential Diagnoses for Air Sacculitis
1. Bacterial organisms, including *Chlamydia*
2. Fungal organisms (commonly *Aspergillus* spp.)
3. Canary pox
4. Paramyxovirus
5. Parasites

G. Differential Diagnoses for Pneumonia
1. Bacterial organisms, *Mycobacterium* spp.
2. Fungal organisms
3. Viruses
4. Parasites, including *Sarcocystis*
5. Toxins, such as Teflon
6. Allergies

H. Differential Diagnoses for Respiratory Abscesses
1. Bacterial organisms, including avian tuberculosis
2. Fungal organisms
3. Neoplasia, with possible secondary infectious foci

The treatment of respiratory disease depends on the diagnosis of the primary problem and the patient's condition. Excellent nutritional management is required to maintain adequate patient health and prevent secondary bacterial and fungal diseases. The treatment of respiratory disease should be based on the clinical diagnosis, aided by the patient's history and by diagnostic testing. Antimicrobial therapy initially should be based on Gram stain results and then adjusted according to results from sensitivity testing of the organisms cultured.

VIII. Treatment of Respiratory Disease

A. Upper Respiratory Tract Therapy
1. Rhinitis (local or systemic)
    a. Environmental modifications (increased humidity, or nebulization, or both should be provided).
    b. Dietary modifications (the patient's diet should be improved and should include foods high in vitamin A and beta carotene; a vitamin A injection might also be needed).
    c. The nares should be cleaned and flushed once a day for several days[1]
    d. Topical antibiotic medications (e.g., ophthalmic solutions) often are helpful when instilled into each naris.
    e. Nasal plugs should be removed carefully and treated both systemically and topically. The nasal cavity is highly vascular, and

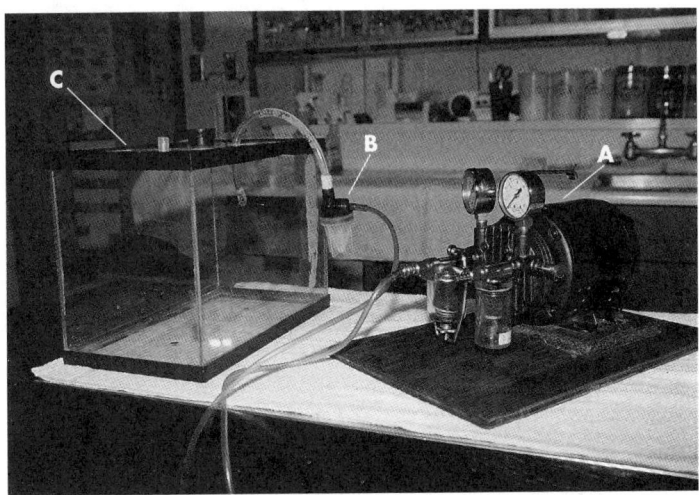

Fig. 3-4 Equipment for nebulization therapy. **A,** Air compressor. **B,** Infant (human) nebulizer. **C,** Enclosed chamber for patient.

plugs and granulomas should be removed in stages to control hemorrhage.
2. Periorbital swelling
   a. Cage rest
   b. Dietary modifications (underlying vitamin deficiencies often are a factor)
   c. Sinus flushing
   d. Nebulization therapy (Fig. 3-4)
   e. Periorbital abscesses may need to be removed while the infectious organism or organisms are treated.

B. Lower Respiratory Tract Therapy
   1. Therapy for lower respiratory tract disease
      a. Environmental support (heat)
      b. O_2 with nebulization therapy
      c. Fluid therapy
      d. Antibiotics (if indicated)
      e. Nutritional support (if the patient is anorexic, nutritional support may be administered through tube feeding. Other means of providing nutritional support are a catheter duodenostomy and total parenteral nutrition.)[7]
      f. Standing radiographs (to determine whether presenting signs are respiratory or nonrespiratory in origin)
   2. Therapy for tracheal disease
      a. See Therapy for dyspnea, above
      b. Inspection of the trachea with a rigid endoscope

c. Removal of a foreign body
   d. Surgical removal of a granuloma (if localized)
   e. Systemic antimicrobial therapy (see Chapter 26)
3. Therapy for subcutaneous emphysema
   a. Removal of air (The disease process often results in a one-way valve effect and cannot be localized.)
   b. Possibly, placement of a tube in a caudal air sac (to alleviate dyspnea of tracheal origin)
   c. Surgical removal of a granuloma (if localized)
   d. Systemic antimicrobial therapy (see Chapter 26)
   e. Placement of a stent to remove air (if the condition is chronic and unresolved)
4. Therapy for lower respiratory tract disease
   a. Systemic antimicrobial and antifungal drugs (based on results of culture and sensitivity tests, aspergillosis testing, Gram stains, and cytologic testing of the aspirate)
   b. Nebulization therapy (improves the function of the mucociliary escalator by using saline to reduce the stickiness of the mucus in order to sweep the foreign material out of the respiratory tree; also can deliver antimicrobial and antiinflammatory drugs directly to the surface of the lungs and to the air sacs)
   c. Topical air sac treatment (may be risky for a debilitated patient)
5. Therapy for choanal atresia
   a. Surgery (recommended treatment)
6. Therapy for neoplasia
   a. Debulking by surgery if possible
   b. Chemotherapy (as indicated by tumor type)

REFERENCES

1. Tully TN, Harrison GJ: Pneumonology. In Ritchie BW, Harrison GJ, Harrison LR, editors: *Avian medicine: principles and applications,* Lake Worth, Fla, 1994, Wingers, pp 556-581.
2. Hillyer EV, Orosz S, Dorrestein GM: Respiratory system. In Altman RB et al, editors: *Avian medicine and surgery,* Philadelphia, 1997, Saunders, pp 387-411.
3. Murray MJ: Diagnostic techniques in avian medicine, *Semin Avian Exotic Pet Med* 6(2):86-95, 1997.
4. Campbell TW: Cytology. In Ritchie BW, Harrison GJ, Harrison LR, editors: *Avian medicine: principles and applications,* Lake Worth, Fla, 1994, Wingers, pp 199-222.
5. Taylor M: Endoscopic examination and biopsy techniques. In Ritchie BW, Harrison GJ, Harrison LR, editors: *Avian medicine: principles and applications,* Lake Worth, Fla, 1994, Wingers, pp 327-354.
6. Tully TN: Avian respiratory disease: clinical overview, *J Avian Med Surg* 9:162-174, 1995.
7. Jenkins JR: Hospital techniques and supportive care. In Altman RB et al, editors: *Avian medicine and surgery,* Philadelphia, 1997, Saunders, pp 232-252.

4

Abnormal Droppings

Louise Bauck

I. Evaluation of Droppings

A. Importance of Evaluation
1. Evaluation: The careful evaluation of droppings is an interesting and vital part of every patient workup. The unique combination of urinary and digestive tract products provides many important clues to states of both normalcy and disease.
2. Hospital staff: The hospital staff should be encouraged to take an interest in evaluating droppings, because the staff sometimes are the first people each morning to take note of what has transpired overnight. Staff members should be trained not to discard droppings while cleaning hospital areas; rather, they should store them in a specific location for professional evaluation if necessary. At the very least, staff members can include written descriptions of droppings in the medical record (Box 4-1).

B. Number and Volume of Droppings
1. Number and volume: The number and volume of droppings are an excellent indicator of food ingestion and are easy for owners to monitor when they are unsure of a bird's true appetite. During the monitoring period the cage papers must be changed frequently to allow good evaluation of the droppings, and loose substrates (e.g., ground corn cobs or shavings) should not be used because they can obscure changes in the droppings.
2. Changes: Changes in droppings may be good indicators of early disease, which can be seen well before the owner notices changes in the bird's behavior. As an important part of preventive health maintenance and monitoring, technicians can train bird owners to recognize a normal dropping. The reception staff consistently should advise bird owners that recent droppings must be collected and brought with the bird for all consultations. It can be explained that this is necessary because it is difficult for a lay person to describe droppings accurately, and transport to the hospital causes stress-induced

> **Box 4-1**
>
> **Description of Avian Pet Droppings**
>
> In the patient's medical record, the hospital staff should include the following information:
> 1. The number and location of droppings, a general evaluation, and an estimate of the volume. For example, are the droppings all in one spot, which indicates inactivity?
> 2. General odor (if any)
> 3. Stool evaluation (based on the freshest droppings plus observations from dried specimens):
>   a. Color (brown, green, black, other, varied)
>   b. Form (water content; stool tubular, coiled, semiliquid, or liquid, or absent)
>   c. Volume (scant, abundant, puffed)
>   d. Matter in stool (e.g., whole seeds, foreign bodies, frank blood or blood clots)
> 4. Urate and urine evaluation (based on freshest specimens plus observations from dried specimens):
>   a. Color of urates or urine (specify some or all)
>   b. Volume of liquid urine associated with the droppings (for possible polyuria or polydipsia); use of special substrates (e.g., wax paper, plastic sheeting) may enhance observation
>   c. Urinalysis findings, if any
> 5. Other items on the cage floor:
>   a. Suspected vomitus
>   b. Evidence of food ingestion (pulverized formulated diets, seed hulls, food in the water dish)
>   c. Molted or chewed feathers (or both), broken pin feathers
>   d. Fresh or dried blood
>   e. Other

changes in the droppings. Droppings produced in the normal environment can be an invaluable diagnostic tool.

C. Anatomy of a Dropping

1. Normal shapes: In most pet birds a normal dropping consists of a coiled or partly coiled stool, a portion of crystalline urates, and some liquid urine (Fig. 4-1). The stool originates in the large intestine but may acquire a coiled shape if stored in the cloaca. The urates originate in the kidneys and travel down the ureters (Figs. 4-2 and 4-3). Urates may be quite discrete, as in parakeets, or more diffuse, as in macaws. The normal amount of liquid urine that accompanies the dropping depends both on the species of bird (e.g., desert-living species, such as the parakeet, produce little liquid urine under most circumstances) and on the amount of water taken in before the dropping is formed. The amount of water taken in, in turn, depends somewhat on the type and volume of food eaten. Normal urine usually is clear, and normal urates usually are snow-white. However, in many birds transient changes in the color of the urates occur as the day progresses, especially when the bird is stressed. Consistent green or yellow staining of the urates or urine is cause for investigation.

2. Color of droppings: In most psittacine species fed a seed-based diet, freshly produced stool usually is a rich, dark green. The stool of finches, canaries, pet quail, and doves is more brown in color (quail and doves also normally have a much higher stool and urate volume

64 *Abnormal Droppings*

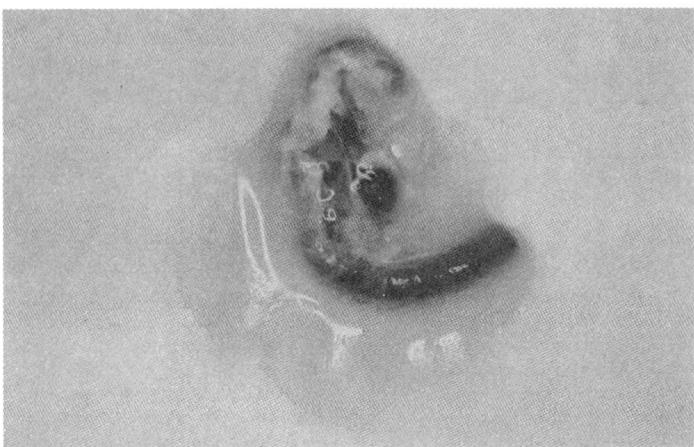

Fig. 4-1 Normal dropping (finch), freshly collected on white paper. Like most normal droppings, this specimen shows a small to moderate amount of liquid urine, snow-white urates, and a well-formed, partly coiled stool.

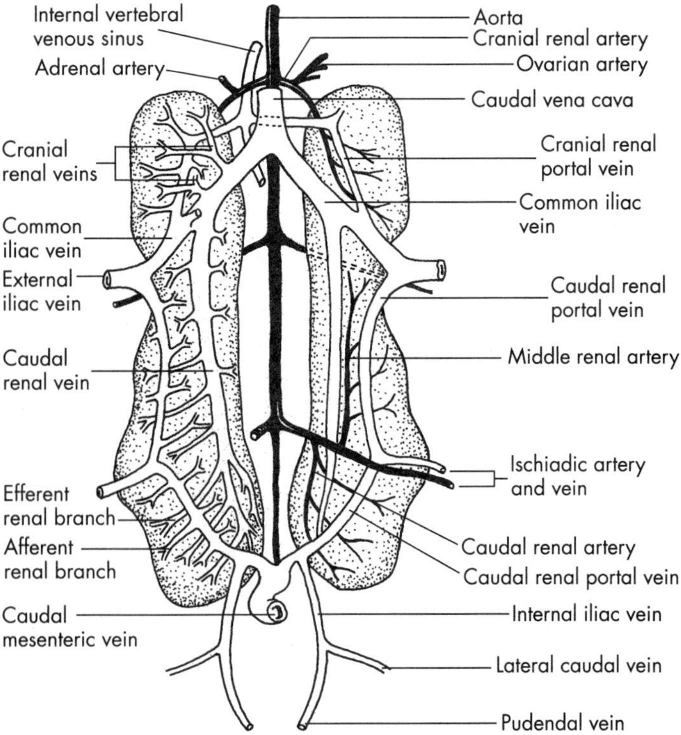

Fig. 4-2 Anatomy of the kidneys, showing major blood vessels and the portal system.

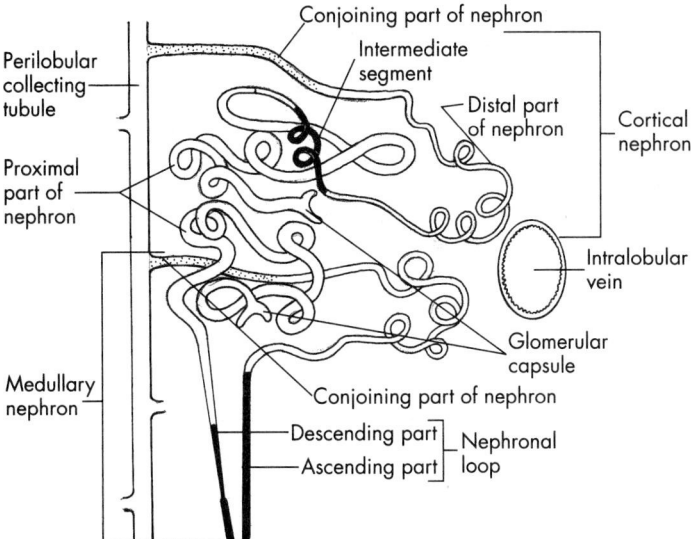

Fig. 4-3 Cross section of a cortical and medullary-type nephrons found in bird kidneys.

relative to their size). Psittacine birds fed formulated diets may also have brown stool when the major portion of the diet is derived from the pellets. Some formulated diets have dyes or coloring agents that can further alter the color of the dropping. Stool color also is affected by changes in normal and abnormal gut flora and by certain foods, such as beets, blueberries, pomegranates, and other fruit. When the normal droppings of a parakeet dry out, the stool may take on a black color; this should not be confused with blackish stools produced by melena or digested blood in the stool, discussed below. Examining a fresh dropping from a healthy parakeet should confirm the initial green color.

II. Common Abnormal Droppings

A. Abnormal Urates
1. Polyuria: The common presentation of polyuria usually is a large volume of clear urine with little or no discoloration. Small amounts of crystalline urates also may be present, and the stool usually is well formed. Polydipsia often accompanies polyuria. The bird should be evaluated for diabetes-like conditions (e.g., insulin or glucagon imbalance), chronic or iatrogenic kidney damage, pituitary disease, and polydipsia associated with formulated diets or excess ingestion of fruit.[1] Polyuria may develop secondary to behaviorally induced polydipsia (see Chapter 8).

2. Reddish, pinkish, or brownish urates or urine, or chocolate milklike urates: Hematuria and hemoglobinuria both may produce pinkish or reddish changes in urate color, as well as other changes. Port wine–colored urine in Amazons, formerly described as Amazon hemorrhagic syndrome,[2] which generally is described as a hemoglobinuria, may be caused by a transient lead or zinc poisoning.[3] Besides lead and zinc poisoning, common problems that produce these urine discolorations are polyomavirus infection[1] and other factors that affect clotting mechanisms or cause hemorrhage, such as vitamin K antagonist poisons.
3. Yellow, lime green, or bright green urates or urine: Biliverdinuria and other causes of bile pigment discoloration are common in birds, and a wide range of problems can affect the liver directly or indirectly. Common problems to investigate include chlamydiosis, herpesvirus hepatitis (Pacheco's disease), fatty liver syndrome, bile duct carcinoma, aspergillosis, psittacine proventricular dilation syndrome, zinc and lead intoxication, and anorexia[4] (see also section C below, Common Abnormal Stool and Urate Combinations).

B. Abnormal Stool

1. Liquid stool (persistent liquidity in species other than lorikeets, toucans, and similar species): True diarrhea may be caused by factors ranging from decreased intestinal transit time to malabsorption problems. Intestinal abnormalities that commonly produce diarrhea without many other changes include bacterial, protozoal, and viral infections; stress; and changes in diet (Table 4-1). Cultures, fecal flotation tests, staining and microscopic examination of stool, and a thorough history usually are needed to determine the most appropriate course of action. In general, the bird's hydration and electrolyte status also should be evaluated immediately.
2. Liquid, blackish stool: A black or dark discoloration in a fresh stool may indicate hemorrhage or intestinal diathesis in the upper part of the gastrointestinal tract.[3] Anorexia or moribund states commonly result in this change, particularly in smaller birds.
3. Brownish stool: Primary coliform intestinal infections may produce a distinct brown change and liquidity in certain affected psittacine species. An abnormal odor has been associated with some bacterial diarrheas. Viral diseases with secondary intestinal infection (e.g., psittacine beak and feather disease and psittacine proventricular dilatation syndrome) also may produce brown diarrheas.
4. Other discolorations of stool (e.g., fresh blood, dyes, food): Fresh blood may be difficult to associate strictly with the stool (see section C, Common Abnormal Stool and Urate Combinations). Dyes and coloring agents from ingested food are common causes of color changes and also may stain urates to some extent. Beets, blueberries, blackberries, pomegranates, and many other fruits are well known for altering color. Formulated diets with artificial colors also may cause changes. Colored newspaper may retroactively stain droppings.

Table 4-1
Possible causes of liquid stool in avian pets

Cause	Effect
Adenovirus	Diarrhea is probably rare, although it has been reported in budgerigars and a variety of large psittacines; hepatitis may accompany the disease.[9]
Aspergillosis and candidiasis	Most severe systemic cases may produce diarrhea.
Atoxoplasmosis	"May cause diarrhea, splenomegaly, and hepatomegaly in canaries.
Avian polyomavirus	Occasionally causes diarrhea, although during the acute phase, gastrointestinal stasis is more common. Melena or blood droppings may follow.
Campylobacter sp.	More common in passeriformes; yellowish diarrhea has been reported.[5]
Change in diet	Any sudden change can cause diarrhea, as can low-fiber foods and high-fat foods.
Chlamydia sp.	Produces a wide range of clinical signs, including diarrhea.
Citrobacter sp.	Has been reported as a cause of diarrhea in passeriformes.[5]
Coccidia	May cause diarrhea in certain passerine birds, also reported in lorikeets.[6,8]
Cryptosporidia and *Microsporidia* spp.	Have been reported in a wide variety of pet birds; death may occur in younger birds. Diagnosis is made by fecal floatation testing, histopathologic tests, or immunoassay.[7]
Escherichia coli	A common cause of diarrhea in young psittacines and passeriformes; the organism often is recovered from normal droppings without having caused clinical signs.
Giardia sp.	Causes diarrhea that often is described as yellowish or pale; may or may not be voluminous; more common in budgerigars and cockatiels.
Hexamita sp.	Has been reported to cause diarrhea in grass parakeets and cockatiels.[6]
Intoxication	Lead, zinc, pesticide, and houseplant intoxication are most commonly associated with abnormal stools.
Malabsorption syndromes	Most of these syndromes cause voluminous stool, but some cause transient diarrhea.
Pacheco's virus	Scant, liquid, or absent stool, which may or may not be hemorrhagic, may accompany discolored urates.
Paramyxovirus	Newcastle strains may have gastrointestinal effects, including diarrhea.
Pseudomonas and *Aeromonas* spp.	Often associated with contamination of water by organic waste; diarrhea or respiratory signs (or both) are common.
Psittacine beak and feather disease	Diarrhea may be seen with acute infection (preweaning) or as a secondary infection with *E. coli* or similar organisms.
Psittacine proventricular dilatation syndrome	Diarrhea probably occurs secondary to immune suppression and concurrent infectious agents or to maldigestion problems.
Salmonella spp.	Can cause diarrhea, hematochezia, melena, and sudden death; a carrier state is possible.
Septicemia	Causes scant, liquid dark stools; acute disease, such as septicemia, may affect gastrointestinal function.
Stress and anorexia	Stress may result in a decrease in the intestinal transit time and anorexia.

5. Malodorous stool: Yeast infections and some bacterial infections may be capable of changing the odor of avian droppings. Gram stains and cultures are recommended.
6. Foreign bodies in the stool (e.g., fiber, foil, metal, plastic): Birds that chew on rope or cloth objects (particularly those made of artificial fibers, such as polyester) may develop fiber obstructive bodies or fiber mats and foreign bodies. Heavy metal toxicosis must be ruled out whenever metal objects are reported in the stool. Plastic and diamonds or other precious stones normally pass out a few days to weeks after ingestion unless they are large enough to cause obstruction.
7. Hulled, whole seeds in stool; foam or mucus on or near dropping: Foam or mucus in or near droppings is more likely to be vomitus associated with droppings than a primary dropping abnormality. Psittacine proventricular dilatation syndrome is a common and important disorder associated with the passage of whole seeds in the droppings. Conditions causing hypermotility, including colibacillosis and salmonellosis, occasionally have been associated with hulled or whole seed in the stools.[10] Pancreatitis and intestinal infiltrative diseases, including neoplasia, also should be ruled out. Foam may be associated with cecal droppings in species with a cecum (see also section C below, Common Abnormal Stool and Urate Combinations).

C. Common Abnormal Stool and Urate Combinations
1. Lime green or bright green urates with liquid or poorly formed stool: Greenish changes in the normally white urates can be very nonspecific. Liver disorders, liver inflammation, and anorexia or other causes of biliverdinuria should be considered. Diarrhea often accompanies some of these disorders, especially chlamydiosis, psittacine proventricular dilatation syndrome, and heavy metal intoxication.
2. Initial lime green urates followed by pink urates; poorly formed stool: This is the classic presentation for many New World psittacines affected with lead or zinc intoxication. The pinkish discoloration is believed to be caused by hematuria or hemoglobinuria that develops secondary to the effects of the heavy metal poisoning. Early in the course of the disease process, green-tinged urates may be seen. Diarrhea (poorly formed stool) and vomiting are both common in heavy metal intoxication.
3. Cream-colored or yellow urates with stool absent: The absence of stool and discoloration of urates is common in birds that are unable or unwilling to eat, often because they are in very acute stages of disease or because changes have been made suddenly in the diet. Primary or secondary ileus also can cause absent stool. Ileus can be seen in cases of obstruction or clostridial infection or in some cases of advanced psittacine proventricular dilatation syndrome.
4. Puffed, popcorn-like or chalky droppings (increased volume of both urates and stool): Since the publication of Petrak's first edition of *Cage and Aviary Bird Medicine* (1982) with its color plates that clearly depict these unusual droppings,[11] the presence of these droppings

in an affected bird has sparked much interest. Many believe that liver or pancreatic dysfunction, or both, may be present. Physical and clinical pathologic investigations may indicate which course of investigation the practitioner should take.

5. Black, scant stool with urates absent: A very abnormal stool suggestive of hemorrhage into the upper intestine and possible evidence of renal failure (absent urates) may be seen in certain avian patients (e.g., finches) in the end stages of infectious disease or with other life-threatening conditions, such as advanced polyomavirus infection.[11]
6. Blackish stool with greenish urates: Although a wide variety of conditions can cause a biliverdinuria and a hemorrhagic diathesis, one example of this type of dropping might be chronic lead poisoning in small psittacines such as cockatiels. Advanced infectious diseases, particularly those that cause liver inflammation, might also be worth investigating.
7. Fresh blood in or around stool or urates: Extracloacal origins, such as a broken pin feather or blood from the nares, oropharynx, or some other bleeding surface that may drip onto the cage floor, must be ruled out.[12] Next, the bird must be examined carefully for any evidence of prolapse, egg laying, papilloma, or other cloacal defects that may cause hemorrhage.[12] Vitamin K disorders, toxicities, and acute and severe infectious intestinal disease (both viral and bacterial) also must be eliminated as possibilities.

REFERENCES

1. Oglesbee B: Polyuria, Proceedings of the Mid-Atlantic Association of Avian Veterinarians, Philadelphia, 1996.
2. Galvin C: Acute hemorrhagic syndrome of birds. In Kirk RW, editor: *Current veterinary therapy,* IX, Philadelphia, 1983, Saunders, pp 617-619.
3. Rich G: Basic history taking and examination. In Rosskopf WG, Woerpel RW, editors: *Pet avian medicine, Veterinary Clinics of North America, Small Animal Practice,* Philadelphia, 1991, Saunders, 21(6): 1135-1145.
4. Bauck L: Abnormal droppings. In *Manual of avian medicine,* Lakewood, Colo, 1995, American Animal Hospital Association, pp 32-35.
5. Gerlach H: Bacteria. In Ritchie BW, Harrison GJ, Harrison LR, editors: *Avian medicine: principles and applications,* Lake Worth, Fla, 1994, Wingers, pp 949-983.
6. Greiner EC, Ritchie BW: Parasites. In Ritchie BW, Harrison GJ, Harrison LR, editors: *Avian medicine: principles and applications,* Lake Worth, Fla, 1994, Wingers, pp 1007-1029.
7. Clubb SL: Cryptosporidiosis in a psittacine nursery, Tampa, 1996, Proceedings of the Association of Avian Veterinarians pp 177-186.
8. Bauck L: Diseases of the finch as seen in a commercial import station, Seattle, 1989, Proceedings of the Association of Avian Veterinarians, pp 196-202.
9. Gerlach H: Viruses. In Ritchie BW, Harrison GJ, Harrison LR, editors: *Avian medicine: principles and applications,* Lake Worth, Fla, 1994, Wingers, pp 862-948.
10. Bauck L: Abnormal droppings and their workup, Philadelphia, 1995, Proceedings of the Association of Avian Veterinarians pp 455-459.
11. Minsky L, Petrak ML: Diseases of the digestive system. In Petrak ML, editor: *Diseases of cage and aviary birds,* Philadelphia, 1989, Saunders, pp 432-448.
12. Johnson Delaney C: *Exotic companion medicine handbook: psittacines,* Loxahatchee, Fla, 1996, Wingers, pp 17-26.

5

Vomiting and Regurgitation

Victoria J. Joseph

I. Anatomy Overview

To understand the factors that may influence a bird to vomit and regurgitate, the clinician should be familiar with the anatomy of the avian species and the differences that may exist between the species.[1-3]

A. Oral Cavity
1. The lips and teeth are absent in birds and are replaced by the cutting edge (tomia) of the horny beak.
2. The choana is a median fissure in the hard palate that connects the oropharynx to the nasal cavity (Fig. 5-1). The palate is ridged in most species.
3. The infundibular cleft is a dorsal midline slitlike opening that receives the right and left pharyngotympanic (eustachian) tubes and sits caudal to the choana. A flap is not present over the opening of the pharyngotympanic tube, which allows the bird to make rapid adjustments in air pressure when quick changes in altitude occur.
4. The tongue is supported by the hyobranchial (hyoid) apparatus and has three functions, which include the following:
    a. Collecting food: A long, narrow tubelike tongue functions as a probe, spear, or capillary tube and appears in nectar, insect, or sap eaters.
    b. Manipulating food: Birds have a nonprotrusible tongue. A short, thick fleshy tongue appears in seed and nut eaters. A rasplike tongue appears in raptors and fish-eaters and filters food particles in waterfowl.
    c. Swallowing food: Caudally directed papillae on the tongue help move food bolus downward and help prevent regurgitation in pelican and domestic fowl.
5. The laryngeal mound has a narrow, slitlike opening into the glottis (rima glottidis) of the larynx. No epiglottis is present. Caudally directed cornified papillae along the choanal slit and dorsal oropharynx aid in swallowing.

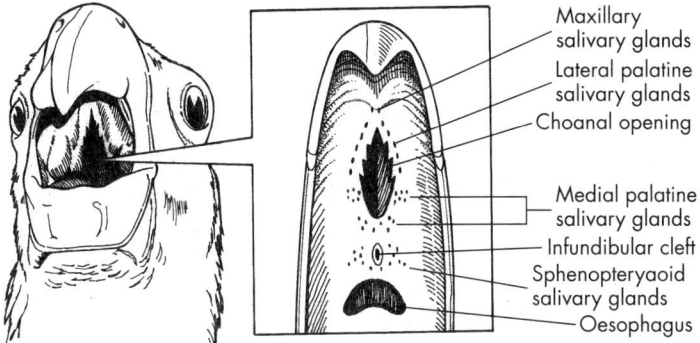

Fig. 5-1 A baby psittacine (parrot) illustrating the choanal opening, infundibular cleft, and salivary glands.

6. The salivary glands develop better in birds that eat dry food and are scattered throughout the oropharynx. The salivary glands secrete mucus, which acts as a lubricant in swallowing. Some birds may have additional secretory organs.
7. Three stages exist in swallowing, including the following:
    a. The tongue moves the food pellets to the palate. The food is held in place by mucous secretions. The choanal opening is reflexively closed.
    b. The tongue rakes the bolus caudally in a rostral-to-caudal movement. The infundibular cleft and glottis close reflexively.
    c. The laryngeal mound carries food toward the esophagus in a rostral-to-caudal movement.

B. Esophagus and Crop
1. The esophagus and crop have thin walls and are distensible (Fig. 5-2).
2. The cervical portion of the esophagus and crop lies on the right side of the neck.
3. The internal surface of the esophagus and crop is folded longitudinally to increase distensibility.
4. The folds in the esophagus and crop are greater in hawks, owls, eagles, and cormorants that eat and store large amounts of food.
5. The esophagus enlarges to form the crop, or ingluvies.
6. The crop is not present in gulls, penguins, cranes, or owls. Hawks have a poorly developed diverticulum.
7. The esophageal sac is an inflatable, croplike diverticulum, or a bilateral symmetrical expansion of the cervical esophagus and may inflate only in male sage grouse, prairie chicken, and great bustard during courtship.
8. Crops that are well developed store food for softening. If the gizzard is full, food may be stored in the crop or in the esophagus if a crop is not present.

72 *Vomiting and Regurgitation*

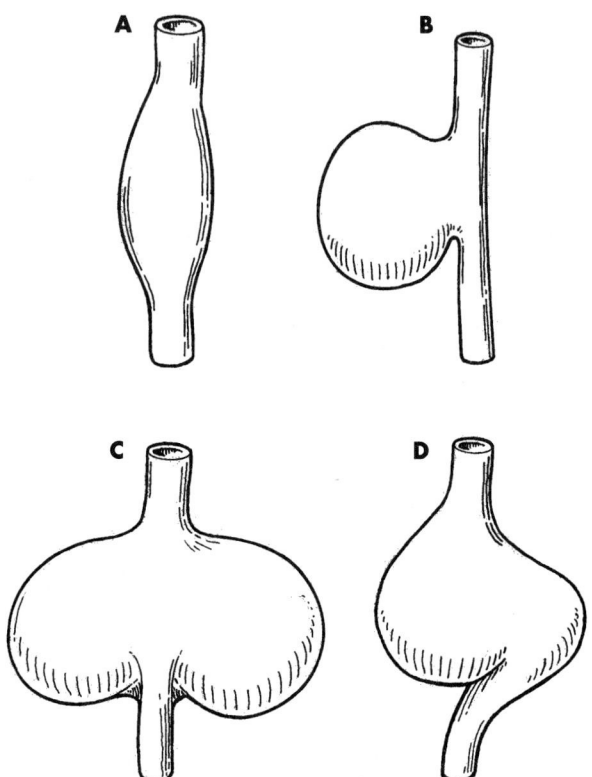

Fig. 5-2 Crops of several species. **A,** Cormorant. **B,** Peafowl. **C,** Pigeon. **D,** Budgerigar.

9. The esophagus and crop transport food in the following stages:
   a. The food enters the crop, which may reduce or inhibit peristaltic contractions. However, if the stomach is empty, food passes directly from the crop to the proventriculus, or the first portion of the stomach.
   b. Chemical digestion does not occur in the crop in most species.
   c. Peristaltic waves propel the food through the gastrointestinal tract.
   d. Retroperistaltic waves are responsible for regurgitation of the food stored in the esophagus or crop.

C. Proventriculus and Ventriculus (Stomach)
  1. Proventriculus (Fig. 5-3)
     a. The proventriculus is the glandular first portion of the stomach.
     b. Large folds are not present, except in fish and meat eaters.
     c. The oxynticopeptic cell secretes hydrochloric acid and pepsin.
     d. The proventriculus and ventriculus mix hydrochloric acid and pepsin.

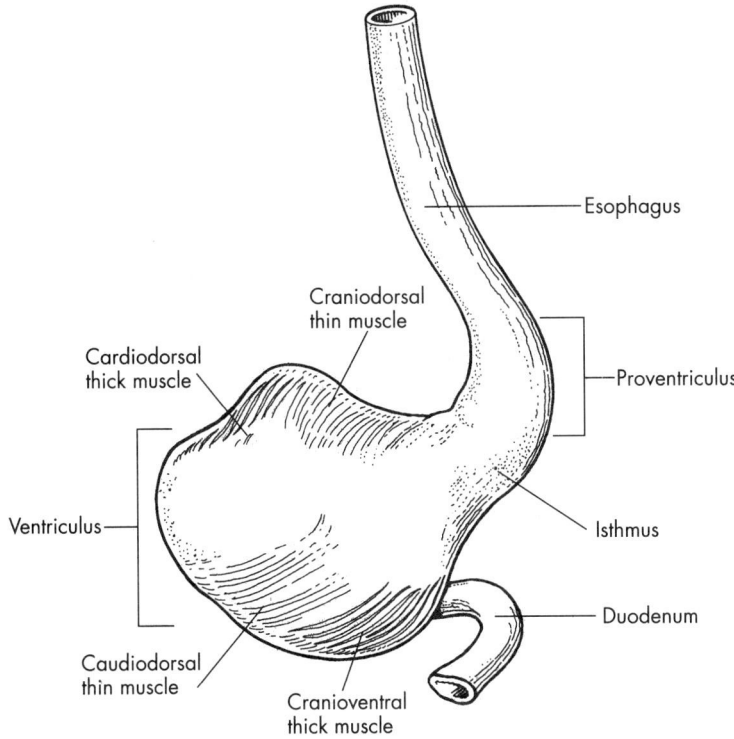

Fig. 5-3 Thoracic esophagus, proventriculus, and ventriculus of a psittacine.

2. Ventriculus (gizzard)
   a. The ventriculus, or muscular portion of the stomach, is most pronounced in insectivores, herbivores, and granivores.
   b. The stomach is saclike in carnivores and piscivores.
   c. Smooth muscle of the ventriculus is rich in myoglobin and asymmetrically arranged into four bands.
   d. The chief cells that excrete protein promote gastric proteolysis and maintain a low pH. The gastric secretions of granivores and omnivores are less acidic than raptors.
   e. The cuticle (koilin) layer is a hardened membrane of a protein-carbohydrate complex overlying the surface epithelium of the ventriculus and aids in mechanical digestion.
3. Mechanical digestion
   a. Contractions propel food in alternate directions between the proventriculus and the gizzard.
   b. Contractions of the stomach in raptors result in the formation of pellets or casts of indigestible portions of the diet (bone, teeth, claws, feathers, and fur). Birds then regurgitate these pellets, or casts.

74 *Vomiting and Regurgitation*

    c. Owls make a pellet with each meal. Hawks may eat more than one meal before regurgitating a pellet.
    d. More bone digestion occurs in the stomach of falconiformes.
    e. Owls swallow their prey and immediately fill the proventriculus and lower esophagus. After several minutes, the food enters the ventriculus.

D. Intestinal Tract
1. Small intestine (Fig. 5-4)
   a. The majority of the pancreas is contained within the duodenal loop.
   b. The duodenum receives the ducts from the liver and pancreas.
   c. Vitelline, or Meckel's, diverticulum is a remnant of the yolk sac and duct, is blind-ended, and divides the jejunum from the ileum.

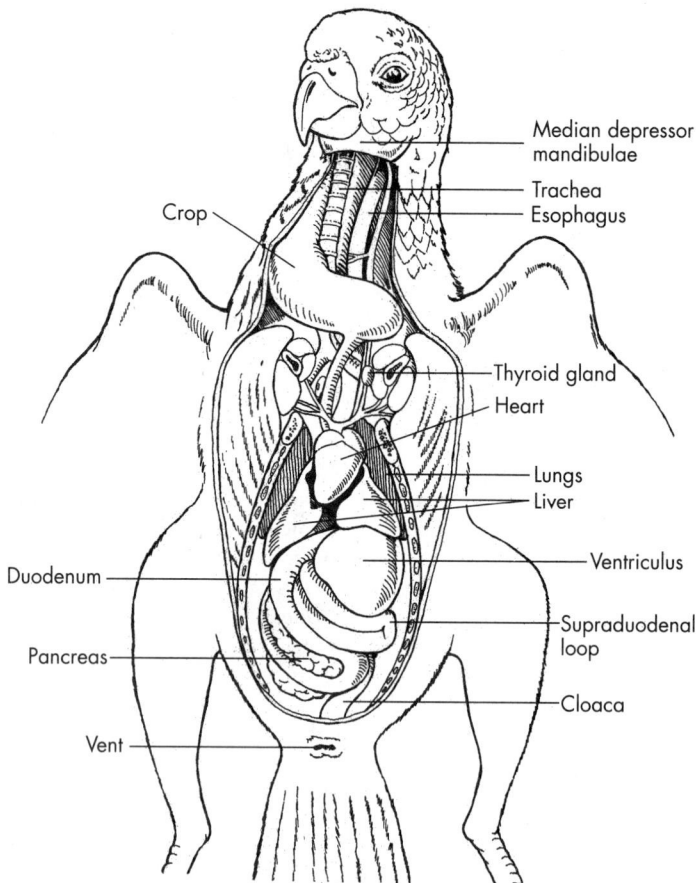

Fig. 5-4 Digestive tract of a psittacine showing the relationship of the various organs.

2. Large intestine (Fig. 5-5)
   a. Paired ceca, if present, are found at the ileorectal junction.
   b. Cecal size and form vary with the species and are large in galliformes and ratites and sacculated in ostrich.
   c. Ceca are reduced or absent in parrots, swifts, and pigeons and are small in the passerines.
   d. Ceca are not present in the red-tailed hawk.
   e. Cecal droppings are easily identified in the great horned owl.
3. Digestion
   a. Chemical digestion and absorption occur in the small intestine.
   b. The ceca break down cellulose by symbiotic bacteria and reabsorb water.
   c. The rectum and cloaca also reabsorb water from gut contents and the ureteral urine (especially in desert birds).

E. **Pancreas and Liver**
   1. Pancreas
      a. Exocrine function provides the chemical phase of digestion in the small intestine with secretions of amylase, lipase, and trypsin, and an increase in pH by HCO_3 secretion.
      b. The cells of the islets of Langerhans provide endocrine function.
   2. Liver
      a. The gallbladder is present in most species and lies on the visceral surface of the right liver lobe.
      b. Pigeons, many parrot species, and the ostrich have no gallbladder.
      c. Bile ducts from each lobe enter the gallbladder or the duodenum directly from the liver.

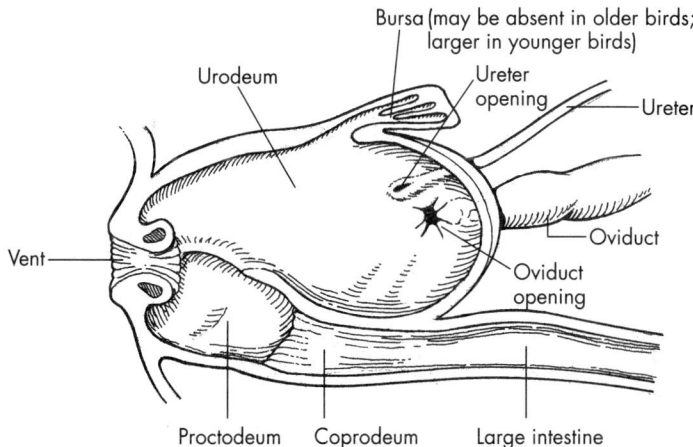

Fig. 5-5 Large intestine, coprodeum, proctodeum, urodeum, and vent.

II. Infectious Agents Causing Regurgitation or Vomiting

Bacterial,[4-9] viral,[6,9-12] parasitic,[13-15] and fungal[16,17] agents may invade the gastrointestinal tract or other organs resulting in regurgitation and vomiting.

A. **Bacteria**
   1. Enterobacteriaceae frequently cause infections in psittacines.
   2. *Aeromonas* spp., *Pseudomonas* spp., *Bordetella* spp., *Vibrio* spp., and *Campylobacter* spp. may colonize the gastrointestinal tract and cause infection.
   3. *Salmonella* spp., *Klebsiella* spp., and *Yersinia* spp. may also cause infections.
   4. *Pasteurella multocida* or fowl cholera affects many species of birds. In waterfowl, diphtheritic enteritis may occur.
   5. Undifferentiated (nonclassified) spirochetes found in the choana and oropharynx of cockatiels and lovebirds may cause regurgitation.
   6. *Mycobacterium* spp. often initially colonizes the gastrointestinal tract and then spread to other organs (liver, lung, and bone). A chronic wasting disease may ensue.
   7. *Chlamydia psittaci* is an obligate intracellular bacterium that is typically ingested or inhaled as a means of infection. Organ systems affected may include the upper respiratory tract, gastrointestinal system, and central nervous system.
   8. Megabacteria are large gram-positive, periodic acid-schiff positive rods and have the following characteristics:
      a. Reported to inhabit the distal end of the proventriculus and ventriculus junction.
      b. May be normal commensal or opportunistic organisms in psittacines, passerines, and Japanese quail.
      c. Signs of infection may include proventricular dilatation, wasting, regurgitation, and soft stool with the passing of whole seed.

B. **Viruses**
   1. Papillomatosis
      a. Papillomatosis is characterized by papilloma-like growths that occur on the mucosa of the alimentary tract.
      b. The etiology of the virus is unknown—no virus is isolated currently.
      c. Papillomatous lesions in New-World psittacines are seen most frequently in the following order: cloaca, glottis, choanal slit, esophagus, oropharynx, ventriculus, and proventriculus.
      d. Lesions may also be found in the crop, liver, pancreas, nasal conjunctiva, and nasolacrimal duct.
      e. Proliferative lesions may interfere with the normal physiologic activities of swallowing, digestion, or defecation.
      f. Chronic weight loss, regurgitation, or vomiting may be present with gastrointestinal lesions.

g. Lesions of the proventriculus, ventriculus, or crop may show signs similar to proventricular dilatation syndrome: regurgitation, poor digestion, and weight loss.
2. Psittacine beak and feather disease virus
   a. The virus is in the Circoviridae family.
   b. Clinicians see feather and beak abnormalities most often.
   c. With acute forms, crop stasis occurs, and with secondary intermittent regurgitation, diarrhea, and central nervous system signs may predominate.
3. Proventricular dilatation syndrome
   a. The suggested etiology is a virus.
   b. Damage to the autonomic ganglia supplying the proventriculus, ventriculus, duodenum, brain, and spinal cord may occur.
   c. Signs of infection may include depression, regurgitation, passage of undigested food, crop impaction, abdominal distention, progressive weight loss, and central nervous system signs (usually more chronic in nature).
4. Pacheco's disease virus
   a. Pacheco's disease is a herpesvirus of psittacines.
   b. Acute fatal hepatitis can occur.
   c. Death is most likely to occur.
   d. Clinical signs of infection may include depression, anorexia, diarrhea, regurgitation, yellow-green urates, or central nervous system signs of acute origin.
5. Duck plague virus
   a. Duck plague virus is a herpesvirus.
   b. Necrosis of the cells lining the alimentary tract can occur.
   c. Clinical signs of infection may include acute onset of vomiting, diarrhea, and then death.
4. Pigeon herpesvirus (PHV-1)
   a. PHV-1 is characterized by infectious esophagitis or inclusion body hepatitis
   b. Signs of infection may include depression, anorexia, diarrhea, vomiting, and central nervous system signs.
7. Falcon herpesvirus
   a. Falcon herpesvirus is associated with depression, anorexia, regurgitation, hepatitis, and death.
   b. The virus is serologically similar to the herpesvirus that is found naturally in pigeons and owls.
8. Polyomavirus
   a. In budgerigars, polyomavirus is associated with hepatitis, abdominal distention, hemorrhage under the skin, feather abnormalities, ataxia, or tremors.
   b. In young nonbudgerigar species, peracute death with no premonitory signs is most common.
   c. Some larger psittacines may show a 12- to 48- hour period of depression, anorexia, regurgitation, diarrhea, dehydration, and subcuticular bleeding before death.

9. Adenovirus
    a. Psittacines show signs of pancreatitis, encephalitis, hepatitis, splenitis, or enteritis.
    b. Pheasants show signs of what is commonly referred to as *marble spleen disease,* diarrhea, vomiting, and anorexia.
    c. Turkeys show signs of hemorrhagic enteritis.
    d. Ostriches show signs of hepatitis, enteritis, and proventricular/ventricular impaction.
    e. Pigeons show signs of polyuria/polydypsia (PU/PD) and vomiting

C. Parasites
1. Trichomoniasis
    a. *Trichomonas gallinae* is the most common species seen clinically.
    b. Other strains of trichomoniasis may cause a latent infection.
    c. Adult birds previously infected may become carriers.
    d. Infections are seen in pigeons, raptors, canaries, finches, budgerigars, owls, and other small passerines.
    e. Squabs are commonly infected during feeding of the crop milk by their parents who can act as carriers.
    f. Protozoa do not survive long outside the host, so strict sanitation reduces the infection rate.
    g. Diphtheritic forms include ulceration in the oral cavity, esophagus, larynx, and pharynx.
    h. Necrotic forms include caseous lesions of the oral cavity and esophagus, and beak lesions may also occur.
    i. Virulent forms include lesions in the liver and abdominal viscera.
    j. Signs of infection may include labored breathing, difficulty swallowing, regurgitation and vomiting, and weight loss.
2. *Giardia*
    a. *Giardia* are protozoal parasites found in the gastrointestinal tract of budgerigars, small passerines, psittacines, and cockatiels.
    b. Signs of infection may include weight loss, chronic diarrhea, and secondary bacterial and yeast infections. Regurgitation and vomiting are rare but can occur.
3. Histomoniasis
    a. *Histomonas meleagridis* is a protozoan spread by eating infected cecal helminth eggs (*Heterakis* spp.) or invertebrate host.
    b. Histomoniasis is seen in captive galliformes and pigeons.
    c. Signs of infection may include loose and blood-tinged feces, liver disease, enlarged cecum, enteritis, and vomiting and regurgitation.
4. *Capillaria contorta* spp.
    a. Crop worm of the mouth, esophagus, or crop may occur in psittacines, passerines, galliformes, and anseriformes.
    b. Eggs are passed in the feces; infection occurs when the bird ingests eggs, contaminated water, or feed.
    c. Weight loss, anorexia, plaques in the oral cavity, frequent swallowing attempts, or regurgitation may be present.

5. *Dispharynx* spp. (*nasuta*)
   a. Gizzard worm is synonymous with *Tetrameres* spp. and *Acuaria* spp.
   b. The parasites are found in the proventriculus or ventriculus of budgerigars, psittacines, passerines, and game birds.
   c. Inflammation of the proventriculus and ventriculus is present.
   d. Signs may include emaciation, regurgitation, ulcers, and vomiting.
6. *Spiropter* (*incerta* spp.)
   a. *Cyrnea* spp. and *Habronema* spp. are synonymous with *Spiropter*.
   b. The parasites are seen in psittacines.
   c. Swelling of the mucous membranes of the proventriculus and degeneration of the ventriculus lining may occur.
   d. Signs may include weight loss, chronic vomiting, and interference of food passage.
7. *Ascarids* parasite
   a. *A. galli, A. columbae,* and *A. platyceri* are types of ascarids that are seen in birds.
   b. Heavy infestations may result in obstructive lesions or perforations.
   c. Signs may include diarrhea, constipation, lethargy, anemia, regurgitation, or vomiting.
8. Cestodes
   a. Cestodes are more common in insect-eating birds than birds feeding on seeds and fruits.
   b. Cestodes require an intermediate host.
   c. Signs may include diarrhea, general debilitation, enteritis, and obstruction in heavy infestations.
9. Trematodes
   a. Trematodes parasitize the gastrointestinal tract, liver, eye, reproductive tract, skin, kidney, blood vessels, respiratory tract, or bile duct.
   b. Signs may include hepatomegaly with hepatic necrosis, anorexia, anemia, weight loss, and diarrhea.
10. *Plasmodium* spp.
    a. *Plasmodium* spp. is a hemoparasite.
    b. The liver, heart, spleen, and lungs are often the target organs.
    c. Signs of infection may include anorexia, depression, dyspnea, or vomiting. Biliverdinuria may be present with liver involvement.

D. Mycoses
1. Candidiasis
   a. *Candida albicans* is the most common isolate.
   b. Opportunistic yeast may be a normal inhabitant of the gastrointestinal tract; however, systemic candidiasis is rare.
   c. Necrotic pseudomembranous lesions may be seen over the mucosa of the tongue, pharynx, and crop. Commissures of the beak and skin may be involved.

d. Signs may include unthriftiness, anorexia, and regurgitation.
  2. Aspergillosis
     a. *Aspergillus* spp. commonly infects the respiratory tract.
     b. Gastrointestinal infections are possible with signs of infection similar to candidiasis.
  3. Mucormycosis
     a. Mucormycosis are synonymous with phycomycosis, zygomycosis, and *Rhizopus* sp.
     b. Mucormycosis are seen in African grey parrots, owls, flamingos, penguins, and canaries who are fed sprouted seed.
     c. Infections may occur secondary to malnutrition, debilitating diseases, and antibiotic therapy.
     d. Signs of infection may include granulomas of the gastrointestinal tract and ventriculus.
  4. Gizzard malfunction syndrome
     a. Fungal mycelia can occur in several species.
     b. Fungal mycelia can penetrate the koilin layer, epithelium, and muscle.

III. Metabolic Dysfunction

Specific organ dysfunction may secondarily result in regurgitation or vomiting in birds. To identify specific organ dysfunction, clinicians often need to perform multiple diagnostic tests and treatment regimens to resolve the symptoms.[8,9,18-20]

A. Liver
  1. Infectious agents
     a. Bacteria including avian tuberculosis and chlamydia
     b. Viruses
     c. Helminths, trematodes, and protozoa
  2. Metabolic factors
     a. Fatty liver syndrome
     b. Iron storage disease
     c. Portal hypertension
     d. Hepatotoxins
     e. Neoplasm

B. Kidney
  1. Vomiting is not as common as with liver disease in birds
  2. Infectious factors as mentioned above
  3. Articular or visceral gout

C. Endocrine System
  1. Thyroid gland
     a. Neoplasm is rare.
     b. Thyroid hyperplasia (goiter) is caused by iodine deficiency and is primarily seen in seed eaters on seed from deficient soils. En-

largement may cause mechanical obstruction in the lower esophagus and tracheal region at the base of the neck. Respiratory stridor and regurgitation are common signs.
   2. Pancreas
      a. An area needing further research
      b. Stool color and consistency changes, PU/PD, and vomiting may all be signs

D. Reproduction
   1. Egg binding may lead to pressure on the gastrointestinal tract and possible necrosis
   2. Peritonitis

E. Electrolyte Disturbances

IV. Toxic Agents

Pesticides and heavy metals are often considered in differentials for toxic conditions in the avian patient who is vomiting and has diarrhea. In addition, many household items, plants, and some foods have toxic principles.[21-25]

A. Pesticides or Insecticides
   1. Organophosphate compounds
      a. Dichlorvos
      b. Malathion, parathion
   2. Organochlorines, including lindane
   3. Carbamates
   4. Organomercurial compounds
   5. Rotenone, phosphorus
   6. Arsenical compounds
   7. Nitrates

B. Household Compounds
   1. Chlorine
      a. Bleaches
      b. Pool chemicals
   2. Aluminum chloride, including deodorants
   3. Detergents
   4. Potassium chloride, including matches
   5. Pine oil disinfectants
   6. Teflon toxicity (polytetrafluoroethylene)

C. Metals
   1. Lead
   2. Zinc
   3. Mercury
   4. Copper

D. Moldy Foods or Seed (Grains, Peanuts, Corn, Cheese, Meats)
1. Mycotoxins
2. Aflatoxin
3. Ochratoxin
4. Trichothecenes

E. Foods and Plants
1. Chocolate, avocados, alcohol
2. Hypervitaminosis D
3. Sodium chloride (salt)
4. Philodendrons, diffenbachia, poinsettia, rhododendron
5. Solanaceae (green berries and roots), yew

F. Pharmaceuticals
1. Levamisole, apomorphine
2. Trimethoprim or sulfadimethoxine, ketoconazole, itraconazole, doxycycline, enrofloxacin

---

V. Miscellaneous Conditions

Vomiting or regurgitation may be explained by a variety of diseases or situations not previously mentioned. The following examples will act as a guide to encourage the veterinarian to consider all possibilities when a bird presents with vomiting or regurgitation.[8,9,19,20]

A. Behavioral
1. Fear or excitement in vultures, pelicans, and penguins
2. Motion sickness
3. Courtship
    a. Bird-to-bird regurgitation
    b. Bird-to-owner regurgitation
    c. Bird-to-shiny object regurgitation

B. Physiologic
1. Psittacines, pigeons, and passerines feed young by regurgitation; overfeeding of formula to psittacine chicks
2. Cast formation by raptors

C. Crop Impaction
1. Galliformes and anseriformes having sudden access to an abundant supply of grass and sprouted seeds
2. Free-ranging Canada geese eating dried oatmeal and soybeans
3. Grit *ad libitum*

D. Crop Stasis
1. Overheated or underheated formula fed to young birds
2. Crop infection due to bacteria and yeast

3. May occur when feeding a liquid formula to granivorous birds, due to lack of mechanical stimulation
4. Idiopathic pendulous crop
    a. Aged geese, poultry, and parakeets
    b. Nestling cockatoos and macaws

E. Foreign Bodies
1. Unweaned psittacines may ingest rubber or metal feeding tubes
2. Fish hooks in waterfowl
3. Bones in the raptor's diet that perforate or obstruct

F. Obstructive Lesions
1. Callous formation following a coracoid fracture
2. Esophageal strictures
3. Neoplasm of the gastrointestinal tract
4. Intussusceptions, volvulus, or ileus
5. Hernia

G. Nutritional
1. Food allergies
2. Vitamin E and selenium deficiency, including degenerative lesions in the smooth muscle of the ventriculus
3. Vitamin A deficiency, including squamous metaplasia of the epithelium

H. Gastric Ulcers

IV. Summary

A bird presenting with regurgitation or vomiting can be a diagnostic challenge for the clinician to identify the inciting agent. Often multiple tests must be given before the clinician can reach a conclusion. This chapter is designed to inform the veterinary student or practitioner of the disease syndromes and environmental conditions that may result in regurgitation or vomiting in birds. The diagnostic test and treatment programs should be based on the category of the problem that exists.

REFERENCES

1. Duke GE: Raptor physiology. In Fowler ME, editor: *Zoo and wild animal medicine*, Philadelphia, 1978, Saunders, pp 225-227.
2. Jones M: Avian digestive system: a review. *Proc Annu Conf Assoc Avian Vet* :419-421, 1995.
3. King AS, McLelland J: Digestive system. *Birds, their structure and function*, London, 1984, Bailliere Tindell: Pitman Press, pp 84-107.
4. Filippich LJ, Parker MG: Megabacteria and proventricular/ventricular disease in psittacines and passerines, *Proc Annu Conf Assoc Avian Vet:* 287-293, 1994.

5. Gerlach H: Bacteria. In Ritchie BW, Harrison GJ, Harrison LR, editors: *Avian medicine: principles and application,* Lake Worth, Fla, 1994, Wingers Publishing, pp 949-965.
6. Gerlach H: Chlamydia. In Ritchie BW, Harrison GJ, Harrison LR, editors: *Avian medicine: principles and application,* Lake Worth, Fla, 1994, Wingers Publishing, pp 984-993.
7. Hall R, Bemis D: A spiral bacterium found in psittacines, *Proc Annu Conf Assoc Avian Vet* :345-347, 1995.
8. Schultz DJ, Rich BG: Gastro-intestinal diseases. In Burr EW, editor: *Companion bird medicine,* Ames, Iowa, 1987, University Press, pp 80-86.
9. Ritchie BW: Select diseases of the alimentary tract, *Proc Annu Conf Assoc Avian Vet* :423-433, 1995.
10. Ritchie BW: *Avian viruses function and control,* Lake Worth, Fla, 1995, Wingers Publishing, pp 135, 140, 173, 190, 207, 230, 322, 440, 441, 447.
11. Shivaprasad HL, Barr BC, Wood LW: Spectrum of lesions (pathology) of proventricular dilation syndrome, *Proc Annu Conf Assoc Avian Vet:* 505-506, 1995.
12. Degernes LA, Love NE, Laughery C: Review of radiographic changes associated with proventricular dilation disease, *Proc Annu Conf Assoc Avian Vet* :209-212, 1996.
13. Burr EW: Digestive tract protozoa. In Burr EW, editor: *Companion bird medicine,* Ames, Iowa, 1987, University Press, pp 129-133.
14. Reuben S, Kumar S: Helminthology. In Burr EW, editor: *Companion bird medicine,* Ames, Iowa, 1987, University Press, pp 135-147.
15. Scott JR: Ascaridiasis and gastrointestinal stasis in a hyacinth macaw, *Proc Annu Conf Assoc Avian Vet* :195-201, 1996.
16. Bauck L: Mycosis. In Ritchie BW, Harrison GJ, Harrison LR, editors: *Avian Medicine: principles and applications.* Lake Worth, Fla, 1994, Wingers Publishing, pp 997-1005.
17. Patgiri GP: Systemic mycoses. In Burr EW, editor: *Companion bird medicine,* Ames, Iowa, 1987, University Press, pp 102-106.
18. Picoux-Brugere J, Brugere H: Metabolic diseases. In Burr EW, editor: *Companion bird medicine,* Ames, Iowa, 1987, University Press, pp 72-79.
19. Lumeij JT: Hepatology. In Ritchie BW, Harrison GJ, Harrison LR, editors: *Avian medicine: principles and application,* Lake Worth, Fla, 1994, Wingers Publishing, pp 522-536.
20. Lumeij JT: Endocrinology. In Ritchie BW, Harrison GJ, Harrison LR, editors: *Avian medicine: principles and application,* Lake Worth, Fla, 1994, Wingers Publishing, pp 582-604.
21. Lumeij JT: Gastroenterology. In Ritchie BW, Harrison GJ, Harrison LR, editors: *Avian medicine: principles and application,* Lake Worth, Fla, 1994, Wingers Publishing, pp 483-519.
22. Dumonceaux G, Harrison G: Toxins. In Ritchie BW, Harrison GJ, Harrison LR, editors: *Avian medicine: principles and application,* Lake Worth, Fla, 1994, Wingers Publishing, pp 1030-1049.
23. Smith A: Zinc toxicosis in a flock of hispaniolan amazons, *Proc Annu Conf Assoc Avian Vet* :447-453, 1995.
24. Atkinson R: Heavy metal poisoning in psittacines and waterfowl, *Proc Annu Conf Assoc Avian Vet* :443-446, 1995.
25. Butler R: Toxins. In Burr EW, editor: *Companion bird medicine,* Ames, Iowa, 1987, University Press, pp 226-230.

6

Abdominal or Coelomic Distention

Michelle Curtis Velasco

Birds do not have a functional diaphragm as do mammals; the thoracic and abdominal spaces are continuous and that space is defined as the *coelom*. Even though veterinarians often describe part of that space as the abdomen, it is anatomically more appropriate to describe it as the coelom. However, for this discussion, both words will be used interchangeably.

Abdominal distention may occur in a multitude of avian diseases and conditions. In mild- to-moderate cases of abdominal distention, the owner or caretaker may not notice the condition because of the feathers covering the abdominal area. Often the first clinical sign noted is the accumulation of fecal material around the vent area. Moistening the feathers over the abdomen or palpation of the area between the caudal edge of the sternum or keel and the vent or pelvic bones will reveal mild to moderate distention. Normally a slight depression should be visible, with the ventral borders of the liver just barely evident on palpation. As abdominal or coelomic distention progresses, the abdominal wall may actually protrude between the legs and hinder proper perching or interfere with respiration. In chronic cases of severe coelomic distention, ulceration of the skin over the abdomen may occur because of contact with the perch or cage bottom.

I. General Considerations

A. Causes of Abdominal or Coelomic Distention
1. Organomegaly (most often from liver enlargement or reproductive organ hyperplasia and neoplasia)
2. Fluid accumulation (ascites, septic or nonseptic peritonitis, serositis)
3. Obesity
4. Egg retention or binding
5. Air sac distention (barrel shaped from pulmonary or air sac disease)
6. Normal anatomic abdominal distention in preweaned neonates
7. Abdominal hernia

II. Etiologies of Abdominal Distention
A. Organomegaly
1. Etiologies of hepatomegaly
   a. Viral diseases (Pacheco's disease, or avian herpesvirus and avian viral serositis, thought to be caused by an EEE virus)[1]
   b. Chlamydiosis (psittacosis and ornithosis)
   c. Bacterial diseases (salmonellosis, colibacillosis)
   d. Parasitic diseases (sarcocystis, flukes, atoxoplasma)
   e. Fatty infiltration and hepatic lipidosis
   f. Neoplasia including bile duct carcinomas, lymphosarcoma
   g. Iron storage disease (mynah birds, toucans, softbills and occasionally psittacines)
2. Etiologies of gastrointestinal tract distention
   a. Ileus
      (1) Neurogenic: A common cause of neurogenic ileus is lead or heavy metal poisoning. Other neurogenic causes include proventricular dilatation disease (PDD) and enteritis such as clostridial enteritis.
      (2) Nonneurogenic: Causes include intestinal lumen occlusion from foreign bodies, enteroliths, or parasites. Stenosis results from tumors, granulomas, and strictures. Compression from extraluminal causes, such as volvulus, intussceptions, and herniation.
   b. PDD
      (1) Also referred to in the literature as *macaw wasting syndrome,* PDS, myenteric ganglioneuritis, psittacine wasting syndrome, proventricular hypertrophy, infiltrative splanchnic neuropathy, and lymphoplasmacytic ganglioneuritis and encephalomyelitis.[2]
      (2) PDD causes segmental distention of the gastrointestinal tract with accumulation of ingesta and often the passage of undigested food in the feces. Recent evidence suggests a viral etiology.[3]
   c. Neoplasias
   d. Impactions
   e. Severe intestinal wall thickening resulting from proliferative diseases such as mycobacteriosis or inflammatory conditions
   f. Cloacal distention resulting from cloacal papillomas or cloacalithiasis
3. Female reproductive organ enlargement
   a. Ovarian enlargement
      (1) Neoplasias
      (2) Cystic ovaries
   b. Oviduct and uterine enlargement
      (1) Egg retention: Eggs can be retained for periods longer than their normal passage rates.

(2) Egg binding: Failure of an egg to pass through the oviduct at a normal rate, including delayed oviposition. Dystocia is defined as the mechanical obstruction of an egg in the caudal reproductive tract.
4. Male reproductive organ enlargement
   a. Testicular enlargement results in abdominal distention most commonly from neoplasia.
   b. Testicular neoplasia is common in the budgerigar and not as common in other species but has been described.
   c. Tumor types reported include Sertoli cell tumors, seminomas, interstitial cell tumors, lymphosarcomas, hemangiomas, fibrosarcomas, and leiomyosarcomas.[4]

B. Abdominal Fluid Accumulation
1. Ascites may be caused by cardiac disease (right-sided or generalized heart failure), hepatic disease, or congestion
2. Peritonitis (septic or nonseptic) commonly causes include yolk peritonitis, intestinal perforation, and bacterial or fungal infections
3. Abdominal effusion resulting from neoplasia
4. Hemoperitoneum
5. Cystic distention of the reproductive tract
   a. Differentiation of the causes of abdominal fluid accumulation may be made by either of the following:
      (1) Analysis and cytology[5] of fluid aspirated by abdominocentesis fine needle aspirates
      (2) Ultrasound in conjunction with other diagnostics

C. Obesity
1. Obesity is common in companion birds fed predominantly oil-seed diets. Amazon parrots, poinus parrots, budgerigars, and cockatiels may present with respiratory distress related to diminished air sac space.
2. Dietary management and gradual increases in activity are required.
   a. Changes in diet should not be taken without close monitoring of food consumption and weight loss.
   b. Obese birds that refrain from eating for even relatively short periods may be susceptible to the development of fatty liver syndrome, which can be fatal if not managed aggressively. Birds with fatty liver syndrome may show lipemia with or without an increase in liver enzyme values.

D. Egg Retention or Binding
1. Causes
   a. Hypocalcemia and other nutritional deficiencies
   b. Oviduct, uterus, or vaginal muscle dysfunction
   c. Excessive egg production
   d. Large misshapen, or softshelled egg or eggs
   e. Age of the hen

f. Obesity
g. Oviductal tumors
h. Oviductal infection
i. Lack of exercise
j. Hyperthermia or hypothermia
k. Genetics

2. Clinical signs: Dystocia can occur from breeding out of season, in first-time layers, or in hens with a persistent functional right oviduct.[6] Torsion of the oviduct may also be a more common cause of dystocia than previously recognized.[7] Small birds (cockatiels, finches, budgerigars, lovebirds) are more likely to show signs of dystocia. Clinical signs include the following:
   a. Depression, lethargy, or shock can occur, depending on the duration of the condition. Often the owner will find the bird on the bottom of the cage instead of the perch.
   b. The owner may note tail wagging, tail bobbing, and straining because of increased abdominal effort by the bird to pass the egg.
   c. The owner may note dyspnea or hyperpnea. Respiratory signs may be due to compression of the air sacs by the retained egg or eggs.
   d. In very severe cases, the birds may have paresis of the pelvic limbs, or the feet may appear bluish-white, indicating vascular compromise.[8]
   e. The pelvic limbs may feel cooler than normal. Large psittacine birds may show lameness, abdominal distention, respiratory distress, depression (usually of a sudden onset), abdominal straining, and sometimes sudden death.
   f. Ratites rarely show signs other than persistent reproductive behavior and a cessation of egg laying.[9]
   g. Intervention can include the following:
      (1) Aggressive therapy is warranted, particularly in the smaller avian species because of their limited reserves and high energy requirements.
      (2) Medical or surgical intervention in the very small birds may be necessary as quickly as 1 to 2 hours to prevent death.

E. Air Sac Distention
1. Causes of air sac distention include the inability of air to be dispelled from the caudal pulmonary air sacs (caudal thoracic and abdominal) and can produce a one-way valve effect.
2. Emphysema has been described most commonly in birds with chronic pulmonary aspergillosis and may be determined radiographically. Potentially any disease that causes dramatic increased in respiratory effort, such as pulmonary fibrosis, can result in this emphysematous condition.

F. Normal Anatomic Distention in Preweaned Neonates
1. Causes include high-volume, high-fluid diets of nestling psittacines.
2. Clinical signs
   a. Distention of the entire gastrointestinal tract

b. Abdominal organs appear to protrude into the abdominal space
   c. As the bird approaches fledging, the abdomen tightens and the breast musculature enlarges in preparation for flight

G. Abdominal Hernia
   1. Causes
      a. Protrusion of abdominal fat or organs secondary to congenital defects
      b. Weakness of the musculature from excessive reproductive activity
      c. Polyostotic hyperostosis

III. Diagnostic Testing

A. Workup Based on Preceding Differential Diagnoses
   1. Gentle palpation and visualization
      a. The abdomen in a normal, adult, non–egg-laying bird should be flat and slightly concave with the edge of the ventriculus just barely palpable under the edge of the sternal border on the left side of center.
      b. With liver enlargement, ascites, proventricular or ventricular distention or displacement, egg development, egg-related peritonitis, or mass formation, the abdomen may appear distended, doughy, and convex.
      c. Excessive digital pressure when fluid is present in the peritoneal space may result in air sac rupture and subsequent asphyxiation from fluid rushing into the lung.
      d. Gentle palpation under the ventral edge of the sternum should not be painful. A painful response to this procedure may indicate hepatitis.[10]
      e. In a well-muscled, low–body-fat canary or finch, the abdominal musculature is nearly transparent; moistening the feathers over the abdomen will allow excellent visualization of the intestinal tract and possibly the liver (if enlarged).
   2. Complete blood count, blood chemistry profiles including bile acids
   3. Radiology including contrast studies
      a. High- or ultra-detail film, short exposure times
      b. Tabletop techniques tend to enhance detail to provide the quality necessary for interpretation and diagnosis
      c. Human mammography films and techniques have been modified to produce films with excellent detail for avian radiology.[11]
   4. Ultrasonography
      a. Very useful in cases of coelomic disease with distention, because air sacs may be compressed or less prominent.
      b. Imaging may be done with the bird away and in dorsal recumbency using the cutaneous window between the pelvic bones and the sternum.
      c. A 6.5 MHz electronically focused sector fingertip probe or 7.5 MHz annular array probe may be used.

d. Ultrasonic findings correlated well with radiographic findings of the abdomen, with genital tract diseases being the most common finding.[12]
e. Ostriches and other ratites are suited anatomically for imaging using ultrasound with a 5.0 MHz convex phased-array ultrasound transducer.[13]

5. FNA (fine needle aspiration) or abdominocentesis
   a. Performed on patients that are adequately restrained, either chemically or physically.
   b. Each patient must be prepared aseptically.
   c. A fine needle (22-gauge or smaller) is inserted on the midline, just caudal to the sternum, directed slightly to the right to avoid the ventriculus (Fig. 6-1).

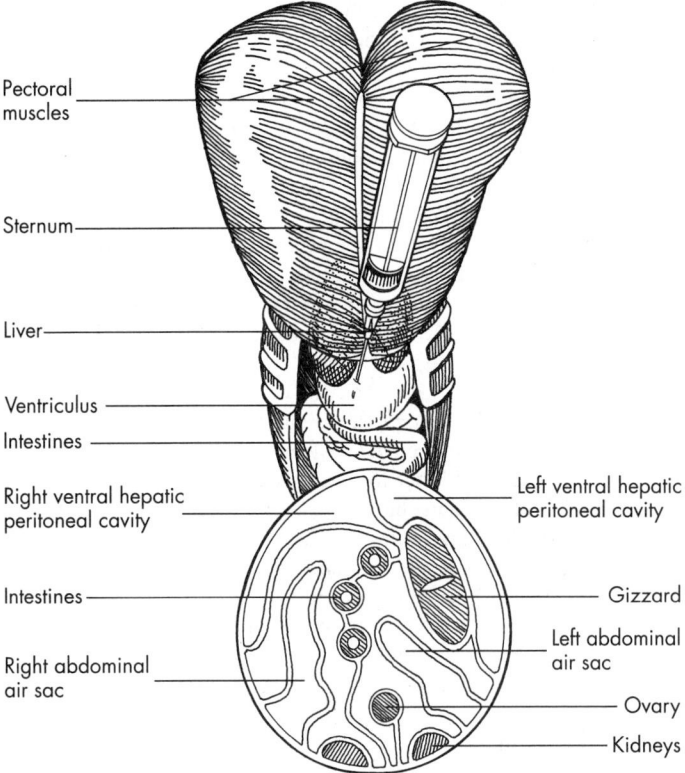

Fig. 6-1 Abdominocentesis (FNA). Drawing combines ventral view of abdomen with abdominal cross section showing placement of needle (22-gauge or smaller) inserted on the midline and slightly to the right to enter the right ventral hepatic peritoneal cavity.

d. May be performed with the help of ultrasound for guiding the needle into pockets of fluid. Fluid can then be aspirated into the syringe for analysis.
e. Samples for exfoliative cytologic examination should be prepared immediately, because cellular morphology will be altered if the sample is allowed to sit for any length of time.[14] Samples can also be cultured.
6. Other diagnostic tests related to diseases affecting the organs of the coelom
   a. Test for the following:
      (1) Chlamydiosis
      (2) Aspergillosis
      (3) Sarcocystis
      (4) Serum protein electrophoresis
      (5) Blood lead, serum, or plasma zinc levels and avian tuberculosis testing (serology and acid-fast stain)
7. Endoscopy
   a. Performed with caution in cases of abdominal distention after radiology and other tests have established the degree of air sac compression and the presence or absence of fluid in the abdominal cavity.
   b. Ultrasonography may help to determine fluid accumulation in the abdominal or coelomic space as well as the possibility of pericardial effusion.
8. Exploratory surgery and biopsy of an enlarged organ
   a. May be performed by exploratory surgery or endoscopy
   b. Liver biopsy (FNA or other) may be the most accurate way to diagnose a variety of hepatic conditions such as hemochromatosis and hemosiderosis, lipidosis, bile duct carcinoma, lymphosarcoma, or hyperplasia. However, anesthesia and endoscopy should be viewed with extreme caution with significant amounts of abdominal fluid accumulation and pericardial effusion.

IV. Workup Plan for Abdominal Distention

A. Physical Examination to Determine if the Bird is Stable or Unstable
1. If the bird is unstable and has respiratory compromise, the clinician should request the following:
   a. Warmed O_2 cage
   b. Nebulization
   c. Standing lateral radiograph to determine if there is primary or secondary respiratory involvement
      (1) Primary respiratory involvement: See Chapter 3.
      (2) Secondary respiratory involvement: Determine which organs are involved and if ascitic fluid is present, which can be determined by the following:
         (a) Radiographs

(b) Ultrasonography
(c) Coelomic aspiration and fluid collection with cytologic examination and culture (aerobic or anaerobic and fungal)
2. If the bird is unstable and does not have respiratory compromise, the clinician should determine if the following are involved:
   a. Liver
   b. Gastrointestinal tract
   c. Spleen
   d. Kidney
   e. Reproductive tract
      (1) Based on the organs involved, the clinician may be able to localize to a disease or condition; treatment should be based on diagnosis or to provide supportive care (see Treatment section).
      (2) The clinician often needs to support birds with secondary respiratory involvement as if the bird has primary respiratory involvement.
      (3) Treatment for secondary respiratory involvement includes O_2 cages, +/− nebulization, warmed cages, and +/− Lasix
3. If the bird is stable, the clinician should proceed with the appropriate diagnostics to determine the cause of abdominal distension.

V. Treatment

Treatment of a bird with a distended abdomen may involve emergency supportive care in severe cases, particularly if the clinical signs include respiratory distress or compression of other critical organs. A warmed incubator with O_2 supplementation may help stabilize the patient so that diagnostics can be performed. Once the cause of the coelomic distention has been defined to the organ system or fluid type involved, specific treatment can then be initiated. Excessive fluid in the coelomic cavity may be aspirated partially to provide immediate relief if respiration is compromised. The removal of large amounts of ascitic fluid at one time may cause the bird to go into shock.

A. Liver Enlargement
Treatment for liver enlargement is based on the etiology of the disease process.
1. Fatty infiltration and hepatic lipidosis
   a. Nutritional support is essential for the successful management in cases of lipidosis.
   b. Gavage feeding with high carbohydrate formulas (Carbofuel, Hepatic Aid), lactulose (Chronulac, Enulose) and parenteral fluid supplementation are helpful.
   c. Correction of the diet and a lower-fat, formulated diet, or a diet tailored to the individual species should be attempted while the patient is supplemented with an appropriate formula (as described previously) and hospitalized if possible.

2. Chlamydiosis should be treated with supportive care and doxycycline for 45 to 60 days
3. Bacterial diseases
   a. Initial treatment should be based on Gram stain results from feces or fluid from the abdomen
   b. Treatment modified from culture and sensitivity results
   c. Avian tuberculosis testing can be performed from serologic examination or fecal sample and organ, biopsy, or FNA fluid aspirate using acid-fast stain
4. Parasitic diseases should be treated based on fecal and zinc sulfate test results and sarcocystis titer or biopsy results
   a. Sarcocystis: Pyrimethamine and sulfa combinations
   b. Flukes: Praziquantel or chlorsulon
   c. Atoxoplasma: Possibly no treatment or, for the tissue form, primaquine; pyrimethamine; sulfachlor-pyrazine for oocysts shedding
5. Hemochromatosis
   a. Reported commonly in mynahs, toucans, and occasionally in psittacines.
   b. Treatment involves phlebotomy with the removal of 1% of the body weight in blood volume on a weekly basis for periods up to a year.[15]
   c. Dietary modification with reduced iron intake is useful, as well as the addition of tea to the drinking water, which may inhibit iron uptake.
   d. A liver biopsy should be performed and stained with Prussian blue.
   e. Biopsies taken at regular intervals will help to monitor progress.[16]

B. **Egg Binding or Dystocia**
   1. General considerations
      a. Paramount in the treatment of egg binding is the return of the patient to a physiologically normal state.
      b. Removal of a retained egg should not be attempted until the patient has been stabilized and is not in shock.
   2. Medical therapy
      a. In noncomplicated cases in which the bird is minimally depressed, the bird may respond to the addition of supplemental heat with increased humidity; injectable calcium; vitamins A, D, and E; and selenium. The bird should also have ready access to food and water.
      b. Depressed or patients in shock may require subcutaneous, intravenous, and IO crystalloids; antibiotics; and possibly short-acting steroids.
      c. Dystocia for longer than 1 hour in a small bird, or more than 3 to 5 hours in a larger bird, indicates the need for more aggressive intervention. Historically oxytocin has been given at a dose of 0.01 to 0.1 ml one time. The clinician should increase uterine contractions and aid in passage of a nonadhered egg in the uterus

if the uterovaginal sphincter is relaxed. *However, it may actually complicate shock.* The use of PGE_2 gel (Prepidil Gel, the Upjohn Co., Kalamazoo, Mich) applied to the uterovaginal sphincter at a dose of 0.2 mg/kg will induce oviposition within 5 to 10 minutes in most species if the egg is not adhered to the oviduct.[17]

3. Surgical intervention
   a. Aspiration of egg contents by syringe via the vaginal opening or abdominal paracentesis may allow the egg to collapse and the shell to pass on its own.
   b. Hysterotomy and laparotomy is indicated in cases where the hen is weak, or the egg may be outside the oviduct. It is also indicated in uterine torsion cases.

REFERENCES

1. Ritchie B: *Avian viruses, function and control,* Lake Worth, Fla, 1995, Wingers Publishing, p 397.
2. Gregory C: Proventricular dilatation disease. In Ritchie B: *Avian viruses, function and control,* Lake Worth, Fla, 1995, Wingers Publishing, p 439.
3. Gregory C: Proventricular dilatation disease. In Ritchie B: *Avian viruses, function and control,* Lake Worth, Fla, 1995, Wingers Publishing, p 439.
4. Orosz SE, Dorrestein GM, Speer BL: Urogenital disorders. In Altman RB et al, editors: *Avian medicine and surgery,* Philadelphia, 1997, WB Saunders, pp 614-644.
5. Campbell T: Cytology. In Harrison GJ, Harrison LR, editors: *Clinical avian medicine and surgery,* Philadelphia, 1986, WB Saunders, p 251.
6. Hochleithner M, Lechner C: Egg binding in a budgerigar caused by cyst of the right oviduct, *Assoc Avian Vet Today* 2:136-138, 1988.
7. Harcourt-Brown N: Torsion and displacement of the oviduct as a cause of egg-binding in four psittacine birds, *J Avian Med Surg* 10:262-267, 1996.
8. Romagnano A: Avian obstetrics, *Semin Avian Exotic Pet Med* 5(4):180-188, 1996.
9. Speer B: Diseases of the urogenital system. In Altman RB et al, editors: *Avian medicine and surgery,* Philadelphia, 1997, WB Saunders, p 634.
10. Harrison GJ, Ritchie BW: Making distinctions in the physical examination. In Ritchie BW, Harrison GJ, Harrison LR: *Avian medicine: principles and applications,* Lake Worth, 1994, Wingers Publishing, pp 167-168.
11. Smith BJ, Smith SA: Radiology. In Altman RB et al: *Avian medicine and surgery,* Philadelphia, 1997, WB Saunders, p 170.
12. Rosenthal K et al: Ultrasonic findings in 30 cases of coelomic disease, *Proc Annu Conf Assoc Avian Vet* 303, 1995.
13. Blue-McLendon A: Ultrasound determination of yolk sac size in ostrich chicks, *Proc Annu Conf Assoc Avian Vet* 1-26, 1995.
14. Murray MJ: Diagnostic techniques in avian medicine, *Seminars in Avian and Exotic Pet Medicine* 6:47-54, 1997.
15. Campbell TW: Cytology. In Harrison GJ, Harrison LR, editors: *Clinical avian medicine and surgery,* Philadelphia, 1986, WB Saunders, p 251.
16. Lemeij JT: Hepatology. In Ritchie BW, Harrison GJ, Harrison LR, editors: *Avian medicine: principles and applications,* Lake Worth, Fla, 1994, Wingers Publishing, pp 523-537.
17. Hudelson RS: A review of the mechanisms of avian reproduction and their clinical applications, *Semin Avian Exotic Pet Med* 5(4):189-198, 1996.

7

Avian Dermatology

April Romagnano and Darryl J. Heard

I. Introduction

Avian dermatology is still in its infancy as a science. In many patients the diagnosis is uncertain. Consequently, many avian dermatologic problems have traditionally been grouped under *idiopathic feather picking*. Regardless of the suspected cause, the approach to an avian dermatology problem should be systematic. Avian dermatology is often frustrating for both the owner and clinician, and many problems may take a long period to resolve.

II. Physical Examination

history - environment - ō
humans/pets - PE

Examination of a bird with a dermatologic problem begins with the following:
1. Gathering a detailed history
2. Observing the bird in its environment
3. Observing the bird's interactions with its owner and other animals
4. Performing a detailed physical examination

Many of these elements have been described in other sections (see Chapter 1).

A. History *Where - how long - cage - ō birds - diet -*
 Information to obtain from the owner includes the following:
 1. Origin of the bird
 2. Length of ownership   *Supplements - changes*
 3. Housing arrangements   *noticed by owner -*
 4. Exposure to other birds *treated by ō vet*
 5. Consumption of a normal diet and supplements
 6. Owner's evaluation of presenting condition including the following:
    a. Changes in food and water consumption
    b. Droppings

c. Environment
d. Behavior
7. Previous treatment by owner or another veterinarian

B. Visual examination
1. Observe the bird from a distance:
    a. Does the bird show any abnormal behaviors?
    b. The bird may not show these behaviors in a novel environment such as the examination room.
2. Observe the cage contents:
    a. See Chapter 1, Chapter 8
    b. Diet
    c. General care
    d. Presence of feathers
    e. Possibility of ectoparasites

C. Interaction with Owners
Note quality and amount of time spent with owner:
1. Has this changed recently?
2. Has the household undergone any other changes?
3. Is the bird handled? Does it obey basic commands, such as step-up?
4. Is the bird living in an adequate size cage, and where is it located?
5. Can the bird view the family in their living areas, such as the kitchen and den?

D. Physical Examination
1. General
2. Feathers
    a. General view of feathers
    b. Examine all feathers, including underfeathers
    c. Pattern of feather loss and abnormality
3. Skin

III. Feathers

A. Anatomy and Physiology
The anatomy and physiology of feathers are well described by Pass[1], King and McLelland[2] and in the definitive text of Lucas and Stettenheim.[3]
1. Arrangement
    a. Skin is incompletely covered by feathers in most birds.
    b. Definitions:
        (1) Ptilosis = Plumage arrangement (size, shape, etc.)
        (2) Pterylae = Feather tracts
        (3) Pterylosis = Arrangement of feather tracts on the skin
        (4) Apteria = Bare spaces between the feather tracts
2. Contour feathers (Figs. 7-1 and 7-2)
    a. Major surface feathers
    b. Divided into flight and body feathers

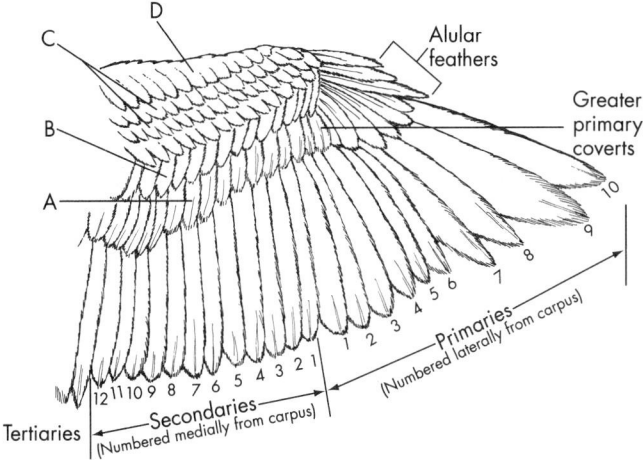

Fig. 7-1 A dorsal view of the wing. The primaries are numbered from inside out; the secondaries are numbered inward. Starting from closest to the flight feathers outward, the coverts are named as follows: **A,** Greater secondary coverts. **B,** Median secondary coverts. **C,** Lesser secondary coverts. **D,** Marginal coverts.

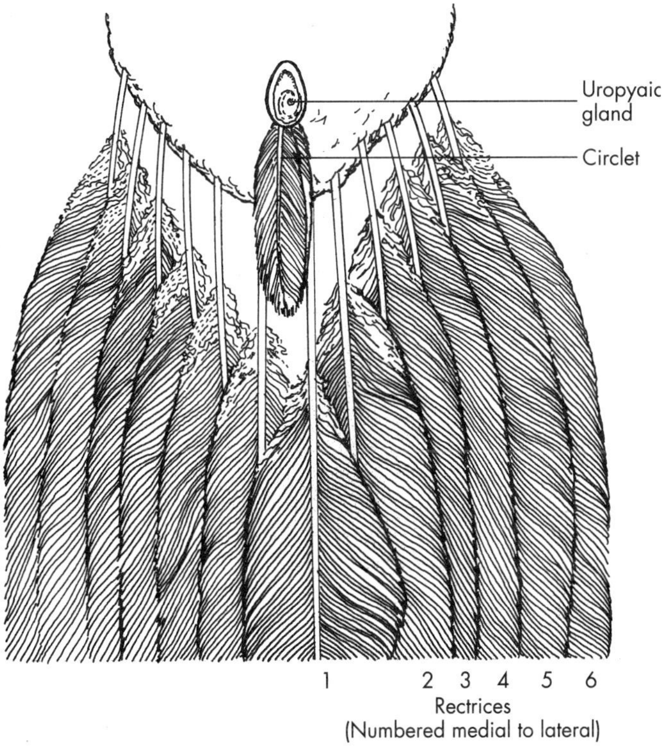

Fig. 7-2 A dorsal view of tail feathers in a partially plucked uropygium. Feathers are numbered from the center out to right and left.

c. Definitions:
   (1) Remiges: Wing flight feathers, 20 to 24 in number
   (2) Primaries: Wing flight feathers attached to the carpus, the carpometacarpus, and the phalanges. They are numbered 1 to 10 from the carpus outward.
   (3) Secondaries: Wing flight feathers inserting on the ulna are numbered inward, 1 to 10 or 14.
   (4) Rectrices: Tail flight feathers, 12 in number
d. Coverts cover the bases of the remiges and rectrices either dorsally or ventrally. They also cover the external ear canal.
3. Semiplume feathers (Fig. 7-3)
   a. Long rachis and entirely plumaceous vane
   b. Occur either along contour feather tracts or in feather tracts of their own
4. Down feathers (Fig. 7-4)
   a. Small, fluffy, wholly plumaceous feathers with very short or absent rachis
   b. Natal down is present in neonates
   c. Definitive down is present in immature and adult plumage

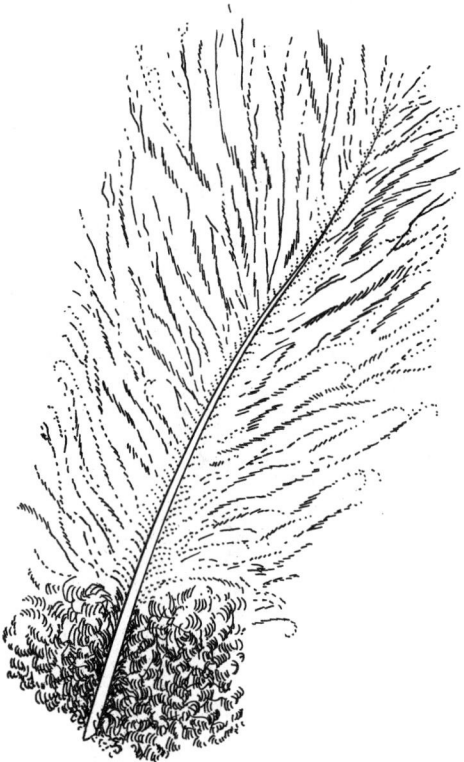

Fig. 7-3 A semiplume feather.

5. Powder down feathers
   a. Specialized down feathers
   b. Disintegrate producing powder (keratin) that is spread through the feathers as a preening and waterproofing agent
6. Hypopennae (afterfeathers) (Fig. 7-5)
   a. Small feathers projecting from the distal umbilicus of plumaceous and pennaceous feathers
   b. Highly variable in form
   c. In emu and cassowaries, afterfeather is as long as the main feather

Fig. 7-4 A down feather.

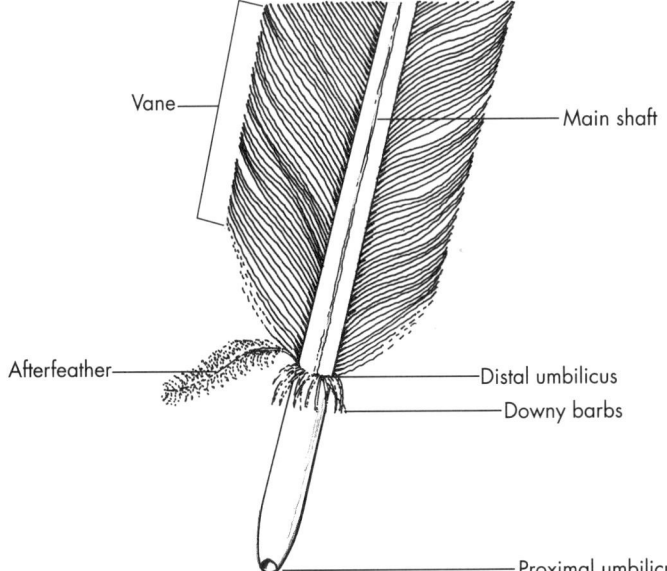

Fig. 7-5 An afterfeather, or hypopennae, seen at the proximal part of a contour feather.

100 *Avian Dermatology*

7. Filoplumes (Fig. 7-6)
    a. Fine hairlike feathers with a long calamus and a tuft of barbs on the tip
    b. Present in all feather tracts
8. Bristle (Fig. 7-7)
    a. Stiff and have a tapered rachis with no barbs except at the proximal end
    b. Usually found on the head (eyelids, nares, and mouth)
9. Feather structure (Fig. 7-8)
    a. Typical contour feather composed of the following:
        (1) Shaft
        (2) Vane
        (3) Afterfeather
    b. Shaft composed of the following:
        (1) Calamus: Tubular, unpigmented end of mature feather within the feather follicle
        (2) Rachis: Shaft above the skin
        (3) Proximal or inferior umbilicus: Circular opening in embedded tip of calamus

Fig. 7-6 A filoplume feather.

(4) Dermal papilla: Small mound of pulp inside the calamus contiguous with dermis of follicle
(5) Dermal papilla covered by a layer of living epidermal cells: Contributes to next feather when molting occurs
(6) Growing feather calamus filled with pulp: Loose reticulum of mesoderm and axial artery and vein.
(7) As feather matures, vessels degenerate and pulp dies: They are resorbed back to the umbilicus.
(8) Chambers: A series of thin epidermal partitions divides the calamus into a sequence of isolated chambers.
c. Barbs and barbules (Fig. 7-9)
(1) Rachis bears two series of slender but stiff filaments, the barbs, at about 45 degrees to the main shaft.
(2) Each barb also carries two series of even finer filaments called *barbules*, again at 45 degrees. Barbules of adjacent barbs cross each other at 90 degrees.
(3) The barbules that slope towards the free tip of the feather (the distal barbules) carry tiny hooks, which loosely engage the proximal barbules.

Fig. 7-7 A bristle feather.

102 *Avian Dermatology*

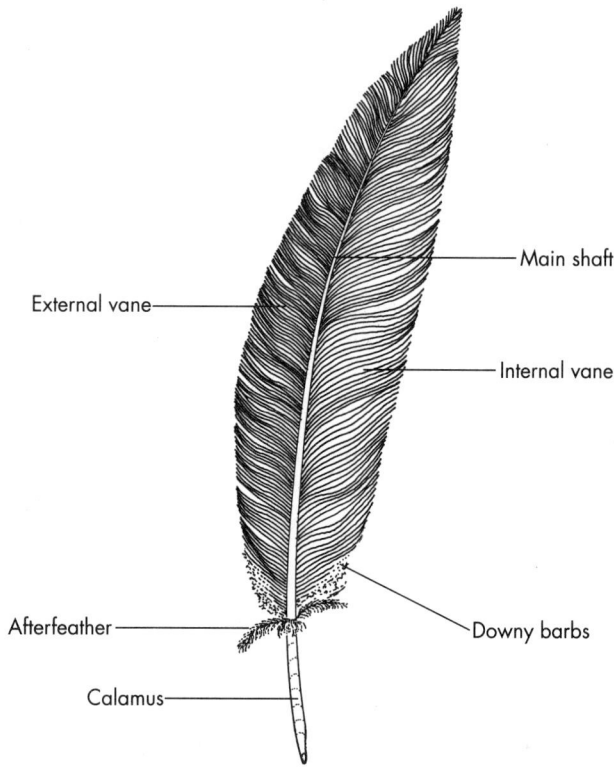

Fig. 7-8 The parts of a flight feather.

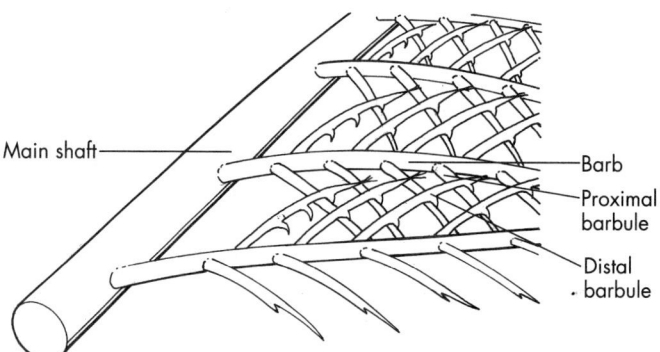

Fig. 7-9 Interlocking barbules of the barbs of a flight feather.

Handwritten annotations at top:
Feather color — produced by: body (melanins), food absorption (carotenoids), pigment (porphyrins), structure (feather surface/viewing angle)

(4) At the base of contour feather the barbs are not hooked together but form downy barbs with a fluffy or plumaceous part. Most of the feather, in which barbs are united, is firm and constitutes the pennaceous part.
  d. Mature follicle
    (1) Follicle is a cylindric pit in the skin.
    (2) Wall of follicle consists of epidermis and dermis.
    (3) Follicle is lined by epidermis.
10. Feather coloration
  a. Determined by pigments (melanins, carotenoids, and porphyrins) and structure
  b. Melanins *(produced by body)*
    (1) Produced by melanocytes that migrate into epidermis during development
    (2) Melanin granules taken up by epidermal cells at a precise time in their development
    (3) Make feathers denser and more resistant to wear and photochemical degradation
  c. Carotenoids (carotenes and xanthophyll) *(absorbed from food)*
    (1) Synthesized by plants
    (2) Absorbed from food and taken up by cells
    (3) Cellular uptake genetically determined
  d. Porphyrins (red and green pigments produced by cells)
  e. Structural colors:
    (1) Product of physical properties of feather surfaces that modify or separate the components of white light
    (2) Iridescent colors determined by viewing angle
11. Molting
  a. Replacement of feathers is typically staggered.
  b. Molt
    (1) Triggered by change in day length
    (2) Usually occurs after breeding

B. Differential Diagnoses: Feather Loss, Picking, and Abnormalities

Feather loss or absence of feathers indicates an imbalance between production and loss. Feather picking can be a behavioral disorder, or it can be due to peripheral neurologic pruritus, infectious pruritus, neoplastic pruritus, trauma, or pain resulting from any irritating disease state. Feather abnormalities reflect damage either during feather growth or after.

1. Normal absence of feathers
  a. Most avian species have areas of skin without feathers called *apteria*.
  b. Many birds also have anatomic bare patches of skin (e.g., the top of the head in cockatoos and the brood patch in breeding birds).
2. Bacterial
  a. Bacterial folliculitis and pulpitis are rare.
  b. Clinicians often find difficulty in differentiating between primary and opportunistic invaders.

c. Etiology
   (1) Gram-negative (e.g., *Aeromonas, Pseudomonas* bacteria)
   (2) Gram-positive (e.g., *Staphylococcus, Streptococcus* bacteria)
   (3) Mycobacteria (very uncommon)
   (4) Role of anaerobic bacteria (unknown)
d. Clinical signs
   (1) Perifollicular and follicular reddening, swelling, and necrosis with crust formation[4]
   (2) In Fischer's lovebird,[4] *Aeromonas* infection associated with acute onset of discomfort and pruritus localized to the ventral surface of the wings
e. Histology
   (1) Characterized by a heterophilic inflammatory infiltrate with or without bacteria[5]
   (2) Mycobacterial granulomatous dermatitis characterized by lump or multiple lumps comprising large macrophages, containing acid-fast bacteria and a variable number of heterophils and plasma cells[5, 6]

3. Congenital and genetic
   a. Feather duster disease, a lethal genetic disorder in budgerigars. Affected birds display gigantism and long filamentous feathering (Fig. 7-10)

Fig. 7-10 A budgerigar with the lethal genetic feather duster disorder. (Courtesy Dr. JR Baker)

b. Straw feathering, a lethal hereditary disorder in canaries. Feathers do not emerge from their shafts, giving the appearance of straw.
   c. Feather cysts are common in canaries but do occur in many other species of passerines and psittacines (Fig. 7-11).
4. Endocrine and hypothyroidism
   a. Very uncommon; only well-documented case was in a scarlet macaw.[7] We have diagnosed a white-fronted Amazon parrot with hypothyroidism, and a conure with hypothyroidism and concurrent bacterial folliculitis.
   b. Thyroxine is important in the initiation of molting. Normal T4 values range from 0.1 to 1.1 ug/dl in Amazons to 0.2 to 4.3 µg/dl in lovebirds (see Chapter 16).[8]
   c. Clinical signs
      (1) Thickened, often fat-laden, skin
      (2) Nonpruritic feather loss
      (3) Molting ceases
      (4) Obesity

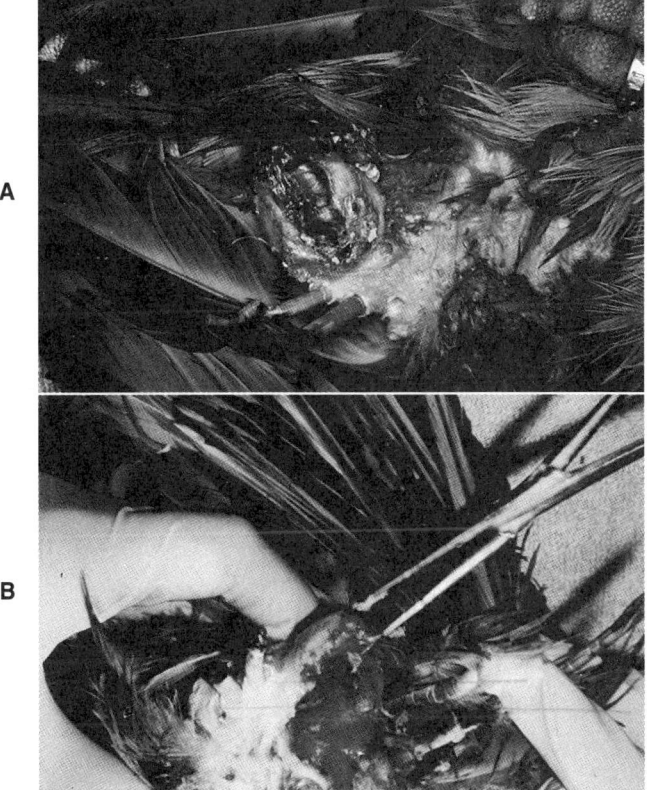

Fig. 7-11 A Bufon macaw with a large tail feather cyst. **A** and **B,** Surgical removal of this cyst.

5. Fungal[5]
   a. Uncommon
   b. Etiology
      (1) *Candida* spp. have been associated with feather and skin lesions in gallinaceous birds, especially with feather loss in the vent area, and with some pet avian species in the head area.[8]
      (2) *Aspergillus* spp. have been associated with feather and skin lesions especially in pigeons.[9]
      (3) Dermatophytes can cause patchy feather loss, especially on the head, neck, and breast; dermatophytes include *Microsporum gallinae* (fowl, canary, duck, quail, pigeon, turkey), *Microsporum gypseum* (fowl, parrot), *Trichophyton verrucosum* (fowl, canary), and *Trichophyton* spp. (robin, European blackbirds).[10]
   c. Clinical signs
      (1) Follicles are usually involved, leading to gross swelling.[5]
      (2) *Candida albicans* is characterized by powdery white crust-like accumulations around the feather follicles.[10]
   d. Histology includes pleocellular inflammatory infiltrate within the follicle and associated epidermis and dermis; fungal hyphae present in lesion give it specificity.[5]
6. Immunologic
   a. Controversial
   b. Eosinophils do not play the same role in allergic reactions in birds as they do in mammals.[11]
   c. Mast cells and basophils can release histamine, but a full allergic response is not well characterized.
   d. IgG, not IgE, appears to be the immunoglobulin involved in anaphylactic responses in birds.[11]
   e. Etiology
      (1) We have suspected several cases of insect bite hypersensitivity in Amazon parrots and macaws housed outdoors in Florida. The bites are typically in nonfeathered areas, such as the cheek patches.
      (2) Glenn Olsen has several documented cases of insect bite hypersensitivity in whooping and sandhill crane legs (unpublished data) (Fig. 7-12).
7. Nutritional (vitamin A deficiency)
   a. Etiology: Poor nutrition; seed diets low in vitamin A
   b. Clinical signs: Skin affected by hypovitaminosis A is dry, flaky, pruritic, and has poor wound healing. These dermatologic changes may lead to feather picking. See 9 below.
   c. Histology: Biopsy shows extensive hyperkeratinization.
   d. Diagnosis: Clinical signs, history, diet, and response to treatment.
8. Parasitic (see Chapter 20)
   a. Lice: Rare in psittacines, noted only in heavy infestation.[8] Pigeons have lice, which produce holes in the feathers.

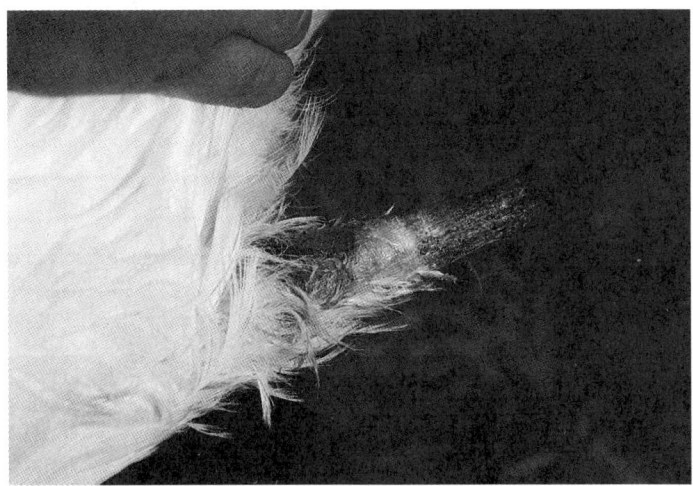

Fig. 7-12 Insect bite hypersensitivity in a whooping crane leg. (Courtesy Dr. GH Olsen)

    b. *Myialges* spp.: Mange mite causes pruritus and feather picking in the face of grey cheeked parakeets. It also causes flaky skin and pruritus in Amazons.[8]
    c. *Knemidokopte* spp.: Cause proliferation of tissues of the beak, legs, feet, and cloaca. Feather loss and honey-combed masses appear. *Knemidokoptes* organisms also cause pruritus of head and neck.
    d. *Giardia* spp.: Hypersensitivity reaction causes feather picking, especially in cockatiels. Ulcerative dermatitis results in cockatiel and lovebird wing web picking.
9. Feather picking and self-mutilation syndromes
    a. Feather picking syndromes and self-mutilation
       (1) Moluccans: Keel feather and muscle mutilation
       (2) Lovebirds: Wing web area mutilation
       (3) Cockatiels: Feather mutilation syndrome
       (4) African greys: Incessant split keel
    b. Feather picking etiology is multifactorial
       (1) Behavioral: Common in African greys, cockatoos, and conures, central nervous system psychosis, loneliness, boredom, change in barbering of mate
       (2) Peripheral neurologic pruritus
       (3) Infectious pruritus
       (4) Neoplastic pruritus
       (5) Trauma
       (6) Pain
10. Mate trauma
    a. Most commonly male cockatoo attacks on females
       (1) Beak and head

(2) Wings, chest, and legs
(3) Cannibalism may occur
b. Cannibalism is most common in lovebirds and red vented and major Mitchell cockatoo pairs
11. Viral
 a. Etiology
  (1) Circovirus[12]
  (2) Polyomavirus
 b. Clinical signs of circovirus of parrots (psittacine beak and feather disease [PBFD])[12]
  (1) Abnormalities depend on species and on stage of feather development when viral infection occurs.
  (2) Chronic form occurs in mature birds. Dystrophic feathers stop growing shortly after emerging from the follicle, and their number increases with each successive molt. The powder down feathers, located over the hips in some species, are typically the first to show signs of dystrophy. The disease later progresses to the body contour feathers followed by changes in the primary, secondary, tail, and crest feathers. Primary feathers are usually the last to manifest the disease. Feather dystrophy and loss are roughly symmetric. Changes include feather sheath retention, pulp cavity hemorrhage, fracture of the proximal rachis, and failure of developing feathers to exsheathe. Short clubbed feathers, deformed curled feathers, stress lines within vanes, and circumferential constrictions may be observed (Fig. 7-13).
  (3) The acute form (French molt, an Australian term, describes both the acute form of PBFD and polyoma virus in young budgerigars) usually occurs in young birds during their first molt, which replaces the neonatal down. Feather abnormalities may be observed in chicks within 25 to 40 days following inoculation. Feather changes are characterized by sudden alterations in developing feathers including hemorrhages, necrosis, fractures, bending, or premature shedding of diseased feathers. Neonates that develop clinical lesions while the majority of feathers are still in the developmental stage exhibit the most severe feather abnormalities. They may appear totally normal initially, but subsequently develop 80% to 100% feather dystrophy within 1 week after feather changes are first observed. After contour feathers mature, feather changes may be limited to developing flight and tail feathers (Fig. 7-14).
 c. Clinical signs of circovirus of doves and pigeons[13]
  (1) Reported in three laughing turtle doves (*Streptopelia senelgalensis*) in Western Australia; feather loss on the wings, tail, and body.
  (2) Most flight and tail feathers were absent; birds otherwise normal.

Fig. 7-13 Feathers of a cockatoo with PBFD. (Courtesy Dr. GH Olsen)

   d. Clinical signs of polyomavirus of parrots[14,15]
      (1) Feather abnormalities are common in budgerigars but occur in large parrots as well.
      (2) Clinical signs depend on age and condition of bird at infection.
      (3) Hatchlings may have reduced formation of down and contour feathers.
      (4) Juveniles that survive early infection may exhibit symmetric feather abnormalities characterized by dystrophic primary and tail feathers, lack of down feathers on the back and abdomen, and lack of filoplumes on the head and neck.
      (5) Birds with dystrophic flight feathers may be unable to fly and have been called *runners*. Progressive feather changes in these birds are also referred to as *French molt*.
      (6) Subcutaneous hemorrhaging is common.
   e. Histology of circovirus (PBFD) of parrots[12]
      (1) Necrosis and inflammation are common within feathers and are usually diffuse and severe. Necrosis varies from mul-

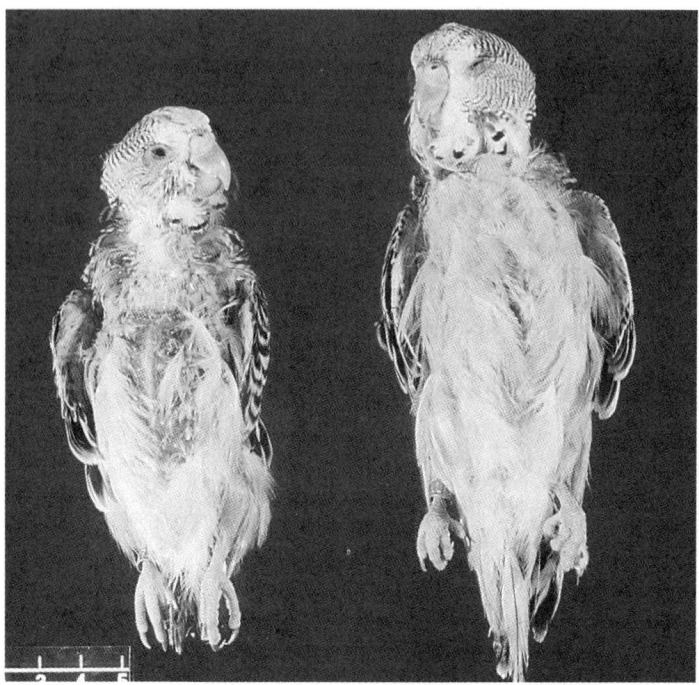

Fig. 7-14 Two budgerigars with the French molt caused by PBFD virus. (Courtesy Dr. JR Baker)

tifocal involvement of the basal feather epithelium to diffuse involvement of the pulp cavity. Inflammatory cell infiltrates are variable in composition (heterophils > macrophages > lymphocytes > plasma cells > giant cells).
(2) Follicular epithelium lesions are less frequent and much less severe.
(3) Characteristic findings include basophilic to amphophilic nuclear and cytoplasmic inclusions both within hematoxylin and eosin-stained tissue sections. Nuclear inclusions appear glassy and are confined to feather and follicular epithelial cells. Multiple, globular cytoplasmic inclusions are confined to macrophages in the feather epithelium, follicular epithelium, pulp cavity, or feather sheath.
f. Histology of circovirus of doves and pigeons[14]
(1) Changes identical to those seen in psittacines
(2) Focal necrosis of feather epidermal cells and the presence of macrophages containing intracytoplasmic inclusion bodies of different sizes are seen. Amount of necrosis differed between feathers as did the presence of macrophages. In some feathers the pulp was infiltrated with heterophils, and in others lymphocytes and plasma cells were present.

(3) Focal necrosis of follicular epidermal cells was also seen, but the extent was much less than in adjacent feather epidermis.
    g. Histology of polyomavirus of parrots
        (1) Produce large clear basophilic or amphophilic intranuclear inclusions only

C. Diagnostic Tests for Feather Loss, Picking, and Abnormalities
1. Biopsy including histopathology, as described previously, is often definitive.
2. Feather pulp cytologic examination
    a. Pluck blood feathers: Squeeze shaft to express liquid from core and make multiple touch preps onto slides.
    b. Cytologic examination
        (1) Stain with Diff-Quik, Wright's, Giemsa, etc.
        (2) Examine for presence of inflammatory cells.
        (3) Examine for presence of inclusions.
    c. Gram stain
        (1) Record bacteria and yeast type and character.
    d. Culture and sensitivity
3. Feather follicle biopsy
    a. Histopathology
        (1) Note cells present, cells infiltrating into feather follicle, or feather pulp.
        (2) Note any cell changes.
4. Complete blood count
5. Plasma biochemical panel
6. Heavy metal screening
    a. Rule out zinc toxicity, which has been implicated in feather picking.
7. PBFD test: DNA probe, a PCR test, unclotted whole blood preferred. Positive tests should be repeated in 90 days. Healthy birds can clear infection.
8. Polyomavirus test: DNA probe, a PCR test, swab test preferred. Indicates if bird is shedding virus. Whole blood test also available, as is serology.

D. Therapy: Feather Loss, Picking, and Abnormalities
1. Nonspecific
2. Specific: Based on etiology, may be protracted
    a. Bacterial
        (1) Treat systemically as per culture and sensitivity
    b. Fungal
        (1) Best treated systemically. Treatment is usually protracted.
        (2) Itraconazole for aspergillosis
        (3) Diflucan, once daily, good tissue and central nervous system penetration
        (4) Amphotericin B only fungicidal antifungal

c. Viral
   (1) Aggressive supportive treatment
   (2) Herpes virus treated with topical and systemic acyclovir
d. Parasitic
   (1) Remove parasite, clean environment.
   (2) Topical treatment with 5% carbaryl powder for lice and other ectoparasites
   (3) Treat systemically for mites with ivermectin, and for *Giardia* with metronidazole
e. Nutritional
   (1) Improve diet
   (2) Reduce or eliminate fatty seed portions
   (3) Increase fruit and vegetable portions
   (4) Convert carefully and slowly to pelleted and extruded bird diets.
   (5) Administer vitamins, as needed, parenterally (e.g., vitamin A)
f. Behavioral including feather picking and self-mutilation
   (1) Psychotropic drugs must be used with great care and only in cases in which medical causes of feather picking and self-mutilation syndromes have been ruled out.
   (2) Birds treated with psychotropic drugs must be strictly monitored with regular chemistry panels, because the majority of these drugs are eliminated by the liver.
   (3) The following drugs have been studied in birds with mixed results.[8] We recommend consulting the current literature for updated dosage and complication information before using these medicines.
       (a) Fluoxetine 1.0 to 2.0 mg/kg PO, BID
       (b) Haloperidol 0.02 to 0.20 mg/kg PO, BID
       (c) Clomipramine 0.5 to 1.0 mg/kg PO, BID
       (d) Naltrexone 1.5 mg/kg PO, BID

IV. Skin

A. Anatomy

The anatomy of the skin is well described by King and McLelland,[2] and in the definitive text of Lucas and Stettenheim.[3]

1. Structure
   a. Usually thinner and more delicate than mammalian
   b. Attached to muscles at several sites
   c. Extensive skeletal attachments (e.g., wings and feet)
2. Epidermis
   a. Deep living layer and outer dead layer
   b. Feathered epidermal areas about 10 cells thick divided equally between living and dead layers
   c. Living cell layer divided into the following:
      (1) Basal layer adjacent to dermis: Replaces cells lost at the surface

(2) Intermediate layer: Contains enlarged polygonal cells characterized by desmosomes and merges with the transitional cell layer

(3) Transitional layer (stratum corneum): Cornified layer of horny dead cells that contain mainly keratin and keratin-bound substances

3. Cutaneous glands
   a. Uropygial gland
      (1) Main cutaneous gland
      (2) Bilobed, located on dorsal surface of tail base
      (3) Well developed in some aquatic species
      (4) Absent in ostrich, emu, cassowary, bustards, frogmouths, and many pigeons, woodpeckers, and parrots
      (5) Usually drained by a pair of ducts, one duct for each lobe, each duct opening through a single narrow, median, nipple-like papilla.
      (6) The papilla is usually bare, except for a tuft of down feathers at the tip, the uropygial wick, in most species.
      (7) Produces a lipoid sebaceous secretion composed of a combination of sudanophilic secretory granules and cell fragments
      (8) Preening spreads secretions over feathers to provide waterproofing and possibly some antimicrobial function.
   b. External ear glands secrete a waxy material composed of masses of desquamated cells.
   c. Cells of the epidermis contain lipoid sebaceous secretions, which are presumably released all over the skin.
   d. No sweat glands
   e. Incubation patches
      (1) Dermis of abdomen in most birds
      (2) Modified during brooding period to form the incubation (brood) patch or patches
      (3) Dermis thickens, becomes vascular, and feathers are lost
      (4) Single median patch in grebes, pigeons, and many passerines
      (5) Occurs in both sexes in some species and in only one sex in others (depending on the normal incubation behavior of the species)

B. Clinical Signs and Differential Diagnoses
1. Small to large swellings
   a. Traumatic
      (1) Hematomas: Small or large swelling, pink-red to green-brown in color. A localized collection of extravasated usually clotted blood. May overlie other traumatically induced pathology, such as a fracture.
   b. Congenital
      (1) *Complexus muscularis:* Normal muscular swelling at the back of a neonatal chick's neck, also known as the *pipping muscle.* Can be confused with a swelling.

(2) Tarsus pad: Young woodpeckers, toucans, and barbets have a pad present on the back of the tarsus, which is molted when the bird leaves the nest. The pad can be confused with a swelling.
c. Viral
  (1) Papillomavirus[16]
    (a) Cutaneous papillomatosis reported in a Timneh grey parrot
    (b) Proliferative cutaneous lesions seen only on the head, particularly on the palpebrae, at the beak commissures, and on the skin contiguous with the lower beak
    (c) Histopathology includes hyperplasia and hyperkeratosis of epidermis with pronounced folding.
    (d) Nuclei in stratum corneum stained positively for papillomavirus structural antigens by the peroxidase-antiperoxidase technique used to detect antigens of mammalian papillomaviruses.
  (2) Poxvirus
    (a) Dry cutaneous form occurs in many species (see Chapter 17).
    (b) Dry form manifests itself as nodular and papular lesions on nonfeathered skin around the eyes, beak, nares, and distal to the tarsometatarsus. The dry form is most common in passerines.
    (c) Intracytoplasmic inclusions, Bollinger bodies, pathognomonic.
    (d) Diagnosis is by histopathology, tissue viral culture, or feces detection by electron microscopy or culture.
  (3) Herpesvirus[17]
    (a) Dry proliferative lesions on the toes of cockatoos can also be found on the feet and legs. These lesions are common in Moluccan, sulfur crested, and umbrella cockatoos but are rare in black cockatoos.
    (b) Herpes lesions are dry, gray to white, proliferative, and occasionally crusty. Lesions are limited to the extremities and are not life-threatening.
    (c) Histopathology of lesions similar to squamous papillomas but intranuclear inclusion bodies were detected that, on electron microscopy, consisted of herpesvirus particles.
d. Bacterial
  (1) Abscess: Typically contain *Staphylococcus, Aeromonas,* or *Mycobacterium* spp. (see below); can be of any size; a localized collection of pus in cavities formed by disintegration of tissues; typically occurs as a result of a localized infection, foreign bodies, and trauma; can be ulcerative.
  (2) Mycobacteriosis

(a) Dermatitis and tubercle formation in the skin associated with mycobacteriosis are rare in avian species but have been noted in Amazon parrots, blue and gold and green wing macaws, and ostriches.
(b) Distinct lump or lumps or raised area or areas of dermatitis is noted. Can be ulcerative.
(c) Focal granulomatous dermatitis resulting from mycobacterial infection appears as a lump or multiple lumps comprising large numbers of macrophages and a variable number of heterophils and plasma cells. Acid-fast bacteria are found in the macrophages.[5]

e. Fungal
  (1) Candidiasis can occasionally cause focal raised lesions but is typically generalized and can become ulcerated. Eclectus parrot, cockatiels, and Amazon parrots have had lesions on the head.

f. Parasitic
  (1) Knemidokoptes: Swellings noted around the cere and face (see previous section). The mites tunnel into the proliferative tissue, giving it a honeycomb appearance.
  (2) Microfilaria: Rare. May produce localized swellings in the skin. We have seen it in eclectus parrots, and it has been reported in the hocks of an Amazon parrot.[9]

g. Neoplasia
  (1) Many neoplasias affect avian skin; only a few more common types are listed. Xanthomas are also included in this section, but they are not neoplastic masses.
  (2) Lipomas: Benign adipose cell tumors of varying size. Common in budgerigars, cockatiels, Amazon parrots, caninde macaws, rose-breasted cockatoos, and slender-billed cockatoos. Abdomen, hips, breast, and neck are common sites. Can be ulcerative.
  (3) Fibrosarcomas: Malignant fibroblastic and mesenchymal cell tumors. Common in budgerigar and cockatiel legs and wings. These may also be ulcerative.
  (4) Liposarcoma: Malignant adipose cell tumors. Common in budgerigars and cockatiels. Found on the face, neck, and uropygial gland.
  (5) Squamous cell carcinoma: Malignant squamous cell tumor; common in budgerigars and cockatiels; found on the face, legs, and uropygial gland; can be ulcerative.
  (6) Pseudolymphoma: Benign papules reported in the cheek patches of a blue and gold macaw.
  (7) Xanthomas: Nonneoplastic yellowish nodules or plaques caused by an accumulation of cholesterol and lipids; often found over other pathology, such as lipomas; especially in budgerigars, cockatiels, Amazon parrots, caninde macaws, and rose-breasted cockatoos; can be ulcerative.

h. Nutritional
  (1) Hypovitaminosis A (see previous section): Affected skin dry, flaky, pruritic, and has poor wound healing; can be ulcerative; extensive hyperkeratinization can cause swellings.
2. Ulcerative lesions. Chronic ulcerative dermatitis (CUD) is characterized by septicemia, edema, hyperemia, and ulceration of the skin. As described previously, CUD has been associated with tumors and is also associated with xanthomas, abscesses, and mycobacteriosis, among other conditions. The most common CUD lesions are discussed.[9]
  a. Prepatagial CUD: Wing web area, most common in lovebirds (lovebird dermatitis). Giardiasis and hypovitaminosis E have been implicated. Love birds, cockatiels, grey-cheeked parakeets, cockatoos, Amazon parrots, and eclectus are affected. Lesions are extremely pruritic. Patagium may also be affected.[9]
  b. Proventer CUD: Keel area, common in African greys and heavy Amazon parrots. Lesions result from hard landings on hard surfaces. Keel skin and musculature may bruise or immediately split open. Area ulcerates and lesions are protracted. Chewed feathers or too short feather clips implicated.[9]
  c. Postventer CUD: Area between cloaca and tail, common in cockatiels and rose-breasted and Umbrella cockatoos. Lesions result from hard landings on hard surfaces; one author believes malnutrition is also implicated.[9] Proventer skin typically splits on impact. Pygostyle may fracture. Area ulcerates and lesions are protracted. Chewed feathers or too short feather clips implicated.[9]

C. Therapy
1. Nonspecific
2. Specific: Based on etiology, may be protracted. CUD lesions are especially difficult to treat and require intense management to avoid septicemia.
  a. Traumatic: Clean area. Eliminate exposure to source of trauma.
  b. Congenital: None needed for normal complexus muscle.
  c. Viral
    (1) Aggressive supportive treatment
    (2) Herpes virus treated with topical and systemic acyclovir
  d. Bacterial: Clean and debride lesion. Treat systemically as per culture and sensitivity
  e. Fungal
    (1) Best treated systemically (treatment is usually protracted)
    (2) Itraconazole for aspergillosis
    (3) Diflucan, once daily, good tissue penetration
    (4) Amphotericin B only fungicidal antifungal
  f. Parasitic
    (1) Remove parasite, clean environment
    (2) Topical treatment with 5% carbaryl powder for lice and other ectoparasites

(3) Treat systemically for mites with ivermectin, and for *Giardia* bacteria with metronidazole
g. Neoplastic
   (1) Complete surgical excision
   (2) Various chemotherapeutic therapies have been tried: intravascular combination and oral[8,9]
h. Nutritional
   (1) Improve diet.
   (2) Reduce or eliminate fatty seed portions.
   (3) Increase fruit and vegetable portions.
   (4) Convert carefully and slowly to pelleted and extruded bird diets.
   (5) Administer vitamins as needed parentally (e.g., vitamin A).
i. CUD
   (1) Prepatagial CUD: Clean and debride area, manage as an open wound. If skin defect is large and open, apply Duoderm to freshened tissue margins and replace in 1 to 2 weeks. Debride as needed. Treat systemically based on culture and sensitivity. Treatment often protracted.
   (2) Proventer CUD: Clean and debride area, manage as an open wound if ulcerated and not split open. If open, apply Duoderm to freshened tissue margins and replace in 1 to 2 weeks. Debride as needed. Treat systemically based on culture and sensitivity. Falls, and therefore keel splits, frequently recur, so manage bird's environment, flight, and activity.
   (3) Postventer CUD: Clean and debride area, manage as an open wound if ulcerated and not split open. If open, use radiography and assess skeletal damage. Next, apply Duoderm to freshened tissue margins and replace in 1 to 2 weeks. Debride as needed. Treat systemically based on culture and sensitivity.

V. Feet

A. Anatomy and Physiology
1. Claws
   a. The horny claw: Encloses the terminal phalanx of each digit of the foot, which is made up of a strongly keratinized dorsal plate forming the dorsal ridge and lateral walls of the claw, with a softer ventral plate forming the sole of the claw. The dorsal plate grows faster than the ventral plate leading to curved claws.
   b. Spurs: Caudomedial surface of the tarsometatarsal region of most *Phasianidae* spp., including the domestic fowl and turkey, are well developed and pointed in the male but small in the female.
   c. Scales: Most species' scales cover lower part of hindlimbs instead of feathers. The nonfeathered skin of the foot is called *podotheca*. Scales are raised areas of highly keratinized epidermis separated by folds of less keratinized skin.

d. Foot pads
   (1) Thickening of the skin on the plantar surface of the foot, specialized to withstand compression.
   (2) Young woodpeckers, toucans, and barbets have a pad present on the back of the tarsus, which is molted when the bird leaves the nest.

B. Clinical Signs and Differential Diagnoses
1. Pododermatitis
   a. A general term for inflammatory or degenerative conditions of the avian foot. Pododermatitis is also known as *bumblefoot* and can range from a mild redness or swelling to a chronic deep infection to bony changes without or with osteomyelitis.[9] Pododermatitis is the most important and common problem of the feet of birds. Although most common in raptors, pododermatitis also occurs in heavy bodied psittacines, especially Amazon parrots and hyacinth macaws.
   b. Classifications of pododermatitis[9]
      (1) Grade 1: Shiny pink areas noted on plantar foot surfaces. Minor peeling or flaking of skin, feet, and legs.
      (2) Grade 2: Smooth thin areas on plantar foot surfaces. Subcutaneous tissue visible through thinned extra skin.
      (3) Grade 3: Ulcerative lesions on plantar foot surfaces. Peripheral calluses may form.
      (4) Grade 4: Necrotic tissue present within ulcer. Pain and mild lameness result.
      (5) Grade 5: Cellulitis surrounds the area of necrosis. Digits or foot may also be edematous. Tendons and metatarsal pads may become infected. Severe lameness results.
      (6) Grade 6: Necrotic tendons visible as swollen digits. Ruptured tendons, bony ankylosis, and digit dysfunction result.
      (7) Grade 7: Osteomyelitis
   c. Pododermatitis is a multifactorial disease and may include one or more of the following:
      (1) Nutritional: Hypovitaminosis A, high fat diets, etc.
      (2) Environment: Inactivity, perches of improper size, wooden dowels, sandpaper perches, etc.
      (3) Trauma: Self-induced (e.g., nail punctures) or bite wounds can lead to opportunistic infection.
      (4) Systemic disease: Can see inflammation at the site of injurious foot lesion.
2. Amazon foot necrosis
   a. Amazon foot necrosis is characterized by lesions on the feet, as well as the legs and wings. Because self-mutilation is part of the syndrome, another name for Amazon foot necrosis is *psittacine mutilation syndrome.*
   b. Black, red, ulcerative, blistered, and necrotic skin is typically noted on toes and feet. The lesions of the wings and legs are gangrenous, blackened, and necrotic.

c. The skin necrosis is of unknown etiology and variable. Mutilation follows an original lesion that tends to recur.
   3. Herpesvirus lesions
      a. Common on the feet, and especially the toes, of cockatoos.
      b. Proliferative and crusty lesions are typically dry and grayish.
      c. Herpes virus has been associated with these lesions.

C. Diagnostic Tests
   1. Biopsy
      a. Histologic examination
         (1) Hematoxylin and Eosin stain (H & E)
         (2) Gram's stain
         (3) Fungal stains
         (4) Acid-fast stains
      b. Culture and sensitivity
         (1) Aerobic culture
         (2) Anaerobic culture
         (3) Fungal culture
         (4) Mycobacterial culture
         (5) Viral isolation
      c. Electron microscopic examination
   2. Skin scraping
      a. Direct
      b. Oil immersion
   3. Touch preparations
      a. Cytologic examination
   4. Radiographs
      a. Detailed radiographs are performed whenever involvement of the deeper tissues is suspected:
         (1) Pain, swelling, and reddening
         (2) Lameness
         (3) Ulceration

D. Treatment
   1. Pododermatitis
      a. Husbandry, perches, substrate, and caging
      b. Improve diet, especially vitamin A
      c. Debride and clean lesions, trim nails
      d. Treat topically with antibiotics, ointments, DMSO, etc
      e. Treat systemically
      f. Bandage with padding
   2. Amazon foot necrosis
      a. Husbandry, perches, and cage
      b. Improve diet, especially vitamin A
      c. Debride and clean lesions
      d. Treat topically with aloe vera gel
      e. Treat systemically secondary infections
      f. Bandage with antibiotics, Telfa, and Vetwrap
      g. Place collar on bird

3. Herpesvirus
    a. Clean lesions
    b. Treat topically with acyclovir
    c. Treat systemically with acyclovir

VI. Beak (see Chapter 17)

A. Anatomy and Physiology
1. The horny beak (rhamphotheca) is the hard, keratinized epidermal structure covering rostral parts of the upper and lower jaws. The rostrum maxillaris (upper beak) and the rostrum mandibularis (lower beak) are all inclusive terms for keratin, soft tissue (dermis), and bony structures of the indicated beak. The keratin of the maxilla is called the *rhinotheca*. The keratin of the mandible is called the *gnathotheca*.
2. The beak is usually formed by a single sheath. The rhinotheca grows in a cranialventral plane and the gnathotheca grows, more quickly, in the cranial dorsal plane.
3. Histologically the beak is composed of bone, dermis (closely attached to periosteum of the jaw bones), epidermis, a transitional layer, and the keratinized epidermis (horn or rhamphotheca).
4. Epidermis modified—stratum corneum very thick, cells of stratum corneum contain free calcium phosphate and oriented crystals of hydroxyapatite in addition to other components—give hardness.
5. Despite hardness, horny tissue of beak is normally lost by wear and is continuously being replaced. Complete replacement takes approximately 6 months in adult psittacines.

B. Clinical Signs and Differential Diagnoses
1. Trauma
    a. Mate trauma: Male cockatoos typically attack females around the beak, face, and head. Trauma ranges from minor beak puncture to complete avulsion.
    b. Trauma: Walls, ceiling fans, other birds or pets.
2. Nutritional
    a. Diets low in vitamin A; in damaged beak tissue, healing may be prolonged.
    b. Diets low in calcium; beak may be soft as calcification is inhibited.
    c. Diets high in fat; hepatic lipidosis is reported to cause overgrowth of the beak.
3. Neoplasia
    a. Fibrosarcoma of the beak (see previous section). Invasive, malignant mass that usually incorporates both the beak and the cere.
    b. Neoplasia of the liver
        (1) Adenocarcinoma: Similar to hepatic lipidosis; neoplasia of the liver can contribute to overgrowth of the beak.

4. Congenital
   a. Prognathism: Pug beak—suspect etiologies include genetics, incubation, and hand-feeding technique. Most common in cockatoos.
   b. Lateral deviations of the maxilla: Scissor-beak—suspect etiologies include genetics, incubation, and hand-feeding technique. Most common in macaws.
   c. Mandibular compression deformity: Incorrect hand feeding and cleaning technique. The lateral sides of the mandible are compressed inward from excessive finger tip pressure. Most common in macaws.
5. Bacterial
   a. A common sequela of beak trauma, primarily secondary infections.
6. Fungal
   a. Aspergillosis can cause primary beak necrosis. Infections tend to be protracted and may involve bony tissue. We have treated fungal beak lesions that respond to topical amphotericin B in cockatoos and African greys.
   b. Candidiasis has been reported in the lower beak of cockatiels and ostriches.[8]
   c. A common sequela of beak trauma, a secondary infection.
7. Viral
   a. PBFD: Beak involvement occurs less frequently than feather involvement; more prevalent in certain species (sulfur-crested, rose-breasted, little corellas, and Moluccan cockatoos); initially the rhamphotheca at the distal portion of the beak separates from underlying tissues; as disease progresses, beaks may elongate or have longitudinal-to-transverse delaminations or fractures; palatine necrosis also may develop; deformities, fractures, necrosis, and sloughing of beak may be observed

C. Diagnostic Tests
1. Trauma
   a. Biopsy
      (1) Histologic examination
      (2) H & E
      (3) Gram's stain
      (4) Fungal stains
   b. Culture
      (1) Aerobic culture
      (2) Anaerobic culture
      (3) Fungal culture
   c. Radiographs
      (1) If bony involvement suspected
2. Nutritional
   a. Analyze diet
3. Neoplasia
   a. Biopsy

        (1) Histologic examination
        (2) H & E
    b. Electron microscopy
  4. Congenital
    a. Physical examination
  5. Bacterial
    a. Culture and sensitivity
    b. Biopsy
        (1) Histologic examination
        (2) H & E
        (3) Gram's stain
  6. Fungal
    a. Culture and sensitivity
    b. Biopsy
        (1) Histologic examination
        (2) H & E
        (3) Gram's stain
        (4) Fungal stains
  7. Viral
    a. PBFD, DNA probe, whole blood test
    b. Isolation

D. Treatment
  1. Trauma
    a. Address wounds, clean, debride, and treat with topical wound dressings.
    b. Remove damaged and necrotic beak tissue, using a Dremel drill.
    c. Treat as an open wound.
    d. Treat systemically as per culture and sensitivity.
  2. Nutritional
    a. Improve diet, administer vitamin A.
    b. Remove damaged and necrotic beak tissue with a Dremel drill.
  3. Neoplasia
    a. Attempt surgical excision.
    b. Chemotherapeutics may be helpful.
    c. Remove damaged and necrotic beak tissue with a Dremel drill.
  4. Congenital
    a. Physical therapy or beak prosthesis as needed
  5. Bacterial
    a. Treat as per culture and sensitivity; topically and systemically.
    b. Remove damaged and necrotic beak tissue with a Dremel drill.
  6. Fungal
    a. Treat as per culture and sensitivity; topically and systemically.
    b. Remove damaged and necrotic beak tissue with a Dremel drill.
  7. Viral
    a. Supportive treatment
    b. Remove damaged and necrotic beak tissue with a Dremel drill

REFERENCES

1. Pass DA: Normal anatomy of the avian skin and feathers, *Seminars in Avian and Exotic Pet Medicine* 4:152-160, 1995.
2. King AS, McLelland J: Integument. In King AS, McLelland J, editors: *Birds: their structure and function,* Philadelphia, 1984, Bailliere Tindall, pp 23-42.
3. Lucas AM, Stettenheim PR: Avian anatomy, integument, *Agriculture Handbook* 362, Washington, DC, 1972, Department of Agriculture.
4. Rosskopf WJ et al: Treatment of feather folliculitis in a love bird, *Mod Vet Pract* 64(11):923-924, 1983.
5. Schmidt RE: Use of biopsies in the differential diagnosis of feather picking and avian skin disease, *Proc Assoc Avian Vet,* 1993, pp 113-115.
6. Pass DA, Perry RA: Granulomatous dermatitis in peach-faced lovebirds, *Aus Vet J* 64(9):285-287, 1987.
7. Oglesbee BL: Hypothyroidism in a scarlet macaw, *J Am Vet Med Assoc* 201:1599-1601, 1992.
8. Altman RB et al: *Avian medicine and surgery,* Philadelphia, 1997, WB Saunders, pp 540-562.
9. Ritchie BW, Harrison GJ, Harrison LR: *Avian medicine: principles and applications,* Lake Worth, Fla, 1994, Wingers Publishing, pp 607-639.
10. Perry RA, Gill J, Cross GM: Disorders of the avian integument, *Vet Clinics North Am: Small Anim Pract* 21(6):1307-1327, 1991.
11. Johnson-Delaney C: The avian immune system and its role in disease, *Proc Assoc Avian Vet,* 1989, pp 20-27.
12. Latimer KS: An updated review of psittacine beak and feather disease, *J Assoc Avian Vet* 5(4):211-220, 1991.
13. Pass DA, Plant SL, Sexton N: Natural infection of wild doves (*Streptopelia senegalensis*) with the virus of psittacine beak and feather disease, *Aus Vet J* 71(9):307-308, 1994.
14. Ritchie BW: Avian polyomavirus: an overview, *J Assoc Avian Vet* 5(3):147-153, 1991.
15. Bernier G, Morin M, Marsolais G: A generalized inclusion body disease in the budgerigar (*Melopsittacus undulatus*) caused by a papovavirus-like agent, *Avian Dis* 25:1083, 1981.
16. Jacobson ER et al: Papilloma-like virus infection in an African gray parrot, *J Am Vet Med Assoc* 183 (11):1307-1308, 1983.
17. Lowenstine LJ: Diseases of psittacines differing morphologically from Pacheco's disease, but associated with herpesvirus-like particles, *Proc 31st Western Poultry Dis Conf,* 1982, pp 141-142.

8

Behavior Problems in Pet Parrots

Liz Wilson

Behavior Problems

Defined as any behavior that threatens parrot's pet potential.

I. Common Behavior Problems

A. Biting
1. Oddly enough, the term *biting* needs to be clarified. Biting does *not* include a human being touched by a bird's beak.
2. A good judgment of the true severity of a bite is encompassed in the question, "How much did you bleed?"

B. Excessive Screaming
1. Note the use of the word *excessive*. Parrots are not quiet animals by nature and cannot be expected to be.
2. Normal behaviors, depending on species, would include intermittent periods of vocalization during the day, as well as periods of intense vocalization morning and evening.
3. Screaming nonstop for hours at a time would obviously be considered excessive.
4. This problem can be difficult, mainly because the behavior is generally rewarded by the owners, who will do almost anything (e.g., give it a treat, let it out of the cage) to get the bird to stop—thereby rewarding the behavior.
   a. This problem is often exacerbated by a limited time frame.
   b. Owners may not seek help until an ultimatum has been set by family members, neighbors, landlord, or local police.

C. Excessive Territoriality
   1. The bird defending its own area is a normal instinct–but one that can get out of hand.
   2. Territoriality is necessary for survival and propagation in the wild.
   3. Territoriality is excessive when the owner can't exert control over the bird in its cage vicinity or play areas.
   4. Territoriality becomes extreme when even the parrot's preferred person cannot safely put his/her hand in the cage to feed.

D. Overbonding or the One-Person Bird
   1. This is defined as a bird incapable of being separated comfortably from its preferred person.
   2. This animal generally cannot be handled by anyone else.
   3. Overbonding is often exacerbated by certain types of owners.

E. Overdependency
   1. Defined as a bird that is incapable of amusing itself when left alone.
   2. Characterized as a problem because it is often the precursor to other behavior problems, such as:
      a. Phobias
      b. Feather destruction
      c. Incessant screaming
      d. Separation anxiety
      e. Inability to adapt to change
      f. Rigidity in eating habits

F. Plucking or feather picking
   1. Physical etiology
   2. Psychologic etiology

G. Phobias
   1. The bird acts terrorized when approached, acting very much like a totally wild import. It often injures itself in an effort to escape.
   2. Consequently, avian veterinarians see these parrots as emergencies with such problems as broken blood feathers.

II. General Etiologies of Behavior Problems

A. Physical Problems
   1. Inadequate food and/or water
   2. Cage problems
      a. Too small
      b. Location inappropriate–gregarious bird in a room by itself, nervous bird in the middle of a high traffic area, dominant bird too high, nervous bird too low
      c. Nonstimulating environment without toys and other enrichments–leads to boredom.

3. Sleep deprivation: Because parrots are equatorial animals, they get 12 to 13 hours of darkness year round in the wild. Consequently, companion parrots probably need 10 to 12 hours of sleep in captivity.[1]
   a. In some homes, a "sleep cage" can be ideal. This is a small, spartanly equipped cage that is placed in a room unoccupied by humans at night. Human sleep movements might disturb the sleep of a prey animal such as a parrot, leading to sleep loss.

B. Medical Problems
1. Malnutrition
2. Bacterial, fungal, and/or systemic disease
3. Allergies

C. Psychologic Problems of the Owner
1. The owner misunderstands what a parrot *is* and *is not*.
   a. A parrot *is*:
      (1) Genetically a wild animal, despite having been born domestically. A tiger born in a zoo is *still* a tiger.
      (2) Not inclined to perceive humans as superior beings.
      (3) A loud, boisterous, highly social creature with a tendency to chew and make large messes.
   b. A parrot *is not*:
      (1) A person with feathers
      (2) A dog with feathers
      (3) A surrogate child
2. The owner has misconceptions of a parrot's normal behavior.
   a. Whose problem is it?[2]
      (1) Some behaviors, such as chewing, are normal for parrots and are therefore not the parrot's problem.
      (2) It is the owner who perceived chewing as a problem, therefore it is the owner's problem.
      (3) The owner has problematic relationships with other humans (e.g., marital problems).
         (a) For example, some owners have an investment in a parrot's aggressiveness.
         (b) May encourage aggressiveness with the belief, "I'm the only one who can handle him—he bites everyone else."
3. Abuse, either aggressive or benign
4. Neglect

D. Lack of Control by the Owner
1. This is by far the most common etiology for behavior problems in parrots.
2. The most common mistake the owner makes is to let the parrot do whatever it wants, whenever it wants.

III. Basic Premises for Behavior Modification

A. Training Is Necessary
1. To teach a parrot to be a good pet, the owner must have a dominant position with the parrot.
   a. To get a dominant position, the owner must assume a decision-making role, no longer allowing the parrot to decide major issues, such as the following:
      (1) Whether or not it will come out of (or off of) the cage
      (2) Whether or not it will go back into the cage
      (3) Whether or not it will get on or off the owner's shoulder
      (4) Whether or not it is allowed to bite, scream, etc.
   b. The owner must first establish himself/herself as *flock leader*—this will allow the owner to outrank the parrot.
2. To become *flock leader*, the owner must establish his/her rank by the use of clear, consistent controls. An easy and effective technique for establishing controls is through a training technique called *dominance training*.[3]
3. No matter which behavior problem(s) the owner is encountering in a parrot, *dominance training must be taught first.*
   a. Successful behavior modification depends on the following:
      (1) Lesson times are upbeat and fun, and *always end on a positive note.*
      (2) The owner does his/her best to always reward positive behaviors and ignore (when possible) negative behaviors.

B. Realistic Time Frames
1. Behavior problems do not develop overnight.
2. Behavior problems are not corrected overnight.
3. To properly change a parrot's unwanted behavior, the owner must be:
   a. Clear
   b. Consistent
   c. Patient
4. To properly change a parrot's unwanted behavior, the owner must also change his/her behaviors that created or exacerbated the bird's behavior.
5. The primary obstacle to an improvement in the parrot's behavior is the *owner's impatience.*
6. The results will definitely justify the time invested.

IV. Miscellaneous Parrot Characteristics of Note

A. Parrots: Genetically Wild Animals
1. Domestics are only a couple of generations from the wild, because the larger species have been breeding routinely in captivity for only

about 20 years. Therefore, they have no idea how to respond to "the environment of our living room." Notable exceptions to this genetic wildness: budgerigars, cockatiels, and some species of lovebirds.[4]

2. Instinctive behaviors are generally blocked by captivity. Unless taught by the owner how to adapt to this, the parrot will improvise alternative behaviors.
   a. Alternative behaviors are called *displacement behaviors*.[5]
      (1) Common negative displacement behaviors include biting and obsessive cage territoriality.
      (2) Displacement behaviors are not all bad—a popular example of a positive displacement behavior would be a parrot bonding to a human.

B. Accidental Rewards for Bad Behaviors

1. Humans are often incorrect as to what a parrot may perceive as negative feedback and will therefore actually *reward* behaviors that they are trying to eliminate.
2. Supposed punishments that most parrots actually enjoy include:
   a. The drama reward[6]
      (1) Most parrots *love* drama, especially the drama of humans yelling.
      (2) Yelling at a parrot to reprimand it (e.g., for screaming or biting) is therefore perceived by the parrot as positive feedback.
   b. "Beak-wrestling"—a common, old-style reprimand for biting in which owners were instructed to grab the parrot's beak, shake it, and yell, "NO!" at the same time.
      (1) Beak wrestling among parrots has now been found to be play behavior between peers within a flock. With a bonded pair of parrots, beak-wrestling appears to be associated with sexual foreplay.
      (2) Parrots also enjoy the drama reward of a human yelling.
      (3) This reprimand is therefore perceived by the parrot as positive feedback.
   c. Altitude vs. attitude
      (1) When a flock roosts, the height at which a parrot sits is directly correlated to that individual's rank in the flock.[7] The higher the bird, the higher the bird's rank. Height is equated with safety.
      (2) When a human places a parrot above eye level, the human is telling the parrot nonverbally that the bird outranks the human. Extremely common situations that place a bird too high:
         (a) Cage top playgrounds
         (b) High perches
         (c) The owner's shoulder, which is especially dangerous, not just because of the superior attitude and therefore dominance of the bird.
         (d) Extremely vulnerable parts of the owner's anatomy (eyes, ears, nose, lips) are subject to severe damage from the

parrot's beak. This type of injury will also *permanently damage* the parrot-human bond.
- (3) Simply placing the parrot no higher than chest level to the owner will automatically increase the owner's rank in the flock.
- (4) The bird can be lowered in a variety of ways:
  - (a) Lower the cage.
  - (b) Lower the perches within the cage.
  - (c) Do not allow the parrot access to the top of its cage.
  - (d) Remove the highest perches from tall climbing "trees."
  - (e) If the bird cannot be lowered, raise the people—place a footstool or small ladder next to the cage, thereby raising smaller humans to a position of higher rank.
- d. Fluidity of flock hierarchy
  - (1) It appears that the hierarchical structure of a psittacine flock is more fluid than that of familiar mammals such as canines.
    - (a) Because of this, the flock leader must deal with constant challenges from flock members.
    - (b) The average psittacine will constantly test controls.
  - (2) Parrot owners must therefore expect constant challenges to their authority from their birds.
    - (a) The owner can *never* back down from a command once given.
    - (b) By relinquishing control to the parrot, even momentarily, he/she risks losing rank completely within the flock.

V. Dominance Training

A. Purpose: Establishing the Human(s) as Higher in Rank than the Parrot
1. Accomplished by the owner teaching the parrot to respond to simple commands.
2. Consistent use of commands and parrot acquiescence to them establishes the bird in a subservient position.

B. Position of Dominance
1. The owner's status as flock leader places him/her in a position of dominance.
2. The owner can now implement various behavior modification techniques to deal with specific problem behaviors.

VI. Implementing Dominance Training

A. Daily Training Sessions of 10 to 15 Minutes

B. The Owners' Approach to Training
1. Positive attitude

2. Relaxed, nonaggressive body language
3. Positive facial expression

C. Location of Training: Neutral Territory and Neutral Perch
Definition: An area and perch in/on which the parrot has spent no time and therefore has no agenda or territory established.
1. No new perch need be purchased—the back of a chair works very well.
2. Neutrality of the training area is critical to the success of training.
3. Training area cannot be within sight of cage or established play areas.
4. The height of the training perch should keep bird's head no higher than chest level to the owner.
5. In neutral territory, the owner is the *only familiar thing*. This is critical to successful training. Reserve the area for training only.

D. Commands to Be Taught: Up and Down
1. Up is defined as "Step onto the human's hand NOW."
2. Down is defined as "Step off the human's hand onto a perch NOW."

E. Training Procedure
1. Step one: teaching the basic commands
   a. The owner places the bird in a neutral room on a neutral perch.
   b. Holding the hand flat and open (as if to shake hands) and slightly higher than the perch, the owner pushes firmly, but gently, on the parrot's lower abdomen, saying "Up" at the same time.
   c. Owner's arm should be bent, with the elbow close to his/her waist; this discourages the bird from running up the owner's arm to the shoulder.
   d. When the bird steps on (rather than losing its balance), the owner smiles and praises the bird.
   e. The owner places the bird back on a neutral perch with "Down" command, then owner smiles and praises the bird.
   f. Owner should also step the bird from one hand to the other, saying "Up" each time and rewarding the positive responses with praise and smiles. This technique is known as "laddering" and will be used again later in another context.
   g. *Important note: These are commands, not requests.* The owner's voice should reflect this by being quiet and authoritative, not questioning.
2. Step two: foot holding[8]
   a. A parrot that is trained to have its foot held while sitting on a hand is a bird that can be controlled in almost any type of situation.
      (1) This is especially important with a parrot that instantly jumps for the owner's shoulder.
      (2) Shouldering cannot be tolerated during training sessions, and is not recommended any other time. (See IV. Miscellaneous Parrot Characteristics of Note, Altitude vs. attitude, p 128.)

b. The owner steps the parrot onto the hand with an "Up," then puts his/her thumb down on one or both of the bird's feet. The owner smiles and praises the bird.
c. When/if the bird gets agitated about the foot holding, the owner steps it to the other hand with the "Up" command. Then the owner puts a thumb on its feet, smiles, and praises the bird.
d. The duration of foot holding is gradually lengthened, a few seconds at a time if necessary, until the bird is comfortable with having its feet held for prolonged periods.
3. Step three: moving the training perch
   a. Once the bird follows the "Up" and "Down" commands perfectly for several lessons, the owner should start moving the training perch out of the training area.
   b. By moving the perch a few inches, then using "Up" and "Down," the owner teaches the parrot that these commands work everywhere—not just in the teaching area.
   c. The perch should be moved at whatever speed the bird is comfortable with. (Note: This is not a race!)
   d. If the bird starts resisting the commands, the owner should:
      (1) Move the training perch back to an area where the bird performed perfectly, and start the moving process again.
      (2) Slow down a little in the speed at which the perch is moved.
   e. Taking as much time as is necessary, move the training perch into the area that houses the parrot's cage and personal play areas.
   f. The owner should expect some initial resistance from the psittacine when in sight of its own territory. In this case, the owner should:
      (1) Move the perch back out of sight of the cage and work on the command again until perfection is reached, and
      (2) Be patient.
   g. Once the cage (and/or play area) is reached, the owner should do the following:
      (1) Put the parrot in the cage with the "Down" command.
      (2) Immediately pick up the bird with the "Up."
      (3) Repeat this procedure several times.
   h. This procedure should also be followed with the top of the cage and any other play areas the bird might have.
   i. The owner has taught the bird that commands now work all around the bird's cage and play areas.
   j. From then on, the owner must *always* remove the bird from the cage with the "Up" command and return it with the "Down."
      (1) The owner must *never* simply open the door and walk away, letting the bird decide when it will come out of the cage.
      (2) The owner can take the bird out with the "Up," place it on top of the cage, and let it go in and out as it wishes; but the initial coming out must be on the owner's command.
4. Once training has begun, any other humans who wish to handle the parrot *must* use the commands in the same manner, so all handling is consistent with the training.

5. The other two commands
   a. "No"–The owner generally already uses this command multiple times daily, to little avail. The owner will be delighted to find that once he/she gains the rank of flock leader, the word "No" becomes a power word.
   b. "Okay"–This is the command for the owner to use when the parrot is determined to do something, no matter what. The command gives the parrot permission to do what it wishes to do. Use this command to make the decision become the owner's, *not* the parrot's, and to maintain control.

VII. Specific Techniques for Reprimand

These reprimands are effective only *after* the implementation of dominance training. *Important note: Under no circumstances is violence an acceptable reprimand.*

A. Results of Use of Violence with a Parrot
1. The bird is seriously injured.
2. The parrot responds to the owner's violence with more aggression. According to Doane, "Violence begets violence."[9] The parrot will also probably never trust the owner again.

B. The "Evil Eye"
1. Definition: An *extremely* dirty look, delivered to a parrot whenever the bird needs correction in its behavior. (This is the technique used by disapproving grandmothers everywhere, especially when small grandchildren act up in church.) Used properly, the evil eye will stop a child in his or her tracks, and it has precisely the same effect on parrots.
2. The evil eye is used alone or in conjunction with another reprimand technique, Laddering.

C. Laddering
1. Stepping a parrot from one hand to the other with the "Up" command with a positive focus is integral to the process of dominance training. In dominance training, it is initiated when the owner says "Up" with a friendly tone of voice, smiling and giving the bird positive feedback each time the command is followed.
2. Laddering as a reprimand is done differently:
   a. The owner's voice is quiet, but decidedly unfriendly as he/she says "Up."
   b. The owner gives the parrot the full force of the evil eye.
   c. The owner steps the bird from one hand to the other several times with the "Up" command.
   d. In this manner, the parrot is reminded in a completely unaggressive manner that it is lower in rank than the human. Note:

The parrot's behavior will change drastically in response to this reprimand; it has a strong effect.

D. "Little Earthquake"[10]
Effective reprimand for a parrot that bites the hand on which it is sitting. The owner instantly moves his/her hand just enough to cause the bird to lose its balance for a second, without causing the bird to fall to the floor. Done consistently, the parrot will learn not to bite when sitting on a hand.

VIII. Dominance Training as Applied to Specific Behavior Problems

Once dominance training has been fully implemented, the owner can begin to work on specific behavior problems.

A. Biting
1. When a bird bites, the owner should *instantly* reprimand the parrot with a sequence of techniques:
   a. Evil eye–Give the parrot the full benefit of a hideous look.
      (1) The owner's facial expression must express *exactly* his/her displeasure.
      (2) The owner should *not* show his/her physical (or emotional) hurt.
   b. The ladder: Saying "Up" in a quiet but extremely unfriendly tone of voice–the owner steps the parrot from one hand to the other several times.
   c. Forgiveness
      (1) The bird has responded correctly to the ladder commands.
      (2) The owner smiles and praises the bird.
      (3) The incident is over and no grudges are held by the owner.

B. Excessive Screaming
1. If the bird is screaming in the same room as the owner:
   a. The owner gives the bird the full force of the evil eye.
      (1) The owner says, "No."
      (2) The owner's voice is quiet but *extremely* displeased.
      (3) Used properly, the evil eye should silence the bird instantly.
   b. The owner then smiles and praises the parrot for responding. The incident is over and no grudges are held by the owner.
2. If the bird is screaming in another room:
   a. The owner does *absolutely nothing*. If the owner enters the room to reprimand the bird, that will result in positive feedback for the parrot for screaming. The parrot will realize that it made the owner appear on command.
   b. The owner waits patiently until the parrot stops screaming momentarily.

c. The owner now enters the room, smiling and praising the parrot for being quiet.
   3. The owner *always* makes the point of rewarding a parrot when it is being quiet. When the parrot is not noisy, the owner calls out to the bird from wherever he/she is, telling the parrot how good it is for being quiet. In this way, the parrot learns that attention is received when it behaves, not when it is vocalizing excessively. This is a radical change, because most parrots are ignored when they are quiet and yelled at when they are loud.

C. Excessive Territoriality
1. Territoriality is defined as a parrot being protective of any area it perceives as its territory (e.g., its cage and/or play areas). "Normal" territorial behavior would entail a parrot driving strangers (e.g., non-flock members) away from its territory.
2. Excessive territoriality may manifest in degrees of severity:
   a. Mild territoriality includes aggressive posturing when someone approaches the protected area.
   b. This generally escalates until the bird lunges at people walking by the protected area.
   c. The end result is a bird that does not allow anyone to feed it and clean the cage without the humans having to take defensive measures.
3. Excessive territoriality is instinctive behavior that has gotten out of control. It is illogical that the bird's preferred person (therefore, the bird's "mate") be driven away from the bird's territory.
4. Excessive territoriality is eliminated if the owner pays special attention to step three of dominance training, moving the training perch. The parrot is taught that "Up" and "Down" commands work everywhere—especially in the cage area. By always requiring the parrot to step "Up" before leaving the cage, the owner maintains control over the cage area.
5. Should the parrot stage a small mutiny and refuse the "Up" command from within the cage, the owner is instructed to do the following:
   a. Close the door to the cage and walk away.
   b. Return in a short time (approximately 5 minutes).
   c. Repeat the "Up" command, requiring the bird to step on before being allowed out of the cage.
   d. Because the parrot generally wishes to exit the cage, it will eventually acquiesce to the owner's command.
   e. It is critical that the owner understand the importance of controlling the initial cage exit.
6. Maintaining strong controls around the cage should block any further displays of cage dominance.

D. Overbonding or the One-Person Bird
1. This bird appears incapable of being comfortable if separated from its preferred person.
2. Other members of the owner's family cannot generally handle this bird unless the primary person is out of town for several days. When

the owner is home, other family members usually get bitten if they get too close.
3. Overbonding can be counteracted, to a degree, in the following manner:
   a. The preferred person takes the parrot into the training room, places it on the training perch, then leaves the room.
   b. A nonpreferred person then enters the training room.
   c. This person trains the bird in exactly the same manner as the preferred person. In this way, the nonfavored person establishes himself/herself in a position of dominance with the parrot.
   d. The nonpreferred person now chooses one of the parrot's favorite food, games, or toys. *From then on,* the parrot must go to *that specific person* to receive the favored item. For example, if the parrot loves sunflower seed treats, then the nonpreferred person becomes the designated sunflower person. If the parrot wishes to get a sunflower seed treat, it must go to this specific individual. This individual has now gained value in the parrot's eyes.
4. The preferred person *cannot tolerate* any aggression shown by the bird towards nonpreferred people.
   a. If the bird shows aggression towards another, the preferred person should *instantly* do the following:
      (1) Give the bird the evil eye and say "No" in a quiet, but exceedingly displeased voice.
      (2) Get up and leave the room.
   b. In this situation, the preferred person should *never*:
      (1) Grab the bird—or the bird will learn to bite others to get picked up by their favorite person.
      (2) Yell at the bird—remember the drama reward (see Accidental Rewards for Bad Behaviors p 128).
      (3) Laugh, a powerful reinforcer.
   c. The recipient of the aggression should immediately do the following (see Biting p 133):
      (1) Say "No"; ladder the bird several times while giving it the evil eye.
      (2) When the bird does as it is told, the human should smile, verbally reward it, and forgive the bird.
      (3) The incident is over, and no grudge is held by the human.
5. Overbonding in a new bird can be prevented if the owner does the following:
   a. All the people who wish to interact with the bird sit in a circle. The parrot is passed from one person to the next with the "Up" command.
   b. Each person talks to the bird, plays with it and pets it, then passes it to the next person with the "Up" command. This game should be played frequently (every week or two) for the remainder of the bird's life.
6. Overbonding may be exacerbated, consciously or otherwise, by the preferred person. He/she often claims to want to fix this problem, but unconsciously that isn't always the case. Aggression towards

others may unconsciously be encouraged by the preferred person. He/she may talk obsessively about how attached the bird is and worries about "what will happen" if he/she dies. Some will go so far as to put it in their wills to euthanize the bird in the event of their death, because the bird "can't live without them." Consciously or not, overbonding is being rewarded. This parrot's behavior will not change until the human's behavior is changed.

E. Overdependency

Generally speaking, captive parrots exhibit various characteristics that can be classified as either dependent or independent. Either category has positive and negative aspects. A parrot that is too independent would probably not be considered a good pet. This is equally true of the bird that becomes *too dependent*.

1. Positive and negative characteristics of independency seen in captive parrots[11]
   a. Positive independent characteristics include the willingness of the bird to do the following:
      (1) Play happily by itself
      (2) Explore new territories
      (3) Meet new people or encounter new situations
      (4) Try new foods
   b. Negative independent characteristics of the bird include the following:
      (1) Reluctance or resistance to petting or touching
      (2) A "leave-me-alone" attitude
      (3) General inability to adapt to life in the human environment
   c. Overly independent birds are often wild-caught tamed birds or domestic-bred babies who were not properly socialized. They are therefore more "wild" than "pet." They may appear to be lonely and isolated.
2. Positive and negative aspects of dependency seen in captive parrots
   a. Positive aspects
      (1) Willingness to be hugged, snuggled, and cuddled
      (2) Craving interaction and attention by using attention-getting techniques such as vocalizing, displaying, or dancing
      (3) Obvious pleasure at being reunited with flock members
   b. Negative aspects of overdeveloped dependence may produce the following:
      (1) Feather destruction
      (2) Incessant screaming
      (3) Separation anxiety
      (4) Inability to adapt to change
      (5) Unwillingness to play with toys
3. The overdependent parrot is incapable of amusing itself during periods of isolation.
   a. Methods for encouraging negative bird dependency
      (1) The owner(s) carried it around all the time.

- (2) When people came home, the bird was immediately released from its cage and carried around some more.
- (3) Whenever it appeared anxious, the bird was picked up and reassured.
- (4) The bird was not trained or socialized—it was simply loved.
  b. Results of encouraging bird dependency
    - (1) The newness of parrot ownership wore off.
    - (2) The owner(s) started leaving the bird in the cage more.
    - (3) The bird *expects* to be let out of the cage when people come home, because that is what it was taught to expect.
  c. When not interacting with humans, the bird generally does one or more of the following:
    - (1) Simply sits in the cage like a lump
    - (2) Climbs around the cage obsessively, exhibiting high anxiety levels
    - (3) Screams to be let out
4. Steps to correct the problem of overdependency
  a. Establish a relationship of nurturing dominance (see Implementing Dominance Training p 129)
  b. *Do not* let the parrot out of the cage when it screams. To deal with this behavior, see Screaming p 133.
  c. Provide the flock greeting: When the owners return home, they should greet the bird in the following manner[12]:
    - (1) Go to the cage first.
    - (2) Take the bird out of the cage with the "Up" command and greet it.
    - (3) Give it a hug and a scratch and a treat to eat.
    - (4) Put it back in the cage with the "Down" command and close the cage door.
    - (5) Ignore the parrot for awhile.
  d. Teach the bird to amuse itself. Teach it to play with toys by letting the parrot watch the owner play with a toy. Whenever the bird touches it, the owner should reward the bird *briefly*. The owner should call out to it, telling it what a good bird it is and smile at it from across the room, or from a doorway.
    - (1) Encourage curiosity by walking the bird around the house, looking in different rooms—touch different objects and talk to the bird about them in an animated voice.
    - (2) Whenever the bird is doing *anything* by itself, the owner should reward.
    - (3) While the bird is caged, the owner can stop at cage and play briefly with a toy without looking at the bird or acknowledging it.
    - (4) The owner can also sit near the caged bird, positioning their body facing slightly away and play happily with a toy while ignoring the bird.
    - (5) The owner is therefore demonstrating the behavior that is expected of the bird.

F. **Feather Destruction**[13]

It is one of avian medicine's ironies that uneducated parrot owners often totally disregard evidence of life-threatening disease, yet they rarely miss a case of feather damaging behaviors. The avian veterinarian is then presented with one of the most frustrating situations in avian medicine.

1. Before a case of feather destruction is considered "behavioral," the various possible medical etiologies must be ruled out. Medical etiologies include the following (in alphabetical order)[14]:
   a. Allergies (e.g., food allergies)
   b. Endocrine imbalances, especially related to thyroid hormone alterations
   c. Environmental factors (e.g., low humidity, low light, inability to bathe, inadequate rest)
   d. Infectious dermatitis (bacterial, viral, and/or fungal)
   e. Malnutrition (both through diet and a lack of exposure to unfiltered light)
   f. Parasites, both external and internal
   g. Systemic disease (e.g., liver disease, air sacculitis[15])
   h. Toxins (e.g., exposure to cigarette smoke and/or nicotine on human hands)
2. A full medical work-up should include the following:[14]
   a. A detailed history and physical examination
   b. Fecal (direct and flotation) and a skin scraping (or biopsy)
   c. Hematology and serum chemistries
   d. Radiographs
   e. Bacterial cultures, preferably choanal and cloacal and feather pulp
   f. Skin biopsy
3. Behavioral work-up/history
   a. If tests are within normal limits, the clinician can do one of two things:
      (1) Refer the client to a parrot behavior consultant
      (2) Do a behavioral work-up themselves
   b. In addition to the extremely detailed questions of a behavioral history (see Behavior Questionnaire, Addendum), various questions specific to plucking need to be asked, such as the following[16]:
      (1) When does the bird pick? When the owner is there? When the owner is watching? When the owner is absent?
      (2) How much does the bird pick? Only a little, all of the coverlets, only flight feathers, large bare areas, body totally bald, etc.?
      (3) Where on its body does the bird pick? Just the tail, all over, wings and tail, etc.?
4. "Normal" vs. "abnormal" behaviors involving feather destruction
   a. "Normal" etiologies include the following:
      (1) Nesting birds preparing to clutch. This type of plucking is generally limited to feathering the nest and is seen only during breeding season. Creation of a "brood patch" on the lower abdomen is a common behavior in poultry, but is rarely reported in psittacines.

(2) Reproductive frustration. This is a favorite catch-all category for pluckers.
 (a) Assuages the owner's guilt
 (b) Provides multitudes of free or inexpensive birds for avicultural breeding stock
 (c) Many pluckers placed in breeding situations continue to pluck, and many pluck their mates as well. Obviously, sexual frustration was not the source of the problem in these birds.
 (d) An increased potential for feather-destructive behaviors in the offspring of parent birds who pluck.
 (e) Plucking from reproductive frustration would generally be seasonal, localized to certain areas of the body (e.g., clavicular region or tops of the wings), and seen only in sexually mature parrots.
(3) Exaggerated preening or feather grooming—often seen in young parrots who were either incubator-raised and had no parents to learn from, or were raised in sight of other parrot species. Preening may be instinctive, but the *finesse* may be learned from the adult birds. These young birds may not actually learn how to preen properly. They may, therefore, overpreen and accidentally damage their feathers. This is evidenced by tattered-looking feathers, even just after a molt.

If the lack of finesse is compounded by watching another species of parrot, these young parrots may actually learn incorrect preening techniques for their specific type of feathers, (e.g., an African grey learning to preen like an Amazon).

b. "Abnormal" etiologies for feather plucking
 (1) A control device—a powerful tool, because some owners will do almost anything to stop their birds from plucking. The parrots learn this very quickly, then use plucking to "punish" the owners whenever their desires are blocked.
 (2) An attention-getting device: When the bird plucks, it gets attention. So it learns to pluck whenever it wants attention.
 (3) Boredom: Parrots are extremely intelligent and prone to boredom if not occupied. Working owners often leave parrots alone for long periods. Parrots need interesting and challenging toys to occupy them during these periods. Four basic categories of toys: climbing toys, chewing toys, foot toys, and puzzle toys.[17]
 (a) Climbing toys include plastic chains, ladders, swings, the cage itself.
 (b) Chew toys include wood, raw pasta, nontoxic twigs with bark (herbicide- and insecticide-free), empty paper towel and toilet paper rolls (no perfume). *Note*: Owners should slit these cardboard rolls lengthwise for small species such as lovebirds and grey-cheeked parakeets so that they do not get caught inside and suffocate.

(c) Food toys encourage manual dexterity and include purchased toys, small chucks of corn on the cob, and nuts left in their shells.
(d) Puzzle toys are very important and include parrot-style music boxes, puzzle boxes that hold food treats, C-clamps that suspend toys, and the cage itself (hence all the padlocks and birds named *Houdini*).
   (i) Generally speaking, parrots need only four toys at a time, one from each category. This gives them lots of room to play.
   (ii) Toys should be moved around the cage frequently and rotated in and out of the cage weekly at most. This keeps interest levels high and reduces boredom to a minimum.

(4) Stress
   (a) Extrinsic or intrinsic
      (i) Extrinsic sources from outside the bird (environment, noise, climate, etc.)
      (ii) Intrinsic sources stemming from the bird itself
   (b) Impossible to quantify
   (c) "What is perceived as stressful to a bird may not appear so to its caretaker."[18]
   (d) Environments that are too quiet may cause low-level stress to caged birds.
      (i) When the jungle is quiet, it means that there is a predator stalking.
      (ii) Radios and/or televisions on timers set to turn on and off can relieve this situation.

(5) Environmental change
   (a) Many parrot owners try valiantly to protect their animals from change, believing that change is stressful to parrots and that parrots must be protected from it.
   (b) Change is definitely stressful to birds *that are unaccustomed to it*.
   (c) However, change is *inevitable* in the long lives of parrots.
   (d) Consequently, owners should accustom their parrots to change as part of their early socialization.
   (e) Owners should teach their parrots that change is interesting and nonthreatening.
   (f) Then, *when* change happens, the birds will adapt easily.
   (g) After all, who keeps everything the same in the wild?

(6) Psychologic disturbances
   (a) Diagnosis of choice when all else is ruled out.
   (b) Province of the parrot behavior consultant.
   (c) Little information is available regarding parrot behavior in the wild.
      (i) The offspring of most of the larger species stay with their parents long after they are "food independent."

(ii) While they are with their parents, they must be learning various things—these are their survival skills.

IX. Survival Skills, Socialization, and the Young Parrot

A. Survival Skills
1. Food location and procurement
2. Predator avoidance
3. Adaptability in the face of change
4. The nuances of social interactions within the flock
5. The proper response to the flock leader
6. Knowledge about the individual's rank within the flock

B. Socialization: the Process of Learning Survival Skills[19]
1. Early socialization is the training that domestic-bred parrots lack.
   a. Because they are sentient beings, it is not unreasonable to assume that young parrots sense this lack.
   b. Parrots that lack survival skills do not survive in the wild.
   c. Parrots seem to sense this lack at the time in their life that they would be becoming more independent in the wild: at this age (from 6 to 24 months, depending on species).
   d. During this period a preponderance of the really serious behavior problems appear:
      (1) Phobias
      (2) Excessive territoriality
      (3) Biting
      (4) Excessive screaming
      (5) Feather destruction
      (6) Food rigidity (refusal to eat a variety of foods)
      (7) Behavioral rigidity (refusal to accept any environmental change)
   e. These are all displacement behaviors that are manifestations of stress.

X. The Role of Behavior Modification with the Feather-Destructive Parrot

Because teaching may be what is lacking with domestics, teaching is needed to fill in the gaps.

A. Implementation of a Training Program (see Implementing Dominance Training p 129)
1. Owners teach their parrots to follow simple commands.
   a. By requiring the bird to acquiesce to these commands, the owner is established in a dominant role.
   b. The owner has now assumed the role of flock leader.

2. Once in that position of authority, the owner can enclose the parrot in a framework of consistent controls. The parrot is no longer confused about its rank in the flock. It is clearly subservient. It is therefore protected and secure. Stress is removed.
3. Consistency is further maintained in the following manner: all people interacting with the parrot must use the same rules and commands at all times.
4. It is often helpful for each interactive person to have training sessions with the bird.

B. The owner's reaction to the parrot when it damages feathers should be *no reaction at all.* It is especially important not to give the parrot a drama reward. If the owner *has to do something,* he/she should say absolutely nothing, get up, and leave the room.

XI. Drug Therapy and the Feather-Damaging Parrot

Various psychoactive drugs can have a positive effect in some cases of feather destruction—e.g., haloperidol (Haldol: Henry Schein). However, it is unreasonable to assume that the animal can be kept on drug therapy forever. As Dr. Richie and Dr. Harrison state (on the subject of haloperidol), it "should be used in conjunction with behavioral modification to correct the inciting cause of the destructive behavior."[20] The author has found this combination to be extremely effective.

XII. Phobias

A phobia is an exaggerated, usually inexplicable and illogical, fear of a particular object, class of objects, or situation.[21]

A. Behaviors of the Phobic Parrot
1. The phobic parrot may act terrorized when approached.
   a. Its terror often focuses on a specific human (e.g., the owner's husband) or group of humans (e.g., all men).
   b. It may act very much like a totally wild import, despite being domestically bred and hand-raised.
   c. Often there is no known etiology for this behavior. Owners may report the behavior started "overnight" with no precipitating factors. However, overly-aggressive handling during veterinary appointments has been implicated in a number of cases.
   d. The phobic bird *must* be differentiated from the abused bird, and the bird with a medical problem.

e. Avian veterinarians often see these parrots as emergencies. This bird often injures itself in an effort to escape the object of its fear—for example, broken blood feathers are common.
f. Overdependence is often the precursor to phobic behaviors, including the following:
(1) Phobic feather picking
(2) Separation anxiety
(3) Inability to adapt to change

B. Characteristic Age for the Occurrence of Phobias
1. Phobias often appear at the onset of adolescence.
    a. Theories regarding these cases generally attribute this behavior to two possible causes:
        (1) A lack of early socialization resulting in a confused, frightened parrot.
        (2) The onset of sexual behavior in a confused, unsocialized parrot.
2. In the wild, this animal would be becoming more independent from its parents, thereby leaving its parents' protection (see Survival Skills, Socialization, and the Adolescent Parrot p 141).
3. Onset of sexual behavior in a confused, unsocialized parrot
    a. In many parrots, the beginning of hormone behaviors causes behavior changes. This is what Linden calls "sexual **IM**maturity."[22] In an effort to gain control, many adolescent parrots become increasingly aggressive. However, if an individual has a fundamentally submissive personality, it may vent the confusion of adolescence via phobias, instead.
        (1) With no warning that the owner recognizes, this individual essentially "wakes up phobic."
        (2) The object of the bird's terror may be the owner himself/herself.
        (3) However, with more insight, owners often comment that there were signs of problems developing, such as the parrot becoming more uneasy with new people and/or objects such as toys and the parrot becoming more rigid in its eating habits. The bird often chose less variety and exhibited nervousness when new food items were offered.

C. Dealing with the Phobic Parrot
Phobic birds should *immediately* be referred to an extremely experienced parrot behavior consultant. I believe this type of bird can be *ruined—possibly permanently—if mishandled*. Phobics can be helped with experience and exquisite patience.

Addendum

Examples of Questions Used in a Behavioral Work-Up

Management Questions:
1. How frequently does the bird bathe?
2. Is it exposed to unfiltered sunlight or full spectrum light, and if so, for how long?
3. Is it exposed to smokers? Do they wash hands before handling the bird?
4. What is the size and location of the bird's cage?
5. What is the placement of the cage within the room, specifically indicating window placement, door placement, existence and placement of skylights, and human flow patterns through the room? Note: diagrams are useful to derive this information.
6. How many hours of *sound* sleep does the bird get at night? Note: sound sleep starts *after* the family vacates the room in which the parrot is housed (see Sleep Deprivation p 125).

Diet Questions:
One needs to ask owners to estimate what percentage of the parrot's daily food *consumption* is in which of the following categories:
1. Base diet
   a. What kind (e.g., seed mix or pellets)?
   b. If pellets, what brand(s)?
   c. What volume?
2. Vegetables
   a. What kinds?
   b. Cooked or raw?
   c. What volumes of each consumed?
3. Fruits
   a. What kinds? (A tremendous nutritional difference exists between, for example, grapes and mango.)
   b. What volumes of each?
4. Complex carbohydrates
   a. What kind? (e.g., pasta, pizza, popcorn)?
   b. How much?
5. Animal protein
6. Junk food—what kinds and how much?
7. Other—have the owner explain.
8. Vitamins—if offered, how much, how often, in food or water? Note: birds consuming a sufficient percentage of high-quality pelleted foods should not be given vitamins.
9. When asking about diet, it is critical to get a feel for volumes consumed. As the owner of an obese Amazon said, "But I only feed *nine grapes a day.*"
10. If when asked what their parrot eats, the owner answers "Everything"—that often means that the bird eats *everything the owner eats* (and the odds are good that the owner never eats things like vegetables).

Behavioral Questions:
1. Parrot's weaning age and history
   a. Who weaned the parrot?
   b. If the owner did, ask these questions:
      i. Were you experienced at hand feeding and weaning?
      ii. If so, how did you learn?
      iii. If not, how were you taught?
      iv. How many feedings a day was the baby bird being given when brought home?
   c. How old was the parrot when it was weaned?
   d. Were there any problems associated with weaning?
      i. If so, what were they? (Explain in detail.)
      ii. If the problems were medical, get a detailed history.
2. Amount of attention the parrot gets per day
   a. Direct attention (in physical contact with the bird), getting 100% of owner's attention.
   b. Ambient attention (owner and bird are physically close). For example, the owner is petting the bird while watching television.
   c. Indirect attention (bird is out of cage and owner doing something else in the same room with the bird, talking to the bird but not touching)
3. Are the wings clipped?
   a. If not, and the bird is getting aggressive, suggest a wing clip for now.
   b. The owner can always let them grow out again, later—once the problem is resolved.
4. Approximately how many toys are in the bird's cage?
   a. Are the toys rotated?
   b. How often?
5. Does the bird play with them by itself?
6. Does the bird play with them with the owner?
7. Number of hours a day the parrot is out of its cage?
8. Is the bird supervised when out of the cage?
9. Does the bird expect to be out when the owner is home? How can the owner tell?
10. Is the bird allowed on the owner's shoulder?
11. Is the bird afraid of new situations, objects, or people? If so, have the owner explain.
12. Is the bird picky about foods? If so, what kinds?

Questions to Ask About a Specific Behavior Problem[23]:
1. What is the behavior problem and how serious is it?
2. Are you concerned that you may have caused the problem? (Ask the owner to explain.)
3. Do you feel guilty about this problem? (Ask the owner to explain.)
4. Have you considered giving up this bird? (Ask the owner to explain.)
5. What is the duration of the problem in terms of days, months, etc.?
6. What is the frequency of the problem in terms of hours, days, weeks, etc.?
7. Does any event or behavior routinely occur immediately before the problem begins?

8. Does any event or behavior routinely occur immediately after the problem ceases?
9. Is there a change in the bird's body postures before or during the inappropriate behavior?
10. What was the age of the bird when it started showing signs of the problem?
11. Has the bird's general behavior changed in any way since the onset of the problem?
12. Was there a change in the household or an event that seemed to precipitate the problem?
13. Is there a time of day when the behavior seems more or less intense? If so, what is usually happening at that time?
14. Is there a person or pet around which the behavior seems more or less intense? If so, who and why?
15. What is the general attitude of the bird while performing the behavior? (e.g., Does it seem frightened, angry, etc.?)
16. What kinds of things, if any, will interrupt the behavior once it has started (e.g., noises, treats, etc.)?
17. What does the owner do when the behavior begins?
18. Is a particular location associated with the behavior?
19. Has the frequency or the intensity of the occurrence of the behavior changed since this problem started? If so, how and when?
20. In the owner's opinion, does the parrot look "mean" or "friendly?"[24]
21. Did the owner realize the amount of responsibility parrot ownership entailed when he/she first obtained the bird?
22. Does the owner think the bird is pretty or does he/she prefer the way most other parrots look?
23. Is the owner happy with his/her selection of this particular parrot species?
24. Is the owner's spouse jealous of the time he/she spends with the parrot?
25. Who likes the parrot better—the owner or the spouse?
26. Has the owner been under a lot of stress lately? If so, have the owner explain.
27. Why does the owner think this is happening with their parrot?

REFERENCES

1. Wilson L: Sleep: How much is enough for a parrot? *The Pet Bird Report* 43:60-62, 1999.
2. Doane BM: How do behavior problems develop? *My parrot, my friend,* New York, 1994, Howell Book House, pp 110-155.
3. Blanchard S: Don't spoil your hand-fed birds, *Bird Talk,* 8:47, 1990.
4. Murphy J: Breeding the truly domesticated parrot, *Pet Bird Report* 5:44-47.
5. Blanchard S: Understanding pet parrot behavior, *Birds USA* 3:62-68, 1991.
6. Blanchard S: Games parrots play, *Bird Talk* 9:48-53, 1991.
7. Athan MS: The importance of being tall, *Guide to a well-behaved parrot,* Hauppauge, NY, 1993, Barrons, pp 64-66.
8. Hubbard J: Hand taming, *The New Parrot Training Handbook,* San Jose, Calif, 1997, Parrot Press, pp 51-53.
9. Doane BM: Domestication and discipline, *My parrot, my friend,* New York, 1994, Howell Book House, pp 92-109.

10. Davis C: Behavior seminar, Exotics and family pet showcase, Expo USA, April 9, 1995.
11. Linden P: Dependence & independence: finding the right balance, *Pet Bird Report* 3:38, 1993.
12. Blanchard S: Personal communications, 1993.
13. Wilson L: Non-medical approach to the behavioral feather plucker, *Proc Annu Conf Assoc Avian Vet,* 1996, pp 3-9.
14. Rosenthal K: Differential diagnosis of feather-picking in pet birds, *Proc Annu Conf Assoc Avian Vet,* 1993, pp 108-112.
15. Tully TN, Harrison GJ: Pneumonology. In Richie BW, Harrison GJ, Harrison LR, editors: *Avian medicine: principles and application,* Lake Worth, Fla, 1994, Wingers Pub, pp 607-633.
16. Csaky K: Feather plucking survey, *The Pet Bird Report* 4-1:40, 1993.
17. Foushee D: Play therapy for parrots, *The Pet Bird Report* 23:30-32, 1995.
18. Worell AB, Faber WL: The use of acupuncture in the treatment of feather picking in psittacines, *Proc Annu Conf Assoc Avian Vet,* 1993, pp121-126.
19. Blanchard S: Understanding pet parrot behavior, *Birds USA,* Irvine, Calif, 1991, Fancy Pub, pp 62-68.
20. Richie BW, Harrison GJ: Formulary. In Richie BW, Harrison GJ, Harrison LR, editors: *Avian medicine: principles and applications,* Lake Worth, Fla, 1994, Wingers Publishing, pp 457-476.
21. *Merriam-Webster's collegiate dictionary,* Springfield, Mass, 1993, Merriam-Webster.
22. Linden PG: Early avian development: a breeder's view of important behavioral stages, *Proc Annu Conf Internatl Avicult Soc,* Jan, 1996.
23. Adapted from interview form of the Behavior Clinic at the Veterinary Hospital of the University of Pennsylvania, Dr. Karen Overall, Director.
24. (Questions #20-25) Doane BM: How do I work with my bird?, *My parrot, my friend* New York, 1994, Howell Book House, pp 174-203.

9

Neurologic Signs

Simon R. Platt and Tracy L. Clippinger

I. Neurologic Syndromes
A neurologic syndrome is a cluster of clinical signs, caused by disease processes, which occur together, characterize a particular disease, and allow localization of defects within the nervous system.[1] The disease processes may affect any of the following:
  Central nervous system (CNS), including the brain and spinal cord
  Peripheral nervous system (PNS), including cranial nerves, spinal cord nerve roots, spinal nerves, peripheral nerve branches, and the neuromuscular junction, or organ systems and body processes that have an integral relationship with nervous system function

A. Multifocal Syndrome

B. Brain Syndromes
  1. Cerebral
  2. Hypothalamic
  3. Midbrain
  4. Cerebellar
  5. Pontomedullary
  6. Vestibular

C. Spinal Cord Syndromes
  1. Cervical
  2. Cervicothoracic
  3. Thoracolumbar
  4. Lumbosacral

D. Neuropathic Syndrome

II. Differential Diagnosis
A. Neurologic Disease

B. Myopathy

C. Behavioral Disorders

D. Weakness

III. General Considerations
A. Diagnostic Evaluation
1. Purpose: To identify the location, extent, and etiology of disease with the aid of ancillary tests
2. Diagnosis
   a. Signalment, history, and clinical picture
   b. Avian neurologic examination[2,3]
      (1) Observation
         (a) Mentation and behavior
         (b) Posture and attitude
         (c) Movement and gait
         (d) Involuntary limb and body motion
      (2) Cranial nerve evaluation
      (3) Palpation
      (4) Postural reactions
         (a) Conscious proprioception
         (b) Hopping
         (c) Drop and flap reaction
         (d) Extensor postural thrust reaction
      (5) Spinal (segmental) reflexes
         (a) Vent sphincter reflex
         (b) Pedal flexor (withdrawal) reflex
         (c) Patellar (stretch) reflex
         (d) Wing withdrawal reflex
      (6) Cutaneous sensation
         (a) Touch perception
         (b) Nociception
   c. Database
      (1) Complete blood count
      (2) Biochemistry panel
      (3) Serologic tests
         (a) Infectious disease screening
         (b) Toxologic screening
   d. Imaging techniques
      (1) Survey or selective radiography
      (2) Myelography
      (3) Computed tomography (CT)
      (4) Magnetic resonance imaging (MRI)
      (5) Nuclear imaging (Technetium-99m)
   e. Ancillary testing

(1) Cerebrospinal fluid (CSF) analysis
(2) Electrophysiologic evaluation
   (a) Electromyography (EMG)
   (b) Nerve conduction velocity (NCV) studies
   (c) Evoked responses (visual, auditory)
   (d) Electroencephalography (EEG)
(3) Histopathology
   (a) Biopsy of muscle and/or nerve
   (b) Postmortem examination

B. Treatment and Prognosis

Treatment and prognosis for diseases of the nervous system are dependent upon the disease location and pathologic process. Therapy must support continued neurologic function and repair of damaged structures.

1. Address underlying cause and tailor therapy to specific etiology.
   a. Electrolytes, minerals, and/or vitamins
   b. Antidotes
   c. Antibiotics, antifungals, preexposure vaccination, antiparasitics
   d. Antiinflammatories
2. Remove source.
   a. Physically remove the offending item by appropriate procedure: bathing, lavage, surgery, or endoscopy.
   b. Decrease absorption by binders and protectants.
   c. Enhance elimination by catharsis and diuresis
3. Control neurologic signs
   a. Antiinflammatory therapy
   b. Anticonvulsant therapy (see Chapter 10)
4. Provide supportive care.

C. General Causes

Many of the same disease processes that elicit cerebral seizure activity cause neurologic signs elsewhere in the CNS and PNS.[3-6] Please refer to Chapter 10.

1. Metabolic disorders
2. Intoxications
3. Inflammatory/infectious diseases (neuritis, myelitis, encephalitis, meningoencephalitis, encephalomyelitis)
   a. Bacterial sepsis or abscess: *Listeria monocytogenes, Streptococcus* sp., *Chlamydia psittaci, Salmonella* sp., *Pasteurella multocida, Mycoplasma* sp., *Clostridium* sp.
   b. Fungal systemic infection or granuloma: *Aspergillus* sp., *Dactylaria gallopava, Cladosporius, Mucomyces*
   c. Viral systemic infection or residual damage: Newcastle disease virus (Paramyxovirus), duck plague virus (also known as duck enteritis virus, herpesvirus), duck viral hepatitis (Picornavirus), avian encephalomyelitis virus (picornavirus), polyomavirus, reovirus, eastern and western equine encephalitis

viruses (togavirus), proventricular dilatation disease (unconfirmed virus).[7]
- d. Parasitic presence or aberrant migration: *Baylisascaris* sp., *Toxoplasma, Sarcocystis* sp.; schistosomiasis - *Dendritobilharzia* sp., Filariasis - *Chandlerella quiscali*
4. Trauma
5. Neoplasia
6. Anatomic, congenital, and familial disorders

IV. Multifocal Disease

A. Characteristics
1. Lesion present at more than one site
2. Infectious diseases and intoxications most commonly affect both the CNS and PNS.
3. Degenerative storage diseases are rare occurrences that usually affect multiple systems or the neurologic system in multiple locations.[1]

V. Brain Syndromes

A. Cerebral Syndrome
1. Characteristics
   a. Damage to the cerebral cortex (telencephalon) affects intellectual, learned, and sensory activities dependent upon the lesion location.
   b. The frontal lobe processes environmental information and influences fine motor ability.
   c. The parietal lobe processes input from cutaneous sensation.
   d. The occipital and temporal lobe deal with visual and auditory information, respectively.[1,8]
2. Clinical signs and neurologic examination[1-3]
   a. Altered mental status ranges from hyperexcitability and seizures to apathy, depression, and coma.
   b. Behaviorial changes include recognition failure and loss of trained habits.
   c. Pleurothotonus: Altered posture in which the head and trunk twist toward the side of the lesion
   d. Adversion: Altered attitude in which the head turns toward the side of the lesion and the bird circles toward the side of the lesion
   e. Abnormal behavioral movement (circling, head pressing, compulsive pacing)
   f. Altered depth of respiration and recurrent apnea (Cheyne-Stokes respiratory pattern) possible
   g. Cranial nerve function is generally normal but vision may be impaired if involvement of the optic tectum or optic lobe exists.
   h. Postural reactions usually range from depressed to absent in the contralateral limbs.

B. Diagnosis
1. Physical and neurologic examinations localize lesion to a region of the brain.
2. History and diagnostic testing, including blood work, survey radiographs, CSF analysis, electrophysiologic testing (EEG, visual evoked response), and advanced imaging techniques (CT, MRI) identify the etiologic agent or disease process.

C. Treatment
1. General principles (see Treatment and Prognosis p 150)
2. Refer also to Chapter 10

D. Causes
1. Trauma
2. Hydrocephalus
3. Encephalitis
4. Neoplasia
5. Toxicity
6. Degenerative
7. Developmental
8. Metabolic i.e., hepatic encephalopathy
9. Vascular disorders

VI. Hypothalamic Syndrome

A. Characteristics
1. Disease and/or damage to the hypothalamus (ventral portion of the diencephalon) affects autonomic visceral body functions and regulation of endocrine activity (pituitary gland).
2. Modulation of appetite, sexual activity, sleep-wake cycle, body temperature, blood pressure, and emotions occurs in the hypothalamus. The olfactory and optic cranial nerves are located in the diencephalon division.[1,8]

B. Clinical Signs and Neurologic Examination[1-3]
1. Altered mental status includes disorientation and lethargy.
2. Hyperexcitability and seizures may occur.
3. Abnormal behavioral movement (circling, head pressing, compulsive pacing) possible
4. Alterations in autonomic activity
    a. Abnormal temperature regulation (hyperthermia, hypothermia, poikilothermia)
    b. Abnormalities in appetite (hyperphagia and obesity, anorexia and cachexia)
    c. Endocrine disturbances (polyuria and polydipsia associated with pituitary lesions)
    d. Normal movement and gait

e. Optic nerve (cranial nerve [CN] II) damage - ipsilateral
   (1) Impaired vision may manifest if CN II (optic nerve) and/or the optic chiasm is/are affected.
   (2) Pupils will be dilated with weak or absent responses to light stimulation.

C. Diagnosis
1. Physical and neurologic examinations localize lesion to a region of the brain.
2. History and diagnostic testing, including blood work, survey radiographs, CSF analysis, electrophysiologic testing (visual evoked response), and advanced imaging techniques (CT, MRI) identify the etiologic agent or disease process.

D. Treatment
1. General principles (see Treatment and Prognosis p 150)
2. See also Chapter 10

E. Causes
1. Pituitary tumor
2. Granulomatous mass
3. Parasite presence or migration
4. Abscess
5. Trauma
6. Infarction

VII. Midbrain Syndrome

A. Characteristics
1. Damage to the midbrain (mesencephalon) is relatively uncommon because of its protected central location.
2. The reticular activating system passes through the mesencephalon and is responsible for the maintenance of an alert status dependent upon the environmental input received and processed.
3. The oculomotor and trochlear cranial nerves are located in the mesencephalon division.[1,8]

B. Clinical Signs and Neurologic Examination[1-3]
1. Altered mental status: sleepy, depressed, stuporous, or comatose
2. Opisthotonus: Altered posture (head thrown back) with rigid extension of all limbs
3. Voluntary movement may be weak or absent in the wing and leg contralateral to a unilateral lesion or in all four limbs with an extensive lesion.
4. Hyperventilation may be seen in some patients.
5. Cranial nerve deficits may occur on the same side as a unilateral lesion
   a. Ipsilateral oculomotor nerve (CN III) damage

(1) Ventrolateral strabismus
(2) Mydriasis
(3) Ptosis
(4) Depressed or absent pupillary light response (may be difficult to assess because pupillary constriction in birds is under both voluntary motor and intrinsic parasympathetic control)
   b. Contralateral trochlear nerve (CN IV) damage
      (1) Dorsomedial strabismus (may be impossible to detect in those birds with circular pupils, because the sclera is often not visible to serve as a reference point)
6. Bilateral pupillary miosis may be seen initially in birds with severe cranial trauma that diffusely involves the midbrain. This miosis gradually changes to fixed dilation.
7. Palpation may reveal increased tone in contralateral limbs or in all limbs.
8. Postural reactions deficient in limbs on contralateral side or in all limbs
9. Spinal reflexes increased in limbs on contralateral side or in all limbs

C. Diagnosis
1. Physical and neurologic examinations localize lesion to a region of the brain.
2. History and diagnostic testing, including blood work, survey radiographs, CSF analysis, and advanced imaging techniques (CT, MRI) identify the etiologic agent or disease process.

D. Treatment
1. General principles (see Treatment and Prognosis p 150)
2. See also Chapter 10

E. Causes
1. Thiamine deficiency
2. Cranial trauma
3. Hydrocephalus
4. Tumors
5. Infectious disease
6. Infarction

VIII. Cerebellar Syndrome

A. Characteristics
1. The cerebellum (dorsal metencephalon) is important for coordinating and reinforcing actions.
2. Dysmetria (inability to regulate rate, range, and force of movement) and dysequilibrium (inability to balance and maintain posture) occur in cerebellar disease.[1,8]

B. Clinical Signs and Neurologic Examination[1-3]
  1. Mental status, behavior, spinal reflexes, and sensation unchanged
  2. Posture and attitude
     a. Broad-based stance at rest
     b. Truncal swaying when walking
     c. Opisthotonus (if severe lesion of the rostral lobe)
     d. Vestibular signs (if rare lesion of the flocculonodular lobe)
  3. Voluntary movement and gait
     a. Exaggerated limb responses on initiation of a movement (hypermetria)
     b. Jerky or clumsy limb movements
  4. Involuntary intention tremors (head, eyes)
  5. Cranial nerves not affected specifically
     a. The menace response may be affected ipsilaterally or bilaterally.
     b. Vision is not impaired.
  6. Postural reactions are intact but reflect incoordination and dysmetria.

C. Diagnosis
  1. Physical and neurologic examinations localize lesion to a region of the brain.
  2. History and diagnostic testing, including blood work, survey radiographs, CSF analysis, and advanced imaging techniques (CT, MRI) identify the etiologic agent or disease process.

D. Treatment
  1. General principles (see Treatment and Prognosis p 150)
  2. See also Chapter 10

E. Causes
  1. Neoplasia
  2. Inflammation
  3. Trauma
  4. Toxicity
  5. Developmental hypoplasia
  6. Vascular infarction

IX. Pontomedullary Syndrome

A. Background Information
  1. Diseases of the pons (ventral metencephalon) and medulla oblongata (myelencephalon) produce multiple cranial nerve deficits, influence motor and sensory pathways, and alter vital body functions (respiration, blood pressure, heart rate).
  2. This division of the brainstem contains major ascending and descending tracts and the nuclei of cranial nerves V-XII.[1,8]

B. Clinical Signs and Neurologic Examination[1-3]
1. Mental depression (damage to the ascending reticular activating pathways) which can progress to stupor and coma
2. Behavior, posture, and attitude affected secondarily
3. Voluntary movement may be weak or absent in the wing and leg ipsilateral to a unilateral lesion or in all four limbs in an extensive lesion. Ataxia may be present in ambulatory animals.
4. Altered respiration if extensive damage
   a. Irregular and apneic (apneustic ventilation)
   b. Rapid and shallow (central neurologic hyperventilation)
   c. Slow and shallow (central alveolar hypoventilation)
5. Multiple cranial nerve deficits are present.
   a. Trigeminal nerve (CN V) damage
      (1) Diminished beak strength for mastication, jaw strength
      (2) Decreased palpebral reflex (unlike mammals, CN V supplies the orbicularis oculi muscle in birds)
      (3) Decreased sensation to the head including the eye, nasal cavity, beak, skin and floor (taste) of the oropharynx
   b. Abducent nerve (CN VI) damage[9]
      (1) Protrusion of the third eyelid
      (2) Medial strabismus
   c. Facial nerve (CN VII) damage (difficult to assess in birds as a result of minimal facial musculature)
      (1) Reduced tone in the muscles of facial expression
      (2) Decreased taste sensation from the maxillary and mandibular rhamphotheca
      (3) Reduced parasympathetic stimulated secretions of the lacrimal gland, the microscopic salivary glands, and the nasal glands
   d. Vestibulocochlear nerve (CN VIII) damage
      (1) Impaired hearing (thus, impaired hunting abilities)
      (2) Nystagmus: horizontal, rotatory, or vertical if medulla oblongata damaged
      (3) Impairment of normal physiologic nystagmus (absent if bilateral)
      (4) Vestibular signs
         (a) Head tilt, often toward the side of the lesion
         (b) Dysequilibrium causing a loss of balance, falling, rolling towards the side of the lesion
      (5) Possible ipsilateral hypotonia and contralateral hypertonia
   e. Damage to the glossopharyngeal nerve (CN IX), vagus nerve (CN X), spinal accessory nerve (CN XI), and hypoglossal nerve (CN XII) may be difficult to discriminate as their distal nerve fibers undergo considerable anastomoses in birds.[8,9]
      (1) Dysphagia/regurgitation with absent or depressed gag reflex (CN IX, X, XI, XII)
      (2) Poor prehension, deviated tongue, lingual atrophy, dysphonia (CN IX, XII)
      (3) Taste perception decreased or lost (CN VII, IX)

(4) Visceral signs such as inspiratory dyspnea and tachycardia (CN X)
(5) Poor neck movement (CN X, XI)
6. Palpation may reveal normal or increased muscle tone in all limbs.
7. Postural reactions deficient in limbs on ipsilateral side or in all limbs
8. Spinal reflexes intact

C. Diagnosis
1. Physical and neurologic examinations localize lesion to a region of the brain.
2. History and diagnostic testing, including blood work, survey radiographs, CSF analysis, electrophysiologic testing (brainstem auditory evoked responses -CN VIII), and advanced imaging techniques (CT, MRI) identify the etiologic agent or disease process.

D. Treatment
1. General principles (see Treatment and Prognosis p 150)
2. See also Chapter 10

E. Common Causes
1. Cranial trauma
2. Encephalitis
3. Neoplasia,
4. Inflammation

X. Vestibular Syndrome

A. Characteristics
1. Both central (medulla oblongata of the myelencephalon) and peripheral vestibular components (vestibular portion of CN VIII and more commonly, the vestibular receptors of the membranous labyrinth) are necessary for equilibrium to maintain appropriate attitude.
2. Peripheral vestibular disease is more common than central disorders.
3. Paradoxical vestibular disease is due to lesions in the caudal cerebellar peduncle and/or the flocculonodular lobe of the cerebellum.[1,8]

B. Clinical Signs and Neurologic Examination in Peripheral Vestibular Disease[1-3]
1. Posture: Possible exaggerated extensor tone of the contralateral limbs accompanied by decreased extensor tone in the ipsilateral limbs
2. Ipsilateral head tilt
3. Falling or rolling
4. Nystagmus
    a. Horizontal with the fast phase away from the side of the lesion
    b. Possible depression or absence of the physiologic nystagmus
    c. May also be rotatory
5. Flying or walking in tight circles

6. Strabismus
    a. Ipsilateral
    b. Possible ventrolateral direction if the neck is extended
7. Preservation of strength
8. Loss of righting reactions

C. Clinical Signs and Neurologic Examination in Central Vestibular Disease[1-3]
1. Possible altered mental status: depression, stupor, or coma
2. Nystagmus
    a. Vertical, horizontal or rotatory
    b. Can be positional, or changing direction with different positions of the head
3. Ipsilateral hemiparesis
4. Clinical evidence of other cranial nerve dysfunction
5. Ipsilateral postural reaction deficits

D. Clinical Signs of Paradoxic Vestibular Disease[1-3]
1. Head tilt contralateral to the side of the lesion
2. Strabismus contralateral to the side of the lesion
3. If the lesion is in the caudal cerebellar peduncle, it may cause ipsilateral postural reaction and proprioceptive deficits.

E. Diagnosis
1. Physical and neurologic examinations localize lesion to a region of the brain.
2. History and diagnostic testing, including blood work, survey radiographs, bullae radiography, and advanced imaging techniques (CT, MRI) identify the etiologic agent or disease process.

F. Treatment
1. General principles (see Treatment and Prognosis p 150)
2. See also Chapter 10

G. Causes
1. Cranial trauma
2. Infectious disease (especially of the middle and inner ear)
3. Toxicity (metronidazole in dogs and cats but as yet unreported in birds)
4. Nutritional deficiencies
5. Neoplasia

XI. Spinal Cord Syndromes

A. Cervical Syndrome
A lesion in the spinal cord caudal to the brainstem and cranial to the brachial plexus produces clinical signs that reflect damage to the external white matter pathways.[10]
1. Clinical signs and neurologic examination[2,3,10]

a. Mentation, behavior, and cranial nerves unchanged
   b. Posture and attitude suggest neck pain
      (1) Abnormal posture with the beak held toward the ground
      (2) Splinting of neck muscles
   c. Movement and gait
      (1) Variable signs from weakness to spastic paralysis of all four limbs (tetraparesis or tetraplegia)
      (2) Occasionally cause motor deficits in the limbs of one side of the body alone (hemiparesis or hemiplegia)
      (3) Ataxia in ambulatory birds (knuckling, crossover legs)
   d. Palpation
      (1) Muscle tone is normal to increased in all limbs.
      (2) Severe lesions may cause marked increases in muscle tone that cause extensor rigidity (termed clasp-knife if this hyperextension gives away suddenly to forced flexion).
      (3) No segmental muscle atrophy in any of the limbs
   e. Postural reaction deficits in ipsilateral limbs or in all limbs
   f. Spinal reflexes intact or increased in all limbs
      (1) Wing withdrawal reflex and pelvic limb reflexes are intact to exaggerated.
      (2) Vent reflex is intact to exaggerated.
   g. Sensation
      (1) Variable losses of pain perception in the neck caudal to the lesion and in all limbs
      (2) Complete loss of pain perception is extremely rare, because a lesion this severe would be accompanied by respiratory failure.
   h. A lesion within the center of the spinal cord (tumor or traumatic necrosis) produces wing signs more commonly than limb signs, because motor tracts to the wings are situated more centrally than those to the legs.
2. Diagnosis
   a. Physical and neurologic examinations localize lesion to a region of the spinal cord.
   b. Diagnostic testing, including survey radiographs, CSF analysis, myelography, and EMG, identify the etiologic agent or disease process.
3. Treatment of spinal cord syndromes depends upon the severity and location of the lesion.
   a. Cage rest
   b. Medical therapy[10]
      (1) Therapy should be initiated as soon as possible after spinal cord injury is recognized.
      (2) Corticosteroids may reduce inflammation and scavenge free radicals. Birds on steroid therapy must be closely monitored for decreased resistance to systemic infection. Administer parenteral rapid-acting corticosteroids, such as prednisolone sodium succinate (30 mg/kg initial dose; 15 mg/kg IV/IO/IM q8 hr ξ (24-48 hr) if onset is within 8 hr.

(3) Pain relief may be indicated. Consider buprenorphine (0.01 to 0.05 mg/kg IM q12 hr) or butorphanol (1-2 mg/kg IM q4-12 hr).
(4) Direct therapy at underlying cause.
c. Surgical intervention may be indicated in cases of vertebral instability or persistent spinal cord compression. Bird size may hinder technique.
4. Common causes of spinal cord syndromes: trauma (spinal fracture), neoplasia, myelopathy (nutritional or toxin-induced disease), myelitis (infectious or inflammatory disease), vascular disorder

B. Cervicothoracic Syndrome
1. Characteristics
   a. A lesion in the spinal cord within the cervical intumescence produces clinical signs that reflect damage to the central gray matter pathways.
   b. The spinal nerves arising from this region constitute the brachial plexus, innervate the thoracic body wall and wings, and include the pectoral, medianoulnar, bicipital, ventral propatagial, axillary, and radial peripheral nerves.[1,2,8,10]
2. Clinical signs and neurologic examination[2,3,10]
   a. Mentation, behavior, and cranial nerves unchanged
   b. Posture affected secondarily by muscle tone
   c. Movement and gait
      (1) Weakness to paralysis in one wing (monoparesis or monoplegia), both wings, ipsilateral limbs, or in all four limbs
      (2) Ataxia in ambulatory birds
   d. Palpation
      (1) Muscle tone is decreased (flaccid) in one or both wings.
      (2) Segmental muscle atrophy in wings (muscles associated with peripheral nerve arising from affected spinal cord segment)
      (3) No pelvic limb atrophy
   e. Postural reaction deficits in all limbs, especially the wings
   f. Spinal reflexes
      (1) Depressed or absent wing withdrawal on one or both sides
      (2) Normal to exaggerated leg and vent reflexes
   g. Sensation
      (1) Variable losses of pain perception caudal to the cervical intumescence
   h. Urinary incontinence possible
   i. Traumatic avulsion of the brachial plexus mimics the cervicothoracic syndrome, but limits clinical signs to the affected wing only.
      (1) Monoplegia/paresis and muscle atrophy of one wing
      (2) Postural reactions depressed and spinal reflexes absent in the affected wing but normal in all other limbs
      (3) Horner's syndrome (miosis; inconsistent ptosis, enophthalmos, nictitans prolapse) may occur if the vagosympathetic trunk is damaged.

XII. Thoracolumbar Syndrome

A spinal cord lesion caudal to the cervical intumescence and cranial to the lumbar intumescences produces clinical signs that reflect damage to the external white matter pathways.[10]

A. Clinical Signs and Neurologic Examination[2,3,10]
1. Mentation, behavior, and cranial nerves unchanged
2. Posture and attitude
    a. Kyphosis (back slightly arched)
    b. Schiff-Sherrington posture (recumbency of the bird with rigid extension of the wings)
3. Movement and gait
    a. Spastic weakness or paralysis of the legs (spasticity associated with increased extensor muscle tone)
    b. Ataxia in ambulatory birds
4. Palpation
    a. No segmental muscle atrophy in wings or legs
    b. All muscles related to nerves of the spine caudal to a lesion may develop generalized atrophy with long-term paralysis.
5. Postural reaction deficits in legs
6. Spinal reflexes
    a. Normal wing withdrawal
    b. Normal or exaggerated leg reflexes
        (1) The leg undergoing testing may exhibit clonus (rapid contraction and relaxation).
        (2) The contralateral leg may exhibit a crossed extensor reflex.
    c. Normal or exaggerated vent reflex
7. Sensation
    a. Variable losses of pain perception caudal to the lesion
    b. Variable enhanced perception at or just cranial to the lesion

XIII. Lumbosacral Syndrome

A lesion in the spinal cord within the lumbar intumescence (lumbar, sacral, pudendal plexus) produces clinical signs that reflect damage to the central gray matter pathways. The spinal nerves arising from this region constitute the lumbosacral plexus, innervate the legs, cloaca, vent, and tail, and include the femoral, obturator, ischiatic, pudendal, pelvic, and coccygeal peripheral nerves.[1,2,8,10]

A. Clinical Signs and Neurologic Examination[2,3,10]
1. Mentation, behavior, cranial nerves unchanged
2. Posture affected secondarily by decreased muscle tone
3. Movement and gait
    a. Flaccid weakness to paralysis of the legs and tail

4. Palpation
   a. Muscle tone decreased in the legs if the lesion is caudal to the lumbar and/or sacral plexus
   b. Muscle tone decreased in the vent sphincter if the lesion is caudal to the pudendal plexus
   c. Muscle tone decreased in the tail if the lesion is caudal to pudendal plexus and caudal spinal nerves
   d. Segmental denervation atrophy occurs in the muscles of the leg, cloaca, and vent.
5. Postural reaction deficits in the legs
6. Spinal reflexes are altered depending upon the level of the lesion
   a. Normal wing reflexes
   b. Normal to exaggerated leg and/or vent reflexes if lesion cranial to lumbosacral and/or pudendal plexus, respectively
   c. Depressed or absent pedal withdrawal reflex if the lesion is caudal to the lumbar plexus
   d. Depressed or absent patellar reflex if the lesion is caudal to the sacral plexus
   e. Depressed or absent vent sphincter reflex if the lesion is caudal to the pudendal plexus
7. Sensation
   a. Variable losses of pain perception in the legs and vent area dependent upon the level and extent of the spinal cord lesion
8. Fecal and urinary incontinence may be apparent.
9. A space-occupying mass within the kidney or pelvic canal (renal tumor or egg) may place pressure upon the lumbosacral plexus and its ischiatic nerve, thus mimicking the lumbosacral syndrome and causing a depressed or absent withdrawal reflex in the legs.[11,12]

XIV. Neuropathic Syndrome

The neuropathic syndrome, classically known as lower motor neuron disease, refers to disease of peripheral and sometimes cranial nerves. Disease course often has a gradual and cumulative effect, but may be self-limiting.[1]

A. Nerves Affected in Various Combinations
1. Mononeuropathy
2. Polyneuropathy (usually bilaterally symmetric)

B. Neuromuscular Junctionopathies Mimicking Diffuse Polyneuropathies
1. Botulism

C. Clinical Signs and Neurologic Examination[1-3]
1. Mentation and behavior unaffected unless diffuse disease
2. Posture and attitude unaffected unless diffuse disease
3. Movement and gait

a. Paresis or paralysis of limb or head muscles
b. Ataxia
4. Involuntary tremors and muscle fasciculations occur occasionally.
5. Cranial nerve dysfunction is uncommon.
6. Palpation
   a. Reduced or absent muscle tone (hypotonia, atonia, flaccidity)
   b. Neurogenic muscle atrophy
      (1) Fibrosis
      (2) Contractures that limit joint motion
7. Postural reaction deficits if sensory nerve involvement
8. Spinal reflexes reduced or absent
9. Cutaneous sensation affected if sensory nerve involvement
   a. Variable loss of pain (hypalgesia or analgesia)
   b. Variable loss of sensation (hypesthesia or anesthesia)
   c. Abnormal sensitivity over specific feather tracts (paresthesia)
   d. Self-mutilation

D. Common Causes
1. Metabolic and nutritional deficiencies
   a. Hypoglycemia[13] (see Chapter 10)
   b. Hypocalcemia[14,15] (see Chapter 10)
   c. Hypoxia
      (1) Vascular disorders can induce acute occlusion of oxygen supply to the spinal cord and/or spinal nerves and arrest all aerobic metabolic processes (e.g., cerebrovascular accidents, ischemic infarction, thromboembolic disease, atherosclerosis, idiopathic)[12,16,17]
      (2) Clinical signs: an "aura-like" behavior while going into a semiconscious state, ataxia, paresis, seizures, acute death
      (3) No antemortem diagnostic techniques have been described. MRI and CT scans may be useful in diagnosing these lesions.
      (4) Supportive treatment
   d. Nutritional deficiencies
      (1) Hypovitaminosis E[18,19]
         (a) Deficiency in tocopherol (vitamin E) commonly causes encephalomalacia and muscular dystrophy. A variety of age, classes, and species have been affected. Piscivorous birds fed an unsupplemented diet of frozen fish are particularly at risk because of breakdown of this fat-soluble vitamin during storage. Hypovitaminosis E is possibly the cause of cockatiel paralysis syndrome.[4,20]
         (b) Clinical signs: vestibular deficits, cerebrocortical signs, weakness, electrocardiographic abnormalities
         (c) Diagnosis may be presumptive.
            (i) Feedstuff, serum, and tissue analysis
            (ii) Response to vitamin E supplementation
            (iii) Treat with oral or injectable vitamin E (Chapter 26)
      (2) Hypovitaminosis B_1 (thiamine deficiency)[21]

(a) Deficiency in thiamine causes a polyneuropathy with peripheral nerve myelin degeneration.
(b) Clinical signs: "star-gazing"
(c) Diagnosis is presumptive based on a rapid (within hours) response to treatment.
(d) Treat with parenteral and oral vitamin B_1 administration (1 to 3 mg/kg IM q7d) and correct diet.

(3) Hypovitaminosis B_2 (riboflavin deficiency)[4,22]
(a) Deficiency in riboflavin causes a demyelinating peripheral neuritis with concurrent axonal edema and occasional neuromuscular endplate degeneration. The ischiatic and brachial nerves are commonly affected.
(b) Clinical signs: "curled toe paralysis" (poultry, nestling budgerigars)
(c) Treat with parenteral and oral vitamin B complex and correct diet.
(d) Prognosis is guarded, because the changes are often irreversible

(4) Hypovitaminosis B_6 (pyridoxine deficiency)[6]
(a) Deficiency in pyridoxine
(b) Clinical signs: "jerky" gait, running gait with wing flapping, seizures, death
(c) Treat with parenteral and oral vitamin B complex and correct diet.

2. Intoxication[23]
   a. Heavy metals
   (1) Lead[24,25]
   (a) Most common causes of toxicity of birds
   (b) See also Chapters 10, 14.
   (2) Zinc ("new cage syndrome")[26]
   (a) Direct irritation of gastric mucosa may occur. Enzyme systems throughout the body, particularly those of the pancreas, are damaged. Exposure occurs by ingestion of zinc source, including galvanized wire and clips, white rust from wire, powder coated cages, pennies (minted after 1982), and Desitin ointment.
   (b) Clinical signs: ataxia, weakness, depression, lethargy, anorexia, vomiting, polyuria, diarrhea, weight loss
   (c) Diagnosis
      (i) Serum or plasma levels
         a) Levels above 200 μg/dl (2 ppm) suggestive
         b) Blood must be collected in plastic containers or in specifically designed trace mineral (royal blue stoppers) glass tubes. Avoid rubber stoppers and grommets on collection devices.
      (ii) Postmortem tissue levels (pancreas, kidney, liver)
      (iii) Cage, metal, and/or paint samples containing zinc > 800-1000 ppm considered dangerous

(d) Treatment
  (i) Prevent further absorption of toxin
  (ii) Remove any metallic object visualized in digestive tract.
  (iii) Enhance passage of small particles by gavage administration of warm liquified peanut butter/mineral oil (2:1) mixture.
  (iv) Chelation therapy
    a) Edetate calcium disodium injection (Calcium versonate: 3M Pharmaceuticals; 25-40 mg/kg IM BID ω 5 days, rest ω 3 to 5 days, repeat prn)
    b) Dimercaptosuccinic acid (DMSA, Aldrich; 25 to 35 mg/kg PO BID ω 5 days, rest × 2 days, repeat 3 to 5 weeks)
  (v) Reduce zinc level of wire by scrubbing the wire with a brush and mild acidic solution (vinegar).[27]
b. Botulism[28,29]
  (1) Botulism ("limberneck") occurs most often in waterfowl after ingestion of organic matter (plant or animal tissue) or maggots that are contaminated with *Clostridium botulinum* spores or concentrated toxin, respectively. The exotoxin blocks acetylcholine release at the endplates of all motor nerves, causing peripheral neuropathy.
  (2) Clinical signs: paresis, paralysis, green diarrhea with vent pasting. The condition classically results in generalized lower motor neuron paralysis and parasympathetic dysfunction and shows an ascending course affecting vent and legs initially and later progressing to wing, neck, and head effects.
  (3) Diagnosis
    (a) History
    (b) Identification of the toxin in the stomach contents, serum, or plasma
    (c) Mouse protection test wherein only inoculated mice that are not protected by an antitoxin die
  (4) Treatment
    (a) Eliminate the toxin from the gastrointestinal tract
    (b) Protect bird from environmental dangers (drowning)
    (c) Antitoxin not commercially available and unreliable efficacy
    (d) Provide supportive care (feeding and physical therapy) during 2- to 3-week recovery period
c. Pesticides
  (1) Organophosphates and carbamates such as carbaryl, chloropyrifos, diazanon, malathion, and dichlorvos are acetylcholine inhibitors, to which birds are 10 to 20 times more susceptible than mammals.
  (2) Clinical signs: tremors, ataxia, weakness, decreased proprioception, paralysis, seizures, respiratory distress, progressive bradycardia, diarrhea

(3) Diagnosis of acetylcholinesterase inhibition is usually based on clinical signs and the results of a cholinesterase assay of blood, plasma, serum, or brain tissue. A decrease of 50% from normal is diagnostic.[30]

(4) Treatment is aimed at reversing the toxin's effects on the muscarinic or nicotinic nervous system receptors with atropine (0.2 to 0.4 mg/kg IV, IM, SQ q 8-24 hr or prn) or diphenhydramine (1 mg/kg IM), respectively. *Protopam* chloride (2-PAM; 10-20 mg/kg IM) may be used only within the first 1 to 2 days after exposure for organophosphate intoxication.

3. Inflammatory/infectious diseases[3,4,12]
   a. Infectious agents may cause direct tissue effects or secondary inflammatory effects because of their presence.
   b. Bacterial sepsis or abscess, fungal systemic infection or granuloma, viral systemic infection or residual damage, parasitic presence or aberrant migration
   c. Clinical signs (dependent upon infectious agent or involved organs): peripheral nerve dysfunction secondary to infiltrates, torticollis, opisthotonus, diffuse neuropathy exacerbated by excitement, seizures, acute blindness (*Toxoplasma gondii*)
   d. Diagnosis
      (1) Culture fluid or tissue sample
      (2) Serology for antibody and/or antigen titer
      (3) Cytologic evaluation of fluids and/or impression smears
      (4) Histopathology and immunohistochemical staining of tissues
   e. Tailor treatment towards etiology.

4. Trauma
   a. Compression resulting from impact, laceration resulting from bone fractures, and/or chemical irritation resulting from injection may injure the peripheral nerve, causing varying degrees of damage. Neurapraxia denotes transient dysfunction (often, paralysis) in the absence of structural changes. Relatively good preservation of internal architecture despite damage to the axon and subsequent peripheral degeneration defines axonotmesis. Neurotmesis describes functional transection or disruption of all essential internal structures.[1-3,10] The brachial plexus (spinal nerve roots C6-T2) is the most common location for traumatic avulsions.[31] The neuropathic syndrome in this region may include a Horner's syndrome (miosis inconsistent feature of this syndrome compared with mammals).
   b. Clinical signs: dependent upon affected nerve but generally include acute loss of motor and sensory function, followed by muscle atrophy. Flaccid paralysis of the affected limb, with loss of sensation distal to the elbow, occurs in the case of a brachial plexus avulsion.
   c. Diagnosis
      (1) History, clinical signs, neurologic examination (level of sensation and pain perception particularly important)
      (2) Electrophysiologic evaluation[2]

(a) EMG: Spontaneous activity may be present within a week after denervation.
(b) NCV: Action potential may be absent with complete nerve transection. Velocity and amplitude may be altered in axonotmesis or neurotmesis.
d. Treatment
(1) Rest and supportive care with continuous neurologic monitoring
(2) Surgery (correction of compression or entrapment, reanastomosis)
(3) Physical therapy to prevent joint and muscle contracture
(4) Prognosis is determined by the level of nerve injury, the length regenerating axons must travel to successfully reinnervate their target, and the alignment of connective tissue supportive structures in the transected cord. Damage may be transient (2 to 4 weeks) or permanent. Prognosis is poor if there has not been improvement within 4 weeks.

5. Neoplasia[4,11,12]
  a. Glioblastoma multiforme, choroid plexus tumors, schwannomas, astrocytomas, undifferentiated sarcomas, and hemangiomas have all been recorded in the nervous system of birds. Neoplastic cell proliferation may destroy or compress peripheral nerves.
  b. Abdominal masses
    (1) The pelvic nerves pass through the renal parenchyma; therefore renal neoplasia can cause a neuropathic syndrome. Renal tumors have a poor prognosis, and surgical resection may not be possible.
    (2) Ovarian masses and tumors can compress the renal parenchyma.
  c. Clinical signs: unilateral paresis (with possible accompanying abdominal distention) progressing to paralysis
  d. Diagnosis
    (1) Physical and neurologic examinations localize lesion to a peripheral nerve.
    (2) History and diagnostic testing, including survey radiographs and advanced imaging techniques (CT, MRI), identify the mass. Histopathology is usually needed to confirm neoplastic process.
  e. Treatment dependent upon tumor location and type
    (1) Chemotherapy
    (2) Surgical excision

6. Congenital
  a. Lafora body neuropathy[32]
    (1) A suspected degenerative defect in intracellular metabolism causes accumulation of indigestible material (cytoplasmic inclusions) within neurons and a subsequent disruption in cellular function.
    (2) Clinical signs: myoclonus, seizures
    (3) Postmortem diagnosis

REFERENCES

1. Braund KG: *Clinical syndromes in veterinary neurology,* ed 2, St Louis, 1994, Mosby.
2. Clippinger TL, Bennett RA, Platt SR: The avian neurologic examination ancillary neurodiagnostic techniques, *J Avian Med Surg* 10:221-247, 1996.
3. Jones MP, Orosz SE: Overview of avian neurology and neurological diseases, *Sem Avian and Exotic Pet Med* 5:150-164, 1996.
4. Bennett RA: Neurology. In Ritchie BR, Harrison GJ, Harrison LR, editors: *Avian medicine: principles and application,* Lake Worth, Fla, 1994, Wingers, pp 721-747.
5. Lyman R: Neurologic disorders. In Harrison GJ, Harrison LR, editors: *Clinical avian medicine and surgery,* Philadelphia, 1986, WB Saunders, pp 486-490.
6. Hasholt J, Petrak ML: Diseases of the nervous system. In Petrak ML, editor: *Diseases of cage and aviary birds,* Philadelphia, 1992, Lea & Febiger, pp 468-477.
7. Ritchie BW: *Avian viruses: function and control,* Lake Worth, Fla, 1995, Wingers.
8. King AS, McClelland J: *Birds—their structure and function,* ed 2, London, 1984, Baillière Tindall, pp 233-314.
9. Orosz SE: Principles of avian clinical neuroanatomy, *Semin Avian and Exotic Pet Med* 5:127-139, 1996.
10. Chrisman CL: *Problems in small animal neurology,* ed 2, Philadelphia, 1991, Lea & Febiger, pp 177-205.
11. Neumann U, Kummerfeld N: Neoplasms in budgerigars, *Avian Pathol* 12:353-362, 1983.
12. Shivaprasad HL: Diseases of the nervous system in pet birds. A review and report of diseases rarely documented, *Proc Assoc Avian Vet,* 1993, pp. 213-222.
13. Walsh MT: 1985. Seizuring in pet birds, *Proc Assoc Avian Vet,* 1985, pp 121-128.
14. Hochleithner M: Convulsions in African grey parrots, *Proc Assoc Avian Vet,* 1989, pp 78-81.
15. Rosskopf WJ, Woerpel RW, Lane R: The hypocalcaemia syndrome in African greys: an updated clinical viewpoint with current recommendations for treatment. *Proc Assoc Avian Vet,* 1985, pp 129-131.
16. Clubb SL, Karpinski L: 1992. Aging in macaws, *Proc Assoc Avian Vet,* 1992, pp 83-86.
17. Joyner KL: Encephalitis, proventricular and ventricular myositis and myenteric ganglioneuritis in an umbrella cockatoo, *Avian Dis* 33:379-381, 1989.
18. Campbell TW: Hypovitaminosis E: its effects on birds, *First Intl Conf Zoo & Avian Med,* 1987, pp 75-77.
19. Schmidt R: Systematic survey of lesions from animals in a zoological collection. 1. Central nervous system, *J Zoo Wildl Med* 17:8-11, 1986.
20. Harrison GJ: Preliminary work with selenium/vitamin E responsive conditions in cockatiels and other psittacines, *Proc Assoc Avian Vet,* 1986, pp 257-262.
21. Lowenstine LJ: Avian nutrition. In Fowler M, editor: *Zoo and wild animal medicine,* ed 2, Philadelphia, 1986, WB Saunders, pp 201-212.
22. Johnson WD, Storts RW: Peripheral neuropathy associated with dietary riboflavin deficiency in the chicken I. Light microscopic study, *Vet Pathol* 25:9-16, 1988.
23. LaBonde J: Avian toxicology, *Vet Clin North Am Sm Anim Pract* 21:1329-1342, 1991.
24. Mautino M: Avian lead intoxication, *Proc Assoc Avian Vet,* 1990, pp 245-247.
25. McDonald SE: Lead poisoning in psittacine birds. In Kirk RW, editor: *Current veterinary therapy IX,* Philadelphia, 1986, WB Saunders, pp 713-718.
26. Van Sant F: Zinc and clinical disease in parrots, *Proc Assoc Avian Vet,* 1997, pp 387-391.
27. Reece RL: Zinc toxicity (new wire disease) in aviary birds, *Aust Vet J* 63:199, 1986.
28. Harrison GJ: Clostridium botulinum type C infection on a game fowl farm, *Proc Am Assoc Zoo Vet,* 1974, pp 221-224.

29. Foreyt WJ, Abinanti FR: Maggot-associated type C botulism in game farm pheasants, *J Am Vet Med Assoc* 177:827-828, 1980.
30. Porter SJ: Organophosphate/carbamate poisoning in birds of prey, *Proc Am Assoc Zoo Vet*, 1992, pp 176-177.
31. Graham D: Surprises at necropsy, *Assoc Avian Vet Today* 2:192-195, 1988.
32. Britt JO, Paster MB, Gonzales C: Lafora body neuropathy in a cockatiel, *Comp Anim Pract* 19:31-33, 1989.

10

Seizures

Tracy L. Clippinger and Simon R. Platt

I. Seizure

A. Definition
1. A sudden, uncontrolled, transient alteration in behavior
2. Characterized by a change in motor activity, consciousness, sensation, or autonomic function
3. Accompanied by a paroxysmal electrical discharge from the brain defines a seizure or convulsion[1,2]

B. Phases of the Seizure Event
1. Prodrome: Behavior change preceding the seizure event
2. Aura: Actual beginning of the seizure
3. Ictus: Manifestations of the seizure
4. Postictal period: Recovery to normal behavior following the seizure

C. Types of Seizures
1. Generalized seizures
   a. Manifested by symmetric and synchronous clinical signs, which are classically tonic-clonic motions accompanied by salivation, urination, and defecation
   b. Supported by diffuse, symmetric dysrhythmia on electroencephalogram
   c. Classified by level of consciousness
      (1) Mild (presence of consciousness)
      (2) Severe (complete loss of consciousness)
   d. Associated with acquired or congenital cause
      (1) Metabolic disorder
      (2) Intoxication
      (3) Primary epilepsy
2. Partial seizures
   a. Manifested initially by focal signs that may secondarily become generalized

b. Supported by localized dysrhythmia on electroencephalogram
c. Associated with acquired cause
    (1) Encephalitis (inflammation/infection)
    (2) Trauma
    (3) Neoplasia

II. Differential Diagnosis

A. Neurologic Disease
1. Extracranial
    a. Metabolic disorder
        (1) Alteration of homeostasis
        (2) Deficiency state
        (3) Accumulation of by-products
    b. Intoxication
2. Intracranial
    a. Inflammatory disease
    b. Trauma
    c. Neoplasia
    d. Vascular disorder
    e. Degenerative condition
    f. Congenital and familial disorders
        (1) Malformation
        (2) Primary epilepsy

B. Other Conditions
1. The following may cause transient changes in consciousness, behavior, or motor activity but are not accompanied by cerebral paroxysmal electrical discharges:
    a. Narcolepsy
    b. Syncope
    c. Behavioral disorders
    d. Transient vestibular attacks
    e. Weakness

III. General Considerations

A. Neuronal Excitability of the Brain
1. Varies among individual birds
2. Depends upon the reception and balance of excitatory and inhibitory synaptic activity
3. Seizures result whenever one of the following states occurs:
    a. Imbalance between inhibitory and excitatory activity at the neuronal synapse
    b. Abnormal neuronal cell membrane
    c. Altered internal neuronal cell metabolism

B. Diagnosis
1. Signalment, history (seizure log), and clinical picture
2. Neurologic examination[3]
3. Database
   a. Complete blood count
   b. Biochemistry panel
   c. Serologic tests
      (1) Infectious disease screening
      (2) Toxologic screening
4. Imaging techniques
   a. Survey or selective radiography
   b. Computed tomography (CT)
   c. Magnetic resonance imaging (MRI)
   d. Nuclear imaging (Technitium-99m)
5. Ancillary testing
   a. Cerebrospinal fluid analysis
   b. Electroencephalography (EEG)
6. Postmortem examination (histopathology)

C. Treatment
1. Address underlying cause, particularly for extracranial seizure disorders.
2. Remove source.
   a. Physically remove the offending item by appropriate procedure: bathing, lavage, surgery, or endoscopy.
   b. Decrease absorption by binders and protectants.
   c. Enhance elimination by catharsis and diuresis.
3. Control seizures.
   a. Diazepam limits the spread of seizure activity and elevates seizure threshold.[1]
      (1) Bolus (0.5 to 1.0 mg/kg IV or IO) to arrest status epilepticus.
      (2) Repeat diazepam two to three times over several minutes as needed.
   b. Barbiturates (phenobarbital, pentobarbital, and primidone) depress repetitive electrical activity of multineuronal networks.[1] Efficacy of therapy is determined by achievement of consistently high concentrations in the brain. Goal of therapy is reduction of frequency and severity of seizures.
      (1) Administer parenteral anticonvulsant to gain seizure control.
         (a) Phenobarbital (1 to 2 mg/kg IV/IO or IM) q 6 to 12 hr or to effect
         (b) Pentobarbital (5 to 15 mg/kg IV/IO) slowly to effect. Pentobarbital stops refractory seizures due to its anesthetic properties. Hypoventilation occurs commonly with pentobarbital; therefore intubation and positive pressure ventilation are recommended.
      (2) Convert to oral maintenance therapy within 48 hours.
         (a) Phenobarbital (1 to 10 mg/kg PO) q 8-12 hr is standard therapy.

(b) Primidone, through its metabolites phenobarbital and phenylethyl malonic acid (PEMA), may be an alternative anticonvulsant. Primidone usage (125 mg/day in water source) has been reported in an Amazon parrot.[4]

(3) Adjust to lowest dosage required to control seizure activity.

(4) Monitor serum anticonvulsant level.
   (a) Phenobarbital (itself and as metabolite of primidone) reaches steady state in 7 to 21 days in small mammals.
   (b) The therapeutic range in small mammals for phenobarbital is 15 to 40 µg/ml serum and for primidone is 5 to 15 µg/ml.[1]

(5) Monitor serum liver enzymes and liver function every 3 to 6 months. PEMA, a metabolite of primidone, may cause greater hepatoxicity than other anticonvulsant metabolites.

c. Gaseous anesthesia may be warranted in seizuring birds to allow diagnostic sampling and medication. General anesthesia diminishes neurologic activity by reducing metabolic rate and, thus, cerebral demands for glucose and oxygen.

d. Steroids and diuretics may decrease cerebrospinal fluid production.

e. Steroids may reduce central nervous system (CNS) edema and inflammation. Birds on steroid therapy must be closely monitored for decreased resistance to systemic infection. The effect of steroids on the gastric mucosa is unknown.

IV. Causes

A. Metabolic Disorders

A metabolic disorder is an acquired disease that develops when another organ system fails to provide components or constituents required for cerebral processes or fails to remove substances that hinder cerebral processes.

1. Hypoglycemia
   a. Depletion of cerebral glucose reserve does not allow adequate energy for brain metabolic processes.
      (1) Transient juvenile hypoglycemia
      (2) Starvation or malnutrition
      (3) Sepsis
      (4) Hepatogenous hypoglycemia
   b. Clinical signs: Depression, general weakness, seizures
   c. Diagnosis: Blood glucose level falls below 100 mg/dl or below half the normal value for the species.[5]
   d. Treatment
      (1) Administer intravenous 50% dextrose (1 to 2 ml/kg slowly to effect) for immediate relief of clinical signs, while the underlying cause of the hypoglycemia is being determined and corrected.

(a) Tissue damage may occur with perivascular leakage or subcutaneous administration as dextrose solutions greater than 2.5% are hypertonic.
(b) Dextrose alone should not be used in a dehydrated bird as it will compromise acid-base balance.
(2) Administer oral glucose-containing solution directly or by gavage to alert bird. Rub small amount of glucose-containing substance on mucous membranes to depressed or obtunded bird if parenteral access is unavailable.
(a) 10% dextrose
(b) Karo syrup and honey

2. Hypocalcemia
   a. Inadequate calcium is present for maintenance of stable cellular membranes.
      (1) Depletion secondary to egg production
      (2) Hypoparathyroidism
         Parathyroidectomized birds have seizures when serum calcium falls below 5.0 mg/dl.[6]
      (3) Renal disease
      (4) Intestinal malabsorption
      (5) Idiopathic in African grey parrots[7,8]
   b. Clinical signs: Muscle fasciculations, tremors, seizures
   c. Diagnosis
      (1) History often reveals poor diet consisting mainly of whole seeds (deficient in calcium, vitamin A, and vitamin D_3)[9]
      (2) Blood calcium level falls below 5.0 to 6.0 mg/dl.[6,8]
   d. Treatment
      (1) Administer parenteral calcium.
         (a) Intravenous: 10% calcium gluconate solution (0.5 to 2.0 ml/kg)
         (b) Intramuscular: Calcium glycerophosphate/lactate solution (Calphosan: Glenwood-Palisades Co.; 0.5 to 1.0 ml/kg q 8-24 hr)
         (c) Oral: Calcium glubionate syrup (Neocalglucon: Sandoz Pharmaceuticals Corp.; 1 to 6 ml/kg directly q 12-24 hr or 1 to 5 ml in 30 ml water daily)
      (2) Corticosteroids are contraindicated, because they increase urinary excretion and decrease intestinal absorption of calcium.
      (3) Ensure proper diet with adequate calcium in proper balance with phosphorus and sufficient levels of vitamin D_3. Eliminate high fat seed.
      (4) Check serum calcium q 2-4 months.
      (5) Prognosis depends upon severity of damage to the parathyroid glands and is more favorable if treatment occurs before complete degeneration of the glands.

3. Hypoxia
   a. Vascular disorders can induce acute occlusion of oxygen supply to the brain and arrest all aerobic metabolic processes.

(1) Ischemic infarction[10]
   (a) Thromboembolic disease
   (b) Idiopathic
(2) Atherosclerotic reduction in carotid artery lumen (reported in Amazon parrots[10,11] and an umbrella cockatoo[12])
  b. Clinical signs: An "aura-like" behavior while going into a semiconscious state, ataxia, sudden onset of blindness, paresis, seizures
  c. No antemortem diagnostic techniques have been described.
  d. Supportive treatment
4. Nutritional deficiencies
  a. Seizures may occur in a bird receiving inadequate levels of vitamins and nutrients to support essential oxidative energy pathways, and thus, cellular life and processes. It may be difficult to distinguish one particular missing element.
    (1) Calcium, phosphorus, vitamin D_3 imbalance (see hypocalcemia discussion)
    (2) Thiamine (vitamin B_1) deficiency[13] (reported in piscivorous birds secondary to degradation of frozen fish by thiaminase and raptors secondary to pure red meat diet)
    (3) Pyridoxine (vitamin B_6) deficiency[10]
  b. Clinical signs: Ataxia, dysmetria, gait abnormalities (i.e., pyridoxine deficiency produces a jerky walk that progresses to rapid, clonic-tonic head and leg movements), paresis/paralysis, seizures
  c. Diagnosis
    (1) Feedstuff analysis
    (2) Serum and tissue analysis
  d. Treatment
    (1) Ensure high quality diet
    (2) Provide supplemental vitamins and minerals
5. Hepatic encephalopathy
  a. Cirrhotic liver is too diseased to clear toxic substances produced in metabolic and gastrointestinal processes, which can damage the cerebral cortex.[1,14]
    (1) Hepatic lipidosis
    (2) Mycotoxicosis[15]
    (3) Hemochromatosis
  b. Clinical signs: Depression, behavioral disorders, ataxia, seizures
  c. Diagnosis
    (1) Alterations in liver enzyme levels signal hepatic cell damage.
    (2) Metabolite measurement indicates status of hepatic function.
      (a) Serum bile acids
        (i) Evaluates enterohepatic circulation and the extraction of bile acids following their emulsifying role in fat digestion
        (ii) Signals hepatopathy whenever spontaneous and/or postprandial concentrations elevated
      (b) Ammonia tolerance test

(i) Evaluates catabolism of protein or exogenous ammonia
(ii) Signals hepatopathy whenever postprandial plasma concentrations elevated
(iii) Reported use in *Toco toucan* with hemochromatosis[16]
- (c) Consider parallel testing in asymptomatic bird of same species to enhance diagnostic value.
- (d) Evaluation of bile acids appears to be a safer and more reliable test of hepatic function.
- d. Treatment
  - (1) Direct therapy toward underlying cause of hepatic failure.
  - (2) Lactulose syrup may indirectly lower blood ammonia concentrations through its usage by intestinal bacteria.[17]
    - (a) Hydrolysis creates acid environment that promotes nondiffusible form of ammonium.
    - (b) Hydrolysis provides carbohydrate source that generates lower amounts of ammonia waste.
    - (c) May promote mild osmotic diarrhea.
  - (3) Neomycin sulfate or metronidazole may indirectly decrease the formation of ammonia by reducing the quantity of bacteria that are potent generators of ammonia in the colon.[17]
  - (4) Low-protein, high-carbohydrate, high-quality protein diet with a vitamin supplement may provide symptomatic relief while the underlying hepatopathy is corrected.[16]
6. Uremic encephalopathy
   a. Severely diseased kidney is unable to excrete nitrogenous byproducts (cause direct damage to neurons) or to maintain electrolyte balance.[18, 19]
   b. Clinical signs: Stupor, seizures, emesis
   c. Diagnosis
      (1) Hyperuricemia signals inadequate renal tubular secretion.
      (2) Glucosuria suggests damaged renal absorption.
   d. Treatment
      (1) Fluid diuresis
      (2) Supportive care
7. Acid-base and osmolality disorders
8. Hyperthermia

B. Intoxications
1. Lead[20,21]
   a. Peripheral weakness results from demyelination of nerves and block of presynaptic transmission by competitive inhibition of calcium. Seizures result from diffuse perivascular edema, accumulation of cerebrospinal fluid, and necrosis of nerve cells.[21] All body systems may be affected, with particular concern to hematopoietic, digestive, and nervous systems.
   b. Clinical signs: Lethargy, depression, weakness, circling, ataxia, paresis, paralysis, blindness, seizures, diarrhea, polyuria (hemoglobinuria in some species)

c. Diagnosis
   (1) History of exposure. Common sources (lead-based paint, lead shot, solder, fishing sinkers, linoleum, putty, ceramics and automobile exhaust) shown to contribute to cumulative lead concentrations in body tissues[22]
   (2) Hypochromic regenerative anemia
   (3) Whole blood analysis
   (4) Radiographic suggestion of metal density
d. Treatment
   (1) Prevent further absorption of toxin.
      (a) Remove any lead object visualized in digestive tract.
      (b) Enhance passage of small particles by gavage administration of warm liquified peanut butter/mineral oil (2:1) mixture.
   (2) Chelation therapy
      (a) Edetate calcium disodium injection (Calcium versonate: 3M Pharmaceuticals; 25 to 40 mg/kg IM BID $\times$ 5 days, rest $\times$ 3 to 5 days, repeat prn)
      (b) Penicillamine (Cuprimine: Merck & Co.; 55 mg/kg PO BID)
      (c) Dimercaptosuccinic acid (DMSA, Aldrich; 25-35 mg/kg PO BID $\times$ 5 days, rest $\times$ 2 days, repeat cycle 3 to 5 weeks)
2. Organophosphates and carbamates
   a. Two types of neurologic syndromes and corresponding clinical signs exist:
      (1) Acute toxicity wherein seizures result from excessive stimulation of acetylcholine (parasympathetic) receptors resulting from toxin binding and subsequent inactivation of acetylcholinesterase[23]
         (a) Clinical signs may include tremors, respiratory distress, progressive bradycardia, ptyalism, diarrhea, and seizures.
      (2) Delayed toxicity wherein organophosphate ester induces primary axonal degeneration of the central and peripheral nervous systems, with secondary myelin degeneration[24]
         (a) Clinical signs may include ataxia, weakness, decreased proprioception, and paralysis.
      (3) Binding of acetylchloinesterase considered irreversible with organophosphate intoxication and slowly reversible resulting from spontaneous decay with carbamate intoxication.
      (4) Birds considered 10 to 20 times more susceptible to toxicity than mammals.[24]
   b. Diagnosis
      (1) History of exposure
      (2) Acetylcholinesterase concentration[20,25]
         (a) Plasma, serum, and/or brain tissue level
         (b) Decreased concentration by 50% of normal level considered diagnostic
         (c) Parallel testing in bird of same species
      (3) Identification of toxic substance

c. Treatment
  (1) Atropine (0.2 to 0.4 mg/kg IV, IM, SQ q 8-24 hr or prn) blocks the muscarinic signs (miosis, salivation, bradycardia, emesis, diarrhea).
  (2) Diphenhydramine (1 mg/kg IM) blocks the nicotinic signs (fasciculations, weakness).
  (3) Pralidoxime chloride, 2-PAM (Protopam: Wyeth-Ayerst; 10 to 100 mg/kg IM) may be used only within the first 1 to 2 days after exposure for organophosphate intoxication.
3. Other chemical intoxicants for which no antidote is available
  a. Excess amounts of some chemicals may cause excessive stimulation of the CNS and may produce dose-dependent seizure activity.
    (1) Chlorinated hydrocarbons
    (2) Strychnine
    (3) Metaldehyde
    (4) Therapeutic agent overdosage
      (a) Dimetridazole (no longer available in United States)
      (b) Metronidazole
    (5) Domoic acid neurotoxin (produced by marine diatoms)[26]
    (6) Chocolate (theobromine and caffeine constituents)

C. Inflammatory/Infectious Diseases (Encephalitis, Meningoencephalitis, Encephalomyelitis)

1. Any infectious agent or disease process may cause inflammation to the cerebrum by primary destruction or invasion, ischemia or vasculitis, edema or swelling, and direct toxic effects of inflammatory cells on neurons, axons, or myelin. A subacute progressive disease process may produce systemic signs of illness and may be accompanied by various neurologic signs, including seizures.
   a. Bacterial
     (1) Acute septicemia
     (2) Chlamydiosis[27]
   b. Fungal
     (1) Granuloma formation in cerebrum
     (2) Aspergillosis
   c. Viral
     (1) Paramyxovirus (velogenic viscerotropic Newcastle's disease)[28]
       (a) Clinical signs are variable depending upon the virulence of the strain and the species of bird affected.
       (b) Respiratory, digestive, and neurologic signs
       (c) Neurologic signs include depression or hyperexcitability, vestibular signs, muscle tremors, paresis or paralysis, and seizures.
     (2) Herpesvirus (duck viral enteritis)[17,28]
       (a) Highest mortality in naive waterfowl exposed to asymptomatic carriers
       (b) Clinical signs include lethargy, ataxia, seizures, photophobia, serosanguinous nasal discharge, and hemorrhagic diarrhea.

(3) Picornavirus (duck viral hepatitis)[17,28]
   (a) Highest mortality in ducklings
   (b) Clinical signs include lethargy, opisthotonos, seizures, and death.
(4) Equine encephalomyelitis viruses[10,28]
   (a) Mosquito vector
   (b) Often innocuous to indigenous wild bird population
   (c) Clinical signs include torticollis, opisthotonos, tremors, staggering, and seizures
  d. Parasitic
    (1) Aberrant larval migration
    (2) Toxoplasmosis
 2. Diagnosis
  a. Identification of organism through culture or histology
  b. Serologic evidence of antigen itself or of antibody response
  c. Cerebral fluid analysis
 3. Treatment
  a. Antibiotics, antifungals, preexposure vaccination, antiparasitics
  b. General supportive, antiinflammatory, and anticonvulsant therapy

D. Trauma
 1. Concussive force to cerebrum may cause direct neuronal injury or residual damage resulting from scar formation.
 2. Clinical signs: Altered consciousness and seizures
 3. Diagnosis
  a. History of trauma
  b. Radiographic visualization of skull fractures suspicious
  c. Evidence of hemorrhage at site of injury
 4. Treatment for acute head trauma
  a. Cage rest in cool environment
  b. Administer parenteral rapid-acting corticosteriods, such as prednisolone sodium succinate (30 mg/kg initial dose; 15 mg/kg IV/IO/IM q 8 hr × 24-48 hr). Consider prophylactic use of gastrointestinal protectant agents during steroid treatment period.
  c. Elevate head 30 degrees in recumbent patients.
  d. Hyperventilate with oxygen delivery in intubated patients.
  e. Consider diuretic (mannitol: 250-1000 mg/kg IV slowly over 30 minutes +/− furosemide: 0.15-2 mg/kg SQ q 4-12 hr) or investigational agents (DMSO=dimethyl sulfoxide as a 10% solution: 100-1000 mg/kg IV slowly over 45 minutes; caution low safety margin) if deterioration in mentation and cranial nerve function.

E. Neoplasia
 1. Seizures result from destruction or compromise (compression, distortion, or insufficient blood supply) of vital neuronal elements.
  a. Both primary and secondary (metastatic) lesions have been identified in rare cases.[10,18,29]
  b. Pituitary adenoma in young male budgerigars produces a classic syndrome of neurologic signs: somnolence and seizures that are

accompanied by incoordinated wing-flapping and clonic leg-twitching and followed by a period of unconsciousness.[30]
   c. Enlarged thyroid glands may indirectly cause cerebral congestion secondary to occlusion of venous drainage (jugular stasis - reported in budgerigars[10])
 2. Clinical signs vary dependent upon tumor location and size, and may include inability to perch, incoordination, tremors, and seizures.
 3. Diagnosis
   a. Imaging techniques
 4. Treatment
   a. Supportive care and anticonvulsant therapy
   b. Steroid therapy may provide temporary relief
   c. Surgical excision ideal but may not be practical

F. **Anatomic, Congenital, and Familial Disorders**
 1. Hydrocephalus
   a. Seizures may occur in a bird with malformation of the skull, expansion of cerebral ventricles, and/or increased amount of cerebrospinal fluid that compresses the cerebrum (reported in a cockatoo,[18] African grey parrots,[29,31] and a yellow-collared macaw[29]).
   b. Clinical signs vary: dementia, irritability, ataxia, seizures, blindness
   c. Diagnosis
     (1) Antemortem diagnostics include ultrasonography, CT, and MRI.[1]
     (2) Typically diagnose from postmortem examination
   d. Treatment
     (1) Symptomatic seizure control
     (2) Prednisone (0.5 mg/kg PO q 12-48 hr) may encourage decreased production of cerebrospinal fluid,[32] but must be used cautiously in long-term situations for avian patients.
 2. Lafora body neuropathy
   a. A suspected degenerative defect in intracellular metabolism causes accumulation of indigestible material (cytoplasmic inclusions) within neurons and a subsequent disruption in cellular function (reported in cockatiels[17,33]).
   b. Clinical signs: Myoclonus, seizures
   c. Postmortem diagnosis
 3. Primary epilepsy
   a. Primary epilepsy is characterized by recurrent seizures arising from a nonprogressive, nonpathologic intracranial process.
     (1) Inherited biochemical defect (reported in chickens,[34] red-lored Amazon parrots,[35] and greater Indian Hill mynahs[17])
     (2) Acquired cerebral insult and residual brain damage
   b. Clinical signs: Intermittent and varying seizure activity
   c. Diagnosis
     (1) Normal diagnostic tests yield diagnosis by exclusion
     (2) EEG
       (a) Paroxysmal bursts of diffuse electrical activity during seizure

(b) Recognize limitations: EEG is difficult in any species; limited knowledge of normal and limited availability of equipment and trained personnel to make and evaluate the EEG; and the EEG may be normal between seizures[9]
d. Treatment through anticonvulsant therapy (see Treatment p 172)

REFERENCES

1. Chrisman CL: *Problems in small animal neurology,* ed 2, Philadelphia, 1991, Lea & Febiger, pp 177-205.
2. Schunk KL: Seizure disorders. In Morgan RV, editor: *Handbook of small animal practice,* New York, 1992, Churchill Livingstone, pp 243-250.
3. Clippinger TL, Bennett RA, Platt SR: The avian neurologic examination and ancillary neurodiagnostic techniques, *J Avian Med Surg* 10:221-247, 1996.
4. Harrison GJ: Long term primidone use in an Amazon, *Assoc Avian Vet Newsletter* 4:89, 1982.
5. Walsh MT: Seizuring in pet birds, *Proceedings of the Association of Avian Veterinarians,* 1985, pp 121-128.
6. Ryan T: Calcium in the avian species. *Comp Anim Pract* 19:38-40, 1989.
7. Hochleithner M: Convulsions in African grey parrots, *Proceedings of the Association of Avian Veterinarians,* 1989, pp 78-81.
8. McDonald LJ: Hypocalcemic seizures in African grey parrots, *Can Vet J* 29:928-930, 1988.
9. Rosskopf WJ, Woerpel RW, Lane R: The hypocalcaemia syndrome in African greys: an updated clinical viewpoint with current recommendations for treatment, *Proceedings of the Association of Avian Veterinarians,* 1985, pp 129-131.
10. Hasholt J, Petrak ML: Diseases of the nervous system. In Petrak ML, editor: *Diseases of cage and aviary birds,* Philadelphia, 1982, Lea & Febiger, pp 468-477.
11. Clubb SL, Karpinski L: Aging in macaws, *Proceedings of the Association of Avian Veterinarians,* 1992, pp 83-86.
12. Joyner KL, Kock N, Styles D: Encephalitis, proventricular and ventricular myositis and myenteric ganglioneuritis in an umbrella cockatoo, *Avian Dis* 33:379-381, 1981.
13. Lowenstine LJ: Avian nutrition. In Fowler M, editor: *Zoo and wild animal medicine,* ed 2, Philadelphia, 1986, WB Saunders, pp 201-212.
14. Center SA: Pathophysiology and laboratory diagnosis of liver disease. In Ettinger SJ, editor: *Textbook of veterinary internal medicine,* ed 2, Philadelphia, 1989, WB Saunders, pp 1421-1478.
15. Wyatt RD, Colwell WM, Hamilton PB, Burmeister HR: Neural disturbances in chickens caused by dietary T-2 toxin, *Appl Microbiol* 26:757-761, 1973.
16. Spalding MG et al: Hepatic encephalopathy associated with hemochromatosis in a toco toucan, *J Am Vet Med Assoc* 189:1122-1124, 1986.
17. Bennett RA: Neurology. In Ritchie BR, Harrison GJ, Harrison LR, editors: *Avian medicine: principles and application,* Lake Worth, Fla, 1994, Wingers, pp 723-747.
18. Jones MP, Orosz SE: Overview of avian neurology and neurological diseases, *Semin Avian Exotic Pet Med* 5:150-164, 1996.
19. Lumeij JT: Nephrology. In Ritchie BW, Harrison GJ, Harrison LR, editors: *Avian medicine: principles and application,* Lake Worth, Fla, 1994, Wingers, pp 223-245.
20. LaBonde J: Avian toxicology, *Vet Clin N Am Sm Anim Pract* 21:1329-1342, 1991.
21. McDonald SE: Lead poisoning in psittacine birds. In Kirk RW, editor: *Current veterinary therapy IX,* Philadelphia, 1986, WB Saunders, pp 713-718.

22. Lyman R: Neurologic disorders. In Harrison GJ, Harrison LR, editors: *Clinical avian medicine and surgery,* Philadelphia, 1986, WB Saunders, pp 486-490.
23. LaBonde J: Two clinical cases of exposure to household use of organophosphate and carbamate insecticides, *Proceedings of the Association of Avian Veterinarians,* 1992, pp 113-118.
24. Mohan R: Dursban toxicosis in a pet bird breeding operation, *Proceedings of the Association of Avian Veterinarians,* 1990, pp 112-114.
25. Porter SJ: Organophosphate/carbamate poisoning in birds of prey, *Proc Am Assoc Zoo Vet,* 1992, pp 176-177.
26. Work TM et al: Epizootiology of domoic acid poisoning in brown pelicans (*Pelecanus occidentalis*) and Brandt's cormorants (*Phalacrocorax penicillatus*) in California, *J Zoo Wildlife Med* 24:54-62, 1993.
27. Gerlach H: Chlamydia. In Ritchie BR, Harrison GJ, Harrison LR, editors: *Avian medicine: principles and application,* Lake Worth, Fla, 1994, Wingers, pp 985-1006.
28. Ritchie BW: *Avian viruses: function and control,* Lake Worth, Fla, 1995, Wingers.
29. Shivaprasad HL: Diseases of the nervous system in pet birds. A review and report of diseases rarely documented, *Proceedings of the Association of Avian Veterinarians,* 1993, pp 213-222.
30. Mathey WJ et al: Brain tumors in young male budgerigar breeders, *Proc 33rd West Poult Dis Conf,* 1984, pp 103-105.
31. Wack RF, Lindstrom JG, Graham DL: Internal hydrocephalus in an African Grey parrot (*Psittacus erithacus timneh*), *J Assoc Avian Vet* 3:94-96, 1989.
32. Munana KR: Disorders of the brain. In Morgan RV, editor: *Handbook of small animal practice,* Philadelphia, 1997, WB Saunders, pp 230-251.
33. Britt JO, Paster MB, Gonzales C: Lafora body neuropathy in a cockatiel, *Comp Anim Pract* 19:31-33, 1989.
34. George DH, Munoz DB, McConnell M, Crawford RD: Megalencephaly in the epileptic chicken: a morphometric study of the adult brain, *Neuroscience* 39:471-477, 1990.
35. Rosskopf WJ, Woerpel RW, Lane R: Epilepsy in red-lored Amazons (*Amazona autumnalis*), *Proceedings of the Association of Avian Veterinarians,* 1985, pp 141-145.

11

Straining and Reproductive Disorders

Michael D. Doolen

I. Tenesmus

A. Egg Binding
  1. Signalment
     a. Single female pets that are strongly bonded to their owner or to inanimate objects, such as mirrors.
        (1) Budgerigars, cockatiels, lovebirds, and finches present most commonly, larger species present occasionally.
     b. Females that are allowed to produce multiple clutches without rest periods between clutches, especially if the eggs or immature young are pulled from the nest to increase production.
     c. Immature females that are breeding for the first time, or older birds.
     d. Females with a history of laying malformed or soft-shelled eggs.
     e. Females with a history of previous reproductive problems.
  2. Diagnosis is based on confirmation of the presence of an egg by palpation or radiographic films, along with historical and clinical evidence of egg binding.
     a. History
        (1) Recent egg laying.
        (2) Behavioral changes (e.g., nest-building activity, masturbation, regurgitant feeding, increased territorial aggression, increased intimate interaction with owner or mate).
        (3) Fewer droppings and increased volume of droppings, especially in the morning.
        (4) Marginal nutrition
           (a) Low vitamin A level results in increased thickening of mucosal epithelium of oviduct and thickened mucus.
           (b) Low vitamin E and selenium levels may result in decreased muscle tone.
           (c) Hypocalcemia causes decreased muscle tone.
           (d) Overall marginal nutrition

b. Clinical findings
   (1) Distended abdomen (abdomen normally is concave); an egg may be evident on abdominal palpation unless it is not well calcified.
   (2) Wide-based stance.
   (3) Droppings stuck to feathers around the vent because the bird is unable to lift its tail high enough for droppings to clear.
   (4) In advanced stages: Depression, fluffed feathers, weakness, dehydration, dyspnea, and lameness caused by nerve compression
   (5) Prolapsed oviduct, with or without egg
c. Radiographic films: One or more calcified or partly calcified eggs in the abdominal cavity. Estrogen-induced hyperostosis often is seen as calcification of the medullary cavities of the long bones (femur, tibiotarsus, and ulna).[1] The egg may be oversized compared to the distance between the distal ends of the pubic bones.
d. Clinical pathologic studies: Normally the serum calcium is elevated during egg laying.[2] In advanced stages of egg binding, the serum calcium may be normal or decreased. Protein and cholesterol may be elevated. Other abnormalities usually are attributed to the bird's health status and the amount of stress on the system at the time.

3. Treatment
   a. Stage 1: No depression, still perching and eating normally, no prolapse, straining less than 24 hours.
      (1) Environment is warmed to 29° to 35° C (85° to 95° F).
      (2) The vent and distal oviduct are lubricated with K-Y jelly (petroleum jelly is not used).
      (3) A calcium injection is administered (100 mg/kg given intramuscularly).
      (4) An oxytocin injection is given, or topical PGE_2 (Prepidil gel) is used.
         (a) Oxytocin: 5 to 10 IU/kg given intramuscularly, but only if no physical obstruction is suspected, because oviductal rupture or reverse peristalsis may occur.
         (b) Prepidil: 0.01 to 0.03 ml applied topically to the opening of the shell gland within the cloaca.
   b. Stage 2: Depression, not perching, other severe signs, straining longer than 24 hours, especially if conservative measures have not resulted in oviposition within 24 hours.
      (1) Manual removal of the egg: I prefer to induce anesthesia with isoflurane in most cases before attempting to remove the egg, especially if the bird is fractious or dyspneic. The cloaca is dilated with lubricated, cotton-tipped applicators. Using gentle pressure with the thumb and index finger, the egg is manipulated caudally toward the cloaca. The pressure should be directed from the sides rather than ventrodorsally. Often the vaginal opening into the cloaca is constricted. Gentle dilation of the opening with a cotton-tipped applicator often allows direct visualization of the egg, which is imploded by suction. The shell fragments may be gently removed with forceps or may be left in the shell gland. They usually pass within 24 hours. If the

oviduct has ruptured or if multiple eggs are present, an abdominal exploratory and salpingectomy may be required.
   (2) Percutaneous ovocentesis may be required if the egg cannot be visualized. The needle is directed through the abdominal wall from a lateral approach so that the egg is not pushed dorsally, which may put a dangerous amount of pressure on the kidneys and associated vessels.
   (3) Prolapse reduction is required if the shell gland, with or without the egg, has prolapsed out of the cloaca. If the tissue is not yet necrotic from vascular strangulation, the egg is removed through the vaginal opening, if possible, or through an incision made in a less vascular area of the oviduct. The incision is closed, and the oviduct is replaced within the abdomen. If the vaginal opening can be visualized, the everted oviduct should be reduced to its natural position with a lubricated, cotton-tipped applicator; if the oviduct is simply pushed into the cloaca, it may remain everted in the cloaca.
   (4) Medical treatment as needed
      (a) Fluids (see Chapter 26).
      (b) Dextrose (see Chapter 26).
      (c) Nutritional support: I prefer a rice-based, commercially available formula, Emeraid II, made by the Lafeber Co, Cornell, Ill.
      (d) Antibiotics (in cases of prolapse): I prefer enrofloxacin, 15 mg/kg, given intramuscularly.

B. Cloacal Papilloma
1. The condition is seen primarily in New World psittacines (Amazons, hawk-headed parrots, macaws, and conures); most commonly in imported birds.
2. The etiology is unknown; a viral cause is likely but has not yet been documented.
3. Clinical signs and diagnosis
   a. Straining, malodorous stools, soiled vent, and partial cloacal prolapse with inflamed tissue protruding in more severe cases.
   b. Routine blood test results usually are unremarkable. The white cell count may be elevated because of a secondary infection. Bacterial culture and sensitivity testing is indicated to document these infections.
   c. All physical examinations should include everting the cloaca with a cotton-tipped applicator to examine the mucocutaneous junction, the most common site of papillomatous lesions. Lesions may vary from diffuse or multifocal inflammation to discrete masses with a "raspberry" appearance. These masses may be broad based or pedunculated. A small amount of 5% acetic acid applied to the tissue will turn the lesions white, but a definitive diagnosis is made by histopathologic studies.
   d. In some cases lesions occur in the oral cavity, most commonly at the edge of the opening of the glottis or the choana, causing

upper respiratory signs. Lesions may occur anywhere in the gastrointestinal tract. Internal papillomatosis has been implicated in an increased incidence of bile duct and pancreatic duct carcinomas in affected birds.[3]

4. Treatment
   a. Any secondary bacterial and fungal infections are treated with an appropriate antibiotic or antifungal agent.
   b. Large masses may be debulked surgically, but electrocautery or chemical cautery (silver nitrate), or both, may be the most effective treatment. A hemoclip may be put across the base of some of the lesions before they are excised. It usually is not possible to eliminate the disease, which may recur in times of stress. I have had limited success with a cloacotomy procedure that eliminates diseased tissue more extensively, but strictures of the vent are a common side effect.
   c. It is important to take measures to improve the bird's immune status through nutritional improvement and stress reduction.
   d. Immune therapy has shown promise in preventing recurrence. I use alfa-interferon (50,000 IU/kg given intramuscularly once daily for 5 days, then once weekly for 2 months, then once monthly thereafter), and I have seen a dramatic decrease in the incidence of recurrence with this regimen.

C. Enteritis, Colitis, Cloacitis
1. Possible etiologies (see Chapter 4)
   a. Bacterial
   b. Parasitic
   c. Viral
   d. Fungal

D. Neoplastic Disease
1. Neoplasms may cause straining directly by affecting the lower gastrointestinal tract or oviduct, or indirectly by displacing and crowding the gastrointestinal or reproductive tract.
2. Reproductive tract
   a. Cystic hyperplasia
   b. Carcinoma, adenocarcinoma
3. Other neoplastic disease that may cause straining
   a. Renal adenocarcinoma
   b. Intestinal or cloacal carcinoma and adenocarcinoma
   c. Intestinal leiomyosarcoma

II. Infertility

Infertility may be caused by disease that directly involves the reproductive tracts and by generalized disease conditions, poor nutrition, lack of adequate nesting territory, or behavioral problems[4] (see Chapter 16).

A. Female Infertility
  1. Egg peritonitis
  2. Generalized disease state
  3. Ectopic eggs
  4. Chronic egg laying
  5. Abnormal eggs
  6. Salpingitis
  7. Immaturity
  8. Mate incompatibility

B. Male Infertility
  1. Orchitis
  2. Generalized disease state
  3. Prolapse of phallus in waterfowl
  4. Seasonal changes
  5. Low sperm count or poor quality semen
  6. Immaturity
  7. Mate incompatibility

III. Behavioral Problems Caused by Reproductive Hormonal Pressure

See Chapter 8.

A. Effect of Reproductive Hormones
  1. When reproductive hormones are stimulated, several types of abnormal behavior may be seen. Hormonal and other pharmaceutical treatments usually are only palliative. Most cases require consultation about behavioral modification techniques to achieve long-term improvement.
      a. *Masturbation* is more common in males, but females may also display this behavior. It sometimes results in trauma to the skin and vent. Administration of human chorionic gonadotropin (HCG), 1000 IU/kg given intramuscularly every third day for two to four administrations, often is an effective temporary measure.
      b. *Excessive "feeding" regurgitation* sometimes is seen in males of the smaller species that chronically regurgitate to feed their "mate" (e.g., the owner, a mirror, a toy). In some cases these birds actually lose weight and begin to suffer from malnutrition. These birds often respond to HCG administration (as described above).
      c. *Displacement behaviors* may be seen in birds that strongly desire to reproduce. These birds usually are bonded to the owner and are given all the things at a young age that they would spend years achieving in the wild: a secure flock environment (almost any living thing within the household is considered a flock member), adequate dominance within the flock, nesting territory, good food, and a bonded mate. Unfortunately, the "mate" is usually a human and is not anatomically correct, so the behavior ritual that

should end with babies to raise is interrupted. The bird sometimes then begins to display one or more displacement behaviors. It is interesting to note that the small species mentioned above that have a tendency for chronic egg laying rarely manifest behavioral feather picking. On the other hand, larger species rarely are chronic egg layers, but these species do show behavioral picking.
(1) Three common forms of displacement behavior:
  (a) *Feather picking* is a complex behavior problem with many variable factors interacting in most cases. Each case must be evaluated medically and from a behavioral and environmental perspective. Reproductive hormonal pressure is at least a component in the syndrome in many cases. Many cockatoos and African greys stop picking after being given HCG (1000 IU/kg given intramuscularly every 1 to 4 weeks). This protocol helps in the evaluation of the degree of hormonal stress present versus the degree of stereotypical behavior (habit). If the bird does not stop picking after receiving hormone therapy, the picking is caused by something else or, most commonly, has become a habit.
  (b) *Aggression* often is seen in young male cockatoos and Amazons (2 to 8 years old) and, to a lesser extent, in most other psittacine species that have been allowed to become too dominant and to control too much territory. Although aggression is usually a problem in males, females (especially Amazons, eclectus parrots, and Asian parakeets) also may develop this behavior. These birds sometimes respond to HCG therapy, but behavior modification techniques designed to reduce the dominance and territoriality are recommended.
  (c) *Screaming* may be a response stimulated by a number of factors. Most psittacines bond very closely with their mates, and the two spend all their time together. In the case of a bird that is bonded to a human, when the "mate" is within earshot or eyesight but is not paying attention to the bird, it is only natural for the bird to call out for its mate. This behavior does not seem to respond well to HCG therapy, and behavior modification usually is required, as well as acceptance of what often is natural behavior.

REFERENCES

1. Johnson AL: Reproduction in the female. In Sturkie PD, editor: *Avian physiology*, New York, 1986, Springer-Verlag, pp 403-431.
2. Hochleithner M: Biochemistries. In Ritchie BW, Harrison GJ, Harrison LR, editors: *Avian medicine: principles and applications*, Lake Worth, Fla, 1994, Wingers, pp 223-245.
3. Graham DL: Papillomatous disease: a pathologist's view of cloacal papillomas and then some! *Proceedings of the Association of Avian Veterinarians* pp 141-143, 1991.
4. Speer BL: 1997. Disease of the urogenital system. In Altman RB et al, editors: *Avian medicine and surgery*, Philadelphia, 1997, WB Saunders, pp 614-644.

12

Embryologic Considerations

Glenn H. Olsen

I. Anatomy of the Female Reproductive Tract

A. Yolk Forms in the Ovary
1. Protoplasm with nucleus (called a blastoderm after fertilization or a blastodisc if unfertilized) is found on the outer margin of the yolk.

B. Yolk Passes into the Oviduct (during Ovulation)
1. *Oviduct:* The oviduct is composed of five sections: the infundibulum, magnum, isthmus, uterus (shell gland), and vagina. In most bird species only the left ovary and oviduct are developed. Falconiformes have both ovaries, but usually only the left one is functional.[1] Paired functional ovaries are found only in the brown kiwi.[1]
2. *Ovarian pocket:* The left abdominal air sac tightly encloses the ovary except caudally at the opening to the infundibulum.
3. *Internal laying:* If the infundibulum fails to envelop the yolk, the yolk will enter the coelomic cavity. It can be absorbed within 24 hours, can become walled-off necrotic tissue (Fig. 12-1), or can break apart, causing a generalized peritonitis (egg peritonitis). Internal laying most often occurs in association with the first or last eggs laid in the clutch or season.[1]
4. *Fertilization:* Fertilization occurs in the upper infundibulum (funnel portion). First cleavage occurs 4 to 5 hours after ovulation.
5. Albumin and the *inner* and *outer shell membranes* are laid down in the magnum and isthmus, respectively.
6. *Vaginal sphincter:* The vaginal sphincter marks the division between the vagina and the uterus.
7. *Uterus (shell gland):* The uterus, which is S shaped in galliformes, contains smooth muscle and connective tissue. The egg remains in the uterus for most of its transit time down the oviduct. Most of the calcification process occurs in the uterus.
8. *Plumping:* During the first one fourth of the egg's time in the uterus, fluid is added to the albumin in a process called plumping.

Fig. 12-1 Walled-off, necrotic yolk material removed from the coelomic cavity of a black duck. The infundibulum had failed to envelope the developing egg, which was released into the coelomic cavity.

9. *Calcification:* During the last three fourths of the egg's time in the uterus, calcification of the egg occurs rapidly, forming the hard shell. In chickens, every 15 minutes the uterus withdraws from the blood flow the equivalent of the total normal circulating calcium; thus the process of calcification places an extreme demand for calcium mobilization on the laying female.

C. Obstruction or Egg Binding

Obstruction of the oviduct is a common problem in many species. Causes include necrotic egg material in the oviduct, a broken egg, failure of the uterus to lay down shell material (resulting in a soft-shelled egg), infection in the oviduct, hypocalcemia, selenium or vitamin E deficiency, senility (older birds have more obstructions with no known explanation), excessive egg production, obesity, lack of exercise, heredity, stress, or cystic right oviduct.[2,3]

1. Occurrence: Oviductal obstruction occurs more often in cockatiels, lovebirds, canaries, finches, and quail (Table 12-1) but may occur in any species.
2. Clinical signs: In small species, the bird sometimes is found dead. Other signs are depression, reluctance to fly or perch, staying on the bottom of the cage, drooped wings (canaries), straining, and tail wagging.
3. Diagnosis: History, clinical signs, palpation, and possibly radiography (in large or obese birds) are used in diagnosis. The presence of an egg alone, without other clinical signs, is not diagnostic of egg binding (see Chapter 11).
4. Treatment
    a. Early egg binding: Early egg binding may respond to manual attempts to deliver the egg, in which the egg is gently pushed ven-

Table 12-1
Common egg-laying problems in birds seen in veterinary practice

Condition	African grey	Amazon	Cockatoo	Cockatiel	Budgerigar	Lovebird	Macaw	Canary	Finch	Waterfowl	Quail	Crane	Raptor
Metritis				***	***					***			
Yolk peritonitis				***	***	***	*	*	***	***	***		
Oviductal obstruction			*	***	***			***	***	*	***	*	
Cloacal prolapse			***								**	*	*
Cloacal papilloma		***	***				***						
Lipomatosis				***	***								
Abdominal hernia				***	***			*	**	**			
Neoplasm					***								
Prolonged egg laying				***	***	***		***	***	***			
Ectopic eggs					*					*			
Oviductal prolapse	*			*	*					*	*		
Diabetes mellitus following yolk peritonitis				***				*					

Data from Rosskopf WJ, Woerpel RW: Pet avian obstetrics, First International Conference on Zoological and Avian Medicine, 1987, pp 213-231.
*Rare;   **occasional;   ***common.

trally and caudally. Lubrication of the vent with K-Y jelly may help. If the hen is not in poor condition, additional measures can be taken: an injection of calcium (Calphosan, 0.5 to 1.5 ml/kg given intramuscularly), administration of fluids, and application of PGE_2 (Prepidil Gel) topically to the opening of the shell gland (uterus) within the cloaca, followed by placement of the bird in a warm (29.4° to 35° C [85° to 95° F]), dark, humid area may result in delivery of the egg within 30 minutes.[4] Prepidil is used only if the oviduct is considered intact and no adhesions are present (nonobstructive egg binding). If the egg is not delivered and the bird is not showing distress, the process may be repeated.

b. Complicated egg binding: If the bird is in distress, immediate action should be taken to give supportive care, after which manipulations can proceed to deliver the egg. Isoflurane anesthesia often is used. The vent can be cleaned, lubricated with K-Y jelly, and dilated with a forceps or small speculum. If this fails to deliver the egg, an 18-gauge needle is introduced into the egg, and the contents are removed (ovocentesis), which causes the egg to collapse inward on itself. Usually the remains of the egg are delivered in 1 to 2 days. If the egg is high in the oviduct, the egg is manipulated against the abdominal wall, the 18-gauge needle is inserted through the abdominal wall and oviduct into the egg, the contents are aspirated, and the egg is allowed to collapse in on itself for delivery in 1 to 2 days.[4] The bird is given supportive care, including calcium injections and Prepidil, as outlined above. A laparotomy may be needed to remove egg material, especially if adhesions or a ruptured oviduct is present, or a hysterectomy may be required (see Chapter 24).[4]

II. Early Embryonic Development

A. Rapid Cell Division
1. Division occurs as an egg passes down the oviduct.
2. *Blastula:* At egg laying, the embryo is as large as 60,000 cells in a circular disc called a blastula, or *blastoderm.* Some authors refer to the unfertilized cytoplasmic material on the yolk as the *blastodisc.*[1]
3. *Gastrulation:* Gastrulation is the process by which the blastula folds in on itself to become the initial embryonic gut.

B. Blastula
1. At egg laying, the blastula may be visible to the observer (if the egg is broken open).
2. Special stains can be used to identify the presence of a blastula; this is important to determine whether an egg is fertile.[5]

C. Egg Membranes
The amnion, chorion, and allantois (Fig. 12-2) are living tissue membranes that develop from the embryo.[6]

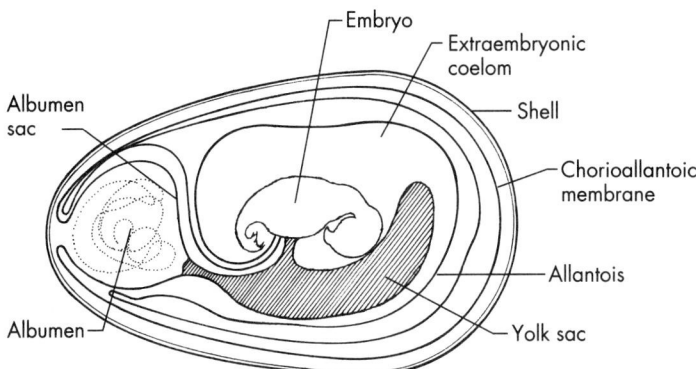

Fig. 12-2 A drawing of the egg membranes.

1. *Amnion* and *chorion:* The amnion and chorion develop as folds of the body wall and encircle the embryo. The amnion forms a fluid-filled sac in which the embryo develops and moves.
2. *Allantois:* The allantois, an extension of the embryonic digestive tract, fuses with the chorion to form the chorioallantoic membrane. This membrane grows to lie under the shell and functions in respiratory exchange through the shell and in calcium mobilization.
3. *Vitteline membrane:* Another pouchlike outgrowth surrounds the yolk material to form the yolk sac. During the last third of incubation and in the first few days after hatching, this sac produces enzymes to digest the yolk material, and blood vessels in the sac function in the absorption and transport of nutrients to the developing embryo. The remnant of the yolk sac forms Meckel's diverticulum (see Chapter 5).
4. The critical development of the egg membranes occurs in the first half of incubation. During this time the embryo is especially susceptible to trauma that can disrupt normal development.

III. Causes of Infertility

A. Causes Directly Related to Parent Birds

1. Age of parent birds: Fertility is diminished in both new breeders and older birds.
2. Muscular, skeletal, or neuromuscular disorders: Any injury or disease that impairs the bird's ability to copulate properly reduces fertility.
3. Metritis, orchitis, or systemic infections: These conditions in either parent bird can reduce fertility.
4. Neoplasms: Testicular and ovarian neoplasms are common in aged budgerigars and sometimes in other psittacines, accounting for 4.3% of reproductive problems in one study.[7] Ovarian or oviductal neoplasms can become quite large; however, when this occurs, usually no eggs are produced. Lymphomatosis can adversely affect testicular

or ovarian function, resulting in infertility. Cystic hyperplasia of the oviduct occurs in budgerigars and gallinaceous birds but generally results in failure to lay eggs rather than in infertile eggs.
5. Genetic factors: Hybrid birds often are infertile. Occasionally genetic factors render a single bird infertile. The fertility characteristics of male birds are inherited.[2]
6. Behavioral factors: Any behavior in either bird that prevents mating leads to infertile eggs. Young, immature birds may be sexually inexperienced. Birds may be improperly imprinted, taken away from parent birds at too early an age, be unfamiliar with the mate, or display mate incompatibility. In monogamous species, mate familiarity (pairing with the same mate yearly) increases male reproductive success and decreases aggression.
7. Obesity: Some psittacines become too obese to breed properly.

B. External Causes
1. Husbandry factors: These include inadequate housing, incorrect nest box or substrate, loose or uncomfortable perches that prevent mating, incorrect photoperiod in birds housed inside, excessive disturbances, incorrect pairing (creating homosexual or incompatible pairs), improper imprinting (bird imprinted on caregiver and not conspecifics), and lack of visual barriers between pairs.
2. Environmental factors: Proper light cycles (usually increasing daylight for most light-sensitive species, proper temperature, humidity, rainfall, and other environmental clues must be present for most species to breed successfully.
3. Nutrition: Psittacines on all-seed diets may be deficient in vitamins A, D_3, and E. Vitamin A is needed for reproductive gland function, vitamin D_3 for proper calcium metabolism, for egg shell formation, and vitamin E for spermatogenesis. A low calcium level in the diet (below 0.3%) prevents or stops egg laying in gallinaceous birds and is presumed to cause similar problems in psittacines. A low calcium level in the diet of cockatiels and other small psittacines may contribute to egg binding.

IV. Incubation

Generally, success in hatching is best achieved by natural incubation. However, various factors, such as the need to multiple clutch a pair of birds, unreliable incubating parents, cannibalism, sickness or death of the parent bird, inclement weather, or the threat of disturbances, all can prompt the aviculturist to turn to artificial incubation.

A. Natural Incubation
1. Parent incubation or incubation by foster parents of the same or related species or by unrelated species with similar incubation characteristics

2. Advantages
    a. Separate facilities are not needed to incubate eggs (or subsequently to raise chicks).
    b. Naturally incubated eggs are not subject to electrical power failures.
    c. Natural incubation may enhance the pair bond of parent birds.
    d. Artificial incubators do not duplicate the temperature gradient from top to bottom of the egg.[8]
 3. Disadvantages
    a. The parent birds may accidentally or deliberately destroy the eggs.
    b. The risk of disease transmission from parent birds, nest material, or other debris is higher.
    c. Predators may be a threat.
    d. Parent birds may abandon the nest at a critical period.
    e. The ability to monitor embryonic development is diminished.[9]

B. Artificial Incubation

Good artificial incubation depends on careful control of several factors.
 1. Temperature: Temperature is the critical factor in artificial incubation. Each thermostat should be carefully calibrated, because normal egg development occurs only within a narrow temperature range. A temperature that is too high or too low can result in death of the embryo. For example, a temperature that is too high by as little as 1° to 1.5° C (2° to 3° F) results in embryo death. The surviving embryos are small and weak and have unhealed navels or exposed yolk sacs. An elevated incubation temperature increases the incidence of scissor bills, curled toes, and wry necks.[6] A slight lowering of the temperature may result in slow development, delayed hatching, and large, soft-bodied chicks. In general, the temperature requirements for commonly incubated species are available in the aviculture literature (Table 12-2). For species for which the incubation temperature is not known, small eggs are incubated at 37.5° C (99.5° F), and large eggs are incubated at 0.25° to 0.5° C lower (37° C to 37.25° C [98.6° to 99.1° F]).
 2. Humidity: Relative humidity requirements vary by species. Humidity is most critical in the first third of incubation.
    a. Too-low humidity dehydrates the embryo, which may lead to embryonic kidney failure, sticky albumin, and a stunted embryo from improper calcium metabolism.
    b. Too-high humidity results in small air cells, soft, edematous chicks at hatch, and open navels or exposed yolk sacs. Late dead embryos can occur if the space in the air cell is too small.
    c. Wet bulb readings: The technique most often used to measure humidity involves taking a reading from a wet bulb thermometer (one in which the mercury bulb is covered with water-soaked gauze) and converting it to a humidity value (Table 12-3).
 3. Recommended temperature and humidity settings (see Table 12-2)[2]
    a. Most psittacines: 37.3° to 37.5° C (99.1° to 99.5° F) and 48% relative humidity

Table 12-2
Incubation requirements for some commonly raised birds

Species	Incubation temperature [°C (°F)]	Relative humidity (%)	Mean number of incubation days
Budgerigar (*Melopsittacus undulatus*)	37.1 (98.7)		18
Cockatiel (*Nymphicus hollandicus*)	37.5 (99.5)	56	21
Moluccan cockatoo	37.2-37.4 (99-99.3)	58-61	29.3
Eclectus parrot (*Eclectus roratus*)	37.2-37.4 (99-99.3)	50-53	28
African grey parrot	99-99.4 (37.2-37.5)	50-53	29.8
Macaw (*Ara* spp)	37.3-37.4 (99.1-99.3)	50-53	25.1-26.5
Hyacinth macaw (*Anodorhynchus hyacinthinus*)	37.2-37.4 (99-99.3)	50-53	26.5
Rose-breasted cockatoo	37.4-37.6 (99.3-99.7)	50-53	21.9
Yellow-naped Amazon parrot (*Amazona ochrocephala avropalliata*)	37.3-37.5 (99.1-99.5)	50-53	27.3
Ostrich (*Struthio camelus*)	36-36.4 (96.8-97.5)	20-25	42-45
Cassowary	36.1 (97)	68	46-49
Rhea (*Rhea americana*)	36.5 (97.7)	62-65	36-44

Data from King AS, McLelland J: *Birds: their structure and function*, London, 1984, Bailliere Tindall; Joyner KL: Theriogenology. In Branson WR, Harrison GJ, Harrison LR, editors: *Avian medicine: principles and applications*, Lake Worth, Fla, 1994, Wingers, pp 748-804; Hochleithner M, Lechner C: Egg binding in a budgerigar caused by a cyst of the right oviduct, *Association of Avian Veterinarians Today* 2:136-138, 1988; Rosskopf WJ, Woerpel RW: Pet avian obstetrics, First International Conference on Zoological and Avian Medicine, 1987, pp 213-231; Brown AFA: *The incubation book*, Surrey, England, 1979, Hindhead; and Olsen GH, Clubb S: Embryology, incubation, and hatching. In Altman RB et al, editors: *Avian medicine and surgery*, Philadelphia, 1996, WB Saunders, pp 54-71.

    b. Cockatiels: 37.5° C (99.5° F) and 56% relative humidity
    c. Galahs and Australian parakeets: 37.1° C (98.7° F)
  4. Turning: Parent birds turn eggs an average of every 35 minutes. In artificial incubation, mechanical turning is required at least five to eight times a day. Inadequate turning leads to early dead, malpositioned, or late dead embryos or to incomplete closing of the ventral body wall.

V. Egg Hygiene

A. Precautions with Eggs
1. Improper handling can kill a fertile egg.
2. The egg is laid at the body temperature of the female bird. Cooling begins immediately, and contents shrink, drawing in air and any contaminants on egg shell.
3. Fresh, clean, dry nesting material must be used. Wet, rotting material is a good source of *Aspergillus* organisms and other infectious agents.
4. The adult bird's cage must be kept clean so that the parent birds do not carry feces or other contaminants into the nest on their feet or feathers.

Table 12-3
Calculation of relative humidity from wet and dry bulb thermometer readings

Wet bulb temperature		Dry bulb temperature													
°C	°F	28	29	30	31	32	33	34	36	37	38	39	40	41
		82	84	86	88	90	92	94	96	98	100	102	104	106
28	82	100	92	84	77	71	65	60	55	50	46	42	39	36
29	84		100	92	85	78	72	66	61	55	51	47	43	40
30	86			100	92	85	78	72	66	61	56	52	48	44
31	88				100	92	85	79	73	67	62	57	53	49
32	90					100	92	85	79	73	68	62	58	53
33	92						100	93	86	79	73	68	63	58
34	94							100	93	86	80	74	69	64
36	96								100	93	86	80	74	69
37	98									100	93	86	80	75
38	100										100	93	87	81
39	102											100	93	87
40	104												100	93
41	106													100

5. Hands must be washed before the eggs are handled, and the collecting basket must be washed and disinfected.

B. Egg Washing
1. Washing of eggs generally is not recommended, because washing removes the naturally protective cuticle, and bacteria can more readily enter the pores of a wet egg. *Removing dried feces by gentle sanding is preferred.*
2. Only eggs with adhering contaminants should be washed.
3. The temperature of the wash water should be 43° C (110° F) to avoid cooling of the egg and contraction of the contents. The egg should be immersed for up to 5 minutes.
4. Disinfectants, such as a 1% povidone-iodine solution, can be added to the wash water. (To make a 1% solution, add one part 10% povidone-iodine to nine parts water.)
5. Eggs also can be disinfected with chlorine dioxide foam.

C. Egg Storage
1. Eggs sometimes must be stored before incubation to synchronize hatching.
2. Storage is recommended at 13° C (55° F) and 75% to 85% relative humidity.
3. Once daily eggs should be warmed to 27° C (80° F) for 5 minutes and then turned.
4. On average, a group of stored eggs has a mortality rate of 2% per day while in storage.

D. Incubator Care
1. Clean and disinfect the incubator before use and after each batch of eggs. Dismantle and clean the incubator with glutaraldehyde solution.
2. Water trays in incubators should be cleaned and disinfected daily.
3. The incubator should be run for 1 month before the first expected use to ensure that the system is functional.
4. The incubator should be tested by culturing periodically.
    a. Surfaces should be swabbed, or an open microbiologic agar plate should be placed in the incubator.
    b. Swabs should be obtained for testing for pscittacine beak and feather disease virus and polyomavirus.

VI. Diagnostic Techniques Useful with Eggs
A. Egg Weights
1. Eggs naturally lose weight because of metabolism and evaporation.
2. Weight loss is affected by temperature, humidity, the porosity characteristics of the egg, air circulation around the egg, and altitude.
3. Each egg should be weighed regularly.

4. Weight loss should be charted. For psittacines, the total weight loss until hatching is 12% to 13%. For most other species, weight loss averages 13% but can range from 11% to 16%.

B. **Candling** (Fig. 12-3)
   1. Candling is a technique for viewing development inside an egg using a light source in a light-tight box with a circular opening, usually with a rubber or soft plastic rim around the opening.
   2. A 40-watt bulb at 4 cm (1½ inches) works well for chicken-size eggs. A lower wattage should be used for smaller eggs. High-intensity lights or high wattage may damage or kill embryos.
   3. Candling allows the following:
      a. Visualization of the air cell and determination of its size.
      b. Determination of fertility status.
      c. Detection of a blood ring, which indicates early dead embryo (see Fig. 12-3).
      d. Identification of cracked eggs, which can be sealed with paraffin or beeswax.

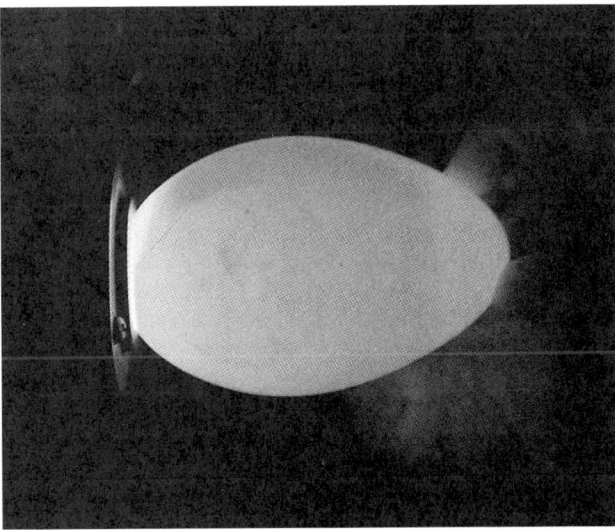

Fig. 12-3 A concentrated, low-wattage light source is used to candle an egg, which allows examination of internal structures. The embryo is in the center of a circle of radiating blood vessels. The pooled circle of blood around the embryo and radiating vessels is called a blood ring, which may be an early indicator of a dead embryo.

e. Determination of vessel and membrane integrity (loss of integrity can signify death of the embryo).
f. Detection of movement, which may be seen in older embryos, indicating a normal, viable embryo.
4. Daily candling is helpful for recognizing stages of development. Otherwise, eggs should be candled every 2 or 3 days. Eggs can be identified by marking gently on the shell with a soft-lead (no. 2) pencil.

C. Floating
1. Floating is a useful alternative for determining viability in dark-shelled eggs and during the last one third of incubation.[9]
2. The egg is floated in warm, mild disinfectant (1% povidone-iodine solution at 43° C [110° F]) is recommended[9]; the observer watches for signs of movement.
3. The egg should not be floated for longer than 1 minute, because aspiration of fluid into the egg or overheating can occur.

D. Egg Necropsy or Breakout
1. Necropsies should be performed on all eggs as soon as it has been determined that the egg is dead.
2. Egg death occurs during three distinct periods:
   a. Early embryonic death occurs at 3 to 5 days' incubation; causes include incorrect temperature, lethal genetic traits, and improper handling of the egg (e.g., jarring, improper storage).
   b. Death during the middle third of incubation often is related to nutrient deficiencies in the egg (vitamins, proteins, or minerals); some types of bacteria and fungi also are causes.
   c. Late incubation death is associated with improper incubation (temperature, humidity, or turning problems), genetics (malpositioning at hatch), or bacterial or fungal disease.
3. Necropsy procedure
   a. The egg records, weight, and candling observations are reviewed.
   b. The egg is opened over the air cell by rotating the point of a small, sharp drill burr (conical burrs from a small hand drill work well [Fig. 12-4]) or by using small, sharp scissors.
   c. Using sterile technique, the examiner carefully removes enough shell to allow visualization of the air cell membrane (Fig. 12-5). The shell and air cell membranes are inspected for abnormalities, and cultures are done if abnormalities are found.
   d. The egg then is opened further, and the air cell membrane is removed. Now the albumin may be cultured.
   e. The internal egg membranes, albumin, allantois, and yolk are examined; and abnormal color or consistency is noted.
   f. The yolk is then cultured.
   g. If no embryo or only a very small embryo is present, the egg contents are gently poured into a sterile Petri dish for further examination. Any abnormalities, hemorrhage, or genetic defects are noted (Fig. 12-6).

Embryologic Considerations **201**

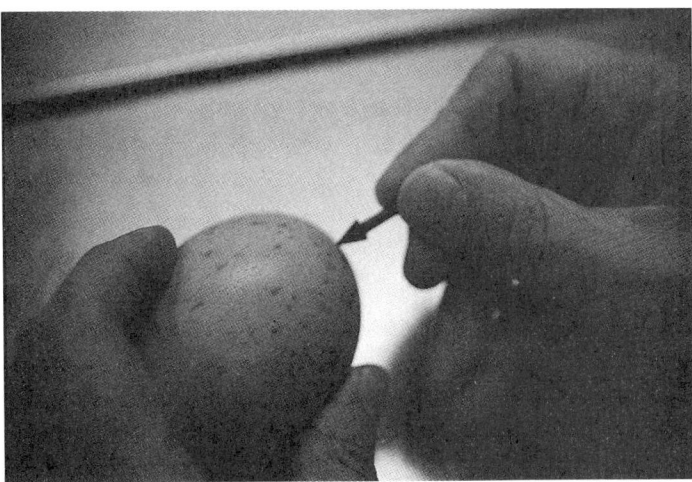

Fig. 12-4 A small, triangular drill burr (Dremel Motor Tool) is held between the thumb and forefinger and rotated back and forth to make an opening into the air cell.

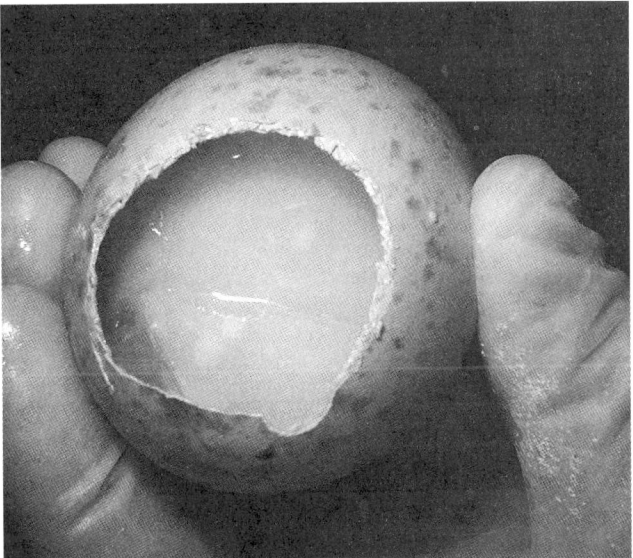

Fig. 12-5 Opening the air cell end of an egg at necropsy using a small, sterile thumb forceps and sterile gloves.

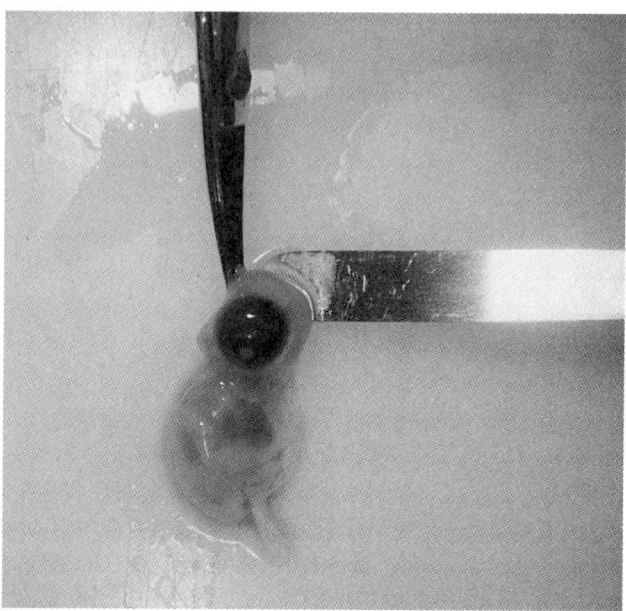

Fig. 12-6 A dead embryo (10-day-old chicken embryo) with a genetic abnormality: only one eye is present (Cyclops embryo).

    h. If a larger embryo is present, the examiner notes the position and condition of the embryo, especially the position of the head, neck, and beak in relation to the body and appendages (used to detect malposition if present).
    i. The embryo is removed carefully from the egg and then weighed and measured.
    j. Larger embryos can then be necropsied using standard necropsy procedures (see Chapter 25). Careful attention is paid to edema or hemorrhage of tissues and to internal organs. Tissues are cultured and then saved in formalin for histopathologic studies.

VII. Causes of Embryo Morbidity and Mortality

A. Inherited Factors
1. Lethal genetic defects may be expressed at several points in egg development but often occur in the first and last thirds of incubation.
2. Genetically lethal traits often arise from recessive genes. The veterinarian should look for patterns in egg deaths; for example, if 1 out of 4 or 1 out of 16 eggs from a pair is found dead with deformity or no discernible cause (the genetically lethal trait may involve metabolic pathways or internal development) then suspect a genetic

factor. If inbreeding is occurring in the flock, the incidence of genetically linked embryo death may be higher.
3. If the dead embryo appears malformed, genetic malformation must be distinguished from that caused by teratogenic factors such as pesticide or exposure to oil (see below).

B. Parental Age
1. Fertility and hatchability change with the age of the parent birds.
2. Both new pairs and old pairs have reduced fertility and hatchability; however, parental age is relative and species specific; for example, a 7-year-old finch pair may be old, but bald eagles breeding at the same age are young.

C. Nutrition of Parent Birds and Dietary Factors
1. Minor deficiencies in a laying female bird are magnified over the breeding season, especially if egg numbers are high.
2. Vitamin A: Marginal deficiencies may lead to reduced hatchability.
3. Riboflavin (vitamin B_2): The most important vitamin for incubation; a large amount is needed in the albumin. Deficiencies result in reduced hatchability, curly toe paralysis, and clubbed down feathers.
4. Pyridoxine (vitamin B_6): This vitamin is important for normal and early chick growth. It also is required for the breakdown and synthesis of protein and may be deficient if the female is fed a high-protein diet. If the diet also is deficient in manganese, perosis and soft bones may result. The embryo may be found dead in the egg with no other apparent cause.
5. Folic acid: Folic acid, which is important for the production of red blood cells, is synthesized by gastrointestinal bacteria. A female bird being treated with antibiotics may have an undetected deficiency that results in early dead embryos because blood does not form properly in the embryos.
6. Vitamin D_3: This vitamin is required for calcium metabolism. Female birds with deficiencies may lay thin-shelled eggs more prone to cracking during incubation, or the embryo may not have received enough vitamin D_3 in the egg for proper mobilization of egg shell calcium, resulting in embryonic death.
7. Calcium: Dietary calcium deficiencies can lead to soft-shelled or thin-shelled eggs. Thin-shelled eggs may crack or break during incubation.
8. Selenium: In amounts exceeding 4 ppm, and especially in the form of selenomethionine, this mineral is teratogenic and causes reduced hatchability of surviving embryos. Deformities seen include ectodactyly, hydrocephaly, microophthalmia or anophthalmia, and beak defects.

D. Infectious Diseases
1. Staphylococci
    a. Avian embryos are highly susceptible to some strains, especially *Staphylococcus aureus* and *S. epidermidis*.

b. Sources of infection include wounds on parent birds or egg handlers. The organisms spread easily in incubators.
   c. Infection can cause embryonic mortality in as little as 48 hours. (Seen as endocarditis and hemorrhage in liver, kidneys, and brain.)
   d. Prevention and treatment of infections in parent birds and handlers are recommended. Adding low dosages of penicillin or tetracycline to feed is not recommended because it leads to development of resistant strains.
2. Streptococci
   a. *Streptococcus faecalis* sometimes invades the ovaries.
   b. The organism is identified in eggs subject to embryonic mortality, and carrier birds are treated or eliminated.
3. Salmonellae
   a. *Salmonella* organisms spread by contamination of the egg in the ovary, by contamination of the egg shell after laying, and by contamination of incubators and egg collection boxes.
   b. Mortality is highest 2 weeks after hatch and can reach 100%.
   c. Postmortem signs include an enlarged, congested liver streaked with hemorrhage and necrotic foci on the liver, lungs, ceca, large intestines, and ventriculus. The spleen is enlarged, and pericarditis is possible.
   d. Spraying eggs suspected of being contaminated with a neomycin sulfate solution before incubation is effective in some cases.
4. *Escherichia coli*
   a. *E. coli* is associated with fecal contamination of egg shells (dirty nests and dirty cages, with birds bringing feces on the feet or feathers into the nest).
   b. In dead embryos the yolk contents appear caseous or watery yellow-brown; the embryo may be edematous.
   c. Omphalitis, or yolk sac infections, may be seen, along with poor weight gain in newly hatched birds.
   d. Fumigation and proper disinfection of the egg shell are important for controlling outbreaks.
5. Chlamydiae
   a. *Chlamydia psittaci* is found in the ovaries and in eggs.
   b. Embryonic mortality occurs 5 to 12 days after laying.
   c. Dead embryos show congested or hemorrhagic yolk sac membranes.
6. Viruses
   a. Herpesvirus can be egg transmitted.
   b. Avian paramyxovirus (Newcastle disease) can enter eggs by direct transmission from the laying bird or by fecal contamination.
   c. Dead embryos may show retarded growth and defects of the neural tube, eye lenses, auditory vesicles, visceral arches, limb buds, and olfactory pits.
7. Mycoplasmas
   a. *Mycoplasma* organisms are transmitted to the egg by an infected female or by infected semen from the male.

b. Lesions are seen on air sacs; also seen are dwarfing, generalized edema, liver necrosis, enlarged spleen, and joint abscesses.
c. Treatment
(1) Tylosin: 0.5 to 1 mg/dose is injected into the air cell before incubation.
(2) The egg is dipped into Tylosin (1000 to 3000 ppm) or gentamicin (400 to 1000 ppm), plus disinfectant (quaternary ammonium, 250 ppm). The egg is warmed to 37° C (98.6° F) and then dipped in a solution maintained at 2° to 8° C (35.6° to 46.4° F) for 5 to 20 minutes.
(3) Removing newly laid eggs from room temperature to 46.1° C (115° F) for 12 to 14 hours inactivates *Mycoplasma* organisms but reduces hatchability by 8% to 12%.
8. Aspergilli
a. *Aspergillus* organisms are transmitted by spores that enter the egg from another contaminated egg or from outside sources. Rapid spread is possible in forced-air incubators.
b. The embryo can die before hatching or may be weak at hatch, having dyspnea, gaping, diarrhea, or possibly nervous system signs.
c. The fungus sometimes can be observed growing on the air cell membrane. Small, yellow-white foci can appear on the embryo's lungs or air sacs, and bronchial plugs are often seen.
d. No treatment is known. Prevention involves eliminating potential sources of contamination, such as decaying plant material.

E. Parasitic Diseases
1. Parasitic infestations in a laying female bird can impair her nutritional status despite an apparently adequate diet. This results in various nutritional deficiencies that appear in the eggs she is laying.
2. Parasites are not normally transmitted on or in the egg itself.

F. Toxins[6]
1. Petroleum products
a. Exposure by parent birds to spilled oil is primarily of concern in wild birds (rehabilitation work) but occasionally occurs in outdoor aviaries.
b. Amounts as small as 1 to 10 µl deposited on the egg shell can result in deformities of the eye, brain, and beak and in embryonic death. This small amount may be carried on the feathers of an incubating bird.
c. Mineral oil can be transferred from the feathers of an incubating bird to its eggs, reducing hatchability by blocking gaseous exchange through the pores in the shell.
2. Insecticides
a. Organophosphorus insecticides can cause embryo deformities, including scoliosis or lordosis, especially in the cervical region; they also can cause micromelia, malformed beak, and abnormal feathering.

b. Exposure to diazinon causes embryonic stunting and incomplete caudal ossification.
3. Herbicides
   a. Herbicides can be as toxic to embryos as insecticides. Exposure occurs through contamination of the egg shell.
   b. Paraquat is highly embryotoxic, resulting in extensive edema of embryonic tissues and anencephaly or exencephaly. Low-level exposure reduces embryonic growth.
   c. Exposure to trifluralin leads to beak defects.
4. Carbon monoxide
   a. Exposure to carbon monoxide can result from automotive exhaust or from a faulty furnace in or near the incubator room.
   b. Exposure at 100 ppm reduces hatchability by 21%; at 200 ppm hatchability is reduced 83%.
   c. Exposure to carbon monoxide causes decreased embryonic growth; hypertrophy of the heart, liver, and spleen, a decrease in hematocrit; and elevation of serum alanine aminotransferase.
5. Nicotine
   a. Nicotine from cigarette smoke causes a reduction in embryonic size and severe skeletal deformities.
   b. The cervical vertebrae are deformed or absent, and the neural tube is malformed or missing.
   c. Other deformities include torticollis, edema, muscular dystrophy, a shortened upper beak, and malformed heart or kidneys.
6. Antibiotics
   a. Chloramphenicol in doses of 0.5 mg/egg reduces growth and development.
   b. Penicillins cause hemorrhage and edema, especially of the head and limbs.
   c. Tetracyclines diminish the size of the embryo, inhibit mineralization of bones, and cause erosion of cartilage.
   d. Sulfa drugs cause degenerative changes in the liver and kidneys, as well as reduced growth, enlarged head, microophthalmia or macroophthalmia, beak hypoplasia, joint ankylosis, and shortened legs.

VIII. Egg Therapeutics

A. Disinfection of Eggs
1. Warmed iodine solution (povidone-iodine 10%, mixed 1:10 with water, warmed to 40° to 43.3° C [104 to 110° F]): the egg is immersed for up to 5 minutes.[6]
2. Gentamicin sulfate, 1000 µg/ml of cold water (4° C [39.2° F]): the egg is removed from a warm incubator (37° C [98.6° F]) and placed in the cold solution.[10]
3. Chlorine dioxide foam sprayed on egg is effective for disinfection.

4. Surgical glue, paraffin, beeswax, and Elmer's white glue all have been used (sparingly) with success to repair cracked eggs.

B. Altering Egg Weights
These techniques are used for eggs that have lost more than 16% or less than 11% of their weight.
1. To increase weight loss:
   a. Lower the humidity in the incubator.
   b. Gently sand the egg or make a small hole over the air cell (hand turn all eggs with altered shells).
2. To decrease weight loss:
   a. Raise the humidity in the incubator.
   b. Cover part (no more than 60%) of the egg shell with paraffin[11] or white glue[12] to reduce evaporative loss.
   c. Injection of sterile lactated Ringer's solution into the small end of the egg (into the albumin) helps correct severe dehydration. The amount is calculated using the egg weight multiplied by the deficit (the percentage difference between the expected 13% weight loss and any actual loss).

C. Antibiotic Injections
1. Antibiotics are injected into eggs to correct bacterial infection detected when other eggs from a clutch or incubator were necropsied and were found to have bacterial infections.
2. A small dental or Dremel drill or needle is used to make an opening in the small end of the egg. After the injection, the opening is sealed with paraffin or white glue.
3. Recommended antibiotics:
   a. Piperacillin (200 mg/ml), 0.02 ml for a macaw egg, 0.01 ml for a cockatoo egg[12]
   b. Tylosin, 2.4 mg plus 0.6 mg gentamicin for a turkey egg (for *Mycoplasma* infections)[13]

IX. Hatching and Malposition

A. Calculation of the Hatch Date
1. The hatch date is calculated from the laying date or the date incubation started (if eggs were stored for any length of time).
2. As the hatch date approaches, the egg is candled daily to monitor air cell and embryo position.
3. The egg is transferred from incubator to hatcher after pipping.

B. Anatomic and Physiologic Changes during Hatching[2,6,13,14]
1. As the embryo grows and develops, the head moves from the narrow end of the egg up toward the air cell in the wide end of the egg, eventually becoming positioned just under the right wing.

2. Allantoic circulation gradually fails to meet the embryo's gaseous exchange needs.
3. The elevated level of carbon dioxide leads to twitching of the neck muscles.
4. During twitching the beak penetrates the air cell, and the embryo gulps air and begins to breathe.
5. As the embryo begins breathing, the lungs begin to function for the first time in air exchange, and the right-to-left cardiovascular shunt closes.
6. In some species, vocalizations (peeping) can be heard for the first time.
7. The elevated carbon dioxide level also causes abdominal muscle contractions, which pull the yolk sac into the abdomen.
8. As the embryo continues to breathe air from the air cell, the carbon dioxide level again rises, to as high as 10%.
   a. The elevated carbon dioxide level leads to further muscle contractions.
   b. During one of the contractions, the beak breaks through the shell, forming a pip hole.
   c. Species vary considerably in the length of the interval between entry into the air cell and pipping (the range is 3 hours to 3 days, generally longer in larger species).
9. Alternating between contractions of the head and neck and of the neck, back, and abdominal muscles, the embryo first chips at the shell, then shifts position slightly before the next head and neck contraction results in further chipping; this process is called "cutting out" (Fig. 12-7).

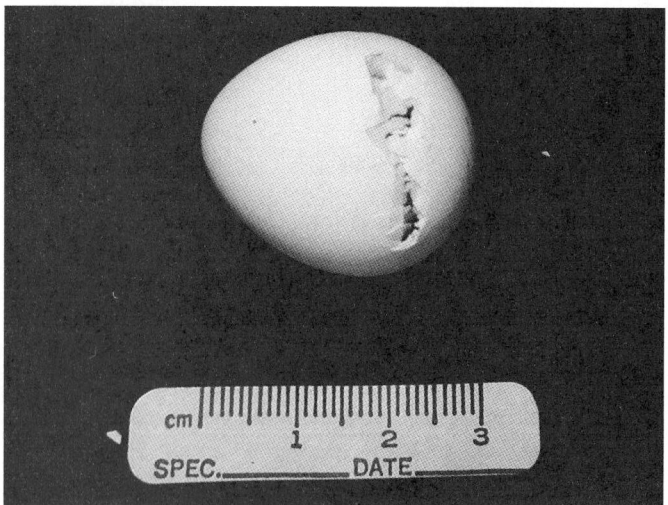

Fig. 12-7 An embryo in the process of cutting out (i.e., the counterclockwise cracking pattern formed in the shell by contractions of the head and muscles and by rotation of the embryo in the egg).

a. Finally, the top of the shell is pushed away, and the chick kicks free.
b. The interval from pipping to kicking free ranges from 30 minutes to 3 days and also is species specific.
10. Newly hatched chicks usually are wet and exhausted and will rest and dry off.

C. Malpositioning
1. Embryos situated in other than the normal position (head under right wing) at the time of hatching are considered malpositioned. Common malpositioning and outcomes are listed in Table 12-4.

D. Assisted Hatches
1. Assisting in hatching should be done only if a problem has been diagnosed. Assisting a normal hatch prematurely can lead to complications, including death of the embryo.
2. The pip-to-hatch interval usually is 36 to 48 hours for most psittacines[6] (for budgerigars it is 12 to 24 hours). An interval shorter than 24 hours or longer than 72 hours may indicate a problem.
3. Chicks with retracted yolk sacs and loss of blood from the vessels in the egg membranes can be safely removed from the egg.
    a. Premature removal, before blood vessels have fully retracted, can lead to hemorrhage and death.
    b. Exposed yolk sacs may rupture or become infected.
    c. Chicks that have rotated at least one-quarter turn during the cutting out process usually can be safely removed from the egg.
4. Techniques for determining if assisting in hatch is required
    a. The normal pip-to-hatch interval has been exceeded (see no. 2 above).
    b. Candling or radiographs show the beak to be in a location other than the air cell.
    c. Chick vocalizations are weak or absent (in species that normally vocalize while still in the egg).
    d. No sign of pipping is apparent 24 to 36 hours after the chick has entered the air cell.
5. Techniques for assisting the embryo to hatch
    a. Just before hatch, the egg is candled and the edge of the air cell is outlined with a soft-lead pencil or fine-tip felt marker.
    b. The area over the air cell is disinfected with a dilute (1%) povidone-iodine solution, and the shell then is opened over the air cell. The burr drill from a motor tool (Dremel) is useful for making such a hole. The drill burr is rotated between the thumb and forefinger (see Fig. 12-4). *Do not use the drill to operate the burr.*
    c. The air cell is entered with a small laparoscope to determine the location of the beak and whether patent blood vessels are present in the shell and air cell membranes.
    d. If the membranes appear white and opaque, they are moistened with a small amount of sterile water or saline solution, applied one drop at a time from a syringe.

Table 12-4
Description and prognosis of malpositioned embryos

Malposition number	Embryo position	Prognosis
1	Head between the thighs	Initially a good prognosis, because this is the early normal position; as time progresses the prognosis becomes poor, and the embryo may die.
2	Head in the small end of the egg	Guarded prognosis; the position is lethal in about 50% of cases; Assistance in hatching improves the chances of survival.
3	Head under the left wing	Poor prognosis; this position generally is lethal to the embryo.
4	Body rotated with the head under the right wing but not pointed at the air cell	Guarded to poor prognosis, because the head will not enter the air cell; may be possible to assist in hatching.
5	Feet over the head	Guarded to poor prognosis, because embryo cannot kick to rotate its body during cutting out; assistance in hatching improves survival.
6	Head over instead of under the right wing	Good prognosis; embryo usually hatches with minimal complications or help.
7	Embryo crosswise in the egg	Poor prognosis; this position often is fatal. It is seen with small embryos and spherical eggs, and other defects commonly are present.

Data from Olsen GH, Clubb S: Embryology, incubation, and hatching. In Altman RB et al, editors: *Avian medicine and surgery*, Philadelphia, 1996, WB Saunders, pp 54-71; and Olsen GH, Duvall F: Commonly encountered hatching problems, *Proceedings of the Association of Avian Veterinarians*, 1994, pp 379-385.

    e. If membrane material must be removed with blood vessels still present, a bipolar forceps or other cautery is used (even a hand-held, battery-operated cautery works well) to prevent hemorrhage.

    f. The membrane is removed from over the beak, mouth, *and* nares to prevent late embryo death by suffocation.

    g. The status of the yolk sac is determined. If the yolk sac is not fully retracted into the abdomen, the opened air cell should be partly covered with flexible, wax-type test tube sealant, or the egg should be placed in a plastic bag with a small opening at one end. These techniques increase the level of carbon dioxide, leading to the muscle contractions required for proper retraction of the yolk sac into the abdominal cavity.

6. Abnormal pip holes
    a. A pip that appears below the air cell is an indication of a possible problem, such as malposition no. 4 or no. 7 (see Table 12-4).
    b. Embryos that make more than one pip hole may be having difficulty and may need assistance.[6]
    c. A pip that appears near the bottom of the egg indicates malposition no. 2.
7. An embryo that enters the air cell normally still may need assistance to hatch; for example, if:

a. The time since entering the air cell (internal pip) is longer than that considered normal and safe for that species (i.e., no external pip has occurred).
b. The time since the external pip is longer than normal and the embryo has not started cutting out.
c. The embryo's vocalizations sound weak.
8. Technique for removing the embryo from the egg
   a. The egg shell is disinfected over the air cell or around the pip hole if the hole is not in the air cell area.
   b. The drill burr is used as described above to make a hole into the air cell.
   c. Forceps or fingertips are used to gently grasp the chick by the beak or head and to pull the head and neck out.[6]
   d. Pulling the head and neck out of the shell frees space for viewing the yolk sac area.
   e. If the yolk sac is visible and no feces are present, the head and neck are put back in the shell as much as possible and covered with stretchable, wax-type test tube sealant, leaving a slit for air exchange.
   f. The embryo must be reexamined every 1 to 3 hours.[6,14]
   g. When the yolk sac has been absorbed into the abdomen or if feces are seen, the chick should be gently pulled from the shell. The clinician should be prepared to cauterize any membrane vessels and attachments to the shell.
   h. Weak chicks can be given small amounts of fluid orally (via a small tube placed into the esophagus, or subcutaneously). Keeping intact membranes moist also helps prevent dehydration.
9. Chicks pipping at or near the narrow end of the egg (malposition no. 2) need constant monitoring. The air cell position should be checked, because in some cases the air cell extends down to the pip site. The pip site should be enlarged to ensure that the chick receives sufficient air exchange (any membrane vessels that start bleeding should be cauterized). The chick is monitored as above, and fluids are given and membranes moistened as described above. After 24 to 48 hours, the hole is enlarged further, allowing the membranes to dry. If the yolk sac has been absorbed, the chick should be allowed or assisted to free itself from the egg.

E. Exposed Yolk Sac
1. This is an emergency that must be handled quickly.
2. If only a small remnant of the yolk sac remains outside the body wall, a ligature can be tied around the exposed yolk sac as close as possible to the body wall (the area is cleaned first with 1% povidone-iodine solution). Generally the exteriorized portion of the sac dries and falls off in a few days.[14]
3. If the chick is newly hatched and the yolk sac remnant is large but clean, the remnant is disinfected with 1% povidone-iodine solution and gently manipulated into the coelomic cavity. After the yolk sac

has been placed in the abdomen, 4-0 or 5-0 nylon or absorbable suture material is used to tie a purse-string suture around the umbilical opening to close the area.[14]

4. If the exteriorized yolk sac remnant is contaminated, necrotic, or too large to go through the umbilical opening into the abdomen, it can be ligated close to the umbilicus, the contents aspirated, and the tissues trimmed (although not too close to the ligature). The area should be kept clean with twice daily applications of 1% povidone-iodine solution.

5. Chicks with exposed yolk sacs often develop infections (omphalitis) in the yolk sac. These chicks are given antibiotics: gentamicin (5 mg/kg once daily), ceftiofur sodium (2.5 mg/kg once daily), or amikacin (8 to 10 mg/kg) and piperacillin (100 mg/kg), both twice a day.

REFERENCES

1. King AS, McLelland J: *Birds: their structure and function,* London, 1984, Bailliere Tindall.
2. Joyner KL: Theriogenology. In Branson WR, Harrison GJ, Harrison LR, editors: *Avian medicine: principles and applications,* Lake Worth, Fla, 1994, Wingers, pp 748-804.
3. Hochleithner M, Lechner C: Egg binding in a budgerigar caused by a cyst of the right oviduct, *Association of Avian Veterinarians Today* 2:136-138, 1988.
4. Rosskopf WJ, Woerpel RW: Pet avian obstetrics, First International Conference on Zoological and Avian Medicine, 1987, pp 213-231.
5. Brown AFA: *The incubation book,* Surrey, England, 1979, Hindhead.
6. Olsen GH, Clubb S: Embryology, incubation, and hatching. In Altman RB et al, editors: *Avian medicine and surgery,* Philadelphia, 1996, WB Saunders, pp 54-71.
7. Keymer IF: Disorders of the avian female reproductive system, *Avian Pathol* 9:405-419, 1980.
8. Gee GF, Hatfield JS, Howey PW: Remote monitoring of parental incubation conditions in the greater sandhill crane, *Zoo Biology,* 14(2): 159-172, 1995.
9. Gabal RR, Mahan TA: Incubation and hatching. In Ellis DH, Gee GF, Mirande CM, editors: *Cranes: their biology, husbandry, and conservation,* Washington, DC, 1996, National Biological Service, pp 59-76.
10. Snogenbos GH, Carlson VL: Gentamicin efficacy against salmonellae and arizonae in eggs as influenced by administration route and test organism, *Avian Diseases* 17:673-682, 1973.
11. Burnham W: Artificial incubation of falcon eggs, *J Wildlife Management* 47:158-168, 1983.
12. Weaver JD, Cade TJ: *Falcon propagation: a manual on captive breeding,* Boise, 1991, The Peregrine Fund.
13. Olsen GH: Problems associated with incubation and hatching, *Proceedings of the Association of Avian Veterinarians,* 1989, pp 262-267.
14. Olsen GH, Duvall F: Commonly encountered hatching problems, *Proceedings of the Association of Avian Veterinarians,* 1994, pp 379-385.

13

Problems of Neonates

Glenn H. Olsen

I. Altricial and Precocial Species

A. Altricial Species
1. Are helpless at hatch, usually naked and with eyes closed
2. Include psittacines, passerines, columbiformes, and raptors
3. Rely on their parents for food, protection, and even, initially, temperature regulation

B. Precocial Species
1. Hatch with eyes open, down covering their bodies, and are able to feed themselves and walk shortly after hatching
2. Include waterfowl, gallinaceous birds, ostriches, and cranes

C. General Neonatal Information
1. Lack the immune system of the adults
2. Are more susceptible to some diseases
3. Congenital and developmental problems, and problems brought on by poor husbandry are commonly seen

II. Rearing Methods

A. Parent-Rearing
1. This term is used when the parent birds incubate the eggs and raise the chicks.
2. Advantages are that the method saves on caretaker time, exposes chicks to adult birds and their behavior, and may produce birds more adapted for wild release.
3. Disadvantages are that parent birds may traumatize, abandon, or fail to care for chicks properly, especially the first or second time the birds rear a clutch, or if unusual disturbances occur. Parent-reared chicks are more difficult to tame; however, by handling parrot chicks for 15 to 30 minutes daily from day 12 onwards, chicks will be as tame as hand-reared.[1]

B. Foster-Rearing
1. Technique of moving eggs or chicks from parent birds to other pairs of the same or different species
2. Useful to increase production by stimulating the parent birds to renest when eggs are removed
3. Disadvantages include foster parents rejecting eggs/chicks (less likely when eggs are fostered for other eggs) and the increased chance of spreading some diseases.

C. Hand-Rearing
1. Includes any techniques in which the chick is raised partially or entirely by humans
2. Advantages include some of those mentioned for foster-rearing (to increase production of parent birds or to remove eggs/chicks from unsuitable parent birds). Other advantages include the production of tame birds for pets, to raise chicks hatched from artificially incubated eggs, and to control diseases spread from parents to chicks.
3. Because common psittacines lay eggs every 2 to 3 days but begin incubating with the first or second egg laid, chicks hatch in the lay order. It may help younger siblings to remove older chicks for hand-rearing.[2]
4. Disadvantages of hand-rearing are the amount of labor involved; the requirement for incubators, hatchers, and nursery facilities; and the possibility of disease spreading through the nursery.

III. Hygiene

A. Methods to Ensure Adequate Hygiene
1. If possible, someone other than the person taking care of adults and parent-reared chicks should handle nursery assignments.
2. If only one person runs the aviary, work with adults only after working in the nursery; shower and change clothes before entering the nursery again.
3. New additions to the nursery should be handled last for the first 7 days or until it can be established that they are healthy.
4. Wash hands between birds or containers to avoid mechanical transmission of pathogens.

B. Food and Water Hygiene
1. *Pseudomonas* spp., *Klebsiella* spp., *Enterobacter* spp., and *Escherichia coli* bacteria and fungus can contaminate diets causing mortality in neonates, but not adults.
2. *Pseudomonas* spp., and to a lesser extent, *Klebsiella oxytoca*, can be contaminants in water, causing neonatal mortalities.
3. Disinfect all feeding utensils and water containers with quaternary ammonium detergent product (especially useful for disinfecting catheter tip syringes)[3] or dilute chlorine bleach. Rinse all disinfected items thoroughly before use.

4. Food hygiene
   a. Most commercial diets are free of contamination.
   b. Contamination comes from improper storage or preparation.
   c. Mix all batches fresh for each feeding; never store premixed diet for another feeding.
   d. Store open dry chick diet in a sealed container in the freezer.
   e. Use a separate clean syringe or feeding utensil for each chick.

C. Nursery Design and Hygiene
   1. Maintain a closed nursery with no contact with adults, and do not bring in chicks from other aviaries.
   2. Test new chicks added to the nursery (cloacal swabs for bacteriology and polyomavirus).
   3. If possible, have separate nursery rooms for older chicks removed from parent-rearing situations and for sick chicks.
   4. Nursery area should have its own sink and counter for preparing diets and washing utensils

IV. General Husbandry Practices to Prevent Neonatal Problems

A. Neonate Environment
   1. Ambient environment
      a. Temperature recommendations may vary with species. For psittacines and other altricial species, maintain the following conditions:
         (1) Recent hatchlings at 92° F to 94° F (33.3 to 34.4°C)
         (2) Older unfeathered neonates at 90° F to 92° F (32.2 to 33.3°C)
         (3) Birds with pin feathers at 85° F to 90° F (29.4 to 33.3°C)
         (4) Fully feathered chicks at 75° F to 80° F (23.9 to 26.7°C)
         (5) By weaning, have chicks at room temperature (68° F, °C).[3]
      b. Humidity, especially for tropical species, is important and should be maintained above 50%.
      c. Watch reaction of neonate: huddling and shivering indicate chilling, whereas panting and holding wings out to side indicate chick is too warm.
   2. Physical environment
      a. Housing: Very young altricial species including psittacines do well in a brooder.
         (1) Commercially available units: best ones have solid state temperature controls.
         (2) Choose models with slow airflow; too high an airflow can dehydrate neonate.
         (3) Older chicks can tolerate temperatures closer to ambient room temperatures and do not need brooders; instead, use pen-top plastic container or small glass aquarium.

B. Diet

1. Diet requirements of neonatal psittacines not established; in cockatiels: 18% to 22% protein, 1% calcium with 2:1 Ca:P ratio.[4]
2. Commercial diets are recommended because of consistent nutrient analysis, freedom from microbiologic contamination, and ease of preparation.
3. Water content: Mix formula very dilute (5% to 10% solids by weight) the first day, increase solids by 2% to 3% per day until diet is 25% to 30% solids by day 7.[3]
4. Feeding times: Initial feeding schedule for most psittacine neonates is every 3 hours starting at 0600.[3] Small species, especially small passerines, need more frequent feedings, every 1 to 2 hours. As a chick grows, feedings can become less frequent until only 2 to 3 per day.
5. Food volume: Psittacine chicks tolerate a feeding approximating 10% to 12% of body weight, less for cockatoos, more for macaws. Some precocial species such as cranes, when fed by esophageal tube, tolerate only 3% to 5% volume by body weight.
6. Supplementing already balanced commercial diets with additional vitamins can lead to hypervitaminosis A or D.[5]
7. Warm food to 38.3° C to 40.0° C (101° F to 104° F) as measured by a thermometer. Use a hot plate or coffee maker to heat formula, not a microwave oven (which usually heats unevenly, leaving extreme hot spots that can burn the crop).
8. Catheter tipped syringes work well for feeding altricial species.
   a. Touching tip of syringe to beak under mandible or at commissure, or even just moving nest container can elicit feeding behavior (moving head up and down, opening mouth, vocalization in some species).
   b. When displaying feeding behavior, glottis and choana slit are closed; can give large bolus of food easily.

C. Weaning

1. Offer a variety of suitable foods (fruits, vegetables, soaked pelleted diet, various seeds for psittacines).
2. Chick will begin by playing with food.
3. Older chicks nearby may stimulate younger bird to eat more quickly.
4. Chicks to be weaned should be on 3 daily feedings. Discontinue midday feeding, followed by morning, then evening feedings.
5. Monitor weight daily; 10% to 15% weight loss seen at weaning even in healthy chicks.[3]

D. Neonate Identification

1. All chicks should have unique identification numbers. This should be posted on the nest container of very young chicks.
2. If a *closed band* is to be used (as required for psittacines by some state laws), this is applied when the chick reaches the pin feather stage. The closed band is placed over the metatarsus by passing it over forward-facing digits 2 and 3 and backward-facing digit 4.[2]

3. Microchip Transponders offer a safer, permanent identification method.

V. Examination of the Neonate

A. History
1. The first step is to learn the current signs.
2. Assess the neonate's previous history.
    a. Health of parents
    b. Previous breeding history of parents; what is the status of their other offspring?
    c. Incubation method and problems during incubation and hatching
3. Assess current husbandry practices.
    a. Nest container, substrate, brooder type, temperature, and humidity
    b. Diet, frequency of feeding, volume fed
4. Review sick chick's records.
    a. All hand-reared chicks should be weighed daily and the weight recorded to track health and development. Best time to weigh is before the morning feeding (crop and gastrointestinal tract empty).
    b. All parent-reared chicks should be weighed when handled for observations, taming, etc.
    c. Compare growth rate history with other neonates at the aviary if possible.
        (1) Growth rates can vary by aviary depending on diet, feeding schedule, husbandry practices, etc.
        (2) Some psittacine species can gain up to 17% daily during the first week post hatching.[6]

B. Physical Examination
1. Attitude, posture, conformation
    a. Before physically picking up the chick, examine its position in the nest.
    b. Newly hatched neonates spend much time sleeping and will sleep in any position.
    c. Prior to weaning age, psittacine chicks will spend much time resting on their hocks and large abdomen.
    d. Chicks, especially the very young, have little muscle mass and large abdomens.
    e. Pectoral muscle mass is almost absent. Evaluate muscle mass over back and hips to assess body condition.
    f. Conformation
        (1) Determine if the size and shape of the neonate is normal for age and species.
        (2) An excessively large head with a small body and limbs can indicate stunting.

2. Skin and feathers
   a. Normal skin of most psittacine and passerines is pink to yellow with some fat and the yellow of the yolk sac sometimes visible under the skin.
   b. Dehydration is indicated by dry, hyperemic skin.
   c. Hypothermia is indicated by white, cool skin and the unresponsiveness of the chick.
   d. Normal feathering
      (1) First feathers to appear in most psittacines and passerines are on the head, tail, and wings.
      (2) Later feathers appear over the body.
      (3) Some species, especially precocial species such as waterfowl, poultry, and cranes, are completely covered by down feathers at hatch. These are later replaced with contour and flight feathers by fledging age.
   e. Abnormal feathering
      (1) Pinched feathers or hemorrhage in the rachis can be signs of PBFD polyomavirus, bacterial infection, trauma, or malnutrition.
      (2) Stress bars or stress marks in feathers indicates systemic disease, antibiotic therapy, or stress at the time the feather was forming.
3. Eyes and ears
   a. In most psittacines and passerines, eyes are closed at hatch and open later.
      (1) Amazon parrot eyes open at 14 to 21 days.[7]
      (2) Cockatoo eyes open between 10 and 21 days.[7]
      (3) Macaw eyes open between 14 and 28 days.[7]
      (4) Clear discharge at the time eyes open is normal.
   b. Ears of old world psittacines open at hatch.
   c. New world psittacine ears open anywhere from 14 to 35 days.
4. Gastrointestinal system
   a. Beak should be examined for deviations, lack of symmetry, brachygnathism (shortened lower beak), prognathism (shortened upper beak), or signs of scissor beak.
   b. Oral cavity should be examined for color, raised white areas (plaques), and signs of trauma (especially if several older chicks are housed together).
   c. Crop capacity in neonates is larger than adults. Check to see if crop is emptying at least once a day
      (1) Palpate for foreign bodies.
      (2) Thickened crop wall or discoloration of tissues over crop area indicates a problem.
   d. In very young naked neonates, it is possible to visualize the yolk sac, and sometimes the liver and a loop of intestines under the skin. Yolk sacs are not visible after day 4 (although full absorption may not occur until day 7).
   e. Droppings are often watery (polyuric) because of the dilute diet fed to neonates.

f. Failure to pass droppings indicates rectal atresia. Pasting of feces around vent area can lead to constipation.
  5. Musculoskeletal system
     a. Look for unusual curvature of the vertebral column (scoliosis, Fig. 13-1, *A* and *B*).
     b. Examine the limbs.
        (1) Look for swelling, fractures, deformities, skin wounds, or discoloration.
        (2) In most species, the legs develop at a faster rate initially than do the wings.
  6. Auscultation (with pediatric stethoscope)
     a. Listen to the heart over each side of the thorax and ventrally for congenital murmurs.
     b. Listen to respiratory sounds over the trachea, dorsal, and lateral thorax.
     c. Normal heart rates 180 to 400 beats per minute and respiration 20 to 60 breaths per minute.[7]

C. Clinical Pathology
  1. Blood collection
     a. Collect a volume equal to or less than 1% of body weight, such as 1 ml in 100 g chick.
     b. Jugular venepuncture is practical in most psittacine and passerine chicks, with small volumes (0.10 ml) being collected from birds weighing only 10 g (use 28 g 1/2 in microfine TB syringe-needle combination).
  2. White blood cell counts higher (20,000 to 40,000) in chicks than adults.
     a. Higher in macaws than cockatoos or eclectus parrots[8-10]
     b. Eclectus parrots and cockatoos go from heterophilia (neonate) to lymphophilia (adult)[8-10]
  3. Red blood cell count, hematocrit, hemoglobin, and mean corpuscular hemoglobin concentration (MCHC) are all low in neonates and increase as the chicks became older.
  4. Serum chemistry values also changed with age.[8-10]
     a. Alkaline phosphatase, creatine kinase, potassium, and phosphorus are initially higher than adult values in macaws, cockatoos, and eclectus parrots decreasing as chicks matured.
     b. In the same 3 species total protein, albumin, globulin, aspartate transaminase, cholesterol, uric acid, calcium, sodium, and chloride values are initially lower than adult values but increase to adult values as the chicks mature.

D. Radiology[11]
  1. Extensive areas of cartilage are present, especially around the ends of long bones in very young chicks. Growth plates close around fledging age.[11]
  2. Soft tissues also appear different.

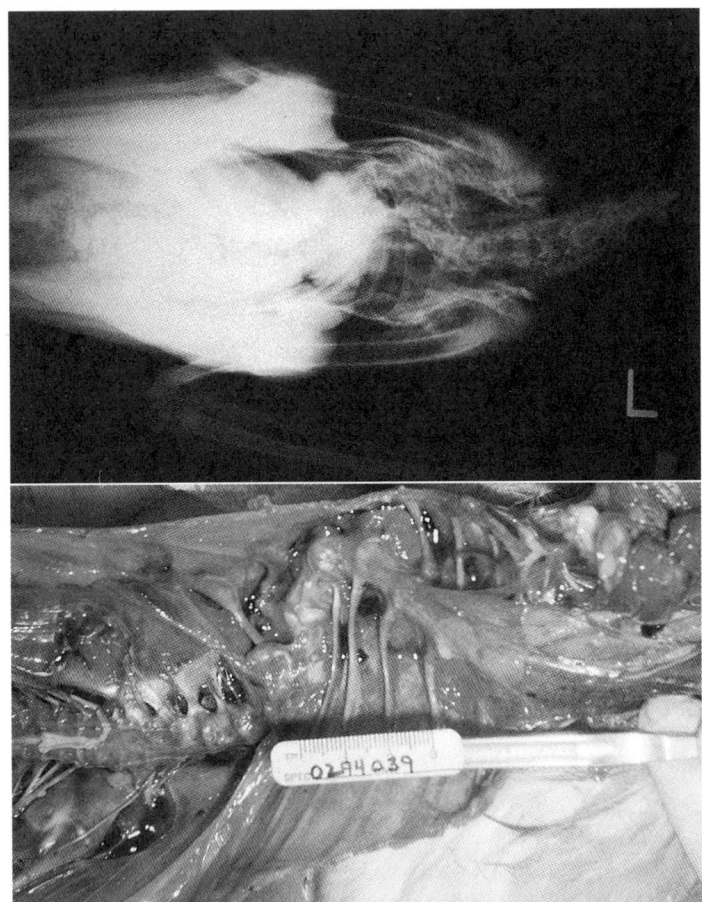

Fig. 13-1 A whooping crane chick. **A,** Radiograph. **B,** Necropsy. Both illustrations show severe curvature of the spinal column (scoliosis). In this species, this trait appears to be a genetically linked recessive abnormality.

a. Liver, heart, and proventriculus appear relatively larger than in adult.
b. Intestines appear large and fluid filled.
c. Little air sac space is seen as compared with radiographs of adults.

E. Microbiology
1. The normal microflora of psittacine chicks is primarily gram-positive: lactobacilli, micrococci, bacilli, corynebacteria, nonhemolytic streptococci, and *Staphylococcus epidermis*.
2. Gram-negative bacteria including *Escherichia coli*, *Klebsiella* spp., *Enterobacter* spp., *Pseudomonas* spp., and *Salmonella* spp. may be pathogens.
3. Some gram-negative bacteria (*E. coli*, *Enterobacter* spp., and *Klebsiella* spp.) are sometimes found in normal healthy psittacine chicks. Finding bacteria without clinical signs of disease is not diagnostic.
4. Raptor, waterfowl, and gallinaceous chicks normally have gram-negative flora, especially in cloacal cultures.
5. Commercial psittacine chick diets often contain brewer's yeast. This can be seen as nonbudding yeast on some smears.
6. Cultures of the crop and cloaca are useful to indicate what bacteria a chick is being exposed to in the environment.
7. Cultures of the choana can be useful to diagnose upper respiratory diseases.

VI. Developmental Problems

A. Yolk Sac Retention, Omphalitis
1. Exposed yolk sac may indicate elevated incubation temperature or premature removal of chick from egg. Exposed yolk sac is ligated, area treated with povidone iodine solution (1%), yolk sac remnants removed (see more complete description in Chapter 12).
2. Retained yolk sac can be due to improper incubation, weak chick, or infectious agents.
    a. Normally, yolk sac absorbed in 5 to 7 days
    b. Yolk contents degrade, leading to toxicity[7]
    c. Treatment is difficult, prognosis poor. Yolk sac removal has been done in larger species but not neonate psittacines.[12]
    d. Aspiration of yolk sac material using a syringe has been successful in some cases.

B. Beak Problems
1. Lateral deviation of the maxilla (scissors beak)
    a. Cause unknown, seen most frequently in macaws
    b. Trim lower beak to force upper beak to slide over to the opposite side. Also, applying finger pressure to the beak twice daily for several minutes will help correct the problem.[14]
    c. After beak has calcified, a dental acrylic ramp is used to apply pressure to correct the problem (see Chapter 17).

2. Mandibular prognathism/maxillary brachygnathism
    a. Causes include developmental and traumatic
    b. Most often seen in cockatoos
    c. Before beak calcifies, this can be corrected by applying finger pressure several times daily. Can use loop of gauze to apply traction at each feeding
    d. After calcification, a dental acrylic prosthesis may be required to help reshape the beak (see Chapter 17, Beak and Oropharynx Problems).

C. Congenital Atresia of the Choana
1. Seen in African grey parrots, in umbrella cockatoos, and a hybrid macaw[15]
2. Sinuses fail to drain through the choana; sinuses fill with fluids. Diagnosis is made by flushing the sinuses and seeing no fluid drain from the choana.
3. No treatment has been described.

D. Malformed Eyelids
1. Seen in cockatiels
2. Malformed eyelids and narrowed eye opening
3. Surgical treatments unsuccessful, but often chicks adapt to their limited vision.[16]

E. Occlusion of External Ear Openings
1. Most often seen in military macaws, other macaw species
2. Macaws hatch with thin membrane over ear canal. Canal should open at 12 to 35 days.
3. Covering over ear canal can be opened surgically if necessary.
4. Occasionally encounter purulent material under covering; will need to culture, flush, and treat with appropriate topical and systemic antibiotics.[3]

F. Congenital Heart Disease
1. Ventricular septal defects, and less commonly, cardiomyopathy, have been diagnosed in psittacine chicks and other species.

VII. Husbandry-Related Problems

A. Gastrointestinal Problems
1. Air in the crop
    a. Usual cause is aerophagia, seen most often in chicks that constantly beg for food. Also feeding too slowly leads to a chick gulping air with food.
    b. Using a bright light source, it is possible to transilluminate the crop area and see the presence of air.
    c. Treatments can be as simple as extending the neck and gently massaging the air out of the crop (burping out). Feeding a sufficient

quantity of food in a reasonably quick manner decreases air gulping and begging. In severe cases, tube feeding may be required.
2. Damage to the pharynx, esophagus, or crop
   a. Damage is often due to using metal tubes or the tip of the plastic feeding syringe, especially if the chick lunges for the food. Burns can result from a single hot meal (hotter than 120°F, 49° C) or several meals between 115° F to 120° F, 46° C to 49° C.[3]
   b. If a puncture occurs, food will move ventrally from the puncture site and produce a large inflammatory area in the tissue.
      (1) Surgically open the area; curet and flush the contaminated tissues.
      (2) Treat with a broad-spectrum antibiotic for 14 days.
   c. Mild burns cause blisters, swelling, and erythema and can be treated with vitamin A and E ointment applied topically plus broad-spectrum antibiotic therapy.
   d. Severe burns can lead to crop or esophageal fistulas.
      (1) Initially give supportive care, because the chick is usually severely debilitated and dehydrated.
      (2) Treatment is surgical excision of necrotic crop and skin, followed by separate closure of each tissue.
      (3) After surgery, feed small amounts frequently, especially if surgery substantially reduced crop size. Use a pharyngotomy tube if esophageal surgery was required.
      (4) Antibiotic therapy for a minimum of 7 days
3. Crop stasis
   a. Causes commonly seen in chicks:
      (1) Problem in the crop resulting from infection, burns, impaction, foreign bodies, cold food, overfeeding
      (2) Intestinal ileus of the distal gut
   b. Mild cases can be treated by adding a small amount of warm water or lactated Ringer's solution to the crop contents and gently massaging the crop.
   c. More severe cases require more aggressive treatment.
      (1) Administer supportive care through intravenous intraosseous or subcutaneous fluids, because chick is often dehydrated.
      (2) Flush crop with warm saline or lactated Ringer's solution using a feeding tube to administer the solution and withdraw the food mixture.
      (3) Gram stain and culture the crop contents. Start antibiotic and antifungal therapy based on the test results, or if general ileus is suspected, start on a broad-spectrum antibiotic (see Chapter 26).
4. Regurgitation
   a. Causes include crop stasis (see above), overfeeding, gastrointestinal infections or blockage, certain medications (doxycycline or trimethoprim-sulfa), or weaning.
   b. Need to differentiate husbandry problems from actual disease
      (1) Some hand-fed psittacines regurgitate frequently when weaning (especially African grey parrots and macaws).

(2) Avoid overfeeding, cut back on the amount fed if neonate is regurgitating frequently.
(3) Avoid using medications known to cause regurgitation unless absolutely necessary.

5. Gastrointestinal blockage and foreign body ingestion
   a. Young birds are extremely curious and may ingest any small object.
   b. Foreign bodies in the crop can be diagnosed by palpation, with the aid of radiographs, or by endoscopy.
      (1) Some objects can be gently maneuvered back up the esophagus.
      (2) The use of a forceps and endoscope allows for the retrieval of many objects.
      (3) Surgical retrieval is relatively simple, requiring an incision over the crop, then suturing each layer in turn on exiting the area.
   c. Foreign bodies in the proventriculus and ventriculus
      (1) Small foreign bodies may cause no problem and may remain here for some time.
      (2) Objects that are larger or potentially harmful (pointed, toxic materials, etc.) should be removed using a flexible forceps or basket and endoscopy, or surgically if needed.
   d. Impaction of the proventriculus/ventriculus
      (1) Associated with grit, bedding ingestion, or large dry bolus of food
      (2) Rehydrate patient and give a mild laxative (psyllium or dioctyl sodium sulfosucccinate)[3]
      (3) If this fails, try giving mineral oil (by feeding tube into crop) followed in 30 to 45 minutes by barium sulfate. Radiograph the patient to assess status of the impaction.[3]
      (4) If more conservative treatments fail, gastric lavage or even surgical removal of the blockage can be attempted, but prognosis is guarded to poor.
   e. Intestinal intussusception
      (1) Seen in young macaws with diarrhea and hypermotile intestines.[17]
      (2) Mild cases may respond to antibiotics and supportive care; more severe cases will require surgery.[17]

B. Musculoskeletal Problems
1. Splay leg or spraddle leg
   a. Chick will have one or both legs deviating laterally from the stifle or hip joints.
   b. Treatment can consist of forming hobbles of elastic tape connecting the tarsometatarsi of the two legs (or sometimes also the tibiotarsi).
   c. Mild cases sometimes respond to placing the chick in a deep cup nest with sufficient padding (tissues work fine) to remove pressure from the legs and keep the chick from standing or pushing.
2. Stifle problems
   a. Stifle luxations are common in young birds, usually resulting from the chick getting its leg caught on something on the cage.

    b. Reducing the luxation is usually not successful, rather, using KE apparatus in the distal femur and proximal tibiotarsus, the joint can be held in a flexed position until fused (3 to 6 weeks).
    c. A Moluccan cockatoo chick with a luxated stifle was successfully treated using a SAM splint (a soft metal splint covered with foam padding).[18]
3. Valgus deformity
    a. A premature closure of one side of the growth plate (usually lateral) of the proximal or distal tibiotarsus leads to bowing of the tibiotarsus and lateral rotation of the femur and/or the tibiotarsus.
    b. The only effective repair is to surgically close the opposite growth plate, or to do periosteal stripping or a wedge osteotomy followed by repair with a KE apparatus. The latter procedure is done only after the long bones have ossified.
4. Crooked toes
    a. In neonates may be caused by improper incubation, improper substrate, or nutrition
    b. Reducing certain dietary proteins and slowing growth has been a factor in reducing toe problems in psittacines[19] and cranes.[20]
    c. Taping the crooked toe to hold it in a normal position is an effective treatment.
        (1) A crooked toe can develop within a few hours, but responds quickly to treatment.
        (2) The tape should remain in place for as long as the toe is crooked, but not more than 48 hours.
        (3) Monitor the chick frequently to ensure that the taping is not causing discomfort or constrictions.
    d. Sometimes several toes on one foot or the entire foot is crooked.
        (1) These cases respond well if a snowshoe is made from thin cardboard (file cards work well) or even x-ray film.
        (2) The shoe is cut to fit the shape of the foot.
        (3) Each digit is taped to the shoe individually, using a low-tack adhesive tape (Fig. 13-2).
5. Toe constrictions
    a. A constrictive ring forms, causing swelling and eventual necrosis of the toe distal to the ring.
    b. Most commonly seen in African grey parrots, eclectus parrots, and macaws. Digits most commonly affected are the outer ones (1 and 4), usually the most distal phalanx.[3]
    c. Sometimes forms as a result of low humidity in brooder or nursery or due to a fracture or trauma of the digits.[21]
    d. Mild constrictive lesions respond to warm water soaks and massage to increase circulation and break down the ring.
    e. More severe cases require incision and some debridement of the constrictive ring of tissue, followed by bandaging the digit using povidone-iodine dressing or DMSO and antibiotic ointment.
    f. Severe cases with necrotic digits require amputation of the necrotic portion of the toe.

Fig. 13-2 A corrective snowshoe made from file card material and attached to the foot with low-tack adhesive tape. This device is useful when multiple toes on one foot are crooked. Usually the device is left on for no more than 48 hours. If after removal the crooked toe condition persists, the device can be reapplied.

VIII. Infectious Problems

A. Bacterial Infections
1. Bacterial infections of the respiratory and gastrointestinal tracts are common problems.
   a. Often isolate *E. coli, Enterobacter* spp., *Klebsiella* spp.[3]
   b. Isolation of potential pathogens, especially if seen only as light growth, in a healthy bird may not require treatment.
2. Transovarian transmission of *Salmonella pullorum, Salmonella enteritidis, Salmonella gallinarum,* and *Mycoplasma* spp. can occur, resulting in sick or dying neonates (see Chapter 12 for other diseases transmitted in the egg).
3. *Bordetella avium* can invade the skeletal muscle of the mandible, resulting in a "lockjaw"-like disease seen most often in hand-fed birds, especially cockatiels.[7]

REFERENCES
1. Clubb K, Clubb SL: Management of psittacine chicks and eggs in the nest. In Schubot RM, Clubb K, Clubb SL, editors: *Psittacine aviculture: perspectives, techniques and research,* Loxahatchee, Fla, 1992, Avicultural Breeding and Research Center, pp 15/1-15/16.
2. Clubb K, Clubb SL: Psittacine neonatal care and handfeeding. In Schubot RM, Clubb K, Clubb SL, editors: *Psittacine aviculture: perspectives, techniques and*

research, Loxahatchee, Fla, 1992, Avicultural Breeding and Research Center, pp 11/1-11/12.
3. Flammer K, Clubb SL: Neonatology. In Ritchie BW, Harrison GJ, Harrison LR, editors: *Avian medicine: principles and applications,* Lake Worth, Fla, 1994, Wingers, pp 805-838.
4. Roudybush TE: Growth, signs of deficiency, and weaning in cockatiels fed deficient diets, *Proceedings Association Avian Veterinarians,* 1986, pp 333-340.
5. Takesita K: Hypervitaminosis D in baby macaws, *Proceedings Association of Avian Veterinarians,* 1986, pp 341-346.
6. Clubb K: Growth rates of hand-fed psittacine chicks. In Schubot RM, Clubb K, Clubb SL, editors: *Psittacine aviculture: perspectives, techniques and research,* Loxahatchee, Fla, 1992, Avicultural Breeding and Research Center, pp 14/1-14/19.
7. Clubb SL: Psittacine pediatric husbandry and medicine. In Altman RB, Clubb SL, Dorrestein GM, Quesenberry K, editors: *Avian medicine and surgery,* Philadelphia, 1997, WB Saunders, pp 73-95.
8. Clubb SL, Schubot RM, Joyner KL: Hematological and serum biochemistry reference intervals in juvenile eclectus parrots (*Eclectus roratus*), *J Assoc Avian Vet* 4:218-225, 1991.
9. Clubb SL, Schubot RM, Joyner KL: Hematological and serum biochemistry reference intervals in juvenile macaws, *J Assoc Avian Vet* 5:154-162, 1991.
10. Clubb SL, Schubot RM, Joyner KL: Hematological and serum biochemistry reference intervals in juvenile cockatoos, *J Assoc Avian Vet* 5:5-16, 1991.
11. Joyner KL: Psittacine pediatric diagnostics, *Semin Avian Exotic Pet Med* 1:11-21, 1992.
12. Cambre RC: 1992. Indications and techniques for surgical removal of the avian yolk sac, *J Zoo Wildlife Med,* 23:55-61, 1992.
13. Joyner KL: Theriogenology. In Ritchie BW, Harrison GJ, Harrison LR, editors: *Avian medicine: principles and applications,* Lake Worth, Fla, 1994, Wingers, pp 748-804.
14. Wolf S, Clubb SL: Clinical management of beak malformations in handfed psittacine chicks. In Schubot RM, Clubb K, and Clubb SL, editors: *Psittacine aviculture: perspectives, techniques and research,* Loxahatchee, Fla, 1992, Avicultural Breeding and Research Center, pp 17/1-17/11.
15. Greenacre CB, Watson E, Ritchie BW: Congenital atresia of the choana in an African grey parrot and an umbrella cockatoo, *J Assoc Avian Vet* 7:19-22, 1993.
16. Buyukminci NC: Eyelid malformation in four cockatiels, *J Amer Vet Med Assoc* 196:1490-1492, 1990.
17. Van Der Heyden N: Jejunostomy and jejunoclocal anastamosis in macaws, *Proceedings of the Association of Avian Veterinarians,* 1993, pp 72-77.
18. Clipsham R: Noninfectious diseases of pediatric psittacines, *Semin Avian Exotic Pet Med* 1:22-23, 1992.
19. Clipsham R: Correction of pediatric leg disorders, *Proceedings of the Association of Avian Veterinarians,* 1991, pp 200-204.
20. Olsen GH, Langenberg J: Veterinary techniques for rearing crane chicks. In Ellis DH, Gee GF, Mirande CM, editors: *Cranes: their biology, husbandry, and conservation,* Washington, D.C., 1996, National Biological Service, pp 95-104.
21. Joyner KL, Abbott U: Egg necropsy techniques, *Proceedings of the Association of Avian Veterinarians,* pp 146-152, 1991.

14

Avian Toxicology

Donita L. Frazier

Overview

Exposure to toxicants is less common in captive birds than in free-ranging birds and mammals. Although information regarding avian toxicoses is limited, compounds that are toxic to mammals should be considered toxic to birds. Because of their size and physiology, birds are more sensitive to some toxicants than are mammals. Sensitivity to toxicants varies widely among different bird species. Most toxicities in captive birds appear as acute illnesses, whereas free-ranging birds often present following chronic exposures to toxicants in contaminated environments. A thorough history from the owner may be the most important information obtained for diagnosis of toxicosis.

I. General Therapeutic Approaches

A. Prevent Further Exposure to the Toxicant
1. Remove birds from contaminated environment or remove source of toxicant exposure.
2. Remove toxicant from feathers, skin, or eyes.
   a. Wash with mild liquid hand-washing detergent. Eyes should be flushed with physiologic saline for 30 minutes.
   b. Rinse thoroughly with large volumes of water.
   c. Caustic alkali (Drano, Clinitest Tablets, "button" batteries) and acid (toilet bowl cleaners, swimming pool cleaners) compounds should be flushed with copious amounts of water. Attempts to neutralize these compounds with vinegar or sodium bicarbonate, respectively, will result in exothermic reactions that may result in more severe damage.
3. Decrease absorption from gastrointestinal tract (GIT)
   a. Gavage crop or proventriculus 3 to 4 times using soft rubber catheter. Proventricular gavages are possible only in birds that

are stable enough to be lightly anesthetized with isoflurane. Placement of an endotracheal tube is essential to avoid aspiration.
  (1) Use saline or activated charcoal slurry and aspirate contents.
    (a) Efficacy of gavage decreases with time after ingestion; unless there is GI stasis, gavage is recommended within 1 to 3 hours after exposure.
    (b) Solid metal objects may necessitate surgery.
  (2) The use of emetics is considered unsafe and ineffective in birds.
b. Use cathartics for toxicants in lower GIT.
  (1) Sodium sulfate or magnesium sulfate (Epsom salt) (0.5 to 1g/kg orally). The magnesium form may result in signs of toxicity.
  (2) Activated charcoal as a slurry
    (a) 2 to 8 g/kg with 1 g charcoal/5 to 10 ml water[1]
    (b) Multiple dose activated charcoal (MDAC) is given by continuous infusion (0.2g/kg/hr) to humans to facilitate removal of toxicants that have been absorbed. It is effective for compounds that undergo enterohepatic cycling or in patients in whom compromised renal function decreases renal clearance of the toxicant.[2]
    (c) Not all compounds are well adsorbed by activated charcoal.[2]
      (i) Compounds that are well adsorbed include atropine, arsenic, barbiturates, camphor, cocaine, digitalis, iodine, malathion, mercuric chloride, morphine, nicotine, oxalates, parathion, penicillin, phenol, phenothiazine, phosphorus, salicylates, selenium, silver, strychnine, sulfonamides
      (ii) Compounds that are not significantly adsorbed include iron, lithium, borates, bromide, potassium, mineral acids, and alkalis.
      (iii) Compounds that are adsorbed to some extent include cyanide, diazinon, DDT, carbamates, ethylene glycol, kerosene, and turpentine.
  (3) Psyllium hydrophilic mucilloid
    (a) Metamucil (Proctor and Gamble, Cincinnati, OH) ½ tsp in 60 ml baby food gruel
    (b) Oil-based cathartics, such as mineral oil, are not recommended because of the risk of aspiration.
    (c) Monitor patients closely; adverse effects of cathartics include dehydration, hypernatremia, hypermagnesemia, hyperphosphatemia, hypokalemia, and hypocalcemia.
    (d) Whole bowel irrigation is used in humans for acute ingestion of metals (iron, lead) and some drugs.
      (i) Gastric intubation and administration of isosmotic electrolyte lavage solution containing polyethylene glycol (Colyte or GoLytely (25 to 40ml/kg/hr) until passage of clear rectal effluent.[3]

(ii) Less evidence of adverse effects than with cathartics
(iii) Contraindications include GI stasis, GI obstruction, GI hemorrhage, GI perforation

B. Institute Supportive Therapy
1. Warm oxygen cage
2. Fluids as needed
   a. Correct acid-base balance if needed
3. Avoid stressing bird

C. Provide Specific Antidotal Therapy, if Available
1. Resources for information
   a. Illinois Animal Poison Information Center, University of Illinois College of Veterinary Medicine, Urbana, IL 1-800-548-2423
   b. Georgia Animal Poison Information Center, Athens, GA 1-404-342-6751
   c. Most states have Poison Control Centers that provide information regarding therapies in humans.

II. Drug and Vitamin Toxicities

A. General Information
1. All drugs, in varying degrees, cause undesirable side effects.
2. Although idiosyncratic reactions occur, adverse drug reactions are most often dose-dependent; this is particularly evident in birds when treatment regimens are extrapolated from protocols for mammals.
3. Adverse drug effects have most often been noted in small birds such as finches and canaries and in softbills.
4. It is essential to weigh small birds daily to avoid improper dosing.
5. Adding medications to water should be avoided whenever possible.
6. Although the most common result is underdosing of birds that refuse to drink unpalatable solutions, polydipsic or heat-stressed birds may be overdosed by drinking more than expected amounts.
7. In all cases of drug-induced toxicity, the drug should be discontinued and symptomatic therapy initiated.

B. Aminoglycosides
1. Nephrotoxic and ototoxic[4,5]
   a. Dose-dependent and perhaps species-dependent
      (1) Greater gentamicin toxicity in rose-breasted cockatoos than in scarlet macaws[6]
      (2) Avoid using gentamicin doses higher than 2.5 to 5.0 mg/kg q8h or q12h[6,7] or amikacin doses higher than 10 to 15mg/kg q8h or q12h.[8,9]
   b. Monitor renal function during treatment
      (1) Polydipsia and polyuria are common during treatment.[8]
      (2) Nephrotoxicity is generally reversible during polyuric stage, irreversible during oliguric or anuric stage.

c. Amikacin is less nephrotoxic than gentamicin.
2. Neuromuscular synaptic dysfunction and paralysis if given by rapid intravenous injection[10]
3. Maintain adequate hydration, saline diuresis if toxicity develops.

C. Amphotericin B
1. Nephrotoxic
   a. May be species-dependent
      (1) Long-term use in raptors was not associated with nephrotoxicity.[11]
      (2) Nephrotoxicity has not been evaluated adequately in all species; therefore, all birds should be closely monitored for polydipsia and polyuria during treatment.
   b. Concurrent use with other nephrotoxic drugs is contraindicated.
   c. Several new lipid formulations that are less nephrotoxic are available in Europe and the United States. Comparative clinical trials have not been conducted in birds; however, use in human patients suggests these formulations are as effective as the parent amphotericin B compound and may be more widely distributed and slowly eliminated from tissues (larger volume of distribution and prolonged half-life). The current limitation to use of the new formulations is cost.
2. Hepatotoxic
3. Acute reactions that commonly occur in mammals after first dose include fever, hypotension, anorexia, vomiting, and tachypnea.
4. Rapid intravenous infusion in mammals may result in hypotension, hypokalemia, arrhythmias, and shock.
   a. Transitory incoordination and mild convulsions reported in birds given rapid intravenous injections of doses greater than 1.0 mg/kg.[11]
5. Anaphylaxis
6. Necrotizing if given subcutaneously or perivascularly, commonly causes phlebitis
   a. Limited aqueous solubility. The manufacturer recommends against dissolving in anything other than water or 5% dextrose. The use of saline or bacteriostatic agents may cause precipitation of amphotericin B.[12]
   b. The recommended dilution is 0.1 mg/ml.
   c. Caution against using as a concentrated flush or saline nebulization; this may instill highly corrosive particulates.
      (1) Severe granulomatous reaction when used as a sinus flush[13]
      (2) Although nebulization of this drug has been recommended, studies are needed to confirm drug delivery by this method.

D. Anesthetics
1. Halothane
   a. In contrast to a stable pattern of tachypnea and low tidal volume induced in mammals, halothane disrupts intrapul-

monary chemoreceptors in birds and causes cardiopulmonary instability and a high incidence of mortality in some avian species (sandhill cranes and small psittacines).[14]
    b. Use isoflurane instead.
  2. Xylazine
    a. When used alone, causes excitation, and in some species convulsions[15]
  3. Ketamine
    a. May produce excitation or convulsions in Gallinules *(Porphyria* spp.*)*, the water rail *(Rallus aquaticus)*, the golden pheasant *(Chrysolophus pictus)*, Hartlaub's turaco *(Tauraco hartlaubi)*, the crowned hornbill *(Tockus alboterminatus)*, vultures *(Gyps rueppellii* and *G. africanus; Torgos tracheliotus, Trygonoceps occipitalis)*, and pigeons *(Columba livia)*.[15]

E. Beta Lactams (Penicillins and Cephalosporins)
  1. These drugs are remarkably nontoxic.
  2. Older cephalosporin, cephaloridine, is nephrotoxic.
  3. Some repositol penicillins contain procaine.
    a. Procaine is an anesthetic.
    b. Intravenous administration is contraindicated because of cardiotoxicity.
    c. Procaine penicillins degrade into procaine and penicillin when heated.
      (1) These drugs must be kept refrigerated.
    d. Adverse reactions described in small birds (finches, canaries, budgerigars, cockatiels)[16] are likely due to procaine.
    e. Penicillin-induced allergic reactions have not been well documented in birds but may occur.[17]
      (1) Treat with epinephrine and O_2.

F. Calcium (See Vitamin D_3)

G. Chloramphenicol
  1. Reversible bone marrow depression in mammals and birds[18,19]
  2. Irreversible non–dose-related bone marrow aplasia in humans[20]
    a. Chloramphenicol should not be dispensed to owners.
    b. It is illegal to use chloramphenicol in birds intended for food.

H. Digitalis Glycosides (Digoxin, Digitoxin, Oleander, Foxglove)
  1. Limited studies in birds; digoxin appears to have varying effects and low therapeutic index in sparrows,[21] parakeets,[21] chickens,[22] Quaker conures,[23] African grey parrots.[24]
    a. Dose in mammals varies allometrically: 0.005 mg/kg for giant breed dogs versus 0.02 mg/kg for toy breed dogs
    b. Recommended dose in birds is 0.02 to 0.05 mg/kg.[21,23]
  2. Vomiting
  3. Bradycardia, arrhythmias, heart block

4. Treat with lavage, charcoal, antiarrhythmics (phenytoin), atropine
5. Resin binding agents used in humans (cholestyramine and colestipol) prevent absorption.

I. Enrofloxacin

1. Damage to developing articular cartilage is well recognized in mammals.
   a. May be dose-dependent and/or species-dependent[25,26]
      (1) Joint deformities reported in offspring of pigeons chronically treated with high doses of enrofloxacin in drinking water (800 ppm)[26]
      (2) Used in many psittacine nurseries without reports of side effects[16]
2. Intramuscular injection causes irritation at site of injection; the intramuscular formulation is safe and efficacious when given orally to birds at 15mg/kg.[16]
3. Polyuria in African grey parrots[27]
   a. Resolved upon discontinuation of the drug

J. Levamisole

1. Regurgitation, ataxia, recumbency, dyspnea, death
   a. Dose-dependent: well tolerated in most birds at 22 mg/kg[28]
   b. Regurgitation common with oral dosing

K. Monensin and Salinomycin[29-32]

1. Feed refusal and growth depression are common in poultry, even at recommended anticoccidial concentrations.
   a. Cream-colored diarrhea is common.
2. Dyspnea, muscular stiffness, or extreme weakness, sternal recumbency, flaccid paralysis
3. Cardiotoxic
   a. Increases movement of Na^+ and K^+ into cells, causing efflux of H^+; Ca^{++} enters cells and damages mitochondria
4. Decreased egg production and hatchability
5. Toxicity appears to vary with the species and dose.
   a. Toxicity is generally due to incorrect mixing of feed or ingestions of feed designed for a less susceptible species by a more susceptible species.
   b. Caution when using in species in which this drug has not been shown to be safe
      (1) Adult turkeys and guinea fowl are especially sensitive.
      (2) Safe in chickens and quail at recommended dose
6. Diagnosis is generally dependent on clinical signs in the flock, because microscopic lesions in heart and skeletal muscle include fragmentation of fibers and intermyofibrillar vacuolation, interstitial infiltration by macrophages and neutrophils.
   a. Histopathologic lesions may be absent in acutely intoxicated birds, and tissue residues decline over time with freezing.

L. Nitrofurans

1. Antibacterial concentrations (MICs) induce widespread systemic toxicity in mammals including nausea, vomiting, hypersensitivity reactions, neurologic dysfunction, and bone marrow suppression.
2. Adrenal cortical alterations reported in birds[33,34]
3. With the exception of topical use, nitrofurans are banned from use in animals intended for food because of carcinogenicity.[35]
4. 1/2 tsp nitrofurazone powder (9.2%) per gallon of drinking water anecdotally reported to be nontoxic to psittacine birds; use should be limited to organisms shown to be sensitive to the dose.[36]

M. Sulfonamides

1. Nephrotoxic
   a. Low solubility, crystalluria
   b. Contraindicated in uricemic or dehydrated birds
2. Hypersensitivity reactions
   a. Hemorrhagic syndrome in gallinaceous birds[37,38]

N. Tetracyclines (Doxycycline, Oxytetracycline)

1. Outdated tetracycline is nephrotoxic.
2. Necrosis at site of intramuscular injection[16,39]
3. GI upset
   a. Regurgitation is not uncommon with oral doxycycline. If this occurs, divide dose and give q12h instead of q24h.
   b. Long-term dosing may alter normal microbial flora.
      (1) Likely contributed to voluminous stools reported in macaws fed pelleted diets containing chlortetracycline[40]
      (2) May lead to overgrowth by *Aspergillus* spp. and other fungal organisms[41]
4. Photosensitization
5. Hepatotoxicity
   a. Less common in mammals with doxycycline; however, hepatotoxicity has been reported with doxycycline administration in macaws.[16]
6. Cardiovascular collapse
   a. Can occur after rapid intravenous injection resulting from propylene glycol carrier

O. Trimethoprim-Sulfonamides

1. Regurgitation is not uncommon in birds.[16]
2. Sulfonamides precipitate in kidneys of dehydrated birds.
3. Injectable formulations may cause irritation and necrosis at site of injection.
4. Rashes, photosensitization, hepatotoxicity, and arthritis have been reported in mammals but not birds.

P. Vitamin A[42]

1. Associated with oversupplementation

2. Osteodystrophy
   a. Thickening of metaphyses, hyperosteoidosis, metaphyseal sclerosis
3. Parathyroid gland hyperplasia

Q. Vitamin D_3[43]
1. Associated with oversupplementation
2. Mineralization of liver, kidneys, stomach, intestines, heart, blood vessels
   a. Macaws and African grey parrots may be particularly sensitive.
3. Hypercalcemia may affect cardiac conduction.
4. Skeletal abnormalities

R. Drug Formulations
1. Procaine
   a. Available as long-acting repositol antibiotic formulations
      (1) Hydrolyzed by nonspecific carboxylesterases in blood and tissues
      (2) Procaine formulations dissociate when heated; these drugs must be kept refrigerated.[44]
   b. Procaine is an anesthetic.
   c. Intravenous administration is contraindicated because of cardiotoxicity.
2. Propylene glycol
   a. Many drugs, when given intramuscularly, can cause focal myositis.
      (1) May be due to propylene glycol or the drug itself[16,39]
      (2) May see elevations in serum AST, LDH, CK
         (a) Significantly greater concentrations of AST and LDH in extrahepatic avian tissues than in mammalian tissues.[45]

III. Metal Toxicity

A. Copper
1. General information
   a. Some birds, including swans, chickens, turkeys, pigeons, and ducks, appear to tolerate higher concentrations than mammals.[46,47]
   b. Sources include copper wire, pennies, antifowling paints, and copper sulfate used to control algae in ponds.
   c. Copper metabolism and toxicity are interrelated with molybdenum and sulfate.
2. Clinical signs and diagnosis[48]
   a. Generalized weakness
      (1) Anemia
   b. Intravascular hemolysis in mammals
   c. Hepatic and renal necrosis
   d. Acute poisoning from a copper salt may result in coagulation necrosis of the intestinal mucosa.

e. Normal copper concentrations in whole blood range from 0.7 to 1.3 ppm in mammals; liver concentrations greater than 150 ppm or kidney concentrations greater than 15 ppm are associated with toxicity in mammals.
   3. Treatment
      a. D-penicillamine
         (1) Dose for mammals: 52 mg/kg/day × 6 days[49]
         (2) CaEDTA is ineffective.
      b. Supportive care

B. Iron
   1. General information
      a. Rare in most birds, associated with cast iron food or water bowls with chipped enamel.
      b. Also available in iron supplement tablets and multivitamins.
      c. In mammals, a high sugar diet has been shown to increase absorption, whereas phosphates reduce intestinal absorption of iron.
      d. A mucosal transfer system has been identified in mammals.
      e. Some birds, including toucans and mynah birds, are predisposed to development of hemochromatosis (iron storage disease) when fed captive diets.[50]
      f. It is generally thought that this is due to adaptive mechanisms for low iron content of food of free-ranging birds.
   2. Clinical signs and diagnosis
      a. Anorexia, emaciation
         (1) Severe, hemorrhagic necrosis of GI mucosa due to direct corrosive effect of iron on tissue documented in mammals[2]
      b. Lethargy
         (1) Cardiovascular collapse, metabolic acidosis (release of H^+ when Fe^{3+} is converted to Fe^{2+}), cyanosis, and fever documented in acute toxicosis in mammals
         (2) Dyspnea may occur secondary to ascites in chronic storage disease.
      c. Hepatic dysfunction
         (1) Hepatic necrosis develops several days after acute exposure in mammals (toxic effect of iron on mitochondria).
         (2) Abdominal distention resulting from hepatomegaly and ascites with iron storage disease
      d. Diagnosis
         (1) Radiographic evidence
         (2) Elevated serum iron
         (3) Elevated WBC count and/or blood glucose in mammals
         (4) Elevated liver enzymes and low plasma proteins with iron storage disease
   3. Treatment
      a. Deferoxamine is used in humans, but safe dose has not been determined in birds.

b. CaEDTA and DMSA chelate iron as well as lead.
   c. Severely dyspneic birds may require abdominocentesis and a diuretic; phlebotomy for hemochromatosis.
   d. Birds predisposed to hemochromatosis should be fed diets low in iron.
   e. Prognosis is poor.

C. Lead
 1. General information
     a. Lead is the most commonly reported avian toxicosis.
     b. Lead is ubiquitous in the environment.
     c. Sources of exposure include paint, linoleum, plaster, caulking compounds, batteries, plumbing materials, galvanized wire, solder, hardware cloth, stained glass, curtain weights, fishing weights, lead shot.
     d. Consider lead toxicosis in any bird with a combination of GI and neurologic signs.
 2. Clinical signs and diagnosis[51-53]
     a. Acute toxicosis is more common in captive birds, whereas chronic toxicities are most common in anseriformes and other free-ranging birds.
     b. Lead toxicosis is more common in diving ducks than in surface-feeding ducks.
     c. In general, young birds are more susceptible than are adults.
     d. GI signs
         (1) Anorexia, chronic weight loss, emaciation[54,55]
             (a) A single No. 4 lead shot in ring-necked ducks resulted in blood concentrations approximately 100 times the accepted toxic threshold in humans; 20 of 23 birds became severely emaciated and 4 died within 2 to 3 weeks.[56]
             (b) An absence of clinical signs noted in mallards dosed with one No. 4 shot and maintained on a balanced diet is likely a result of differences in dose on a body weight basis.[57]
         (2) Regurgitation
         (3) Diarrhea, often bile-green in color[55]
         (4) Crop stasis, esophageal or proventricular impaction resulting from myoneural dysfunction often seen in waterfowl, poultry, pigeons[53,56,58]
     e. Neurologic signs[55,59]
         (1) Head tremors or generalized seizures
             (a) Seizures reported more commonly in cockatiels than in parrots or parakeets[55]
         (2) Behavioral changes
             (a) Depression, circling, lethargy
         (3) Blindness
         (4) Peripheral nerve weakness

- (a) Wing droop, leg paresis, ataxia[55,56,60]
- (b) Antagonist to Ca^{2+} at myoneural junction[61]
- (5) Histopathologically, may see segmental demyelination and CNS edema[62-64]
    - (a) Experimental studies in mallards suggest the CNS is relatively more resistant to the effects of circulating lead compared with mammals; however, peripheral nerves are much more susceptible than the CNS.[65]

f. Renal signs
  - (1) Hematuria
    - (a) Reported more commonly in parrots than in parakeets or cockatiels[55]
  - (2) Hemoglobinuria reported in Amazon and African grey parrots
    - (a) Secondary to intravascular hemolysis[66,67]
  - (3) Renal tubular necrosis[68]
  - (4) Polyuria, polydipsia

g. Necrohemorrhagic enteritis secondary to *Clostridium perfringens* was reported in chronically exposed swans.[52]

h. Diagnosis
  - (1) Radiographs may show metallic densities in GIT.
  - (2) Hematology
    - (a) Hypochromic regenerative anemia[52,59,69]
      - (i) Premature destruction and decreased production of RBCs
      - (ii) Cytoplasmic vacuoles and basophilic stippling (hemoglobin precipitation) of erythrocytes are not always obvious in birds[70]

i. Clinical chemistries
  - (1) May see increased levels of creatinine, uric acid, total protein, aspartate aminotransferase (AST), lactate dehydrogenase (LDH), or creatine phosphokinase (CPK).[71]

j. Blood lead analysis
  - (1) Do not use EDTA tubes; submit whole blood (most lead in blood is in RBCs)[72] in heparinized tubes.
  - (2) If greater than 20 µg/dl (0.2 ppm) plus clinical signs, lead exposure is likely
    - (a) Some birds having characteristic clinical signs but lower levels will respond to treatment.
    - (b) Because of apparent species variability, most avian clinicians will treat with a strong suspicion of lead toxicosis even if blood levels seem within normal range.
      - (i) Positive response to treatment indicates correct diagnosis.
  - (3) Some clinicians consider blood concentrations greater than 50 µg/dL (0.5 ppm), even without characteristic signs to be diagnostic; however, clinically normal birds sometimes have much higher blood lead levels.[70]
  - (4) Tissues (fresh frozen kidney, liver, brain, bone) may also be submitted.

(5) Lack of correlation between blood or tissue concentrations and clinical signs may be due to species differences, duration of exposure, or rate of absorption.
- k. Elevated protoporphyrin levels[52,56,69,70,73] or decreased (-aminolevulinic acid dehydratase (ALAD) activity[56-58,70] may be detected in some, but not all, birds.
  - (1) Altered protoporphyrin or ALAD activities often occur early after exposure and return toward normal in the presence of continuing clinical signs.[56]
  - (2) Protoporphyrin levels should be corrected for hemoglobin content.[69]
- l. Postmortem findings include vascular necrosis, multifocal myocardial degeneration, and renal nephrosis.
3. Therapy
   - a. Remove lead from GIT.
     - (1) Gavage
     - (2) Cathartics
       - (a) Metamucil
       - (b) Sodium sulfate (Glauber's salts) or barium sulfate
       - (c) Activated charcoal (2 to 8 g/kg) as a slurry
         - (i) Inactivated if given concurrently with mineral oil; mineral oil should also be avoided because of risk of aspiration.
     - (3) Surgery, if indicated
   - b. Chelate absorbed lead
     - (1) Make certain lead is out of GIT, because EDTA increases intestinal permeability and may increase lead absorption.[74]
     - (2) Give Ca EDTA 35 mg/kg q12h intramuscularly (it is poorly absorbed when given orally) for 5 to 10 days.[52,75]
       - (a) Some birds may require repeated treatments several days after the first treatment; this allows further equilibration of lead from soft tissues into blood.
       - (b) Oral CaEDTA at 60 mg/kg q8h has been used after injections and resolution of clinical signs.[66]
     - (3) Alternate oral therapies include penicillamine 55mg/kg q12h for 1 to 2 weeks or dimercaptosuccinic acid (DMSA) 25 to 35 mg/kg q12h.[51,58]
     - (4) Evaluate renal function because chelators are nephrotoxic.
       - (a) If birds become polyuric/polydipsic or develop proteinuria or hematuria during treatment, discontinue therapy for 5 to 7 days.
   - c. Supportive therapy
     - (1) Birds with neurologic signs or chronic exposure have poor prognosis.
     - (2) Fluids
       - (a) Lactated Ringer's or 5% dextrose
     - (3) May give corticosteroids to relieve cerebral edema
       Persistent seizures may be treated with diazepam 0.5 to 1.0 mg/kg q8h or q12h as needed.

D. Mercury
1. General information
   a. Present in thermometers, hearing aid and watch batteries, antiquated cathartics and ointments.
   b. Metallic mercury is relatively nontoxic unless it is converted to an ionic form by acids.
   c. Mercury is primarily a problem in free-ranging, fish-eating birds.
   d. Mercury bioaccumulates and adult birds have higher body burdens than young birds.
   e. Seasonal dietary differences and molting influence body burdens.[76]
2. Clinical signs
   a. Inorganic forms of mercury are corrosive.
      (1) Vomiting, diarrhea, ulceration, hemorrhage
   b. Organic compounds damage the kidney.[77]
      (1) Nephrosis, hematuria, oliguria
      (2) Vomiting, diarrhea
      (3) Neurologic disturbances
      (4) Primarily from eating mercury-contaminated seafood
3. Treatment
   a. Lavage and charcoal slurry
   b. d-Penicillamine
      (1) Contraindicated in renal failure because this is the route of excretion
      (2) Alternate treatment is dimercaprol (BAL); 2 to 5 mg/kg q4-12h is human dose

E. Selenium
1. General information
   a. Essential in diet.
   b. Toxicosis is seen primarily in birds in wetlands receiving agricultural drainage from soils high in selenium.[78-80]
   c. May occur with consumption of selenium-contaminated fish in powerplant cooling reservoirs.[81]
   d. Also reported in poultry fed naturally contaminated grains from seleniferous locations.[82]
2. Clinical signs and diagnosis
   a. High incidence of congenital malformations
      (1) Legs, toes, wings, beaks, eyes deformed, rudimentary, or absent[82-84]
   b. Poor reproductive success
      (1) Bioaccumulates in eggs[85]
   c. Acute mortality
      (1) Clinical signs of selenosis in other species include liver damage, enlargement of spleen and pancreas, anemia.
   d. Diagnosis based on clinical signs and selenium concentrations in liver and eggs
      (1) Information regarding normal and toxic concentrations in birds is limited.

(2) Wet weight concentrations in eggs >1 ppm are suggestive of toxicosis; >5 ppm is diagnostic of toxicosis.[86]
(3) Dry weight liver concentration >15 ppm is considered elevated.[79]

3. Treatment
   a. Symptomatic, no antidote available

F. Zinc
1. General information
   a. Birds often present with combined lead and zinc toxicosis.
   b. Sources include galvanized containers, hardware cloth, "bright, shiny" wires, zinc sulfate, pennies, and food and water contamination.
   c. Ingestion of metal objects by ostriches (*Struthio camelus*) is not uncommon.[87]
2. Clinical signs and diagnosis[88,89]
   a. Polyuria, polydipsia
   b. Anorexia, diarrhea, weight loss
   c. Seizures
   d. Histopathology: focal mononuclear degeneration of kidneys, liver, pancreas[90]
   e. Diagnosis
      (1) Palpation and radiographs
      (2) Collect and store samples in glass or plastic without metal or rubber parts.
         (a) Zinc can leach from rubber and artificially elevate blood concentrations.[91]
      (3) Check with laboratory regarding normal concentrations.
         (a) Macaws suffering from zinc toxicoses had blood levels greater than 200 µg/dl.
         (b) Collect pancreas, liver, or kidney for postmortem zinc analysis.
3. Treatment
   a. See treatment for lead toxicity.
   b. Mesh wire should be scrubbed with a mild acid, such as vinegar, or left outside to weather for 1 to 2 months prior to exposing birds.

G. Arsenic
1. Clinical signs[92]
   a. Polyuria and polydipsia
   b. Pruritis, feather picking
   c. Cystic ovaries, egg binding
2. Treatment
   a. Symptomatic
   b. It is particularly important to maintain hydration.
      (1) Diuresis will aid in excretion of arsenic.

IV. Pesticide Toxicity

A. General Information about Pesticides
1. Although some, but not all, pesticide toxicoses are treatable, the best approach for management of pesticide toxicity is client education.
2. An estimated 434,000,000 kg of pesticides are used annually in the United States.
3. Farmers use approximately 3/4 of this amount, with the remainder used around buildings and on lawns and gardens.
4. Of the total amount used, 69% are herbicides, 19% are insecticides, and 12% are fungicides.
5. Approximately 60% of all acreage devoted to crops are treated with pesticides; however, more than 90% of most crops other than forage are treated with pesticides.
6. These grains, fruits, and vegetables have levels of pesticides that are considered to be acceptable for human consumption.
7. Effects of chronic consumption by birds have not been evaluated.
8. An estimated 0.25 to 8.9 birds/hectare are killed by agricultural pesticides in the United States each year.[93]
9. Carbofuran alone is estimated to kill 1 to 2 million birds each year.[94,95]
10. Many birds are killed indirectly by consumption of pesticide-contaminated prey.[96]
11. Many pesticides cause developmental effects at levels lower than those giving rise to detectable parental toxicity.[97]
12. Surprisingly, it has been estimated that losses resulting from pests would increase only 10% if no pesticides were used because of increasing resistance of pests to these chemicals in the United States; however, specific crop losses would range from 0% to nearly 100%.[98]
13. Owners should avoid or minimize pesticide use around birds.
14. Thorough washing of fruits and vegetables with soap and water will remove many superficial pesticide residues.

B. Avicides
1. Most common is alphachoralose. Birds are generally more sensitive than mammals.
2. Clinical signs[99]
    a. CNS stimulation or depression, convulsions, hyperexcitability
3. Treatment
    a. Anticonvulsant therapy

C. Fungicides
1. General information
    a. All fungicides are cytotoxic and most produce positive results in in vitro microbial mutagenicity tests.
    b. It has been estimated that only 11 fungicides may account for 60% of the total estimated dietary carcinogenic risk in humans.[98]
    c. The carcinogenic risk to birds is not known.

d. Unfortunately, a total ban on fungicide use would likely result in an increase in mycotoxin contamination of foods.
e. Mycotoxins are among the most potent carcinogens known, particularly in birds.
f. Organomercurial compounds were commonly used as fungicides on grains, vegetables, cotton, and soybeans until the 1970s. These chemicals are still used in developing countries.

2. Clinical signs of organomercurial toxicity
   a. Leg deformities in young birds
   b. Abnormal egg production
   c. Acute intoxication in mammals is characterized by renal failure and GI signs. Chronic exposures affect all organ systems, but most notably the nervous system.

3. Treatment
   a. Symptomatic, poor prognosis
   b. Poultry that have consumed contaminated grain are unfit for human consumption because of residues.

D. Herbicides

1. General information
   a. Herbicides are the most rapidly growing sector of the agrochemical pesticide business.
   b. They appear to have low acute toxicities in mammals, but contaminants and by-products of herbicide manufacture may be teratogenic and carcinogenic.
      (1) The most widely publicized of these is 2,3,7,8-tetrachlorodibenzo-p-dioxin (TCDD), a contaminant of 2,4-D and 2,4,5-T.
      (2) Contamination of eggs with dibenzodioxins and dibenzofurans also occurs in birds that nest along waterways, especially adjacent to paper and pulp processing plants.[101]

2. Clinical signs depend on chemical
   a. Impaired growth of nestlings, elevated liver enzymes, increase in total plasma thyroxine (T_4) concentrations are seen in raptors with diphenyl ether herbicides (nitrofen, bifenox, oxyfluorfen).[102]
   b. Altricial nestlings are more sensitive to acute effects of diphenyl ether herbicides and bipyridium herbicides (paraquat) than young or adult birds of precocial species.[102-104]
   c. Impaired reproduction seen with chronic dioxin exposure
   d. Contact dermatitis common in mammals
      (1) Herbicides tend to be strong acids, amines, esters, and phenols

3. Treatment
   a. Symptomatic

E. Insecticides

1. General information
   a. Carbamates (includes carbaryl) and organophosphates (includes diazinon, malathion, parathion, DFP, TEPP) are used in insect control and in treating external parasites.

b. There are approximately 25 different carbamates and 200 organophosphates commercially available as thousands of products.
c. These compounds inhibit cholinesterase, resulting in clinical signs typical of cholinergic stimulation.
d. Most organophosphates irreversibly inhibit cholinesterase; therefore, toxicity will persist until new enzyme is synthesized (20 to 30 days in mammals) unless antidotes are given.
e. Carbamates are more readily hydrolyzed from cholinesterase (the rate is still several orders of magnitude slower than acetylcholine hydrolysis).[105]
f. Many organophosphates are bioactivated.
g. Sensitivity varies widely among avian species, presumably resulting from differences in activation by enzymes.[106]
h. The greatest variability in birds appears to be for organophosphates considered to be relatively nontoxic to mammals.
i. The LD_{50}s of these compounds is often hundreds of times lower in birds than in rodents.[107]

2. Clinical signs and diagnosis
   a. Anorexia, vomiting, diarrhea, crop stasis
   b. Bradycardia or tachycardia
   c. Ataxia, muscle fasciculation, inability to fly, flaccid paralysis
   d. Seizures
   e. Delayed neurotoxicity may result in ataxia and paralysis 8 to 14 days after exposure to some organophosphates.[108,109]
      (1) Results from wallerian "dying back" degeneration of large-diameter axons and their myelin sheaths in distal parts of peripheral nerves and long spinal cord tracts.
      (2) Phosphorylation of neurotoxic esterase (NTE) and conversion to a charged form
      (3) Young birds appear resistant to these effects.
   f. Respiratory depression, pulmonary edema
      (1) Death is usually due to paralysis of respiratory muscles.
   g. Teratogenic[110]
      (1) Scoliosis, lordosis, severe edema, incomplete ossification
   h. Behavioral changes
      (1) Affect nesting ecology of gulls[111]
   i. Cholinesterase (ChE) analysis
      (1) Generally must submit normal samples as well as patient samples to laboratory
         (a) Normal brain ChE activity varies among diverse species.[112]
         (b) Whole blood or brain
            (i) Greater than 20% depression of brain cholinesterase activities in birds indicates exposure to cholinesterase inhibitors; birds that die of organophosphate toxicity generally have brain cholinesterase activities depressed 50% or more.[113]

(ii) Blood cholinesterase levels are less reliable than brain because of more rapid rate of recovery of enzyme activity and wide variation in activity in both control and organophosphate-exposed birds.[114] Plasma cholinesterase activity returns to normal within days, whereas brain cholinesterase activity is depressed for one to several weeks, depending on the species.
    j. Identification of insecticide in food, GI contents, liver, body fat, or skin
 3. Treatment
    a. First response should be maintenance of adequate respiratory function.
       (1) Artificial ventilation if respiratory paralysis occurs
       (2) Avoid use of drugs that may depress respiration.
    b. Atropine slowly 0.1 to 0.2 mg/kg intramuscularly as needed until cessation of clinical signs or 0.05 to 0.1 mg/kg intravenously or intramuscularly every 3 to 4 hours[115]
       (1) Atropine counteracts the muscarinic but not nicotinic (muscle fasciculations) signs.
       (2) Diphenhydramine is used to counteract nicotinic signs in mammals.
    c. 2-PAM (pralidoxime chloride) 10 to 100 mg/kg intramuscularly for organophosphate toxicity only
       (1) Give by slow intravenous infusion.
       (2) Of most use if given within the first 24 hours
          (a) Binds covalently to organophosphates either as free circulating toxicant or with the phosphorylated enzyme
          (b) Limited ability to reactivate "aged" cholinesterase
       (3) Use a low dose (10 to 20 mg/kg) if atropine is given.
       (4) Not an effective antidote for carbamate toxicity; may *enhance* toxicity
    d. Activated charcoal if ingested
    e. Diazepam for seizures
 4. Chlorinated hydrocarbons
    a. General information
       (1) Includes DDT, chlordane, dieldrin, lindane, chlordecone, toxaphene, and p-dichlorobenzene
       (2) Chlorinated hydrocarbon insecticides were widely used from 1945 to 1970.
       (3) Their persistence, tendency to move through the food chain, and bioconcentration led to gradual phaseout of most of these compounds in most developed countries.
       (4) Widespread environmental contamination with these compounds persists; and poisoning of native species is not uncommon.[116-118]
       (5) Birds tend to have lower monooxygenase activities than mammals, so are likely to be less efficient in detoxifying organochlorine compounds.[119]

b. Clinical signs
   (1) May be due to the chlorinated hydrocarbon or the petroleum solvent
   (2) Muscle tremors, exaggerated response to stimuli, seizures
      (a) DDT causes nerve membrane to remain in partially depolarized and partially repolarized state.[120]
         (i) Decreases K transport
         (ii) Slows inactivation (closing of Na channels)
         (iii) Inhibits Na^+/K^+ ATPase and Ca^{+2}ATPase, which are essential for repolarization
         (iv) Inhibits calmodulin
      (b) Lindane and cyclodienes (endrin, aldrin, heptachlor, dieldrin, chlordane) primarily prevent complete repolarization in CNS.[121-123]
         (i) GABA antagonist
         (ii) Inhibit ATPases
   (3) Reproductive failure[118, 124]
      (a) Eggshell thinning and breakage
         (i) Inhibition of Ca ATPase and carbonic anhydrase in the avian shell gland (uterus); therefore, decreased ability to secrete calcium and carbonate during eggshell formation[125]
      (b) Failure to lay eggs
      (c) Behavior changes and desertion
      (d) Embryo death, congenital deformities
   (4) Hepatotoxicity, nephrotoxicity, cardiotoxicity
   (5) Diagnosis
      (a) Clinical signs
         (i) Postmortem analysis of liver, kidney, fat, brain; these compounds are very slowly metabolized; they may be mobilized from fat stores during fasting (post-fledgling starvation phase).[126]
c. Treatment
   (1) Supportive
      (a) Diazepam to control seizures
      (b) Epinephrine is contraindicated in patients exposed to chlorinated hydrocarbons because it may induce ventricular fibrillation as a result of chlorinated hydrocarbon sensitization of the myocardium.
   (2) Guarded prognosis: no antidotes exist

5. Pyrethroids
   a. General information
      (1) Account for approximately 30% of worldwide insecticide usage
      (2) Used in agriculture, flea sprays, household and greenhouse sprays.
      (3) Not highly toxic to mammals, and limited information is available regarding their toxicity to birds.
   b. Clinical signs in mammals[127,128]

(1) Allergic reactions
   (a) Contact dermatitis in mammals
   (b) Respiratory distress
   (c) Anaphylaxis
(2) Nausea, vomiting
(3) Muscle fasciculations, seizures, and unconsciousness
   (a) Effects on nerves are similar to organochlorine compounds.
   (b) Readily degraded by nonspecific esterases and mixed function oxidases.
   c. Treatment is symptomatic.

F. Rodenticides: Anticoagulant[129]
1. Sources
   a. Grain-based baits, feed contamination, or consumption of poisoned rodents
2. These agents antagonize the actions of vitamin K in the synthesis of clotting factors (factors II, VII, IX, X).
3. Clinical signs
   a. Weakness, depression
   b. Anorexia
   c. Dyspnea
   d. Petechiation of oral mucosa, bleeding from nares, subcutaneous hemorrhage, anemia
   e. Onset of anticoagulation is delayed several hours after ingestion, with this latency dependent on the half-lives of the clotting factors.[129]
4. Diagnosis
   a. Based on exposure history and clinical signs
   b. Anemia
5. Treatment
   a. Vitamin K_1: 2.5 to 5.0 mg/kg intramuscularly or orally q24h or 0.2 to 2.2 mg/kg q4-q8h
      (1) If warfarin, treat for 10 to 14 days
         (a) Warfarin has a shorter half-life than others.
      (2) If newer compounds (brodifacoum, chlorophacinone) treat for 1 month
   b. Birds should be rechecked in 1 to 2 weeks after treatment to assess recurrence of clinical signs.

G. Rodenticides: Cholecalciferol[129]
1. Clinical signs
   a. Anorexia, vomiting, diarrhea
   b. Polyuria, polydipsia, renal failure
   c. Depression
2. Treatment
   a. Fluids
   b. Activated charcoal
   c. If hypercalcemic

(1) Saline diuresis
(2) Furosemide 2 to 5 mg/kg q8-12h
(3) Mammals treated with calcitonin 4 to 6 IU/kg subcutaneously q2-q3h
d. Corticosteroids
(1) Prednisolone 2 mg/kg orally q12h

H. Rodenticides: Zinc Phosphide[130]
1. Clinical signs
   a. Birds appear to be more sensitive than mammals.
   b. Vomiting, ataxia, convulsions, dyspnea, death
   c. Hypocalcemia, hepatic damage, pulmonary congestion
2. Treatment
   a. Gavage with 5% sodium bicarbonate solution to prevent hydrolysis of zinc phosphide by hydrochloric acid in the stomach and release of phosphine gas.

I. Rodenticides: Strychnine[131,132]
1. Clinical signs and diagnosis
   a. Extreme extensor rigidity and violent tetanic seizures are seen in mammals resulting from antagonism of inhibitory neurotransmitter glycine in spinal cord and medulla.
   b. Birds are generally dead when presented.
2. Diagnosis
   a. Absence of gross or microscopic lesions other than petechial or ecchymotic hemorrhages secondary to trauma
   b. Presence of bait in crop or stomach
   c. Strychnine in liver, kidney, or CNS

V. Plant Toxicoses

A. General Information
1. Rare in birds and are sometimes due to pesticide residues
2. Birds often chew at plants and develop oral irritations but do not ingest enough to cause generalized clinical signs.
3. The most common toxicoses resulting from ingestion consist of GI disturbances (vomiting, diarrhea) resulting from irritation of mucous membranes.
4. Significant differences in sensitivity among different avian species.
5. For all cases of plant toxicoses, treatment is symptomatic.
6. The following plants have been included because clinical, experimental, or anecdotal reports suggest the plants are toxic to birds.
7. All are known to be toxic, to varying degrees, to mammals.
8. This list should not be assumed to be all inclusive.
9. Other books that describe toxicosis in mammals should be consulted when plants not included here are suspected of causing toxicity in birds.

10. A more complete overview of plant-induced toxicoses should be consulted whenever other plants are suspected of causing toxicity in birds.

B. Plant Toxicoses That Present Primarily as GI Toxicity
1. Diffenbachia, calla lily, elephant ear, philodendron (Araceae family)
   a. All parts of the plants are poisonous, active ingredients are resorcinol and calcium oxalates.[133]
   b. Clinical signs
      (1) Severe irritation of mucous membranes; chewing on diffenbachia causes pain.
         (a) Edema and irritation may take weeks to subside.
         (b) Diffenbachia (dumb cane) leaves contain trypsin-like inflammatory protein and have calcium oxalate crystals located in ejector cells throughout surface of leaf.[134]
         (c) Severe dyspnea may develop.
         (d) Severe keratoconjunctivitis may develop if plant juices contact eyes.
      (2) Vomiting, diarrhea
      (3) Philodendrons, but not diffenbachia, cause allergic dermatitis in humans; however, philodendrons appear to be less toxic to mucous membranes.
   c. Antihistamines are useful in humans; use has not been evaluated in birds.
2. Oak (*Quercus* spp.)
   a. Contain tannins; tannins denature proteins and are used to cure or "tan" leather
   b. Anorexia, diarrhea
      (1) Diffuse serosal hyperemia, ulcers, and hemorrhage in small intestine of cassowary that consumed leaves of live oak (*Quercus agrifola*)
   c. Polydipsia, renal failure
   d. Hepatotoxic
   e. May be fatal if large amounts are consumed
3. Castor bean *(Ricinus communis)*:- member of Euphorbiaceae family
   a. Contains Ricin I and Ricin II, toxins that consist of 2 subunits
      (1) Glycosidase (A-chain) that modifies the ribosome and prevents protein synthesis and lectins (B-chain) that bind to cell surface receptors (galactose)
      (2) Presence of a single A-chain molecule within the cytosol is sufficient to kill a cell
   b. Persistent vomiting, diarrhea can be bloody
   c. Necrosis of liver, spleen, lymph nodes, intestine, stomach in mammals
      (1) Carbohydrate groups of ricin bind readily to mannose receptors of cells of the reticuloendothelial system.
   d. Similar signs may be seen with ingestion of precatory beans (prayer beans, rosary peas, or Seminole beads used in bead-

work and rattles) that contain a toxin, abrin, which is similar to ricin.
  4. Poinsettia *(Euphorbia pulcheriama)*
     a. Leaves, stem, and oily white sap are poisonous
     b. Clinical signs
        (1) Irritation, vesication, gastroenteritis, conjunctivitis
     c. Alcohol may be used to solubilize sap and remove it from skin.
  5. Pokeweed *(Phytolacca americans)*
     a. Contains ricin-like lectin and a saponin
     b. Young shoots are eaten by humans after cooking; the root is the most toxic part.
     c. Mature leaves and berries cause ulcerative gastroenteritis.
     d. Acute hemolytic crisis reported in humans, presumably due to saponins.
  6. Rattlebox, wild pea (*Crotalaria* spp.)
     a. Contains pyrrolizidine alkaloid monocrotaline
     b. Naturalized after introduction by Tennessee Valley Authority in 1920s and 1930s as a nitrogen-fixing covercrop in coastal plain and piedmont regions of the southeastern United States
     c. Varying degrees of toxicity in birds
        (1) Turkeys are resistant, chickens intermediate, and quail very sensitive to development of diarrhea and hepatotoxicity.
  7. Coffeebean, rattlebox (*Sesbania* spp.)
     a. Seeds, both green and mature, are much more toxic than leaves.
        (1) Ingestion of a few seeds is lethal to pigeons and chickens.
     b. Toxic principle is a saponin.
     c. Clinical signs include gastroenteritis and erythrocyte lysis.

C. Plant Toxicoses That Present Primarily as Cardiovascular Toxicity
  1. Sensitivity
     a. Relative sensitivity of birds to cardioactive glycosides is unclear.
     b. Although digoxin appears to have a low therapeutic index in some birds,[21-24] one study indicated that clinical signs are seen only when large quantities of plants are ingested by budgerigars.
  2. Lily of the valley *(Convallaria majalis)*
     a. Contains cardioactive glycoside convallatoxin in bulbs
     b. Action resembles digitalis glycosides
        (1) Vomiting, diarrhea
        (2) Cardiac arrhythmias, bradycardia, heart block
        (3) Hyperkalemia
     c. Treatment
        (1) Charcoal slurry
        (2) Antiarrhythmics if needed
        (3) Atropine for bradycardia or heart block
  3. Oleander *(Nerium oleander):* bay laurel[135]
     a. All parts are poisonous; contains glycosides oleadrum and neroside
     b. Clinical signs

(1) Action resembles digitalis glycosides (see lily of the valley)
c. Treatment
(1) See lily of the valley
4. Yew *(Taxus media)*
a. Wood, bark, leaves, and seeds are poisonous
b. Clinical signs
(1) Vomiting, diarrhea
(2) Weakness, convulsion, shock, coma
(3) Deaths result from cardiac or respiratory failure.
(a) Alkaloids, known as taxines, inhibit Ca^{2+} currents in heart.[136]
(b) Lethal doses in mammals and poultry range between 0.1% and 0.5% body weight.[137]
(c) Treatment: see lily of the valley
5. Nightshade *(Solanum nigrum),* Jerusalem or Christmas cherry *(Solanum pseudocapsicum),* potatoes *(Solanum tuberosum)*
a. Alkaloids are most concentrated in skin and sprouts of potatoes, in berries of nightshade and Jerusalem cherries; concentration in leaves is much lower.
b. Several steroidal alkaloids are teratogenic.
c. Steroidal alkaloids produce arrhythmias, bradycardia, heart-block, severe gastroenteritis.
(1) Chronic toxicosis resembling vitamin D intoxication has been reported in mammals. Contains a glycoside of 1,25-dihydroxycholecalciferol[137,138]
(2) Calcification of vascular system (heart and aorta), lungs, kidneys
6. Jimsonweed *(Datura* sp.)[139]
a. Seeds and leaves are toxic.
(1) Contains atropine, scopolamine, hyoscyamine
(2) Parasympatholytic
(3) Tachycardia, convulsions, death
7. Milkweed *(Asclepias* spp.)[140]
a. Cardiac toxins (cardenolides)
b. Weakness, ataxia, seizures, cardiovascular signs
(1) Increased intramyocardial calcium
c. Treatment
(1) Anticonvulsants
(2) Calcium-containing solutions are contraindicated.
8. Avocado *(Persea americana)*
a. Toxic principle unknown
(1) Controversial: some claim it is highly toxic to psittacines, others say it is nontoxic.
b. Reduced activity, inability to perch, fluffing of feathers, labored breathing, rapid death reported in budgerigars.[141]
(1) Subcutaneous edema, hydropericardium, and generalized congestion of tissues noted

D. Plant Toxicoses That Present Primarily as Neurotoxicoses

1. Blue-green algae, Cyanobacteria *(Anabena, Mycrocystis, Aphanizomenon, Nodularia, Oscillatoria* spp.)[136,142]
   a. Neurotoxin, anatoxin A, is a postsynaptic depolarizing agent that affects muscarinic and nicotinic acetylcholine receptors.
      (1) Rapidly lethal from respiratory arrest
   b. Some *Anabena* spp. produce cholinesterase-inhibiting agent, anatoxin-a.
   c. Hepatotoxin mycrocystin
      (1) Cyclic peptides transported by bile acid carriers across intestinal mucosa and into hepatocytes.
      (2) Hepatic degeneration and necrosis
         (a) Secondary to destruction of actin myofilaments in hepatocytes
   d. Clinical signs in ducks include opisthotonos, inability to swim or fly, frequent defecation.
   e. Treatment is symptomatic.
      (1) Control growth of blue-green algae
         (a) Copper sulfate 1mg/L; total alkalinity of pond should be greater than 40 ppm to avoid fish kill.
2. Coffee bean *(Sesbania drumundii)*, cocoa beans *(Theobroma cocas)*, chocolate, tea *(Thea sinensis)*[143]
   a. Contain the stimulants caffeine, theobromine, theophylline
   b. Hyperpyrexia, hypertension, severe CNS stimulation, seizures, tachycardia, diarrhea
      (1) Inhibition of cellular phosphodiesterase causing an increase in intracellular cAMP
      (2) Catecholamine release
      (3) Increased muscle contractility due to increased entry of calcium and inhibition of calcium sequestration by sarcoplasmic reticulum
   c. Give diazepam as needed.
3. Locoweed (*Astragalus* spp.)[144]
   a. Hyperexcitability and locomotor difficulty
   b. In mammals, toxic chemical swainsonine inhibits liver lysosomal and cytosomal cc-D mannosidase and Golgi mannosidase II.
      (1) Results in accumulation of mannose-rich oligosaccharides and abnormal glycoproteins in brain
      (2) Cytoplasmic vacuolation of cerebellar cells
4. Tobacco *(Nicotiana* spp.)[145]
   a. Clinical signs resulting from ingestion of the pyridine alkaloid nicotine; nicotine comprises 97% of the alkaloid content of commercial tobacco
      (1) Vomiting, diarrhea
      (2) Hyperexcitability, muscle fasciculations, seizures
         (a) Nicotine mimics the actions of acetylcholine at all ganglionic synapses and at neuromuscular junctions.
      (3) Rapid death

b. Congenital abnormalities due to pyridino-piperidine alkaloids, including anabasine
   (1) Skeletal malformations, torticollis, malformation of beak, heart, kidneys
   (2) Reduced body weight, muscular dystrophy
c. Clinical signs resulting from exposure to smoke
   (1) Coughing, sneezing, sinusitis
   (2) Conjunctivitis
   (3) Secondary bacterial respiratory infections
d. Pododermatitis may develop in birds handled by people who smoke regularly.
   (1) Nicotine is readily absorbed through skin.

E. Mycotoxins[146-150]
   1. Ergot *(Claviceps purpurea)*
      a. Poorly stored seed, silage, dog food
      b. Ergopeptide alkaloids cause vasoconstriction of arterioles.
         (1) Comb slough in poultry, gangrenous lesions
      c. Lysergic acid derivatives
         (1) Hyperexcitability, seizures
   2. Aflatoxin[147]
      a. Common contaminant of grains, especially those grown in warm, humid environments such as the southeastern United States
         (1) Most commonly in corn, peanuts, cottonseed
      b. Immunosuppression
      c. Lower egg production
      d. Slowed growth, poor feed conversion in chickens
         (1) Inhibition of protein synthesis and inability to mobilize fats
      e. Hepatic necrosis and hepatocarcinogen
         (1) Avian species, such as ducks, are highly susceptible.
         (2) Usually more rapid induction of tumors than seen in mammals
      f. No specific antidote available
         (1) Increased levels of high quality protein and vitamins in diet
         (2) Treat clinical signs of infectious disease aggressively.
   3. Citrinin[151]
      a. Contaminant of feed grains produced by *Penicillium* and *Aspergillus* spp.
      b. Nephrotoxic
         (1) Polyuria, polydipsia
         (2) Degeneration of proximal tubule and distal tubule cells
   4. Zearalenone[152]
      a. Contaminant of grains produced by *Fusarium* spp.
      b. Estrogenic and anabolic effects are species-dependent.
         (1) Zearalenone is metabolized by 3 α-hydroxysteroid dehydrogenase to α-zearalenol and β-zearalenol.
            (a) α-isomer is about three times more estrogenic than β-isomer.
         (2) Potent estrogenic effects in mammals

(a) Signs in mammals generally regress with removal of contaminated feed.
(3) Increased oviduct weight in growing chickens, reduced weight of male broiler comb and testes; precocial development of male turkey poults

VI. Solvents and Vapors

A. General Information
1. Birds are much more sensitive to volatile solvents than are mammals.
2. Clients should be advised that any fumes or smoke can be hazardous to birds, and birds should be removed from the area until the odor is not detectable.
    a. Sources include polytetrafluoroethylene (PTFE) fumes from overheating nonstick cookware or irons (Teflon or Silverstone).[153,154]
3. Pyrolysis of PTFE at temperatures greater than 260° C releases highly toxic carbonyl fluoride, hydrogen fluoride, and perfluoroisobutylene.
    a. Other sources include automobile exhaust, self-cleaning ovens, hair sprays, glues, paints, nail polish, ammonia or strong bleach, mothballs (naphthalene), fluoropolymers from spray starch, furniture polish containing petroleum hydrocarbons or mineral spirits, any chemical spray.

B. Volatile Household Compounds
1. Clinical signs
    a. Most commonly see acute death. Most deaths occur in first 2 to 3 hours postexposure.
    b. Dyspnea, pulmonary congestion, pulmonary hemorrhage
    c. Ataxia, depression, somnolence, narcosis
    d. Concentrations of ammonia and bleach that do not cause acute toxicity can predispose them to infectious diseases.[155]
        (1) Irritation of mucosa of nares, conjunctiva, lower respiratory tract
        (2) Altered lymphocyte function
2. Therapy
    a. Gavage is contraindicated because of the risk of aspiration unless very large amounts are ingested.
    b. Supportive
        (1) Activated charcoal
        (2) Fresh air or oxygen therapy (do not use 100% oxygen)
        (3) Corticosteroids for pulmonary edema
            (a) Some studies indicate corticosteroid therapy may be harmful.[156,157]
        (4) Avoid exciting or stressing birds.
        (5) Prophylactic antibiotics to prevent secondary bacterial bronchopneumonia

C. Environmental Contamination with Hydrocarbons and Petroleum Distillates (Crude Oil, Gasoline, Kerosine)
  1. Clinical signs: acute
     a. Mucous membrane irritation, vomiting
        (1) Enteritis with or without hemorrhage and necrosis[158-160]
     b. Chemical pneumonitis[158,161]
        (1) Acute hemorrhagic necrotizing pneumonia
        (2) Patients who present with coughing have probably already aspirated.
     c. Nephrotoxicity[149-152]
        (1) Hematuria is common
        (2) Renal tubular nephrosis
            (a) Hyaline droplets and casts
     d. Hepatoxicity[159,160,162]
        (1) Elevated liver enzymes
        (2) Fatty liver
        (3) Focal or diffuse hepatic necrosis
     e. Hemolytic anemia[162]
        (1) Phagocytosis of degenerate erythrocytes in liver and spleen
            (a) Enlarged Kupffer cells
        (2) Hemoglobinuria
        (3) Erythroid hyperplasia
     f. Immunosuppression[161,162]
        (1) Decreased resistance to *Pasteurella multocida* infection[161]
        (2) Lymphopenia and lymphocyte depletion of primary lymphoid tissues
        (3) Increased prevalence and severity of lesions in bursa of Fabricius
     g. Cardiac arrhythmias
     h. Adrenal toxicity[163]
        (1) Lipid depletion in steroidogenic cells
            (a) Elevated corticosterone levels in herring gulls and black guillemots following a single oral dose (0.1 to 1.0 ml) of crude oil
        (2) Focal necrosis
     i. Highly toxic to avian embryos and chicks[164,165]
        (1) LD_{50} of Prudhoe Bay oil to chick embryos was found to be 1.3 µl.[166]
        (2) Stunted embryonic growth and increased incidence of abnormal survivors[167]
     j. Impaired osmoregulation
        (1) Inhibition of Na^+/K^+ ATPase in nasal salt gland and altered intestinal mucosa[168]
  2. Clinical signs: chronic
     a. Exposure to petroleum from environmental spills alters the arrangement of feather barbules causing loss of buoyancy and hypothermia.

            (1) Birds become emaciated and dehydrated because they are unable to fly or forage.[169]
        b. Impaired growth
            (1) Result of elevated corticosterone and possibly thyroxine levels[163]
        c. Impaired osmoregulation
        d. Diminished avoidance behavior
        e. Pathologic changes in liver and kidneys
    3. Treatment
        a. Same as for household exposures, but poor prognosis

REFERENCES

1. Osweiler GD et al: Management and treatment of toxicoses. In *Clinical and diagnostic veterinary toxicology,* ed 3, Dubuque, Iowa, 1985, Kendall/Hunt Publishing, pp 52-63.
2. Snodgrass WR: Clinical toxicology. In Klassen CD, editor: *Casarett & Doull's toxicology, the basic science of poisons,* New York, 1996, McGraw-Hill, pp 969-986.
3. Buckley N: Slow release verapamil poisoning: use of polyethylene glycol whole bowel lavage and high-dose calcium, *Med J Austr* 158:202-204, 1993.
4. Fernandez-Repollet E, Rowley J, Schwartz A: Renal damage in gentamicin-treated lanner falcons, *J Am Vet Med Assoc* 181:1392, 1982.
5. Bauck LA, Haigh JC: Toxicity of gentamicin in great horned owls (*Bubo virginianus*), *J Zoo Anim Med* 15:62-66, 1984.
6. Flammer K et al: Adverse effects of gentamicin in scarlet macaws and galahs, *Am J Vet Res* 50:404-407, 1990.
7. Bird JE, Miller KW, Larson AA, Duke GE: Pharmacokinetics of gentamicin in birds of prey, *Am J Vet Res* 44:1245-1247, 1983.
8. Flammer K: Use of amikacin in birds, *Proceed 1st Intl Conf Zoo and Avian Med,* 195-198, 1987.
9. Gronwall R, Brown MP, Clubb SA: Pharmacokinetics of amikacin in African grey parrots, *Am J Vet Res* 50: 250-252, 1989.
10. Bird JE, Walser MM, Duke GE: Toxicity of gentamicin in red-tailed hawks, *Am J Vet Res* 44:1289-1293, 1983.
11. Redig PT, Duke GE: Comparative pharmacokinetics of antifungal drugs in domestic turkeys, red-tailed hawks, and great horned owls, *Avian Dis* 29:649-661, 1988.
12. *Physicians' desk reference,* ed 50, Montvale, NJ, 1996, Medical Economics Data Production Company.
13. van der Mast H et al: A fatal treatment of sinusitis in an African grey parrot, *J Assoc Avian Vet* 4:189, 1990.
14. Pizarro J, Ludders JW, Douse MA, Mitchell GS: Halothane effects on ventilatory responses to changes in intrapulmonary CO_2 in geese, *Resp Physiol* 82:337-348, 1990.
15. Samour JH, Jones DM, Knight JA, Howlett JC: Comparative studies of the use of some injectable anaesthetic agents in birds, *Vet Rec* 115:6-11, 1984.
16. Flammer K: Antimicrobial therapy. In Ritchie BW, Harrison GJ, Harrison LR, editors: *Avian medicine: principles and application,* Lake Worth, Fla, 1994, Wingers.
17. Rosskopf WJ, Woerpal RW, Whittaker D: Anaphylactic reaction in a hawk, *Mod Vet Pract* 64:235, 1983.
18. Rigdon RH, Crass G, Martin N: Anemia produced by chloramphenicol in the duck, *Arch Pathol* 58:85-93, 1954.

19. Rigdon RH, Martin N, Crass G: Consideration of the mechanism of anemia produced by chloramphenicol in the duck, *Antibiot Chemother* 5:38-44, 1955.
20. Yunis AA: Chloramphenicol: relation of structure to activity and toxicity, *Annu Rev Toxicol* 28:83-100, 1988.
21. Hamlin RL, Stalnaker PS: Basis for use of digoxin in small birds, *J Vet Pharmacol Ther* 10:354-356, 1987.
22. Alvarez Maldonado MVZ: Report preeliminar. Digitalizacion en pollos de engorda como metodo preventico en el sindrome ascitico, *Proceed 35th West Poultry Dis Conf,* 1986.
23. Wilson RC, Zenoble RD, Horton CR, Ramsey DT: Single dose digoxin pharmacokinetics in the quaker conure, *J Zoo Wildlife Med* 20:432-434, 1989.
24. Lumeij JT, Ritchie BW: Cardiology. In Ritchie BW, Harrison GJ, Harrison LR, editors: *Avian medicine: principles and application,* Lake Worth, Fla, 1994, Wingers.
25. Bond M et al: Enrofloxacin in neonates, *J Assoc Avian Vet,* 1993.
26. Krautwald ME et al: Further experience with the use of Baytril in pet birds, *Proceed Assoc Avian Vet* 226-236, 1990.
27. Flammer K, Aucoin DP, Whitt DA: Intramuscular and oral disposition of enrofloxacin in African grey parrots following single and multiple doses, *J Vet Pharmacol Ther* 14:359-366, 1991.
28. Ritchie BW, Harrison GJ: Formulary. In Ritchie BW, Harrison GJ, Harrison LR, editors: *Avian medicine: principles and application,* Lake Worth, Fla, 1994, Wingers.
29. Sawant SG, Terse PS, Dalvi RR: Toxicity of dietary monensin in quail, *Avian Dis* 34:571-574, 1990.
30. VanderKop PA, MacNeil JD, VanderKop MA: Monensin intoxication in broiler chicks: is it really so easy to identify? *Can Vet J* 30:823-824, 1989.
31. Hanrahan LA, Corrier DE, Naqi SA: Monensin toxicosis in broiler chickens, *Vet Pathol* 18:665-671, 1981.
32. Harries N, Hanson J: Salinomycin toxicity in turkeys, *Can Vet J* 32:117, 1991.
33. Bartlet AL, Khan FH: Effects of nitrofurans on adrenal cortical tissue in chickens, *J Vet Pharmacol Ther* 13:206-216, 1990.
34. Bartlet AL, Harvey S, Klandorf H: Contrasting effects of nitrofurans on plasma corticosterone in chickens following administration as a bolus or diet additive, *J Vet Pharmacol Ther* 13:261-269, 1990.
35. Ali BH: Some pharmacologic and toxicologic properties of furazolidone, *Vet Res Commun* 6:1-11, 1983.
36. Clubb S: *Assoc Avian Vet Newsletter* 1:1, 1980.
37. Feldman BF, Kruckenberg SM: Clinical toxicities of domestic and wild caged birds, *Vet Clin N Am Sm Anim Pract* 5:653-673, 1975.
38. Gerlach H: Drug hypersensitivity, *J Assoc Avian Vet* 4:156, 1990.
39. Flammer K, Aucoin DP, Whitt DA, Styles DK: Potential use of long-acting injectable oxytetracycline for the treatment of chlamydiosis in Goffin's cockatoos, *Avian Dis* 34: 228-234, 1990.
40. Flammer K, Cassidy DR, Landgraf WW, Ross PF: Blood concentrations of chlortetracycline in macaws fed medicated pelleted feed, *Avian Dis* 33:199-203, 1989.
41. Arnstein P, Eddie B, Meyer KF: Control of psittacosis by group chemotherapy of infected parrots, *Am J Vet Res* 29:2213-2227, 1968.
42. Tang K, Rowland GN, Veltmann JR: Vitamin A toxicity: comparative changes in bone of broiler and leghorn chicks, *Avian Dis* 29:416-429, 1985.
43. Takeshita K et al: Hypervitaminosis D in baby macaws, *Proceed Assoc Avian Vet* 341-345, 1986.
44. Chapman CB et al: The role of procaine in adverse reactions to procaine penicillin in horses, *Austr Vet J* 69:129-133, 1992.

45. Lumeij JT, Westerhof I: Blood chemistry for the diagnosis of hepatobiliary disease in birds. A review, *Vet Quart* 9:255-262, 1987.
46. Frank A, Borg K: Heavy metals in tissues of the mute swan *(Cygnus olor)*, *Acta Vet Scand* 20:447-465, 1979.
47. Pullar EM: The toxicity of various copper compounds and mixtures for domesticated birds, *Austr Vet J* 16:147-162, 1940.
48. Molnar JJ: Copper storage in the liver of the wild mute swan (*Cygnus olor*), *Arch Pathol Lab Med* 107:629-632, 1983.
49. Soli NE, Froslie A, Aaseth J: The mobilization of copper in sheep by chelating agents, *Acta Vet Scand* 19:422-429, 1978.
50. Randell MG, Patnaik AK, Gould WJ: Hepatopathy associated with excessive iron storage in mynah birds, *J Am Vet Med Assoc* 179:1214-1217, 1981.
51. Mautino M: Avian lead toxicosis, *Proceedings of the Association of Avian Veterinarians* 245-247, 1990.
52. McDonald SE: Lead poisoning in psittacine birds. In Kirk RW, editor: *Current veterinary therapy IX, small animal practice*, Philadelphia, 1988, WB Saunders, pp 713-718.
53. Ochiai K et al: Pathological study of lead poisoning in whooper swans (*Cygnus cygnus*) in Japan, *Avian Dis* 36:313-323, 1992.
54. Bailey TA, Samour JH, Naldo J, Howlett JC: Lead toxicosis in captive houbara bustards (*Chlamydotis undulata maqueenii*), *Vet Rec* 137:193-194, 1995.
55. Morgan RV, Moore FM, Pearce LK, Rossi T: Clinical and laboratory findings in small companion animals with lead poisoning: 347 cases (1977-1986), *J Am Vet Med Assoc* 199:93-102, 1991.
56. Mauting M, Bell JU: Experimental lead toxicity in the ring-necked duck, *Environ Res* 41:538-545, 1986.
57. Dieter MP, Finley MT: ALAD enzyme activity in blood, brain, and liver of lead dosed ducks, *Environ Res* 19:127-135, 1979.
58. Degernes LA, Frank RK, Freeman ML, Redig PT: Lead poisoning in trumpeter swans. *Proceedings of the Association of Avian Veterinarians* 1989, 144-155.
59. DelBono G, Bracca G: Lead poisoning in domestic and wild ducks, *Avian Pathol* 2:195-209, 1973.
60. Pokras MA, Chafel R: Lead toxicosis from ingested fishing sinkers in adult common loons (*Gavia immer*) in New England, *J Zoo Wildlife Med* 23:92-97, 1992.
61. Kober TE, Copper GP: Lead competitively inhibits calcium-dependent synaptic transmission in the bullfrog sympathetic ganglion, *Nature* 262:704-705, 1976.
62. Hunter DB, Haigh JC: Demyelinating peripheral neuropathy in a guinea hen associated with subacute lead intoxication, *Avian Dis* 22:344-349, 1978.
63. Bagley GE, Locke LN, Nitingale T: Lead poisoning in Canada geese in Delaware, *Avian Dis* 11:601-603, 1967.
64. Cook RS, Trainer DO: Experimental lead poisoning of Canada geese, *J Wildlife Manag* 30:1-8, 1966.
65. Hunter B, Wobeser G: Encephalopathy and peripheral neuropathy in lead-poisoned mallard ducks, *Avian Dis* 24:169-178, 1980.
66. Rosskopf WJ Jr, Woerpel RW: Heavy metal intoxication in caged birds – parts I and II. In *Exotic animal practice, the compendium collection*, NJ, 1986, Veterinary Learning Systems.
67. Galvin C: Acute hemorrhagic syndrome of birds. In Kirk RW, editor: *Current veterinary therapy VIII*, Philadelphia, 1983, WB Saunders, pp 617-619.
68. Rao PVVP, Jordan SA, Bhatnagar MK: Combined nephrotoxicity of methylmercury, lead and cadmium in Pekin ducks: metallothionein, metal interactions, and histopathology, *J Toxicol Environ Health* 26:327-348, 1989.

69. O'Halloran JO, Duggan PF, Myers AA: Biochemical and haematological values for mute swans (*Cygnus olor*): effects of acute lead poisoning, *Avian Pathol* 17:667-678, 1988.
70. Lumeij JT: Clinicopathologic aspects of lead poisoning of birds: a review, *Vet Quart* 7:133-138, 1985.
71. Fudge AM: Clinical findings in an Amazon parrot with suspected lead toxicosis, *Cal Vet* 36:23-25, 1982.
72. Goyer RA: Toxic effects of metals. In Klaassen CD, editor: *Casarett & Doull's toxicology, the basic science of poisons*, New York, 1996, McGraw-Hill, pp 691-736.
73. Roscoe DE, Nielsen SW: A simple quantitative test for erythrocytic protoporphyrin in lead poisoned ducks, *J Wildlife Dis* 15:127-136, 1979.
74. Rozman KK, Klaassen CD: Absorption, distribution, and excretion of toxicants. In Klaassen CD, editor: *Casarett & Doull's toxicology, the basic science of poisons*, New York, 1996, McGraw-Hill, pp 91-112.
75. LaBonde J: Avian toxicology, *Vet Clin N Am Sm Anim Pract* 21:1329-1342, 1991.
76. Stewart FM, Thompson DR, Furness RW, Harrison N: Seasonal variation in heavy metal lilvels in tissues of common guillemots, *Uria aalge* from northwest Scotland, *Arch Environ Contam Toxicol* 27:168-175, 1994.
77. Snelgrove-Hobson SM, Prasada Rao PVV, Bhatnagar MK: Ultrastructural alterations in the kidneys of Pekin ducks fed methylmercury, *Can J Vet Res* 52:89-98, 1988.
78. Saiki MK, Lowe TP: Selenium in aquatic organisms from subsurface agricultural drainage water, San Joaquin Valley, California, *Arch Environ Contam Toxicol* 16:657-670, 1987.
79. Ohlendorf HM, Hothem RL, Aldrich TW, Krynitsky AJ: Selenium contamination of the grasslands, a major California waterfowl area, *Sci Total Environ* 66:169-183, 1987.
80. Paveglio FL, Bunck CM, Heinz GH: Selenium and boron in aquatic birds from central California, *J Wildlife Manag* 56:31-42, 1992.
81. Lemly AD: Toxicology of selenium in a freshwater reservoir: implications for environmental hazard evaluation and safety, *Ecotox Environ Safety* 10:314-338, 1985.
82. Franke KW, Tully WC: A new toxicant occurring naturally in certain samples of plant foodstuffs, *Poultry Sci* 14:273-279, 1935.
83. Hoffman DJ, Ohlendorf HM, Aldrich TW: Selenium teratogenesis in natural populations of aquatic birds in central California, *Arch Environ Contam Toxicol* 17:519-525, 1988.
84. Ohlendorf HM, Hoffman DJ, Saiki MK, Aldrich TW: Embryonic mortality and abnormalities of aquatic birds: apparent impacts by selenium from irrigation drainwater, *Sci Total Environ* 52:49-63, 1986.
85. Ort JF, Latshaw JD: The toxic level of sodium selenite in the diet of laying chickens, *J Nutr* 108:1114-1120, 1978.
86. Heinz GH, Hoffman DJ, Gold LG: Impaired reproduction of mallards fed an organic form of selenium, *J Wildlife Manag* 53:418-428, 1989.
87. Deeming DC, Dick ACK: Ingestion of metal objects by ostriches (*Struthio camelus*), *Vet Rec* 137:99-100, 1995.
88. Reece RL, Dickson DB, Burrowes PJ: Zinc toxicity (new wire disease) in aviary birds, *Austr Vet J* 63:199, 1986.
89. Van Sant F: Zinc toxicosis in a hyacinth macaw, *Proceedings of the Association of Avian Veterinarians*, 1991, pp 255-259.
90. Howard BR: Health risks of housing small psittacines in galvanized wire mesh cages, *J Am Vet Med Assoc* 200:1667-1674, 1992.

91. Minnick PD, Braselton WE, Meerdink GL, Slanker MR: Altered serum elemental concentrations due to laboratory usage of vacutainer tubes, *Vet Hum Toxicol* 24:413-414, 1982.
92. Vaugh S: Arsenic poisoning in a cockatiel aviary, *Proceed European Assoc Avian Vet,* 1991, pp 258-263.
93. Mineau P: Avian mortality in agroecosystems: 1. The case against granule insecticides in Canada. In Greaves, Smith, Greig-Smith, editors: *Field methods for the study of environmental effects of pesticides,* London: British Crop Protection Council (BCPC) Monograph 40, Thornton Heath: 3-12.
94. Environmental Protection Agency: *Pesticide industry sales and usage: 1988 market estimates,* Washington, DC, 1989, EPA Economic Analysis Branch.
95. Porter SL, Sneed SE: Pesticide poisonings in birds of prey, *J Assoc Avian Vet* 4:84-85, 1990.
96. Henry CJ, Blus LJ, Kolbe EJ, Fitzner RE: Organophosphate insecticide (famphur) topically applied to cattle kills magpies and hawks, *J Wildlife Manag* 49:648-658, 1985.
97. Mineau P, Boersma DC, Collins B: An analysis of avian reproductive studies submitted for pesticide registration, *Ecotox Environ Safety* 29:304-329, 1994.
98. Pimental D et al: 1992. Environmental and economic costs of pesticide use, *Bioscience* 42: 750-760, 1992.
99. Lees P, Pharm Y: Pharmacology and toxicology of alpha-chloralose: a review, *Vet Rec* 91:330, 1972.
100. Ecobichon DJ: Toxic effects of pesticides. In Klaassen CD, editor: *Casarett & Doull's toxicology, the basic science of poisons,* New York, 1996, McGraw-Hill, pp 643-689.
101. Hart LE et al: Dioxin contamination and growth and development in great blue heron embryos, *J Toxicol Environ Health* 32:331, 1991.
102. Hoffman DJ et al: Developmental toxicity of diphenyl ether herbicides in nestling American kestrels, *J Toxicol Environ Health* 34:323-336, 1991.
103. Hoffman DJ, Franson JC, Pattee OH, Bunck CM: Survival, growth, and histopathological effects of paraquat ingestion in nestling American kestrels (*Falco sparverius*), *Arch Environ Contam Toxicol* 14:495-500, 1985.
104. Hoffman DJ et al: Toxicity of paraquat in nestling birds: effects on plasma and tissue biochemistry in American kestrels, *Arch Environ Contam Toxicol* 16:177-183, 1987.
105. Ecobichon DJ: Hydrolytic mechanisms of pesticide degradation. In Geisbuhler H, editor: *Advances in pesticide science, biochemistry of pests and mode of action of pesticides, pesticide degradation, pesticide residues and formulation chemistry. Part 3,* New York, 1979, Pergamon, pp 516-524.
106. Fouts JR, Devereux TR: Hepatic microsomal N-oxidation and N-demethylation of N,N-dimethylaniline in red-winged blackbird compared with rat and other birds, *Life Sci* 17: 819-826, 1975.
107. Hill EF: Acute and subacute toxicology in evaluation of pesticide hazard to avian wildlife. In Kendall RJ, Lacher TE, editors: *Wildlife toxicology and population modeling. Integrated studies of agroecosystems,* Boca Raton, Fla, 1994, Lewis, pp 207-226.
108. Abou-Donia MB, Lapadula D: Mechanisms of organophosphorous ester-induced delayed neurotoxicity: type I and type II, *Annu Rev Pharmacol Toxicol* 30:405-440, 1990.
109. Sprague GL, Bickford AA: Effect of multiple diisopropylfluorophosphate injections in hens: behavioral, biochemical and histological investigation, *J Toxicol Environ Health* 8:973-988, 1981.

110. Hoffman DJ: Embryotoxicity and teratogenicity of environmental contaminants to bird eggs, *Rev Environ Contam Toxicol* 115:39-89, 1990.
111. White DH, Mitchell CA, Prouty RM: Nesting biology of laughing gulls and relation to agricultural chemicals, *Wilson Bull* 95:540-553, 19893.
112. Hill EF: Brain cholinesterase activity of apparently normal wild birds, *J Wildlife Dis* 24:51-61, 1988.
113. Ludke JL, Hill EF, Dieter MP: Cholinesterase response and related mortality among birds fed cholinesterase inhibitors, *Arch Environ Contam Toxicol* 3:1-21, 1975.
114. Fleming WJ: Recovery of brain and plasma cholinesterase activities in ducklings exposed to organophosphorous pesticides, *J Toxicol Environ Health* 7:215-229, 1981.
115. Ritchie BW: Treatment of organophosphate toxicosis in *Columba livia*, *Assoc Avian Vet Today* 1:23, 1987.
116. White DH, Krynitsky AJ: Wildlife in some areas of new Mexico and Texas accumulate elevated DDE residues, *Arch Environ Contam Toxicol* 15:147-157, 1986.
117. Burger J, Viscido K, Gochfeld M: Eggshell thickness in marine birds in the New York Bight - 1970's to 1990's, *Arch Environ Contam Toxicol* 29:187-191, 1995.
118. Kozie KD, Anderson RK: Productivity, diet, and environmental contaminants in bald eagles nesting near the Wisconsin shoreline of Lake Superior, *Arch Environ Contam Toxicol* 20:41-48, 1991.
119. Walker CH, Newton I, Hallam SD, Ronis MJJ: Activities and toxicological significance of hepatic microsomal enzymes of the kestrel (*Falco tinnunculus*) and sparrowhawk (*Accipiter nisus*), *Comp Biochem Physiol* 86C:379-382, 1987.
120. Joy RM: Chlorinated hydrocarbon insecticides. In Ecobichon DJ, Joy RM, editors: *Pesticides and neurologic diseases,* ed 2, Boca Raton, Fla, 1994, CRC Press, pp 81-170.
121. Eldefrawl MES, Sherby SM, Abalis IM, Eldefrawi AT: Interaction of pyrethroid and cyclodiene insecticides with nicotinic acetylcholine and GABA receptors, *Neurotoxicology* 6:47-62, 1985.
122. Matsumura F: *Toxicology of insecticides,* New York, 1985, Plenum Press, pp 122-128.
123. Wafford KA, Sattelle DB, Gant DB: 1989. Noncompetitive inhibition of GABA receptors in insect and vertebrate CNS by endrin and lindane, *Pesticide Biochem Physiol* 33:213-219, 1989.
124. Bogan JA, Newton I: The effects of organochlorine compounds on British birds of prey, *Vet Res Comm* 7:119-124, 1983.
125. Miller DS, Kinter WB, Peakall DB: Enzymatic basis for DDE-induced eggshell thinning in a sensitive bird, *Nature* 259:122-124, 1976.
126. Bend JR, Miller DS, Kinter WB, Peakall DB: DDE-induced microsomal mixed-function oxidases in the puffin *(Fratercula artica)*, *Biochem Pharmacol* 26:1000-1001, 1977.
127. He F, Sun J, Han K: Effects of pyrethroid insecticides on subjects engaged in packaging pyrethroids, *Br J Industr Med* 45:548-551, 1988.
128. He F, Wang S, Liu L: Clinical manifestations and diagnosis of acute pyrethroid poisoning, *Arch Toxicol* 63:54-58, 1989.
129. Katona B, Wason S: Anticoagulant rodenticides, *Clin Toxicol Rev* 8:1-2, 1986.
130. Shivanandappa T, Ramesh HP, Krishnakumari MK: Rodenticidal poisoning of non-target animals: acute oral toxicity of zinc phosphide to poultry, *Bull Environ Contam Toxicol* 23:452-455, 1979.
131. Redig PT: Relay toxicity of strychnine in raptors in relation to a pigeon eradication program, *Vet Hum Toxicol* 24:35-336, 1982.

132. Wobeser G, Blakley BR: Strychnine poisoning of aquatic birds, *J Wildlife Dis* 23:341-343, 1987.
133. Knight TE: Philodendron-induced dermatitis: report of cases and review of the literature, *Cutis* 48:375-378, 1991.
134. McIntire MS, Guest JR, Porterfield JF: Philodendron - an infant death, *J Toxicol Clin Toxicol* 28:177-183, 1990.
135. Shropshire CM, Stauber E, Arai M: Evaluation of selected plants for acute toxicosis in budgerigars, J Am Vet Med Assoc 200:936-939, 1992.
136. Norton S: Toxic effects of plants. In Klaassen CD, editor: *Casarett & Doull's toxicology, the basic science of poisons,* New York, 1996, McGraw-Hill, pp 841-853.
137. Panter KE, Molyneux RJ, Smart RA: English yew poisoning in 43 cattle, *J Am Vet Med Assoc* 202:1476-1477, 1993.
138. Baker DC, Keeler RF, Gaffield W: Toxicosis from steroidal alkaloids of Solanum species. In Keeler RF, Tu AT, editors: *Handbook of natural toxins,* vol 6, New York, 1991, Marcel Dekker.
139. Day EJ, Dilworth BC: Toxicity of Jimson weed seed and cocoa shell meal to broilers, *Poultry Sci* 63:466-468, 1984.
140. Cheeke PR, Shull LR: Glycosides. In *Natural toxicants in feeds and poisonous plants,* Westport, Conn, 1985, Avi Publ Co, pp 192-194.
141. Hargis AM, Stauber E, Casteel S, Eitner D: Avoccado (*Persea americana*) intoxication in caged birds, *J Am Vet Med Assoc* 194:64-66, 1989.
142. Beasley VR et al: Cyclic peptide hepatotoxins from cyanobacteria. In Keeler RF, Tu AT, editors: *Handbook of natural toxins,* vol 6, New York, 1991, Marcel Dekker.
143. Rall TW: Central nervous system stimulants: the methyl xanthines. In Gilman AG et al, editors: *Goodman and Gilman's the pharmacological basis of therapeutics,* ed 7, New York, 1985, MacMillan, pp 589-603.
144. Cheeke PR, Shull LR: Alkaloids. In *Natural toxiocants in feeds and poisonous plants,* Westport, Conn, 1985, Avi Publ Co, pp 142-146.
145. Benowitz NL: Clinical pharmacology of nicotine. *Annu Rev Med,* 37:21-32, 1986.
146. Hoerr FJ: Diseases of poultry. In Calnek BW et al, editors: *Mycotoxicoses,* ed 9, Ames, Iowa, 1991, Iowa State University Press, pp 884-915.
147. Chattopadhyay SK et al: Clinical and biochemical effects of aflatoxin in feed ration of chicks, *Cancer Biochem Biophys* 8:67-75, 1985.
148. Dalvi RR, Ademoyero AA: Toxic effects of aflatoxin B1 in chickens given feed contaminated with *Aspergillus flavus* and reduction of the toxicity by activated charcoal and some chemical agents, *Avian Dis* 28:61-69, 1983.
149. Carnaghan RBA: Hepatic tumors in ducks fed on low level of toxic groundnut meal, *Nature* 208:308, 1965.
150. Windingstad RM et al: Fusarium mycotoxins from peanuts suspected as a cause of sandhill crane mortality, *J Wildlife Dis* 25:38-46, 1989.
151. Glahn RP, Wideman RF: Avian diuretic response to renal portal infusions of the mycotoxin citrinin, *Poultry Sci* 66:1316-1325, 1987.
152. Olsen M, Mirocha CJ, Abbas HK, Johansson B: Metabolism of high concentrations of dietary zearalenone by young male turkey poults, *Poultry Sci* 65:1905-1910, 1986.
153. Wells RE: Fatal toxicosis in pet birds caused by an overheated cooking pan lined with polytetrafluoroethylene, *J Am Vet Med Assoc* 182:1248-1250, 1989.
154. Wells RE, Slocombe RF, Trapp AL: Acute toxicosis of budgerigars (*Melopsittacus undulatus*) caused by pyrolysis products from heated polytetrafluoroethylene: clinical study. *Am J Vet Res* 43:1238-1242, 1982.
155. Klucinski W, Targowski SP: Ammonia toxicity for mammalian and avian lymphocytes from blood, *Immunopharmacology* 8:47-52, 1984.

156. Brown J, Burke B, DaJanias C: Experimental kerosene pneumonia: evaluation of some therapeutic regimens, *J Pediatr* 84:396-401, 1974.
157. Marks MI, Chicoine L, Legere G: Adrenocorticosteroid treatment of hydrocarbon pneumonia in children—a cooperative study, *J Pediatr* 81:366-369, 1972.
158. Beer JV: Post mortem findings in oiled auks dying during attempted rehabilitation. In Carthy, Arthur, editors: *The biological effects of oil pollution on littoral communities,* London, 1968, Field Studies Council, p 123.
159. Khan RA, Ryan P: Long term effects of crude oil on common murres (*Uria aalge*) following rehabilitation, *Bull Environ Contam Toxicol* 46:216-222, 1991.
160. Fry DM, Lowenstein LJ: Pathology of common murres and Cassin's auklets exposed to oil, *Arch Environ Contam Toxicol* 14:725-737, 1985.
161. Rocke TE, Yuill TM, Hinsdill RD: Oil and related toxicant effects on mallard immune defenses, *Environ Res* 33:343, 1984.
162. Leighton FA: Clinical, gross, and histopathological findings in herring gulls and Atlantic puffins that ingested Prudhoe Bay crude oil, *Vet Pathol* 23:254-263, 1986.
163. Peakall DB, Tremblay J, Kinter WB, Miller DS: Endocrine dysfunction in seabirds caused by ingested oil, *Environ Res* 24:6-14, 1981.
164. White DH, King KA, Coon NC: Effects of No. 2 fuel oil on hatchability of marine and estuarine bird eggs, *Bull Environ Contam Toxicol* 21:7-10, 1979.
165. Szaro RC, Alberts PH, Coon NC: Petroleum: effects on mallard hatchability, *J Wildllife Manag* 42:404-406, 1978.
166. Lee YZ, O'Brien PJ, Payne JF, Rahimtula AD: Toxicity of petroleum crude oils and their effect on xenobiotic metabolizing enzyme activities in the chicken embryo in ovo, *Environ Res* 39:1553-1563, 1986.
167. Vangilder LD, Peterle TH: South Louisiana crude oil and DDE in the diet of mallard hens: effects on reproduction and duckling survival, *Bull Environ Contam Toxicol* 25:23-28, 1988.
168. Miller DS, Peakall DB, Kinter WB: Ingestion of crude oil: sublethal effects on herring Gull chicks, *Science* 199:315-317, 1978.
169. Holmes WN: Petroleum pollutants in the marine environment and their possible effects on seabirds. In Hodgson E, editor: *Reviews in environmental toxicology,* Amsterdam, 1984, The Netherlands Elsevier Science Publishers, pp 251-317.

Ophthalmic Disorders

Patricia Bright

This chapter gives the practitioner an overview of the avian ophthalmologic examination, ancillary tests, and information on the diagnosis and treatment of common ophthalmologic disorders. The chapter also provides a brief review of the anatomy of the eye. For a more detailed description of the anatomy and for information regarding physiology, the current literature should be reviewed.

I. Anatomy

A. Eyelids and Adnexa
  1. The lower lid is more mobile and covers more than 50% of the eye when the eye is closed[1,2]
  2. Meibomian glands are absent[1,3,4]
  3. The lacrimal gland is located ventral and lateral to the globe[4]
     a. It is absent in some species.[3]
     b. Ventral and dorsal nasal lachrymal puncta are located in the medial canthus of each eye[3,4]
  4. In precocial birds the palpebral fissure is open at hatching.[3]
  5. In altricial birds, lids are sealed and continue to develop after hatching.[3]
     a. The time when eyelid separation begins varies by species[3,5]
     b. Examples
        (1) Amazons: 14 to 21 days posthatching
        (2) Cockatoos: 10 to 21 days posthatching
        (3) Macaws: 14 to 28 days posthatching
     c. Eyelid separation can take up to 1 week[3]

B. Nictitans
  1. Commonly referred to as the "third eyelid"
  2. It moves over the eye in a dorsonasal to ventromedial direction.[3,6]
  3. It is responsible for passing the tear film over the cornea and removing debris.[3,6]
  4. It should never be surgically removed; removal may result in chronic exposure keratitis.[7]

5. The harderian gland, located at the base of the nictitans, opens into the conjunctival sac and serves as a source of tears.[3,4]

C. Globe
1. The globe and orbit are very large relative to the bird's size[2]
2. The shape of the globe varies by species; there are three basic shapes[1,4,8]
   a. Anterior–posteriorly flattened (most birds, including Psittaciformes)[1,4,8]
   b. Globose or rounded (many diurnal birds, including most raptors)[1,4,6,8]
   c. Tubular (owls and some eagles)[1,4,8]

D. Anterior Segment
1. Cornea
   a. Thinner in birds than in mammals[1]
2. Sclera
   a. Is reinforced by a layer of hyaline cartilage and 8 to 10 rings, or scleral ossicles, immediately posterior to the cornea[1,2,6,8]
3. Anterior chamber
   a. Some anatomic variations exist between species, for example, owls have a deep anterior chamber.[1]

E. Iris
1. Color may vary with age, sex, or species[3] (Table 15-1)
2. Contains striated muscle, which allows for voluntary dilation and constriction of the iris[1-3]
3. Intermittent anisocoria may be normal, especially during periods of excitement[3]
4. Pupillary light responses occur but are difficult to interpret because of voluntary constriction and dilation[1]

F. Lens
1. Shape varies by species, from flat to spherical[1]

G. Posterior Segment
1. Retina
   a. Most birds have good color vision[3]   *discrete area of reflective cells*
   b. Avascular in birds; no tapetum is present[3,9]
   c. Fundus color varies among species but is commonly gray or reddish and may have yellow or gray spots[3,4,6]
2. Optic disc
   a. Oval in shape[3]
   b. Difficult to visualize, because it is hidden behind the pecten[2]
3. Fovea — *cup shaped depression or pit*
   a. Poorly defined or absent in most species[3,4]
      (1) There are some exceptions; macaws and some raptor species have distinct fovea.[1]
      (2) Some raptor and passerine species have two foveae.[1,3,4,6]

avascular — without blood or lymphatic cells.

Table 15-1
Iris Color

Species	Iris color
African Grey[1]	
Juvenile	Dark brown/gray
Adult	Turns yellowish gray then silverish white
Blue and gold macaw[1]	
Immature	Dark brown turning gray
2-3 years old	Color lightens
Adult	Becomes yellow with age
Some species of white cockatoo[1,3]	
Immature (both sexes)	Black
Adult male	Brown
Adult female	Reddish pink
Moluccan cockatoo[1]	
Adult males	Black
Adult female	Dark brown
Amazon parrot[7]	
Young	Brown, become reddish orange with age
Red-tail hawk[3]	
Juvenile	Yellow to gray
By 4 to 5 years of age	Brown
Accipiters[3]	
Nestling	Gray
Juvenile and young adult	Yellow-orange
By 5 years of age	Red

Data from the following sources:
Williams D: Ophthalmology. In *Avian medicine: principles and application*, Lake Worth, Fla, 1994, Wingers, pp 673-694.
Kern TJ: Disorders of the Special Senses. In Altman RB, Club SL, Dorrestein GM, Quesenbury K: *Avian medicine and surgery*, Philadelphia, 1997, WB Saunders, pp 563-582.
Rupley A: *Manual of avian practice*, Philadelphia, 1997, WB Saunders, pp 135-164.

4. Pectin (Fig. 15-1)
    a. Unique to birds, present in all species; it is an outgrowth of the retina over the area of the optic disc.[1,8]
    b. Shape varies by species.[3]
    c. Has pleated "comblike" projections[1,3,9]
    d. Vascular and generally black or brown[1-3]

H. Infraorbital Sinus
1. Infraorbital sinuses are associated with the area surrounding each eye.
2. The sinuses are located <u>rostroventral</u> to each eye and lateral to the nares.[3,7]

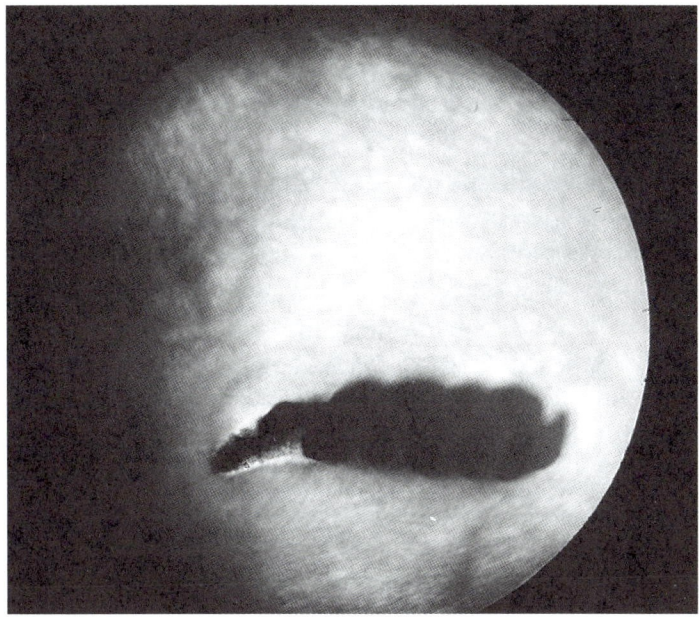

Fig. 15-1 Normal owl pectin. (Courtesy T.J. Kern. From Altman RB et al, editors: *Avian medicine and surgery*, Philadelphia, 1997, WB Saunders.)

3. In many species, the sinuses connect with the cervicocephalic air sac, the upper beak, and pneumatized bone of the skull, and extend around each eye.[3,7]
4. Sinusitis may present as periorbital swelling, conjunctivitis, or intraocular disease.[1]

II. Clinical Examination

A. History
1. Information should be obtained regarding the following:
   a. Medical history and concurrent diseases
   b. Management and husbandry
   c. Current medications and response to treatment
   d. Presenting complaint, including the following:
      (1) Changes in behavior and/or evidence of vision loss
      (2) Changes in the appearance of the eyes or face
      (3) Presence, frequency, and nature of ocular discharge
      (4) Presence of blepharospasm
      (5) Duration and progression of the problem

B. Physical Examination

1. Assessment of vision
   a. Initially the bird should be observed from a distance, preferably in its cage or natural environment.
   b. Determine whether the bird is able to identify and follow a moving object.
      (1) Stand behind the bird and move your hand toward the bird first on the left and then on right to test the vision in each eye.[7]
   c. A menace response is generally not diagnostic in birds.[1,3]
      (1) The response may be absent in normal birds.[3,7]
      (2) A false-positive response can be associated with airflow created by the hand movement[1]
      (3) A positive response may be indicated by the movement of the nictitans rather than eyelid closure.[6]
2. Pupillary light response (PLR)
   a. Voluntary dynamic changes in pupil size, associated with excitement or agitation, makes evaluation of the direct PLR difficult.[1,3,9]
   b. True consensual (indirect) response is absent in birds because of the complete separation of the optic nerves.[3,7]
   c. Inadvertent constriction of the opposite pupil can occur when light passes through the thin osseous septum, which separates the globes caudally.[3,7]
   d. Consensual pupillary light reflexes are believed to be absent in birds.[3,6,8]
3. Anisocoria—Intermittent anisocoria can be normal especially during periods of excitement, because of the presence of striated sphincter muscle in the iris.[3]
4. Neurologic examination: If vision impairment is apparent, and/or neurologic abnormalities exist, a complete neurologic examination is indicated (see Chapter 9).

C. Examination of External Eye and Adnexa

1. Magnification, with a simple head loupe (28 or 30 diopter), or an indirect condensing lens and a point light source, should be used to examine external structures.[1,3,6]
2. Evaluate overall symmetry and position of globes.
3. Note any swellings, discolorations, abnormal growths, and/or ocular discharges.
4. Always obtain samples for microbial culture and sensitivity or virus isolation *before* administering topical anesthetics, medications, or eyewashes; in some cases it may be necessary to administer topical anesthetic before obtaining cytologic samples.[3]
5. Evaluate the function and mobility of eyelids and nictitans.
6. Evaluate the lid margins for any anatomic abnormalities, such as entropion.
7. Corneal surface should be smooth and shiny.
8. Fluorescein stain is indicated in any eye that is red, inflamed, or painful.[10] After staining, evaluate the eye with a Wood's lamp.

9. If blepharospasm is present, or the eye is excessively reddened, use atraumatic forceps to examine the conjunctival fornices and the underside of the nictitans for the presence of foreign bodies and/or parasites.[3]
    a. Instill one to two drops of topical anesthetic (0.5% proparacaine or 0.5% tetracaine HCL) into the eye and wait 1 minute.[3]
    b. Use topical anesthetic sparingly, as overuse can induce systemic toxicity, particularly in small birds.
10. Anterior chamber should be evaluated using a slit lamp.
    a. If a slit lamp is unavailable, the chamber can be evaluated using a penlight, a binocular loupe, or a direct ophthalmoscope at 20 diopters.[1]
    b. Aqueous should be clear; the presence of "aqueous flare" is indicative of anterior uveitis.[1,7,10]

D. Posterior Chamber
1. The small size of the bird's eye can make examination of the posterior chamber difficult.
2. Because of the presence of striated muscle in the iris, mydriasis cannot be achieved with parasympathomimetics.[1,9]
    a. In some larger birds, funduscopic examination may be possible by performing the examination in a darkened room, with a low illumination light source.[1,6]
    b. In smaller birds, mydriasis can be achieved under general anesthesia.[7]
    c. Topical vercuronium bromide solution (4 mg/ml) has been used to induce mydriasis in raptors.[1,3,6]
        (1) The solution is administered three times at 5-minute intervals to one eye only.[1]
        (2) Mydriasis peaks at 1 hour postadministration.[3,6]
        (3) Topical medications should be used sparingly, because overuse can induce systemic toxicity, particularly in small birds.
        (4) A more dilute concentration is recommended in smaller birds.[6]
        (5) Monitor bird for decreased respiratory rates and inspiratory excursions.[6]

E. Ancillary Tests
1. Samples for microbial culture and sensitivity
    a. Always obtain samples for microbial culture and sensitivity or virus isolation *before* administering topical anesthetics, medications, or eyewashes; in some cases it may be necessary to administer topical anesthetic before obtaining cytologic samples.[3]
    b. Use a sterile cotton swab moistened with transport media.[1]
    c. Optimally samples should be obtained from upper fornix to decrease contamination.[1]
    d. Studies indicate that bacteria and fungal organisms can be cultured from clinically normal birds.[1,3]
    e. Clinical disease is more commonly associated with gram-negative organisms.[1]

f. Special media is required for the growth of *Mycoplasma* spp. or *Chlamydia* spp.[1,4,7]
2. Cytologic samples
   a. Samples should be obtained before administering topical medications or eyewashes; however, topical anesthetic will probably be required.[7]
   b. Samples can be obtained by using a sterile moistened swab or the blunt end of a sterile scalpel blade.[3,6]
   c. Examination of samples, stained with Diff-Quik, Gram's, or Wright's stain, may be useful for selecting initial antibiotics until the results of the culture and sensitivity are known.[3,6]
   d. If *Chlamydia* spp. is suspected, Giemsa stain may be useful in evaluating cells for chlamydial elementary bodies.[1]
   e. The identification of fungal hyphae may require the use of methylene blue stain.[7]
   f. The identification of mycobacteriosis may require acid-fast staining.[7]
   g. It is a good idea to make extra slides so that additional stains can be used as needed.
   h. Examine the slide for inflammatory cells (infectious agents, trauma, foreign bodies), eosinophils (allergies), and keratinization (hypovitaminosis A).[7]
3. Fluorescein stain
   a. Fluorescein stain is indicated in any eye that is red, inflamed, or painful.[10]
   b. After staining, the eye should be evaluated with a Wood's lamp.
4. Schirmer tear test
   a. Normal data for most species have not been established.
   b. Tear test strips can be difficult to use in smaller birds.
      (1) Trimming a tear strip to a smaller size (4 mm) will make it easier to use but will yield above-average readings.[1]
      (2) A great deal of variation exists among species; another bird of the same species, or the clinically normal eye in cases of unilateral conditions, is useful for comparison.[3]
5. Schiøtz tonometry
   a. Intraocular pressure results may be unreliable in small birds (birds smaller than Amazon parrots)[1,3]
   b. Little normal data has been established[1,3]
   c. Another bird of the same species, or the clinically normal eye in cases of unilateral conditions, can be used for comparison.[3,6]

III. Diseases of the Globe and Orbit

A. Microphthalmia (Fig. 15-2)
1. Clinical signs
   a. Abnormally small eye or globe[3,8]
   b. Other anomalies may also be present including eyelid malformations, reduction of the palpebral fissure, or retinal dysplasia[3]

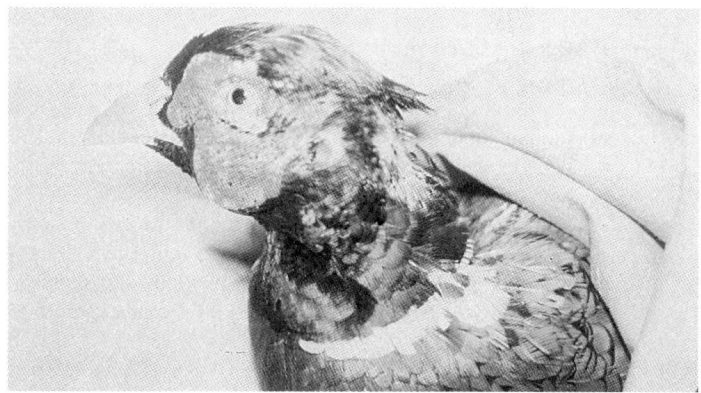

Fig. 15-2 Microphthalmia in a pheasant. (Courtesy Sallie Welte)

    c. Eye is usually visually impaired or blind[3]
2. Differential diagnosis
    a. Congenital malformation
    b. Posttraumatic: may be associated with rupture of the globe, particularly in raptors[3]
3. Treatment
    a. None

B. Phthisis
1. Clinical signs
    a. Abnormally small globe associated with ciliary atrophy [3,6]
2. Differential diagnosis—generally caused by ciliary body atrophy secondary to chronic uveitis or secondary to blunt or penetrating trauma (particularly in raptors)[3,10]
3. Diagnosis
    a. Can be difficult to differentiate from microphthalmia
    b. Inquire about medical history, past and concurrent disease, medications, and response to treatment
4. Treatment—none, eye is usually nonvisual[6,10]

C. Enophthalmia
1. Clinical signs
    a. Recession of the globe into the orbit[7,10]
    b. Enophthalmia must be differentiated from decreased globe size because of microphthalmia or phthisis[7]
    c. Concurrent clinical signs may include prolapse of the nictitans, entropion, and/or ocular discharge[7]
2. Differential diagnosis
    a. Atrophy of periorbital fat secondary to chronic sinusitis (particularly in macaws)[9]
    b. Trauma[7]
    c. Cachexia[9]

3. Diagnosis
   a. Diagnosis is based on clinical signs and history
   b. Obtain a complete history and signalment
      (1) Inquire about the duration and progression of the condition.
      (2) Inquire about medical history, including past or concurrent signs of respiratory disease, medications, and response to treatment.
      (3) Inquire about diet, dietary changes, and changes in eating habits.
      (4) Inquire about potential sources of infectious disease (respiratory infections), such as a history of recent exposure to other birds.
      (5) Inquire about incidences of trauma; birds that are allowed "free flight" are at a higher risk of traumatic injury from flying into windows, walls, ceiling fans, etc.
   c. Perform a complete physical examination
      (1) Evaluate bird for signs of systemic disease.
      (2) Evaluate bird's body condition/body weight.
      (3) Evaluate bird for evidence of sinusitis including periocular swelling, nasal discharge, etc.
      (4) Evaluate the entire bird for evidence of trauma.
      (5) If concurrent disease is suspected, perform diagnostic tests as indicated.
4. Treatment
   a. No treatment is required for the enophthalmia.
   b. Treat concurrent sinusitis, if present (see Chapter 3)
      (1) A twice-daily nasal flush, using a dilute antibiotic solution, and the administration of systemic antibiotics are recommended.[1,4,7,10]
      (2) Antibiotic selection should be based on the results of culture and sensitivity; I have used a 1:10 solution of enenrofloxacin initially while awaiting culture results.

D. Buphthalmos
1. Clinical signs: abnormal enlargement of the eye resulting from increased intraocular pressure or inflammation[7,10]
2. Differential diagnosis
   a. Uveitis[7] (see Intraocular Disease p. 301)
   b. Glaucoma
      (1) Rare in birds[1,7]
      (2) Can occur secondary to trauma[1,7]
   c. Intraocular neoplasia
      (1) Rare in birds[1,7]
3. Diagnosis
   a. Obtain a complete history and signalment.
      (1) Inquire about duration and progression of condition.
      (2) Inquire about medical history, including previous and concurrent diseases, response to treatment and current medications.

(3) Inquire about potential exposure to infectious disease, such as a history of recent exposure to other birds.
(4) Inquire about incidences of trauma; birds that are allowed "free flight" are at a higher risk of traumatic injury from flying into windows, walls, ceiling fans, etc.
b. Perform a complete physical examination.
    (1) Carefully evaluate the bird for evidence of trauma anywhere on the body, particularly on the head.
        (a) Lift the feathers on the head to look for hemorrhage or bruising.[7]
        (b) Palpate the orbital rim area for evidence of fractures and visually examine the sclera for indentations, which suggest a fracture of the scleral rings.[6]
        (c) Examine ears, nostrils, and oral cavity for evidence of hemorrhage.[7]
    (2) Evaluate the bird for evidence of systemic disease or infection.
c. Perform a complete ophthalmologic examination.
    (1) Evaluate overall symmetry and position of the globes.
    (2) Note any swellings, discolorations, abnormal growths, and/or ocular discharges.
        (a) Evaluate the periocular area for evidence of trauma
        (b) Evaluate the eye for evidence of periorbital masses such as abscesses, hemorrhages, or neoplasia.
    (3) Evaluate the eye for function and mobility of the lids.
        (a) Lid margins should meet when eye is closed.
        (b) Nictitans should cover the entire eye; inadequate coverage may lead to "dry eye" or exposure keratitis.
    (4) The cornea should be smooth and shiny: administer fluorescein stain and evaluate the eye for the presence of corneal defects, lacerations, etc.
    (5) Carefully evaluate the eye for the presence of uveitis (see Uveitis), hypopyon, hyphema, and intraocular parasites.
        (a) The eye should be evaluated using a slit lamp.
        (b) If a slit lamp is unavailable, the chamber can be evaluated using a penlight, a binocular loupe, or an ophthalmoscope at 20 diopters.[1,6]
        (c) Aqueous should be clear; the presence of "aqueous flare" is indicative of anterior uveitis.[1,7,10]
    (6) Perform a funduscopic examination.
        (a) Evaluate the posterior chamber for uveitis.
        (b) Evaluate the retina for evidence of trauma, such as retinal hemorrhage, edema, and detachment.
        (c) Because of the presence of striated muscle in the iris, mydriasis cannot be achieved with parasympathomimetics.[1,9]
        (d) In some larger birds, funduscopic examination may be possible by performing the examination in a darkened room, with a low-illumination light source.[1,6]
        (e) Mydriasis can usually be achieved under general anesthesia.[7]

(f) Topical vercuronium bromide solution (4 mg/ml) used to induce mydriasis in raptors[1,3]
   (i) The solution is administered three times at 5-minute intervals to one eye only.[1]
(g) Mydriasis peaks at 1 hour after administration.[3,6]
(h) Topical medications should be used sparingly, because overuse can induce systemic toxicity, particularly in small birds.
(i) A more dilute concentration is recommended in smaller birds.[6]
(j) Monitor bird for decreased respiratory rates and inspiratory excursions.[6]
(k) Evaluate the eye for evidence of intraocular neoplasia.
   d. Perform radiographs if indicated.
      (1) Radiographic changes in the periorbital area may be subtle and difficult to identify.
      (2) Evaluate entire body for evidence of trauma.
         (a) Periorbital trauma may result in fractures of the scleral ossicles.[3]
      (3) Evaluate for evidence of systemic disease, such as organomegaly.
      (4) Evaluate for evidence of neoplasia and/or metastasis.
   e. Ultrasonography (7.5 MHz with coupling gel) may be useful to evaluate the eyes of larger birds for intraocular masses.[6,7]
   f. Perform other tests (hematology, biochemistry, cultures, titers, etc.) as needed to diagnose infectious or inflammatory diseases.
4. Treatment
   a. Provide supportive care as indicated.
   b. If inadequate coverage is provided by the nictitans treatment should be initiated to prevent exposure keratitis (see Keratitis p. 295).
   c. Treat uveitis, if present (see Uveitis p. 301).
   d. The treatment (and prognosis) of neoplasia depends on the type of tumor and its location.[3]
      (1) If orbital neoplasia is present, removal of the eye and the soft tissue of the orbit is recommended.[3]

E. Exophthalmus
1. Clinical signs
   a. Abnormal protrusion of the globe[10]
      (1) Differentiate from buphthalmos and eyelid malformations (abnormally large palpebral opening)
   b. May indicate a space-occupying lesion[1,7,10]
   c. Concurrent signs may include a prolapsed nictitans, strabismus, and/or periocular feather loss.
2. Differential diagnosis
   a. Trauma (common in raptors, often produces secondary retrobulbar abscesses or hemorrhage)[1,7]
   b. Abscess

(1) Periorbital abscess (commonly seen secondary to a chronic upper respiratory tract infection and sinusitis, particularly in cockatiels)[1,4]
(2) Orbital abscess[1,3]
(3) Lacrimal gland abscess (generally identified as a movable mass located at or near the medial canthus)[1,4]; may be accompanied by epiphora and/or lid distortion
  c. Neoplasia
    (1) Includes pituitary tumors (particularly in budgies), optic nerve gliomas, and round cell sarcomas[1,3,7]
  d. Retrobulbar mass
    (1) Hemorrhage/hematoma (uncommon)[3,7]
    (2) Aspergillosis granuloma, mycobacterium, and disseminated cryptococci are uncommon[7]
3. Diagnosis
  a. Obtain complete history and signalment.
    (1) Inquire about duration and progression of condition.
    (2) Inquire about medical history, including past or concurrent signs of respiratory disease, medications, and response to treatment.
    (3) Inquire about incidences of trauma; birds that are allowed "free flight" are at a higher risk for traumatic injury.
    (4) Inquire about potential sources of infectious disease (respiratory infections), such as recent exposure to other birds.
  b. Perform a complete physical examination.
    (1) Evaluate the bird for evidence of respiratory disease.
    (2) Evaluate the bird for evidence of concurrent systemic disease.
    (3) Carefully evaluate the bird for evidence of trauma anywhere on the body (particularly on the head); lift the feathers on the head and examine ears, nostrils, and oral cavity to look for hemorrhage or bruising.[7]
    (4) Palpate the orbital rim area for evidence of fractures and visually examine the sclera for indentations, which suggest a fracture of the scleral rings.[6]
  c. Perform a complete ophthalmologic examination.
    (1) Evaluate overall symmetry and position of globes.
    (2) Note any swellings, discolorations, abnormal growths, and/or ocular discharges.
      (a) Evaluate the periocular area for evidence of trauma such as bruising or lacerations (ocular discharge may lead to facial rubbing).
      (b) Evaluate the eye for evidence of periorbital masses such as abscesses, hemorrhages, or neoplasia.
    (3) Evaluate the eye for function and mobility of the lids.
      (a) Lid margins should meet when eye is closed.
      (b) Nictitan should cover the entire eye.
      (c) Inadequate coverage may lead to "dry eye" or exposure keratitis (see Corneal Ulceration and Keratitis).

(4) Evaluate the eye for evidence of trauma such as hyphema.
(5) Administer fluorescein stain and evaluate the cornea for the presence of corneal defects, ulcerations, or lacerations.
(6) A funduscopic examination should be performed to evaluate the retina for evidence of trauma, such as retinal hemorrhage, edema, and detachment.
   (a) The small size of the bird's eye can make examination of the posterior chamber difficult.
   (b) In some larger birds, funduscopic examination may be possible by performing the examination in a darkened room, with a low-illumination light source.[1,6]
   (c) In smaller birds, mydriasis can be achieved under general anesthesia.[7]
   (d) Because of the presence of striated muscle in the iris, mydriasis cannot be achieved with parasympathomimetics.[1,9]
   (e) Topical vecuronium bromide solution (4 mg/ml) has been used to induce mydriasis in raptors.[1,3,6]
   (f) The solution is administered three times at 5-minute intervals to one eye only.[1]
   (g) Mydriasis peaks at 1 hour after administration.[3,6]
   (h) Topical medications should be used sparingly, because overuse can induce systemic toxicity, particularly in small birds.
   (i) A more dilute concentration is recommended in smaller birds.[6]
   (j) Monitor bird for decreased respiratory rates and inspiratory excursions.[6]
d. Perform radiographs if indicated.
   (1) Evaluate radiographs for evidence of periorbital masses; periorbital masses may be subtle and difficult to identify.
   (2) Evaluate for evidence of trauma including fractures of the ossicles, skull, and limbs.[3]
   (3) Evaluate for evidence of systemic disease.
   (4) Evaluate for evidence of neoplasia and/or metastasis.
e. Ultrasonography
   (1) Ultrasonography (7.5 MHz probe with coupling gel) may be useful to evaluate the eye (in larger birds) for intraorbital or retrobulbar masses and retinal dysplasia.[3,6,7]
   (2) Ultrasound guided aspiration may be required for sampling retrobulbar masses.[3,7]
f. Fine needle aspiration of intraorbital masses[3]
   (1) Samples can be submitted for cytology, culture and sensitivity, acid-fast staining, and/or histology.[3]
   (2) Bird should be placed under general anesthesia.
   (3) Infraorbital masses can be sampled by inserting the needle between the eye and the wall of the orbit.[3]
   (4) For medial lesions the point of insertion should be medial to the nictitans and lateral to the orbit wall.[3]

g. Lacrimal sac abscesses
   (1) Commonly occur secondary to chronic upper respiratory infections
   (2) Identified as movable masses ventromedial to the eye(s)[7]
   (3) May be accompanied by epiphora and/or distortion of the lids
   (4) If caseous material can be obtained (by gently expressing the area over the duct), it should be submitted for culture and sensitivity.[7]
   (5) Samples can be also be stained with Diff-Quik or Gram's stain.[7]
h. Fungal cultures or serology samples (for aspergillosis) may be submitted to laboratories specializing in avian diagnostics.[7]
i. Additional diagnostic tests (such as serum biochemistry, complete cell count, etc.) are indicated if systemic disease is present or suspected.

4. Treatment
   a. Provide supportive care as indicated.
   b. If inadequate corneal coverage is provided by the nictitans, treatment should be initiated to prevent exposure keratitis (see Corneal Ulceration and Keratitis p. 295).
   c. Begin treatment for corneal ulcers if present. (See Corneal Ulceration and Keratitis.)
   d. If periorbital or intraocular inflammation is present, systemic corticosteroids may be administered.
      (1) Topical steroids should *not* be used if corneal ulceration is present.
   e. Remember that topical medications can be absorbed systemically
      (1) Systemic toxicity is possible, particularly in small birds.
      (2) Calculate maximum safe daily dose of the drug, taking into consideration all forms of administration
   f. Abscesses
      (1) Antibiotic selection should be based on the results of a culture and sensitivity.
      (2) It is important to differentiate lacrimal sac abscesses from other types of periorbital abscesses.[1]
      (3) Lacrimal sac abscesses
         (a) Surgical debridement is not recommended because scarring and stricture may result.[1]
         (b) Some lacrimal sac abscesses can be expressed manually.[1,4]
         (c) Cannulate and flush the lacrimal duct(s) twice daily with a dilute antibiotic solution.[1,4]
         (d) Before receiving culture results, use a 1:10 solution of enrofloxacin to sterile saline.
      (4) Other periorbital abscesses
         (a) Surgical debridement and administration of systemic antibiotics is recommended.[1,4]
      (5) A systemic antibiotic should be administered concurrently.
   g. Treat sinusitis and respiratory disease if present.

(1) If sinusitis is present, a twice-daily nasal flush, using a dilute antibiotic solution, and the administration of systemic antibiotics are recommended.[4,10]
(2) Antibiotic selection should be based on the results of culture and sensitivity.
(3) Sinus abscesses usually require surgical debridement and the administration of parenteral antibiotics.[4,10]
h. Periorbital or retrobulbar hemorrhages will spontaneously regress; no treatment is required.[7]
i. Treat systemic disease if present.
j. The treatment (and prognosis) of neoplasia depends on the type of tumor and its location.[3]
(1) If orbital neoplasia is present, removal of the eye and the soft tissue of the orbit is recommended.[3]
5. Prevention
a. Early aggressive treatment of sinusitis with appropriate antibiotics may help prevent the formation of periocular abscesses.[1,4]

F. Flattening of the Globe
1. Clinical signs
a. Decreased depth of anterior chamber[7]
2. Differential diagnosis
a. May occur during handling or restraint.[1,7]
b. May occur in the ventrally positioned eye when the bird is placed in lateral recumbency during general anesthesia.[3,7]
3. Diagnosis
a. Based on history and clinical signs
4. Treatment
a. None; chamber will spontaneously reinflate 15 to 20 minutes after the stimuli are removed.

IV. Diseases of the Eyelid and Periorbita

A. Eyelid Malformations, Abnormalities, and Trauma
1. Clinical signs
a. Signs vary with etiology
b. Secondary periocular dermatitis may be present
2. Differential diagnosis
a. Agenesis
(1) Involves the loss of the eyelid margin; varies from subtle to severe[3]
(2) Developmental abnormality[3]
b. Cryptophthalmos
(1) Congenital fusion of the upper and lower eyelids[1]
(2) Eyelid margins are absent; no palpebral tissue is present[3]
(3) Developmental abnormality[1,3]
c. Ankyloblepharon

(1) Adhesion of the upper and lower eyelid margins[10]
(2) Some or all of the lid margin is present, but the palpebral fissure is absent[3]
(3) Developmental abnormality[3]
   d. Trauma
3. Diagnosis
   a. Diagnosis is based on history, signalment, and characteristic clinical signs.
   b. Agenesis, cryptophthalmos, and ankyloblepharon are first noted at hatching or within the first few weeks of life.
4. Treatment
   a. Agenesis—may require a permanent canthorrhaphy[3]
   b. Cryptophthalmos and ankyloblepharon
      (1) Surgical correction followed by the administration of topical steroids has been suggested but is usually unsuccessful, as skin closure tends to recur[1,3]
   c. Trauma
      (1) Lid tears can be corrected using 5-0 to 7-0 suture material in a half-thickness simple interrupted suture.[6]
      (2) Full-thickness lacerations of the nictitans (and those involving a lid margin) can be repaired using 6-0 to 10-0 suture material.[6]
      (3) Small and partial-thickness lacerations may not require sutures; treat with topical antibiotics.[6]
      (4) The nictitans should never be surgically removed; removal may result in chronic exposure keratitis.[7]
      (5) Care must be taken to avoid exposing the cornea to the suture material and knots.

B. Eyelid Dysfunction
1. Clinical signs
   a. Reduction or absence of the palpebral reflex[3]
   b. Secondary keratitis may be present
2. Differential diagnosis
   a. Facial paralysis (usually secondary to trauma)[3]
3. Treatment
   a. Administer a topical lubricating ointment three to four times a day until paralysis resolves.[3]
   b. If functional loss is permanent, a canthorrhaphy may be necessary to prevent chronic exposure keratitis.[3]
   c. If keratitis is present, initiate treatment (see Corneal Ulceration and Keratitis p. 295).

C. Conjunctivitis
1. Clinical signs (Fig. 15-3)
   a. Signs may include conjunctival hyperemia, edema, ocular discharge, blepharospasm, and photophobia.[1,3,7]
   b. Chronic ocular discharge and/or pruritus may result in periocular feather loss and or ocular trauma due to facial rubbing.[1,3,7]

Fig. 15-3 Conjunctivitis in a house finch. (Courtesy Sallie Welte)

      c. In some cases, signs of upper or lower respiratory disease may be present.[1,3,4]
- 2. Differential diagnosis
  - a. Trauma (e.g., handling, restraint, cage mate aggression, self-induced injury)[1,3,4]
  - b. Environmental allergens, toxins, and irritants (e.g., chemical disinfectants, dust, smoke, ammonia buildup)[1,3,4,7]
  - c. Secondary to either upper or lower respiratory disease and/or sinusitis[1,3,4]
  - d. Foreign bodies
  - e. Vitamin A deficiency[3,4]
  - f. Cockatiel conjunctivitis of unknown etiology[1,3]
  - g. Chlamydiosis, especially in cockatiels and finches[1,3] (see Chapter 3)
  - h. Mycoplasmosis, especially budgerigars, cockatiels, passerines[1,3,7]
  - i. Herpes virus, particularly in Gouldean, zebra, and crimson finches[7]
  - j. Infectious laryngotracheitis, especially canaries[7]
  - k. Paramyxovirus, especially finches[3,7]
  - l. Pacheco's disease[3,7]
  - m. Other bacterial, viral, or fungal infections[1]
  - n. Endophthalmitis secondary to toxoplasmosis (uncommon)[3]
  - o. Cryptosporidiosis particularly budgerigars (uncommon)[1,3,4,7,9]
  - p. Giardia, particularly budgerigars and cockatiels[3,7]
  - q. Parasites (e.g., *Oxyspirura, Ceratospira, Thelazia, Philophthalmus,* and *Setaria*)[1,3,4]
  - r. Secondary to septicemia[1]
- 3. Diagnosis
  - a. Obtain complete history and signalment
    - (1) Inquire about duration and progression of condition.

(2) Inquire about medical history, including past or concurrent diseases, medications, and response to treatment.
(3) Inquire about incidences of trauma.
 (a) Birds that are allowed "free flight" are at a higher risk for traumatic injury from flying into windows, walls, ceiling fans, etc.
 (b) Inquire about incidences of cage mate aggression, etc.
(4) Inquire about exposure to environmental irritants or allergens and evaluate husbandry.
(5) Inquire about potential sources of infectious disease, such as a history of recent exposure to other birds.
(6) Evaluate diet for sufficient levels of vitamin A.
 (a) "All-seed" diets are frequently deficient.[7]
 (b) A diagnosis of hypovitaminoses A is based on characteristic clinical signs, dietary evaluation, and response to treatment (see Periorbital Hyperplastic Lesions p. 289).
b. Perform a complete physical examination.
(1) Carefully evaluate the bird for evidence of trauma anywhere on the body, particularly on the head.
 (a) Lift the feathers on the head to look for hemorrhage or bruising.[7]
 (b) Palpate the orbital rim area for evidence of fractures and visually examine the sclera for indentations, which suggest a fracture of the scleral rings.[6]
 (c) Examine ears, nostrils, and oral cavity for evidence of hemorrhage[7]
(2) Evaluate the bird for evidence of respiratory disease and sinusitis.
 (a) Birds with cryptosporidium may exhibit signs of respiratory disease, including dyspnea, rhinitis, sinusitis, and tracheitis.[7]
(3) Evaluate the bird for evidence of concurrent systemic disease.
c. Perform a complete ocular examination.
(1) Use magnification with a simple head loupe (28 or 30 diopter) or an indirect condensing lens and a point light source[1,3,6]
(2) Evaluate overall symmetry and position of globes.
(3) Evaluate the eye for evidence of trauma; note any swellings, discolorations, abnormal growths, and/or ocular discharges
(4) Evaluate the eye for evidence of periorbital masses such as abscesses, hemorrhages, or neoplasia
(5) Always obtain samples for microbial culture and sensitivity or virus isolation *before* administering topical anesthetics, medications, or eyewashes; in some cases it may be necessary to administer topical anesthetic before obtaining cytologic samples.[3]
 (a) Use a sterile cotton swab moistened with transport media.[1]
 (b) Optimally samples should be obtained from upper fornix to decrease contamination.[1]

- (c) Studies indicate that bacteria and fungal organisms can be cultured from clinically normal birds.[1,3]
- (d) Special media are required for the growth *Mycoplasma* spp. or *Chlamydia* spp.[1,4,7]
- (e) *Mycoplasma* spp. can be difficult to culture; diagnosis may be based on bird's response to treatment
- (6) Cytologic samples should be collected whenever conjunctivitis is present.[3]
    - (a) Samples should be obtained before administering topical medications or eyewashes; however, topical anesthetic will probably be required.[7]
    - (b) Samples can be obtained by using a sterile moistened swab or the blunt end of a sterile scalpel blade.[3,6]
    - (c) Examination of samples, stained with Diff-Quik, Gram's, or Wright's stain, may be useful for selecting initial antibiotics until the results of the culture and sensitivity are known.[3,6]
    - (d) If *Chlamydia* spp. is suspected, Giemsa stain may be useful in evaluating cells for chlamydial elementary bodies.[1]
    - (e) The identification of fungal hyphae may require the use of methylene blue stain.[7]
    - (f) The identification of mycobacteriosis may require acid-fast staining.[7]
    - (g) It is a good idea to make extra slides so that additional stains can be used as needed.
    - (h) Examine the slide for inflammatory cells (infectious agents, trauma, foreign bodies), eosinophils (allergies), and keratinization (hypovitaminosis A).[7]
- (7) Fluorescein stain is indicated in any eye that is red, inflamed, or painful.[10]
    - (a) After staining, the eye should be evaluated for corneal defects with the use of a Wood's lamp.
- (8) Examine the conjunctival fornices and the underside of the nictitans for the presence of foreign bodies and/or parasites.[7]
    - (a) Instill one to two drops of topical anesthetic (0.5% proparacaine or 0.5% tetracaine HCl) into the eye and wait 1 minute.[3]
    - (b) Use atraumatic forceps to examine the conjunctival fornix and the nictating membrane.
    - (c) Topical anesthetic should be used sparingly, because overuse can induce systemic toxicity, particularly in small birds.
- d. When concurrent upper respiratory disease is present, consider submitting samples for bacterial and/or fungal (aspergillus) culture and/or cytology
    - (1) Appropriate samples include sinus aspirates or choanal swabs.[1]

Ophthalmic Disorders **283**

  (2) If choanal swabs are being used to evaluate the organisms present in the sinuses or nasal passages, samples should be taken from the rostral end of the slit.[1]
  (3) Fungal cultures should be submitted to laboratories proficient in culturing avian aspergillosis
 e. Perform radiographs if indicated.
  (1) Radiographic changes in the periorbital area may be subtle and difficult to identify.
  (2) Evaluate entire body for evidence of trauma.
  (3) Periorbital trauma may result in fractures of the scleral ossicles or skull.[3]
  (4) Evaluate for evidence of respiratory changes.
  (5) Evaluate for evidence of systemic disease such as organomegaly.[7]
  (6) Evaluate for evidence of periorbital masses.
 f. Biochemistries and complete cell count are indicated if respiratory or systemic disease is present or suspected.
 g. Fecal samples may be examined for the presence of *giardia* spp. or cryptosporidium oocysts[7] (see Chapter 20).
 h. Perform other tests (cultures, titers, etc.) as needed to diagnose infectious or inflammatory diseases.
4. Treatment
 a. Provide supportive care if necessary.
 b. Initiate treatment for respiratory disease if present.
 c. Initiate appropriate treatment for systemic disease if present.
 d. Eliminate environmental irritants or allergens.
  (1) Eliminate suspected causes, such as chemical disinfectants, secondhand cigarette smoke, ammonia buildup, etc.
  (2) Antihistamines may be useful for treating the initial symptoms.[7]
 e. If cytologic samples are indicative of a bacterial conjunctivitis, begin treatment with topical broad-spectrum antibiotic ointment or solution three to four times a day.[3]
 f. The administration of topical steroids may be indicated.
  (1) Topical steroids should *not* be used if corneal ulceration is present.
  (2) Remember that topical medications can be absorbed systemically.
   (a) Toxicity is possible, particularly in small birds.
   (b) Calculate maximum safe daily dose of the drug, taking into consideration all forms of administration.
 g. Conjunctival parasites
  (1) After administration of a topical anesthetic, parasites can usually be flushed or manually removed from the conjunctiva and/or the underside of the nictitans.[1,3,7]
  (2) Gently express the area surrounding the lacrimal sac to extrude any parasites in the duct[7]
  (3) In the case of severe infection, topical administration of a single dose of ivermectin (200 µg/kg) or 5% carbamate powder[4,7]

(4) Ivermectin can be diluted (1:8) in either propylene glycol or sterile water.[13] Ivermectin will precipitate out of either of these solutions; therefore, thorough mixing before administration is essential.[13]

D. Periorbital Swelling

1. Clinical signs
   a. Focal or general swelling and erythema of the periocular region and/or eyelids.
   b. Other signs may include ocular discharge and pruritus as well as periocular feather loss and trauma resulting from facial rubbing.[1,7]
   c. The presence of strabismus or exophthalmus is suggestive of a space-occupying lesion.[1,7,10]
      (1) Space-occupying lesions include abscesses, hematomas and neoplasia.[7]
2. Differential diagnosis
   a. Periorbital hyperplastic lesions
      (1) Includes *Knemidokoptes* spp., papillomatosis, and hypovitaminosis A (see Periocular Hyperplastic Lesions)[3,7]
   b. Avian pox[3] (see Avian Pox p. 292)
   c. Spontaneous orbital inflation
      (1) Seen in some birds (particularly Amazons) as a an act of aggression or in reaction to stress, handling, restraint, etc.[1,7]
      (2) Will regress when stressful stimuli are removed
   d. Infraorbital sinusitis and rhinitis[1,3]
      (1) Frequently develops concurrent with an upper respiratory infection[1,3,4]
   e. Periorbital or orbital abscesses
      (1) Common cause of periorbital swelling, particularly in cockatiels[1,3,4]
      (2) Common sequela of chronic upper respiratory disease and sinusitis[1,4,9]
      (3) May occur anywhere within or around the orbit[1,4]
   f. Lacrimal gland abscesses
      (1) Develop as movable swellings in the area of the lacrimal duct (ventromedial to the eye)[7,9]
   g. Dermatitis
      (1) Can occur secondary to environmental irritants or allergens, such as dust, cigarette smoke, periorbital insect bites, chemical disinfectants, and ammonia buildup (poor husbandry)[1,3,4,7]
      (2) May occur secondary to bacterial or fungal infections[1,7]
   h. Trauma
      (1) May include periorbital hematomas, lid abnormalities, and cellulitis[7]
   i. Bacterial infections[3]
   j. Fungal infections: Organisms identified include aspergillus and cryptococcus[1,7]

k. Mycobacteriosis[7]
l. Parasitic infections
    (1) Organisms most commonly identified include Thelazia, *Oxyspirura* spp., philophthalmus, *Setaria* spp., ceratospira[1,3,4,7,9]
    (2) *Knemidokoptes* spp. are common in passerines and budgerigars (see Periocular Hyperplastic Lesions)
m. Neoplasia
    (1) Neoplasia of the lids or conjunctiva of birds is uncommon.[3,4,11]
    (2) Infraorbital/retrobulbar neoplasia has been identified in birds and includes tumors of the orbit, pituitary gland, and optic nerve (see Exophthalmia p. 274).[1,3,7]
    (3) Periocular or infraorbital neoplasia may be present with or without globe displacement, exophthalmus, or strabismus[1,7]
3. Diagnosis
   a. Obtain a complete history and signalment.
      (1) Inquire about duration and progression of condition.
      (2) Inquire about medical history, including previous and concurrent signs of respiratory disease, response to treatment, and current medications.
      (3) Inquire about incidences of trauma.
         (a) Birds that are allowed "free flight" are at a higher risk of traumatic injury from flying into windows, walls, and ceiling fans.
         (b) Inquire about incidences of cage mate aggression, etc.
      (4) Inquire about potential sources of infectious agents and parasites, including history of recent exposure to other birds.
      (5) Evaluate diet for sufficient levels of vitamin A; diets composed solely of seed are frequently deficient.[7]
         (a) A diagnosis of hypovitaminoses A is based on dietary evaluation, characteristic clinical signs, and response to treatment (see Periorbital Hyperplastic Lesions-Hypovitaminosis A p. 289).
      (6) Inquire about exposure to environmental irritants or allergens and evaluate husbandry. Birds in outdoor aviaries are more susceptible to insect bites.
   b. Perform a complete physical examination.
      (1) Carefully evaluate the bird for evidence of trauma anywhere on the body, particularly on the head.
         (a) Lift the feathers on the head to look for hemorrhage or bruising.[7]
         (b) Examine ears, nostrils, and oral cavity for evidence of hemorrhage.[7]
         (c) Palpate the orbital rim area for evidence of fractures and visually examine the sclera for indentations, which suggest a fracture of the scleral rings.[6]
      (2) Evaluate the bird for evidence of systemic disease and/or neoplastic disease.

(3) Evaluate bird for evidence of respiratory disease and sinusitis, including periocular swelling and nasal discharge.
c. Perform a complete ocular examination.
  (1) Evaluate overall symmetry and position of globes.
  (2) Note any swellings, discolorations, abnormal growths, and/or ocular discharges.
  (3) Evaluate the function and mobility of eyelids and nictitans.
    (a) Evaluate the lid margins for any anatomic abnormalities, such as lid tears, or entropion.
  (4) Obtain samples for microbial culture and sensitivity if indicated.
    (a) Always obtain samples for microbial culture and sensitivity or virus isolation *before* administering topical anesthetics, medications, or eyewashes; in some cases it may be necessary to administer topical anesthetic before obtaining cytologic samples.[3]
    (b) Use a sterile cotton swab moistened with transport media.[1]
    (c) Optimally samples should be obtained from upper fornix to decrease contamination.[1]
    (d) Studies indicate that bacteria and fungal organisms can be cultured from clinically normal birds.[1,3]
    (e) Special media are required for the growth of *Mycoplasma* spp. or *Chlamydia* spp.[1,4,7]
    (f) Clinical disease is more commonly associated with gram-negative organisms.[1]
    (g) A conjunctival scraping can be used to determine the choice of antibiotics until the results of the culture and sensitivity are known.
  (5) Obtain cytologic samples.
    (a) Samples should be obtained before administering topical medications or eyewashes; however, topical anesthesia will probably be required.[7]
    (b) Samples can be obtained by using a sterile moistened swab or the blunt end of a sterile scalpel blade.[3,6]
    (c) Examination of samples, stained with Diff-Quik, Gram's, or Wright's stain, may be useful for selecting initial antibiotics until the results of the culture and sensitivity are known.[3,6]
    (d) If *Chlamydia* spp. is suspected, Giemsa stain may be useful in evaluating cells for chlamydial elementary bodies.[1]
    (e) The identification of fungal hyphae may require the use of methylene blue stain.[7]
    (f) The identification of mycobacteriosis may require acid-fast staining.[7]
    (g) It is a good idea to make extra slides so that additional stains can be used as needed.
    (h) Examine the slide for inflammatory cells (infectious agents, trauma, foreign bodies), eosinophils (allergies), and keratinization (hypovitaminosis A).[7]

(6) Evaluate the periorbital area for evidence of trauma.
  (a) Palpate the orbital rim area for evidence of fractures and visually examine the sclera for indentations, which suggest a fracture of the scleral rings.[6]
(7) Fluorescein stain is indicated in any eye that is red, inflamed, or painful.[10] After staining, eye should be evaluated for corneal abrasions and ulcerations using a Wood's lamp.
(8) Examine the conjunctival fornices and the underside of the nictitans for the presence of foreign bodies and/or parasites.[7]
  (a) Instill one to two drops of topical anesthetic (0.5% proparacaine or 0.5% tetracaine HCl) into the eye and wait 1 minute.[3]
  (b) Use atraumatic forceps to examine the conjunctival fornix and the nictating membrane.
  (c) Topical anesthesia should be used sparingly because overuse can induce systemic toxicity, particularly in small birds.[3]
d. Respiratory disease
  (1) If concurrent respiratory disease is present, submit samples for bacterial and/or fungal (aspergillus) culture and cytology.
  (2) Appropriate samples include sinus aspirates, choanal swabs, or tracheal washes. If choanal swabs are being used to evaluate the organisms present in the sinuses or nasal passages, take samples from the rostral end of the slit.[1]
e. Fungal cultures should be submitted to laboratories proficient in culturing avian aspergillosis.
f. Lacrimal abscesses
  (1) Commonly occur secondary to chronic upper respiratory infections
  (2) Identified as movable masses ventromedial to the eye(s)[7]
  (3) May be accompanied by epiphora and/or distortion of the lids
  (4) If caseous material can be obtained (by gently expressing the area over the duct), it should be submitted for culture and sensitivity.[7]
  (5) Samples can be also be stained with Diff-Quik or Gram's stain.[7]
g. Retrobulbar abscesses
  (1) If retrobulbar abscesses are present, samples should be taken via fine needle aspirate and submitted for culture and sensitivity (see Exophthalmia p. 274).
h. Radiographs may be indicated, depending on the suspected etiology.
  (1) Radiographs may be useful for identifying the location of periorbital masses
    (a) Radiographic changes of the periorbital area can be subtle and therefore difficult to identify.
  (2) Radiographs should be performed to evaluate the bird for other injuries if trauma is suspected.
    (a) Periorbital trauma may result in fractures of the skull and/or scleral ossicles.[6]

(3) Radiographs may be useful for identifying respiratory changes or signs of systemic disease such as organomegaly.[7]
(4) If neoplasia is suspected, radiographs should be evaluated for evidence of metastasis.
  (a) Ultrasonography (using a 7.5 MHz probe and coupling gel) may be useful to evaluate the eye and orbit of larger birds for neoplasia, abscesses, and hematomas.[6,7]
  (b) If retrobulbar masses are present, samples may be taken via ultrasound guided fine needle aspirate (see Exophthalmia p. 274).
i. A definitive diagnosis of neoplasia requires the examination of histologic samples.
j. Blood samples should be submitted for hematology and plasma biochemistry if clinical signs indicate that respiratory, systemic, or septic disease is present.

4. Treatment
  a. Provide supportive care as needed.
  b. When indicated, the bird should be placed on an appropriate systemic antibiotic as determined by culture and sensitivity.
    (1) Gram staining of a conjunctival scraping can be used initially to determine the choice of antibiotics until the results of the culture and sensitivity are known.
  c. Sinusitis
    (1) If sinusitis is present, a twice-daily nasal flush, using a dilute antibiotic solution, and the administration of systemic antibiotics is recommended.[1,4,7,10]
    (2) Antibiotic selection should be based on the results of culture and sensitivity.
    (3) Sinus abscesses usually require surgical debridement and the administration of parenteral antibiotics.[4,10]
  d. Periorbital abscesses
    (1) Antibiotic selection should be based on the results of a culture and sensitivity.
    (2) Good choices for initial treatment include enrofloxacin or cefotaxime three times daily.[7]
    (3) Recommended treatment is surgical curettage and administration of systemic antibiotics.[1,4]
    (4) It is important to differentiate lacrimal sac abscesses from other types of periorbital abscesses.[1]
  e. Lacrimal sac abscesses
    (1) Surgical debridement is not recommended as scarring and stricture may result.[1]
    (2) Some lacrimal sac abscesses can be expressed manually.[1,4]
    (3) The lacrimal duct(s) should be cannulated and flushed twice daily with a dilute antibiotic solution.[1,4]
    (4) Before receiving culture results, use a 1:10 solution of enrofloxacin to sterile saline.
  f. Neoplasia

(1) Best removed via surgical excision or cryosurgery.[1,3]
(2) A presurgical radiograph should be performed to evaluate for evidence of metastasis.
  g. Environmental irritants and allergens
    (1) Eliminate suspected causes, such as chemical disinfectants, secondhand cigarette smoke, and ammonia buildup.
    (2) Mosquito netting placed over outdoor cages may help to eliminate insect bites.
    (3) Antihistamines may be useful for treating the initial symptoms.[7]
  h. Retrobulbar hematomas will regress naturally; no treatment is required.[7]
  i. Conjunctival parasites
    (1) After administration of a topical anesthetic, parasites can usually be flushed or manually removed from the conjunctiva and/or the underside of the nictitans.[1,3]
    (2) Gently express the area surrounding the lacrimal sac to extrude any parasites in the duct.
      (a) In the case of severe infection, topical administration of a single dose of ivermectin (200 µg/kg) or 5% carbamate powder may be used.[4,7]
    (3) Ivermectin can be diluted (1:8) in either propylene glycol or sterile water.[13] Ivermectin will precipitate out of either of these solutions, therefore, thorough mixing before administration is essential.[13]
  j. Lid tears
    (1) Lid tears can be corrected using 5-0 to 7-0 suture material in a half-thickness simple interrupted suture.[6]
    (2) Full-thickness lacerations of the nictitans (and those involving a lid margin) can be repaired using 6-0 to 10-0 suture material.[6]
    (3) Small and partial thickness lacerations may not require sutures; treat with topical antibiotics.[6]
    (4) The nictitans should never be surgically removed; removal may result in chronic exposure keratitis.[7]
    (5) Care must be taken to avoid exposing the cornea to the suture material and knots.
5. Prevention
  a. Quarantine new birds.
  b. Quarantine sick birds when etiology may be infectious.
  c. Early aggressive treatment of sinusitis with appropriate antibiotics may help to prevent the formation of periocular abscesses.[1,4]

E. Periorbital Hyperplastic Lesions

1. Clinical signs
  a. Hyperplastic epithelial proliferations in the periocular area
  b. Depending on the etiology, similar lesions may be seen in the beak, legs, feet, and vent.[1]
2. Differential diagnosis
  a. *Knemidokoptes* spp. ("scaly face and leg mite")[1,7]

(1) Most commonly identified in budgerigars and canaries, but many other species of psittacines and passerines are susceptible.[1,4,7,9]
(2) Produces characteristic "honeycomb-like" lesions that may be present around the eyes and on the cere, beak, vent, legs, and/or feet.[1,3,7]
(3) Lesions may be scaly or crusty and are highly proliferative.[1,7]
(4) All ages are susceptible, but young birds and inbred individuals are most commonly affected.[15]
(5) Infections may be latent for up to 2 years and a genetic predisposition may exist.[15]
(6) Severe infections may lead to gross deformities and overgrowth of the beak of budgerigars.[15]

b. Hypovitaminosis A
(1) May cause hyperplastic and hyperkeratotic epithelial lesions in the periocular area; signs may be subtle to severe.[1,7,11]
(2) Lesions may include thickening of the conjunctiva and the cornea, and dysplasia of the lacrimal gland.[3,7]
(3) Lesions are generally less proliferative and less extensive than those associated with knemidokoptes.[1]
(4) Other clinical signs may include blepharitis, keratoconjunctivitis sicca ("dry eye"), mild swelling of the periorbital area and conjunctiva, and ocular discharge.[1,7,11]
(5) Bird may show signs of an upper respiratory tract infection, including sneezing and nasal discharge.[1,11]
(6) White hyperkeratotic plaques in the oral cavity and a blunting or swelling of the choanal papillae and/or palatine or choanal abscesses may be present.[1,7]

c. Cutaneous papillomas
(1) Single or multiple "wartlike" epidermal proliferation(s)[7]
(2) Lesions may appear on the periocular skin, eyelids, beak, oral cavity, feet, legs, and/or cloaca.[7]

d. Neoplasia: Neoplasia of the eyelids is rarely reported in birds.[3,4]
e. Avian Pox virus (see Avian Pox p. 292)

3. Diagnosis
a. Obtain a complete history and signalment.
(1) Inquire about the duration and progression of the disease.
(2) Inquire about medical history, including previous and concurrent diseases, response to treatment, and current medications.
(3) Evaluate diet for sufficient levels of vitamin A; diets composed solely of seeds are frequently deficient.[3,7]

b. Perform a complete physical examination.
(1) Examine bird for additional lesions on legs, feet, and vent.
(2) Examine oral cavity and choanae for evidence of hyperplastic lesions, sterile granulomas, abscesses, or cysts; and note any changes associated with the choanal papillae.[7]
(3) Evaluate bird for signs of upper respiratory tract infection.

c. Perform complete ocular examination.

(1) Evaluate overall symmetry and position of globes.
(2) Fluorescein stain is indicated in any eye that is red, inflamed, or painful or if keratoconjunctivitis sicca is suspected.[10]
   d. Initial diagnosis of *Knemidokoptes* spp. can be made based on characteristic clinical appearance.
      (1) A definitive diagnosis of Knemidokoptes is obtained by skin scraping and identification of mites.[7]
   e. A diagnosis of hypovitaminoses A is based on dietary evaluation, characteristic clinical signs, and response to treatment.[1]
   f. A definitive diagnosis of papillomas or neoplasia requires histologic evaluation.[7]
4. Treatment
   a. Supportive care
      (1) Birds with extensive bilateral lesions may have difficulty finding food and water and may require supportive care, including tube feeding.
   b. *Knemidokoptes* spp.
      (1) A single dose of ivermectin (200 µg/kg), administered either topically or orally, and repeated 3 weeks later, is generally effective.[1,4]
      (2) Ivermectin can be diluted (1:8) in either propylene glycol or sterile water.[13] Ivermectin will precipitate out of either of these solutions; therefore, thorough mixing before administration is essential.[13]
      (3) Diluted ivermectin can also be administered intramuscularly; however, toxic reactions and death have been reported in small birds even at recommended doses.[13]
      (4) Mites can be transmitted to cagemates; all birds in the cage should be treated simultaneously.
      (5) Lesions may persist for weeks or months and will recede as new growth occurs on the beak.
      (6) Corrective beak trims may be required.
   c. Hypovitaminosis A
      (1) Parental administration of vitamin A is recommended initially; IM injection of 10,000 to 25,000 IU/300 g of body weight once weekly.[3,4]
      (2) Oral supplementation can also be provided.
      (3) Recommend dietary changes, including the use of formulated diets, to ensure adequate intake of vitamin A.
         (a) Switching birds from an "all-seed" diet to a formulated diet should be done gradually, particularly in small birds such as passerines.
         (b) Fecal output should be monitored to ensure that the bird is eating.
      (4) Consider supplementation with natural sources of vitamin A, including leafy green and yellow vegetables.[11]
      (5) Vitamin supplementation added to the drinking water is *not* recommended.

(6) Treat keratoconjunctivitis sicca and/or corneal ulcers if present (see Corneal Ulceration and Keratitis p. 295).
(7) Surgical debridement of oral and choanal abscesses or cysts is recommended.
  (a) Systemic antibiotics are recommended; choice of antibiotics should be based on culture and sensitivity; initial treatment should be based on cytology.[7]
d. Papillomas
  (1) Surgical removal is indicated if lesions impair the bird's ability to see or eat.
  (2) Complete surgical removal may be difficult depending on location.
  (3) Lesions in the oral cavity may cause dyspnea or dysphagia.
  (4) Lesions may recur if not completely excised.[7]
e. Neoplasia
  (1) Surgical removal (if possible) is recommended; cryotherapy may be useful for tumors that are too large or diffuse to be removed surgically.[1,3]
  (2) A presurgical radiograph should be performed to evaluate for evidence of metastasis.
  (3) Long-term prognosis for malignant tumors is usually poor.[3]

F. Avian Pox

1. Clinical signs
   a. Poxviridae family: Avian pox occurs in three forms—cutaneous, diphtheritic, and septicemia.[1,3,7,14]
     (1) The presenting form depends on the species of the bird and the virulence of the virus.[1,12]
     (2) The different forms can occur alone or in combination.[4]
     (3) The cutaneous form occurs most commonly in blue-fronted Amazons, Pionus parrots, lovebirds, conures, waterfowl, peafowl, quail, pigeons, raptors, mynahs and many other species of passerines—particularly the canary.[1,3,7,14]
     (4) Any age bird can be infected, but young birds are most susceptible.[14]
   b. Cutaneous form ("dry pox")
     (1) Papular lesions develop on the unfeathered skin of the periocular area, beak, nares, vent, and feet.[1,7,12]
     (2) Unilateral or bilateral blepharitis, conjunctivitis, symblepharon, keratitis, progressive corneal ulceration, uveitis, and serous ocular discharge may accompany periorbital lesions.*
     (3) Clinical signs may appear 10 to 14 days postinfection and include ulceration of the lid margins, followed by scab formation, which may seal the lids closed.[1,4,11]
     (4) 10 to 18 days later the scabs begin to fall off.[1,7]

---

*References 1,3,4,7,9,14.

(5) Secondary bacterial (*Escherichia coli*, *Pseudomonas* spp., *Proteus* spp.) or fungal (*Aspergillus* spp., *Candida* spp.) infections are common, and a muco-purulent discharge may be present.[7]
  c. Diphtheric form ("wet pox")
    (1) May accompany or follow the cutaneous form of the disease[4,14]
    (2) Characterized by fibrinous lesions of the oral cavity and upper respiratory tract[3,11,12]
    (3) Infected canaries often develop pneumonia.[4]
  d. Postinfection lesions
    (1) Postinfection lesions may occur, including periocular depigmentation and feather loss, scar tissue formation, eyelid distortions, corneal neovascularization, epiphora resulting from lacrimal duct occlusion, cataract formation, uveitis, and/or phthisis.[1,3,4,9,14]
    (2) Lid margin deformities may result in exposure keratitis, or entropion and corneal abrasions or ulcers.[4]
2. Differential diagnosis
  a. Cutaneous form
    (1) Rule out etiologies that cause other types of proliferative or ulcerated skin lesions.
      (a) Knemidokoptes (see Periorbital Hyperplastic Lesions p. 289)
      (b) Hypovitaminosis A (see Hyperplastic Periorbital Lesions)
      (c) Trauma, including retained foreign bodies[3]
    (2) Other causes of periorbital swelling (see Periorbital Swelling)
3. Diagnosis
  a. Obtain a complete history and signalment.
    (1) Inquire about the duration and progression of the condition.
    (2) Inquire about medical history, including previous and concurrent diseases, response to treatment, and current medications.
    (3) Inquire about potential sources of infection, such as a history of recent exposure to other birds or to vectors (birds housed outdoors).
    (4) Inquire about diet, dietary changes, and changes in eating habits. Evaluate diet for sufficient levels of vitamin A; diets composed solely of seeds are frequently deficient.[7]
    (5) Inquire about instances of cagemate aggression. Horizontal transmission may occur because of traumatic injuries associated with cagemate aggression or territorial behavior.[1,14]
  b. Perform a complete physical examination.
    (1) Examine bird for additional lesions on legs and feet.
    (2) Examine oral cavity for fibrinous lesions.
    (3) Evaluate bird for signs of dyspnea secondary to respiratory lesions.
  c. Perform complete ocular examination.
    (1) Evaluate overall symmetry and position of globes.
    (2) Note any swellings, discolorations, abnormal growths, and/or ocular discharges.

(3) Evaluate the function and mobility of eyelids and nictitans.
(4) Evaluate the lid margins for any anatomic abnormalities, such as entropion.
(5) Fluorescein stain is indicated in any eye that is red, inflamed, or painful.[10] After staining, evaluate eye with a Wood's lamp.
(6) Carefully evaluate the eye for the presence of uveitis (see Uveitis p. 301).
   (a) Evaluate anterior chamber using a slit lamp.
   (b) If a slit lamp is unavailable, evaluate the chamber using a penlight, a binocular loupe, and an ophthalmoscope at 20 diopters.[1,6]
   (c) Aqueous should be clear; the presence of "aqueous flare" is indicative of anterior uveitis.[1,7,10]

d. Tentative diagnosis of avian pox can be made based on history and characteristic clinical signs.[1]
   (1) Asymptomatic carriers exist and infections may be latent for years.[1]
   (2) Intermittent shedding of the virus in the feces may occur.[7]
      (a) Repeated fecal culturing may identify carriers, but negative cultures do not rule out the possibility of infection.[7]
   (3) Definitive diagnosis requires histologic identification of characteristic intracytoplasmic inclusion bodies ("Bollinger bodies").[1,3,4,7,12]

4. Treatment
   a. Provide supportive care as indicated.
   b. Birds with bilateral lesions may have difficulty finding food or water; tube feed and/or administer supplemental fluids as indicated.
   c. Topical administration of a broad-spectrum ophthalmic antibiotic ointment is recommended to prevent secondary bacterial infections.[1]
      (1) If secondary bacterial infection is present, the choice of antibiotic should be based on the results of culture and sensitivity.
      (2) Topical ophthalmic ointments should be administered every 3 to 4 hours, solutions every 2 hours.[3]
      (3) A corneal scraping can be used initially to determine the choice of antibiotics until the results of the culture and sensitivity are known.
   d. In severe cases systemic antibiotics may be indicated.[1]
   e. If intraocular inflammation is present systemic corticosteroid may be administered (see Uveitis p. 301).
   f. Topical steroids may be administered if no corneal ulcers are present.
   g. Remember that topical medications can be absorbed systemically. Toxicity is possible, particularly in small birds.[3]
      (1) Calculate maximum safe daily dose of the drug, taking into consideration all forms of administration.
   h. Initiate treatment for corneal ulceration if present (see Corneal Ulceration and Keratitis).

i. Clean lid margins daily by applying compresses dipped in dilute baby shampoo.[1,3,4,9]
j. Before the formation of scabs on the lid margins, eyes should be flushed daily with a 1:4 dilution of 2% merbromin (Mercurochrome) to eyewash solution.[1,3,4,9]
k. Allow scabs to fall off naturally; manual removal of scabs is traumatic and may exacerbate postinfection lesions.[1,3,4,9]
l. It has been suggested that early administration of parenteral vitamin A may help to limit the severity of the lesions.[1,3]
   (1) Administer IM injection of 10,000 to 25,000 IU/300 g of body weight once weekly.[3,4]
m. Postinfection eyelid distortions may require surgical correction and/or long-term use of ophthalmic lubricants to prevent keratitis.[1]

5. Prevention
   a. Isolate infected birds.
      (1) Direct horizontal transmission may occur in birds because of traumatic injuries associated with cagemate aggression or territorial behavior.[14]
      (2) Virus transmission can also occur via contact with contaminated fomites (food, cages, hand feeding apparatus, etc.); handle infected birds last and wash hands frequently.[3]
      (3) Virus can be shed via epithelial crusts, exudates, skin, feces, and feather quills and remain latent for years.[14]
      (4) Effective disinfectants include hydroxide KOH, 2% sodium hydrochloride NAOH, and 5% phenol.[12]
   b. Primary vector is the mosquito; initiate vector control, such as mosquito netting over outdoor cages, particularly during the late summer and autumn.[3,7,12,14]
   c. Vaccines (see Chapter 17)
      (1) A commercial modified live vaccine is available for canaries (Poximmune-C Biomune, Lenexa, KS).[3]

V. Diseases of the Cornea

A. Corneal Ulceration and Keratitis

1. Clinical signs (Fig. 15-4)
   a. May include mild to severe blepharospasm and/or conjunctivitis, photophobia, and corneal epithelial defects, with or without corneal opacification and leukocytic infiltration, purulent ocular discharge, and signs of intraocular inflammation, including hypopyon and uveitis.[1,3,7]
   b. Corneal vascularization and pigmentation is less common in birds than mammals.
   c. Secondary microbial (including fungal) infections may complicate the condition.
   d. The condition may deteriorate rapidly.

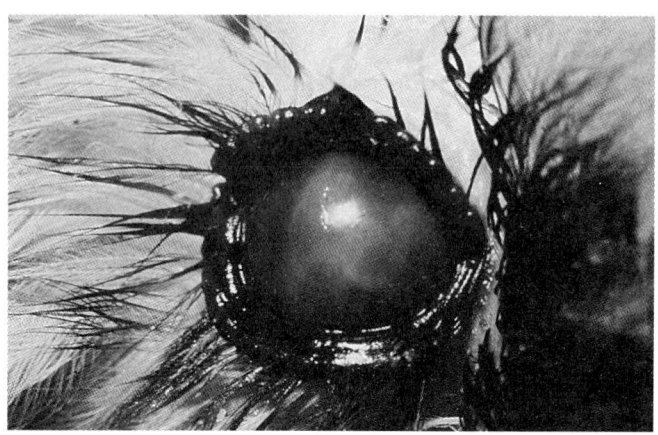

Fig. 15-4 Ulceration and corneal scarring in a fledgling barn owl. (Courtesy Sallie Welte)

2. Differential diagnosis
   a. Traumatic injuries (common in birds)[1,3] because of the following:
      (1) Handling, restraint, or capture
      (2) Foreign bodies
      (3) Lid abnormalities/entropion
      (4) Parasites
   b. Exposure keratitis[3]
      (1) Dysfunction of eyelids or nictitans
      (2) Exophthalmus
      (3) Buphthalmos
   c. Keratoconjunctivitis sicca
      (1) Hypovitaminosis A
   d. Exposure to chemical agents
   e. Thermal injuries
      (1) Can occur secondary to fire or smoke
   f. Chlamydia
3. Diagnosis
   a. Obtain a complete history and signalment.
      (1) Inquire about duration and progression of condition.
      (2) Inquire about medical history, including previous and concurrent diseases, response to treatment, and current medications.
      (3) Inquire about diet.
         (a) Evaluate diet for sufficient levels of vitamin A; diets composed solely of seeds are frequently deficient.[3,7]
         (b) Hypovitaminosis A can lead to a secondary keratoconjunctivitis sicca.
      (4) Inquire about potential sources of parasitic infection such as history of recent exposure to other birds.

(5) Inquire about incidences of trauma. Birds that are allowed "free flight" are at a higher risk of traumatic injury from flying into windows, walls, ceiling fans.
(6) Inquire about housing and husbandry. Note any exposure to environmental toxins, irritants, or allergens such as chemical disinfectants, dust, smoke, ammonia buildup.

b. Perform a complete physical examination.
  (1) Carefully evaluate the bird for evidence of trauma anywhere on the body, particularly on the head.
     (a) Lift the feathers on the head to look for hemorrhage or bruising.[7]
     (b) Palpate the orbital rim area for evidence of fractures, and visually examine the sclera for indentations, which suggest a fracture of the scleral rings.[6]
     (c) Examine ears, nostrils, and oral cavity for evidence of hemorrhage.[7]
     (d) If trauma is suspected, radiographs are indicated.
  (2) Evaluate the bird for evidence of systemic disease.
     (a) Chlamydia-positive birds may initially present with signs of conjunctivitis and keratitis.
     (b) Submit appropriate samples for testing if history and physical examination findings are suggestive of *Chlamydia* spp. (see Chapter 3).

c. Perform a complete ophthalmologic examination.
  (1) Evaluate overall symmetry and position of globes.
     (a) Determine whether buphthalmos or exophthalmos is present (see Diseases of the Globe p. 270).
     (b) Evaluate the periorbital area for evidence of trauma and note any swellings, discolorations, abnormal growths, and/or ocular discharges.
  (2) Always obtain samples for microbial culture and sensitivity or virus isolation *before* administering topical anesthetics, medications, or eyewashes; in some cases it may be necessary to administer topical anesthetic before obtaining cytologic samples.[3]
     (a) Use a sterile cotton swab moistened with transport medium.[1]
     (b) Optimally samples should be obtained from upper fornix to decrease contamination.[1]
     (c) Studies indicate that bacteria and fungal organisms can be cultured from clinically normal birds.[1]
     (d) Special media are required for the growth of *Chlamydia* spp.[1,4,7]
  (3) Cytologic samples are indicated whenever nonhealing progressive ulcers are present.[3]
     (a) Samples should be obtained before administering topical medications or eyewashes; however, topical anesthetic will probably be required.[7]
     (b) Samples can be obtained by using a sterile moistened swab or the blunt end of a sterile scalpel blade.[3,6]

(c) Examination of samples, stained with Diff-Quik, Gram's, or Wright's stain, may be useful for selecting initial antibiotics until the results of the culture and sensitivity are known.[3,6]

(d) It is a good idea to make extra slides so that additional stains can be used as needed.

(e) If *Chlamydia* spp. is suspected, Giemsa stain may be useful in evaluating cells for chlamydial elementary bodies.[1]

(4) Evaluate the function and mobility of the lids.
  (a) Lid margins should meet when eye is closed.
  (b) Nictitan should cover the entire eye; inadequate coverage may lead to "dry eye" or exposure keratitis.[3]
  (c) Evaluate the eyelids for any abnormalities such as entropion.

(5) Corneal surface should be smooth and shiny.
  (a) Corneal lipid deposition can occur secondary to chronic irritation or intraocular inflammation and after the healing of an ulcer.[3]

(6) Fluorescein stain is indicated in any eye that is red, inflamed, or painful.[10]
  (a) The eye should be evaluated for the presence of corneal erosions and ulcerations with the use of a Wood's lamp.

(7) If blepharospasm is present, or the eye is excessively reddened, examine the conjunctival fornices, the underside of the nictitans, and the lacrimal duct for the presence of foreign bodies and parasites.
  (a) Instill one to two drops of topical anesthesia (0.5% proparacaine or 0.5% tetracaine HCl) into the eye and wait 1 minute.[3]
  (b) Use atraumatic forceps to examine the conjunctival fornix and the underside of the nictating membrane.
  (c) Topical anesthetic should be used sparingly, because overuse can induce systemic toxicity, particularly in small birds.[3]

(8) Carefully evaluate the eye for the presence of uveitis, which can occur secondary to corneal ulceration and/or keratitis (see Uveitis).
  (a) The anterior chamber should be evaluated using a slit lamp.
  (b) If a slit lamp is unavailable, the chamber can be evaluated using a penlight, a binocular loupe, or an ophthalmoscope at 20 diopters.[1,6]
  (c) Aqueous should be clear; the presence of "aqueous flare" is indicative of anterior uveitis.[1,7,10]

(9) Perform a funduscopic examination.
  (a) Evaluate the eye for posterior uveitis and evidence of trauma, such as retinal hemorrhage, edema, and detachment.

(b) Because of the presence of striated muscle in the iris, mydriasis cannot be achieved with parasympathomimetics.[1,9]
(c) In some larger birds, funduscopic examination may be possible by performing the examination in a darkened room, with a low-illumination light source.[1,6]
(d) General anesthesia may be required to thoroughly evaluate the fundus; mydriasis can be achieved under general anesthesia.[7]
(e) Topical vecuronium bromide solution (4 mg/ml) has been used to induce mydriasis in raptors.[1,3,6] The solution is administered three times at 5-minute intervals to one eye only.[1] Mydriasis peaks at 1 hour after administration.[3,6]
(f) Topical medications should be used sparingly because overuse can induce systemic toxicity, particularly in small birds.
(g) A more dilute concentration is recommended in smaller birds.[6]
(h) Monitor bird for decreased respiratory rates and inspiratory excursions.[6]

4. Treatment
   a. Identification, treatment, and elimination of the primary cause are essential.
   b. Corneal ulcerations should be treated aggressively, as the potential for rapid deterioration exists.
   c. Corneal ulceration requires frequent reevaluations (every 2 to 3 days) to monitor progression.[3,10]
   d. Corneal ulceration and keratitis require medical treatment.
      (1) Keratitis and noninfected ulcers should be treated prophylactically TID, with a topical broad-spectrum antibiotic ointment or solution.[3]
      (2) Treatment of infected ulcers should be based on culture and sensitivity.
         (a) Initial treatment can be based on the results of a Gram's stain of a corneal scraping.
         (b) Ophthalmic antibiotic ointments should be administered every 3 to 4 hours, solutions every 2 hours.[3]
         (c) Topical steroids should *not* be used if corneal ulceration is present.
      (3) Remember that topical medications can be absorbed systemically.
         (a) Toxicity is possible, particularly in small birds.
         (b) Calculate maximum safe daily dose of the drug, taking into consideration all forms of administration.
      (4) If uveitis is present, systemic corticosteroids are indicated.[7]
      (5) The administration use of anticollagenase may be useful in preventing the progression of deep ulcers.

(a) Acetylcysteine can be sprayed into the affected eye(s) every 3 to 4 hours.[1,7]
(b) To minimize restraint and handling, a 25-gauge needle and tuberculin syringe can be used to spray the solution into the eye(s); this can be done while the bird is in the cage.[1,3]
(6) Topical atropine has little or no effect in birds and is not recommended for the treatment of uveitis.[6]
(7) Cyanoacrylate adhesive may be used for the treatment of deep ulcers to prevent perforation.[3]
  (a) The procedure should be performed with the bird under general anesthesia.
  (b) Dry the cornea with a sterile cotton swab and apply one drop of cyanoacrylate to the cornea; a TB syringe and 27-gauge needle can be used to minimize the amount of cyanoacrylate applied.[3]
  (c) Cyanoacrylate will slough off spontaneously as the epithelium heals.[3]
  (d) Antibiotic treatment should be continued as described previously.[3]
(8) If keratoconjunctivitis sicca (secondary to hypovitaminosis A) is the suspected cause, dietary changes and supplementation with vitamin A should be initiated (see Periorbital Hyperplastic Lesions p. 289).
(9) If lid dysfunction or paralysis is present, long-term treatment with ophthalmic lubricants may be required (see Lid Dysfunctions p. 279).
(10) Eliminate any environmental irritants or allergens.
(11) Eliminate conjunctival parasites.
  (a) After administration of a topical anesthetic, parasites can usually be flushed or manually removed from the conjunctiva and/or the underside of the nictitans.[1,3]
  (b) Gently express the area surrounding the lacrimal sac to extrude any parasites in the duct.
  (c) In the case of severe infection, topical administration of a single dose of ivermectin (200 µg/kg) or 5% carbamate powder may be used.[4,7]
  (d) Ivermectin can be diluted (1:8) in either propylene glycol or sterile water.[13] Ivermectin will precipitate out of either of these solutions; therefore, thorough mixing before administration is essential.[13]
(12) Surgical treatment of corneal ulcers may be required.
  (a) Neither conjunctival nor nictitan flaps are recommended in birds.[1,3]
  (b) Temporary tarsorrhaphy may be useful when exposure keratopathy is the primary cause or when moderate to deep ulcers are present.[1,3,7]
  (c) Partial thickness horizontal mattress or simple interrupted sutures using 5-0 or 6-0 suture material is recommended.[1,3,7]

(d) Care must be taken to avoid exposing the cornea to suture material or knots.
(e) Temporary tarsorrhaphy offers extra protection to the cornea; however, it makes monitoring the progression of the ulcer more difficult.[3]
(f) The administration of both systemic and topical antibiotics is recommended (see Medical treatment of corneal ulcers above).

(13) Corneal lacerations or perforations
   (a) Cornea can be surgically repaired using 7-0 to 11-0 absorbable suture in a simple interrupted pattern.[3]
   (b) The administration of both systemic and topical antibiotics is recommended (see Medical treatment of corneal ulcers above).
   (c) Additional protection may be provided by a temporary tarsorrhaphy.[3]
   (d) In some cases the use of an e-collar may be necessary after surgery to prevent self-trauma.[3]

VI. Intraocular Disease

A. Uveitis

1. Clinical signs
   a. Inflammation may be active or chronic and can range from subtle to severe.
   b. Vision may or may not be impaired.
   c. Chronic uveitis can lead to blindness.
   d. Typical signs associated with anterior uveitis include any or all of the following: aqueous flare, blepharospasm, photophobia, corneal edema, aqueous flare, hemorrhage in the anterior chamber, hypopyon, hyphema, dyscoria, synechiae, miosis and darkening or thickening of the iris.[1,3]
   e. Typical signs associated with acute posterior uveitis include the following: vitreous opacity and retinal changes such as edema, hemorrhage, and/or detachment.[3]
   f. Typical signs associated with chronic uveitis include fibrin clots in the anterior chamber, glaucoma, cataracts, synechiae (anterior or posterior) retinal atrophy, and/or detachment and blindness.[3]

2. Differential diagnosis
   a. Trauma (particularly in raptors)[1,3,4,7]
   b. Corneal ulceration[7] (see Corneal Ulceration and Keratitis)
   c. Keratitis (see Corneal Ulceration and Keratitis)
   d. Intraocular infections
      (1) Can be bacterial, viral, fungal, or protozoal [3,7]
      (2) Infectious agents include reovirus, paramyxovirus-3 and toxoplasmosis.[3,7]
   e. Immune mediated (lens rupture or cataract resorption)[3,4,7]
   f. Septicemia/systemic infections [3,7]

g. Avian pox[7]
h. Ocular nematodiasis/toxoplasmosis[1,3,7]
i. Uveal neoplasia (rare in birds)[1,3,7]

3. Diagnosis
   a. Obtain a complete history and signalment.
      (1) Inquire about duration and progression of condition.
      (2) Inquire about medical history, including previous and concurrent diseases, response to treatment, and current medications.
      (3) Inquire about incidences of trauma. Birds that are allowed "free flight" are at a higher risk of traumatic injury from flying into windows, walls, ceiling fans.
      (4) Inquire about potential sources of infectious disease, including history of recent exposure to other birds.
   b. Perform a complete physical examination.
      (1) Carefully evaluate the bird for evidence of trauma, anywhere on the body, particularly on the head.
         (a) Lift the feathers on the head to look for evidence of hemorrhage or bruising.[7]
         (b) Examine ears, nostrils, and oral cavity for evidence of hemorrhage.[7]
         (c) Palpate the orbital rim area for evidence of fractures and visually examine the sclera for indentations, which suggest a fracture of the scleral rings.[3,6]
         (d) If trauma is suspected, radiographs are indicated.
      (2) Evaluate the bird for signs of systemic disease.
   c. Perform a complete ophthalmologic examination, evaluating the overall symmetry and position of globes.
      (1) Determine whether buphthalmos or exophthalmos is present (see Diseases of the Globe and Orbit).
      (2) Evaluate the periorbital area for evidence of trauma, swellings, discolorations, abnormal growths, and/or ocular discharges.
      (3) Note any periorbital masses such as abscesses, hemorrhages, or neoplasia.
      (4) Examine periocular region for evidence of primary extraocular disease, such as avian pox.
   d. Obtain samples for microbial culture and sensitivity, if indicated.
      (1) Always obtain samples for microbial culture and sensitivity or virus isolation *before* administering topical anesthetics, medications, or eyewashes; in some cases it may be necessary to administer topical anesthetic before obtaining cytologic samples.[3]
      (2) Use a sterile cotton swab moistened with transport media.[1]
      (3) Optimally samples should be obtained from upper fornix to decrease contamination.[1]
      (4) Studies indicate that bacteria and fungal organisms can be cultured from clinically normal birds.[1]
   e. Cytologic samples
      (1) Samples should be obtained before administering topical medications or eyewashes; however, topical anesthesia will probably be required.[7]

(2) Samples can be obtained by using a sterile moistened swab or the blunt end of a sterile scalpel blade.[3,6]
(3) Examination of samples, stained with Diff-Quik, Gram's, or Wright's stain, may be useful for selecting initial antibiotics until the results of the culture and sensitivity are known.[3,6]
(4) The identification of fungal hyphae may require the use of methylene blue stain.[7]
(5) It is a good idea to make extra slides so that additional stains can be used as needed.
f. Evaluate the eye for function and mobility of the lids.
  (1) Lid margins should meet when eye is closed.
  (2) Nictitan should cover the entire eye; inadequate coverage may lead to "dry eye" or exposure keratitis and eventual ulceration.
  (3) Evaluate the eyelids for any abnormalities such as entropion.
g. Corneal surface should be smooth and shiny.
  (1) Corneal lipid deposition can occur secondary to chronic irritation or intraocular inflammation and after the healing of an ulcer.[3]
  (2) Fluorescein stain is indicated in any eye that is red, inflamed, or painful.[10] The eye should be evaluated for the presence of corneal erosions and ulcerations with the use of a Wood's lamp.
h. If blepharospasm is present, or the eye is excessively reddened, examine the conjunctival fornices, the underside of the nictitans, and the lacrimal duct for the presence of foreign bodies and parasites.[7]
  (1) Instill one to two drops of topical anesthesia (0.5% proparacaine or 0.5% tetracaine HCl) into the eye and wait 1 minute.
  (2) Use atraumatic forceps to examine the conjunctival fornix and the underside of the nictating membrane.
    (a) Topical anesthesia should be used sparingly, as overuse can induce systemic toxicity, particularly in small birds.[3]
i. Carefully evaluate the anterior chamber of the eye.
  (1) The anterior chamber should be evaluated using a slit lamp.
    (a) If a slit lamp is unavailable, the chamber can be evaluated using a penlight, a binocular loupe, or an ophthalmoscope at 20 diopters.[1,6]
    (b) Aqueous should be clear, the presence of "aqueous flare" is indicative of anterior uveitis.[1,7,10]
  (2) Evaluate the chamber for the presence of hyphema or hypopyon.
j. Carefully evaluate the lens for evidence of cataract or rupture.
k. A funduscopic examination should be performed.
  (1) Evaluate the eye for posterior uveitis and evidence of trauma, such as retinal hemorrhage, edema, and detachment.
  (2) Evaluate the eye for evidence of uveal neoplasia.
  (3) General anesthesia may be required to thoroughly evaluate the fundus.
    (a) Because of the presence of striated muscle in the iris, mydriasis cannot be achieved with parasympathomimetics.[1,9]

(b) In some larger birds, funduscopic examination may be possible by performing the examination in a darkened room, with a low-illumination light source.[1,6]

(c) Topical vecuronium bromide solution (4 mg/ml) has been used to induce mydriasis in raptors.[1,3,6] The solution is administered three times at 5-minute intervals to one eye only.[1] Mydriasis peaks at 1 hour after administration.[3,6]

(d) Topical medications should be used sparingly because overuse can induce systemic toxicity, particularly in small birds.

(e) A more dilute concentration is recommended in smaller birds.[6]

(f) Monitor bird for decreased respiratory rates and inspiratory excursions.[6]

l. In larger birds, ultrasonography (7.5 MHz probe and coupling gel) may be useful to evaluate the eye for intraocular neoplasia and retinal detachment.[6,7]

m. Full body radiographs are indicated if trauma is suspected.
   (1) Radiographic changes in the periorbital area may be subtle and difficult to identify.
   (2) Evaluate entire bird for evidence of trauma.
      (a) Periorbital trauma may result in fractures of the skull and/or scleral ossicles.[3]
      (b) Evaluate for evidence of systemic disease, such as organomegaly.
      (c) Evaluate for evidence of neoplasia and/or metastasis.

n. Perform other tests (hematology, biochemistry, cultures titers) as needed to diagnose infectious or inflammatory diseases.

4. Treatment
   a. Provide supportive care and treat systemic disease as indicated.
   b. Chronic uveitis can lead to severe consequences, including glaucoma and/or blindness.[10]
      (1) Identification, treatment, and elimination of the primary cause of the uveitis are essential.
   c. The administration of systemic corticosteroids is recommended.[7]
      (1) Monitor bird for polyuria, polydipsia, and immune suppression.[3]
   d. The administration of topical steroids may be indicated.
      (1) Topical steroids should *not* be used if corneal ulceration is present.
      (2) Remember that topical medications can be absorbed systemically and that toxicity is possible, particularly in small birds.[3] Calculate maximum safe daily dose of the drug, taking into consideration all forms of administration.
   e. Corneal ulcerations should be treated aggressively because the potential for rapid deterioration exists[10] (see Corneal Ulcerations and Keratitis).

f. Surgical removal of the lens may be necessary in chronic lens associated uveitis.[3]
g. Enucleation is indicated if uveal neoplasia is present and may be indicated in chronic unresponsive unilateral conditions.[3]
h. Intraocular parasites may require surgical removal.[7]
i. Topical atropine has little or no effect in birds and is not recommended for the treatment of uveitis.[6]

VII. Blindness

A. Clinical Signs
1. Unilateral or bilateral vision loss.
2. Complete vision loss is uncommon in birds.[7]
3. Vision loss may go undetected in caged birds that are handled infrequently (or not at all) as a bird in familiar surroundings may continue to eat and drink normally.

B. Differential Diagnosis
1. Trauma (common cause, particular in raptors, Fig. 15-5)[7]
2. Toxins[1,7]
   a. Heavy metals—especially lead
   b. Organochlorines
   c. Mycotoxins

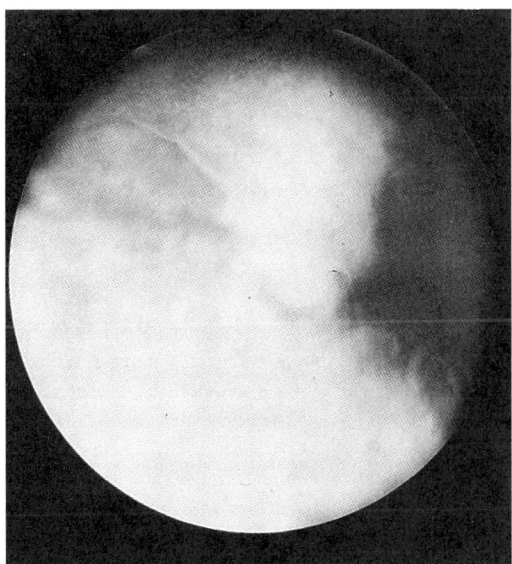

Fig. 15-5 The fundus of a great horned owl, posttrauma. (Courtesy T. J. Kern)

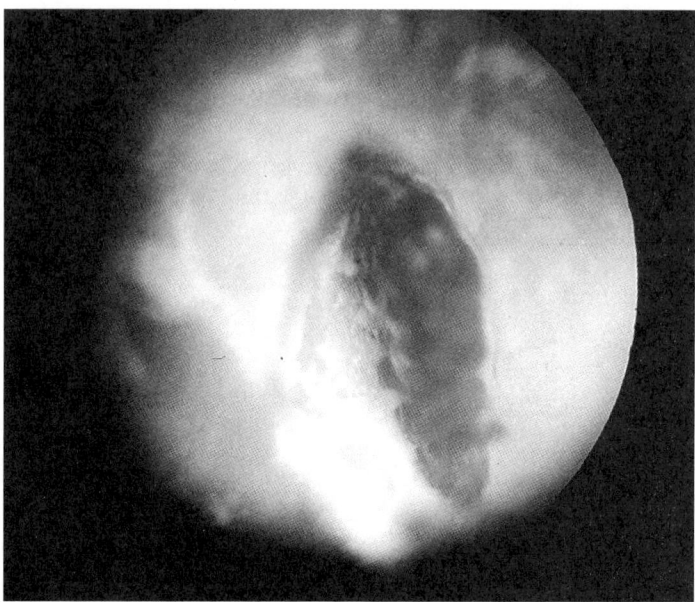

Fig. 15-6 Peri-pectin hemorrhage and retinal edema in a screech owl. (Courtesy T. J. Kern, from Altman RB et al, editors: *Avian medicine and surgery*, Philadelphia, 1997, WB Saunders.)

      d. Hexachlorophene
      e. Cephaloridine
  3. Space-occupying lesions (e.g., pituitary adenomas, particularly in budgies[1,3,7] and in cockatiels, from my experience)
  4. Ocular neoplasia[7]
  5. Developmental disorders[3,7] (see Diseases of the Globe and Orbit p. 270)
  6. Hepatic encephalopathy[7]
  7. Retinal lesions or dysplasia (often associated with trauma, particularly in raptors, Figs. 15-6 and 15-7)[1,3,4,7]
  8. Anterior uveitis (see Uveitis)
  9. Opacities of the cornea or lens (cataracts, Fig. 15-8)
      a. Cataracts can develop secondary to trauma, old age, inflammation, uveitis, nutritional deficiencies, lens luxations, or genetic disorders; or they may be congenital.[1,3,4,7]
  10. Optic nerve lesions
  11. Encephalitis secondary to infections disease, including *Chlamydia psittaci* and toxoplasma (particularly passerines)[1,7]
  12. Vascular accidents or ischemic necrosis (particularly in budgies)[7]

C. Diagnosis
  1. Obtain a complete history and signalment.
      a. Inquire about the duration and progression of the disease.

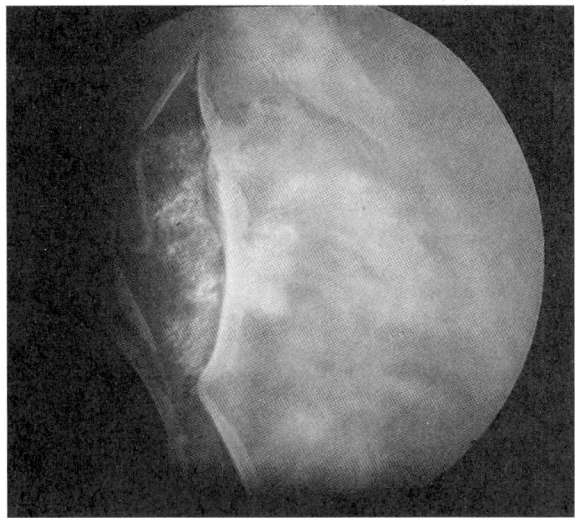

Fig. 15-7 Nasal retinal detachment, species unknown. (Courtesy Sallie Welte)

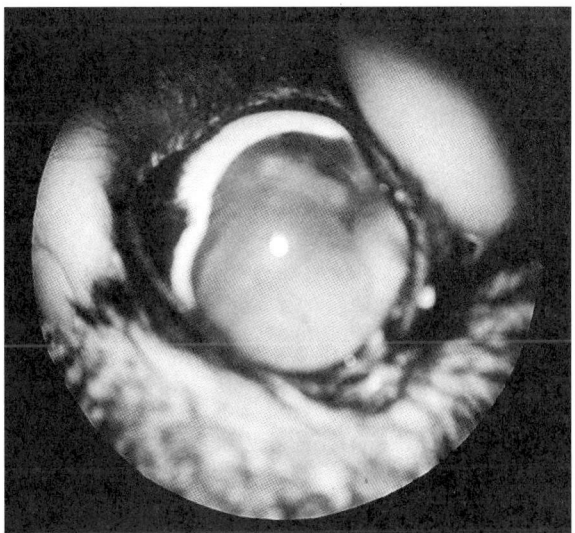

Fig. 15-8 Cataract and hemorrhage in a great horned owl. (Courtesy Sallie Welte)

b. Inquire about changes in behavior, eating habits.
c. Inquire about medical history, concurrent diseases, and response to treatments and current medications.
d. Inquire about incidences of trauma.
    (1) Birds allowed "free flights" in the house are at a higher risk of trauma; common sources of trauma include flying into windows, walls, and ceiling fans.
    (2) Other sources of trauma that can lead to vision loss include chewing on electrical cords and head trauma secondary to being stepped on.
e. Inquire about the potential for toxin exposure.
    (1) Birds that are allowed "free flight" or unsupervised activity outside of the cage are at a higher risk for toxin exposure.
    (2) A variety of toxins can cause blindness, including heavy metal, particularly lead.
    (3) Common sources of heavy metal toxicity include the following:
        (a) Lead paint (older homes)
        (b) Galvanized steel (zinc coating) on new cages
        (c) Antique bird cages and some cages imported from outside the United States
        (d) Some bird toys
        (e) Coating on the backs of mirrors
        (f) Soldering on antique jewelry, lamps, etc.
        (g) Curtain weights, coins, fishing lures
    (4) Evaluate cage and toys for evidence of missing pieces.
    (5) Owner should carefully evaluate areas in the home where the bird plays for potential sources of toxins.
f. Inquire about potential sources of infectious disease, including history of recent exposure to other birds.
g. Inquire about housing and husbandry.
    (1) Inquire about potential exposure to toxoplasma via direct or indirect contact with cat feces.[7]
    (2) Cockroaches and certain other arthropods can act as indirect sources of toxins, contaminating the bird's food or water.[7]
2. Perform a complete physical examination.
    a. Carefully evaluate the bird for evidence of trauma anywhere on the body.
        (1) Lift the feathers on the head to look for hemorrhage or bruising.[7]
        (2) Examine ears, nostrils, and oral cavity for evidence of hemorrhage.[7]
        (3) Palpate the orbital rim area for evidence of fractures and visually examine the sclera for indentations, which suggest a fracture of the scleral rings.[6]
    b. Evaluate the bird for signs of systemic or neoplastic disease.
        (1) Evaluate bird's body weight and condition.
    c. Evaluate feces and urates for evidence of diarrhea, polyuria, and color changes.[7]
        (1) Diarrhea may be present with toxins, toxoplasma, hepatic encephalopathy, and *Chlamydia* spp.[7]

Ophthalmic Disorders **309**

(2) Bright green feces can be indicative of lead poisoning and/or liver disease.[7]
(3) Biliverdinuria can be associated with both toxins and liver disease.[7]
d. Look for evidence of regurgitation such as matting of the feathers around the bill or above the cere.
3. Perform a complete neurologic examination.
   a. Evaluate bird for evidence of neurologic abnormalities.
   b. Inquire about history of ataxia or seizures.
   c. Seizures can be associated with toxins, hepatic encephalopathy, or encephalitis.[7]
   d. Seizures associated with hepatic encephalopathy often occur shortly after the bird has eaten.[7]
   e. Evaluate bird for evidence of nystagmus.
      (1) Nystagmus can be associated with space-occupying lesions (including pituitary adenomas) and with encephalitis,[1,3,7] but not all space-occupying lesions are associated with nystagmus.
4. Perform a complete ophthalmic examination.
   a. Initially the bird should be observed from a distance, preferably in its cage or natural environment.
      (1) Evaluate bird's stance and its awareness of surroundings.
      (2) Determine whether the bird is able to identify and visually follow a moving object.
      (3) Stand behind the bird and move a hand toward the bird, first on the left and then on right to test the vision in each eye.[7]
   b. A menace response is generally not diagnostic in birds.[1,3]
      (1) The response may be absent in normal birds.[3,7]
      (2) A false-positive response can be associated with airflow created by the hand movement.[1]
      (3) A positive response may be indicated by the movement of the nictitans rather than the eyelid closure.[6]
   c. Evaluate overall symmetry and position of globes.
      (1) Evaluate the eye(s) for the presence of exophthalmia (see Exophthalmia p. 274).
      (2) Exophthalmia may or may not be present with space-occupying lesions.
      (3) Evaluate the eyes for the presence of glaucoma; glaucoma can lead to irreversible optic nerve atrophy.[3]
   d. Evaluate bird, particularly a young bird, for evidence of developmental conditions such as microphthalmia, cryptophthalmia[3] (see Diseases of the Globe and Lids p. 270).
   e. Pupillary light response (PLR)
      (1) Voluntary dynamic changes in pupil size, associated with excitement or agitation, makes evaluation of the direct PLR difficult.[1,3,9]
      (2) True consensual (indirect) response is absent in birds because of the complete separation of the optic nerves.[3,7]

(3) Inadvertent constriction of the opposite pupil can occur when light passes through the thin osseous septum, which separates the globes caudally.[3,7]
(4) Consensual pupillary light reflexes are believed to be absent in birds.[3,6,8]
f. Anisocoria
(1) Because of the presence of striated muscle in the iris, intermittent anisocoria is normal in birds; in unilateral disease conditions, a permanent anisocoria may be present.[3]
g. Carefully evaluate the eye for corneal opacities and for cataracts.
h. Evaluate the anterior chamber for evidence of anterior uveitis (see Uveitis p. 301).
(1) Evaluate anterior chamber using a slit lamp.
(2) If a slit lamp is unavailable, the chamber can be evaluated using a penlight, a binocular loupe, and an ophthalmoscope at 20 diopters.[1,6]
(3) Aqueous should be clear; the presence of "aqueous flare" is indicative of anterior uveitis.[1,7,10] Evaluate the pupillary margins and iris for clarity.[1]
(4) Look for evidence of intraocular hemorrhage.
i. Perform a funduscopic examination.
(1) The small size of the bird's eye can make examination of the posterior chamber difficult.
(2) Because of the presence of striated muscle in the iris, mydriasis cannot be achieved with parasympathomimetics.[1,9]
(3) In some larger birds, funduscopic examination may be possible by performing the examination in a darkened room, with a low-illumination light source.[1,6]
(4) Mydriasis can be achieved with the bird under general anesthesia.[7]
(a) Topical vecuronium bromide solution (4 mg/ml) has been used to induce mydriasis in raptors.[1,3] The solution is administered three times at 5-minute intervals to one eye only.[1] Mydriasis peaks at 1 hour after administration.[3,6]
(b) Topical medications should be used sparingly because overuse can induce systemic toxicity, particularly in small birds.[3]
(c) A more dilute concentration is recommended in smaller birds.[6]
(d) Monitor bird for decreased respiratory rates and inspiratory excursions.[6]
(5) Evaluate the retina for evidence of tears, hemorrhage, and dysplasia.
5. Perform full-body radiographs.
(1) Cranial masses (pituitary tumors, retrobulbar hemorrhages, etc.) and other radiographic changes in the periorbital area may be subtle and difficult to identify.

(2) Evaluate entire radiograph for evidence of trauma; periorbital trauma may result in fractures of the skull or scleral ossicles.[7]
(3) Evaluate for the presence of metal density in the soft tissue or GI tract.
(4) The absence of metal in the GI tract does not exclude heavy metal toxicosis.[7]
(5) Evaluate for evidence of hepatomegaly, microhepatia, ascites, and other signs of systemic disease.[6]
(6) Evaluate for evidence of neoplasia and/or metastasis.
6. Orbital ultrasonography (7.5 MHz probe and coupling gel) may be helpful for identifying ocular masses and retinal dysplasia in larger birds.[6,7]
7. Submit blood samples for hematology and biochemistry.
8. Select additional diagnostic tests based on the history, clinical signs, and the results of diagnostic tests performed up to this point.
   a. Submit blood samples to measure blood lead levels or concentrations of other toxins.
   b. Submit samples for *Chlamydia* spp. testing if indicated (see Chapter 3).
   c. Electroretinogram can be used to evaluate retinal function and to differentiate between retinal and central lesions.[7]
   d. Computed tomography and/or magnetic resonance imaging[7]

D. Treatment
1. Provide supportive care as needed including supplemental fluids and tube feeding.
2. Special precautions should be take to prevent the blind bird from injuring itself.
   a. Carefully evaluate the cage. Any sharp objects should be removed or padded; remove any chains and potentially dangerous toys.
   b. If bird is eating, drinking, and maneuvering in a normal manner; perches may be left in the cage.
   c. If bird is ataxic, remove perches and provide a soft substrate such as several layers of non-terry cloth or paper towels.
   d. Remove seizuring birds from cage and placed in a padded aquarium or pet carrier to reduce the risk of leg and wing fractures.
      (1) Place bird in a dark, quiet place.
      (2) Administration of diazepam may help to control seizures.
3. Specific treatments should be based on suspected etiology.
4. Antiinflammatory doses of corticosteroid may be beneficial in treating conditions associated with acute head trauma (including retinopathy and optic nerve neuropathy) and cardiovascular accidents.[1,3,7]
   a. Dexamethasone (2 mg/kg) can be administered IM or IV once or twice daily.
   b. Prolonged use may lead to adrenal suppression and immune compromise; monitor bird for polyuria and polydipsia.
5. Consider chelation therapy in any bird with suspected lead toxicosis, even if the diagnosis has not been confirmed (see Chapter 14).

a. CaEDTA is an effective chelator for lead or zinc.[1]
   b. Administer CaEDTA at 20 to 40 mg/kg, IM, BID-TID.[1]
   c. CaEDTA may cause renal tubular necrosis, discontinue use if polyuria or polydipsia develops.[13]
6. If liver disease is suspected, bird should be placed on a high-quality, low-protein, high-carbohydrate diet; the administration of oral lactulose may be beneficial.[7]
7. If intraocular neoplasia is present, enucleation is recommended.[3]

REFERENCES

1. Williams D: Ophthalmology. In Ritchie BW, Harrison GJ, Harrison LR, editors: *Avian medicine: principles and application,* Lake Worth, Fla, 1994, Wingers, pp 673-694.
2. Dyce KM, Sack WO, Wensing CJG: Avian anatomy. In Dyce KM, Wensing CJG, Sack WO, editors: *Textbook of veterinary anatomy,* Philadelphia, 1996, WB Saunders, pp 772-779.
3. Kern TJ: Disorders of the special senses. In Altman RB, Clubb SL, Dorrestein GM, editors: *Avian medicine and surgery,* Philadelphia, 1997, WB Saunders, pp 563-582.
4. Millichamp NJ: Exotic animal ophthalmology. In Gellat KN, editor: *Veterinary ophthalmology,* ed 2, Malvern, Penn, 1991, Lea & Febiger, pp 680-700.
5. Wissman M: Hatching, neonatology and pediatrics, *Proceedings of the Association of Avian Veterinarians* 1998, pp 391-400.
6. Davidson M: Ocular consequences of trauma in raptors, *Semin Avian Exotic Med* 6:121-130, 1997.
7. Rupley A: *Manual of avian practice,* Philadelphia, 1997, WB Saunders, pp 135-164.
8. Shivaprasad HL: A bird's eye view of avian ocular anatomy and pathology, *Proceedings of the Association of Avian Veterinarians* 1998, pp 273-283.
9. Karpinski LG: Ophthalmology. In Harrison GJ, Harrison LR, editors: *Clinical avian medicine and surgery,* Philadelphia, 1986, WB Saunders, pp 278-281.
10. Slatter D: *Fundamentals of veterinary ophthalmology,* ed 2, Philadelphia, 1990, WB Saunders.
11. Davidson MG: Ophthalmology of exotic pets, *Compend Cont Ed* 7:724-737, 1985.
12. Oglesbee B, Bishop C: Avian infectious diseases. In Birchard SJ, Sherding RG, editors: *Saunders m anual of small animal practice,* Philadelphia, 1994, WB Saunders, pp 1257-1270.
13. Ritchie BW, Harrison G: Formulary. In Ritchie BW, Harrison GJ, Harrison LR, editors: *Avian medicine: principles and application,* Lake Worth, Fla, 1994, Wingers, pp 457-478.
14. Gerlach H: Viruses. In Ritchie BW, Harrison GJ, Harrison LR, editors: *Avian medicine: principles and application,* Lake Worth, Fla, 1994, Wingers, pp 862-948.
15. Greiner EC, Ritchie BW: Parasitology. In Ritchie BW, Harrison GJ, Harrison LR, editors: *Avian medicine: principles and application,* Lake Worth, Fla, 1994, Wingers, pp 1008-1029.

16

Avian Endocrinology

Carol J. Canny and Christal G. Pollock

Hypothalamohypophysial Complex (Hypothalamus and Pituitary Gland)

I. Anatomy

A. Description
   1. The hypothalamohypophysial complex is composed of the hypothalamus, the connective stalk, and the hypophysis (pituitary gland).
   2. The hypothalamus is a small structure that accounts for about 3% of the total brain volume.[1]
   3. The pituitary gland is composed of the adenohypophysis and the neurohypophysis.
   4. The internal carotid arteries supply the hypophysis.
   5. A capillary plexus is formed within the hypothalamus from the internal carotid arteries; the plexus continues as portal veins to the hypophysis.
   6. A second plexus is formed in the pars distalis from the portal vein.
   7. The portal system is divided into two portions, the rostral zone and the caudal zone.
      a. The rostral zone drains the hypothalamus and the median eminence, through which releasing factors are carried.
   8. Innervated by paired hypothalamohypophysial tracts.

II. Adenohypophysis and Neurohypophysis

A. Adenohypophysis
   1. The adenohypophysis comprises two parts, the pars distalis and the pars tuberalis. Unlike mammals, birds do not have a pars intermedia.
      a. The pars distalis is the main component of the adenohypophysis.
         (1) Two regions can be distinguished in the pars distalis: the rostral zone and the caudal zone.

(2) The pars distalis contains seven types of secretory cells and one type of nonsecretory cell.
  b. The pars tuberalis is the region through which the portal vessels pass from the median eminence to the pars distalis.

B. Neurohypophysis
  1. The neurohypophysis is a direct extension of the hypothalamus.
  2. The neurohypophysis has three parts:
    a. The median eminence
    b. The infundibulum
    c. The neural lobe

III. Function of the Hypothalamohypophysial Complex

The hypothalamus and the hypophysis are intimately connected through vascular and nervous connections; for this reason, it is impossible to separate all functions.

A. Hormone Production
  1. Hypothalamus
    a. The hypothalamus produces stimulating or inhibiting chemotransmitters:
      (1) Luteinizing hormone–releasing hormone (LHRH): Stimulates production of luteinizing hormone (LH) and follicle-stimulating hormone (FSH).
      (2) Thyrotropin-releasing hormone (TRH): Stimulates production of thyroid-stimulating hormone (TSH) and growth hormone (GH).
      (3) Growth hormone–releasing hormone (GHRH): Stimulates production of GH.
      (4) Somatostatin: Stimulates production of TSH and inhibits production of GH.
      (5) Prolactin-releasing factor: Exact compound has not yet been elucidated; feedback controls for prolactin are not well understood.
      (6) Suspected corticotrophin-releasing factor (CRF): CRF has not yet been isolated.
  2. Adenohypophysis
    a. The pas distalis has seven types of secretory cells; the hormones produced are either glycoproteins or polypeptides:
      (1) Alpha cell: Produces somatotrophic hormone (STH), also called growth hormone (GH); STH is a polypeptide that regulates growth and influences the release of thyroid hormones.
      (2) Beta cell: Produces FSH, a glycoprotein that stimulates follicular development in the ovaries and seminiferous tubular growth and spermatogenesis in the testes.
      (3) Gamma cell: Produces LH, a glycoprotein that is involved in ovulation in hens and that stimulates androgen hormone synthesis (testosterone) in cocks.

(4) Delta cell: Produces TSH, a glycoprotein that increases the release of thyroxine (T_4) into the plasma while triiodothyronine (T_3) remains relatively unchanged.[2]
(5) Epsilon cell: Produces adrenocorticotropic hormone (ACTH), a polypeptide that controls release of the adrenocortical hormones corticosterone and aldosterone.
(6) Eta cell: Produces prolactin, a polypeptide that stimulates broodiness, nesting behavior, nest protection, and incubation; it also stimulates the production of crop milk in pigeons and inhibits gonadal steroid hormones in both cocks and hens; it may enhance metabolism during migration by promoting hepatic lipogenesis and hyperglycemia.[3]
(7) Kappa cell: Produces melanotropic hormone (MSH), a polypeptide; its physiologic function is unknown, but it may play a role in controlling pigmentation.
b. Nonsecretory cell
(1) Chromophobe cell
3. Neurohypophysis
a. The neurohypophysis produces arginine vasotocin (AVT) and mesotocin (MT).
(1) Arginine vasotocin
(a) Acts on the kidneys and uterus.
(b) Considered the antidiuretic hormone of birds.
(c) Differs from mammalian antidiuretic hormone by one amino acid residue.
(d) Decreases glomerular filtration through reptilian-type nephrons and increases the permeability of the collecting ducts of mammalian-type nephrons.
(e) Is released in response to an increase in plasma osmolality.
(f) In the hen, release causes contraction of the distal oviduct and a subsequent increase in uterine pressure.
(2) Mesotocin
(a) Is similar to the oxytocin hormone of mammals but differs by one amino acid residue.
(b) Actual role is unknown.

IV. Clinical Diseases of the Pituitary

Relatively few conditions associated with diseases of the pituitary have been reported; most documented pituitary disorders are neoplastic conditions.

A. Neoplasia
1. Pituitary adenoma, adenocarcinoma, and carcinoma have been reported primarily in budgerigars and cockatiels.[4,5]
2. Pituitary adenoma also has been reported in one Amazon parrot,[4] a lovebird,[5] and a canary.[6]

B. Dwarfism

1. Dwarfism is a sex-linked, recessive trait in fowl; it also has been reported in the pheasant, black-headed gull, and great crested flycatcher.[7]
2. Chicks have lower plasma T_3 and somatomedin levels but higher levels of GH and T_4.
3. Conversion of T_4 to T_3 via hepatic 5'-monodeiodination does not occur.

C. Diabetes Insipidus

1. Central diabetes insipidus is the result of lack of endogenous antidiuretic hormone (ADH).
2. The condition has been reported in chickens.
3. The primary clinical signs are polyuria and polydipsia.
4. Water deprivation tests are used for diagnosis in small animals.
5. A water deprivation test has been established for racing pigeons.[10]
   a. In racing pigeons, a water osmolality of at least 450 mOsm/kg is expected after 24 hours of water deprivation; this is the normal concentrating ability of the avian kidney.
   b. If the water osmolality does not increase in PU/PD birds, exogenous ADH (AVT) may be given to detect other causes.
   c. A plasma AVT concentration >2.2 pg/ml after 24 hours of water deprivation is considered normal AVT release for the pigeon.[9]

D. Clinical Signs of Pituitary Disease

1. The clinical signs are related to the portion of the hypophysis affected.
2. Causes may include:
   a. An increase in the ACTH level or a decrease in the AVT level.
   b. Changes in sex hormones or the TSH level (or both)
   c. Changes in the GH, TSH, and ACTH levels.
   d. Polyuria and polydipsia
3. Clinical signs include:
   a. Polyuria and polydipsia
   b. Color or pattern change in the feathers
   c. Obesity
   d. Blindness
      (1) May occur as a result of tumor compression, enlargement of associated tissues, and extension of swelling or pressure along the optic nerve
   e. Exophthalmia
      (1) Occurs secondary to extension of a tumor along the optic nerve or to orbit invasion
   f. Alterations in consciousness
      (1) Occur secondary to tumor compression or an increase in intracranial pressure

E. Diagnosis of Pituitary Disease

1. The diagnosis often is suggested by the clinical signs. The diagnostic procedure includes:

a. A complete medical workup, including a complete blood count (CBC), chemistry panel, determination of electrolyte level, and urinalysis
   b. Possibly ultrasound examination, fine needle aspiration, cytologic studies, and magnetic resonance imaging (MRI)
   c. Elimination of other possible causes (e.g., renal or liver disease, infectious diseases, toxicosis, metabolic disorders, diabetes mellitus, and psychogenic disorders)
   d. Examination of the pituitary gland at necropsy if indicated by the clinical signs

F. Treatment
   1. No successful treatment has yet been reported.
   2. Radiotherapy is a possible treatment, because it has proved effective for human and canine pituitary tumors.[11]

Thyroid Glands

I. Anatomy

A. Description
   1. The thyroid glands are paired, oval, dark red glands located on the medial side of the jugular vein at the thoracic inlet.
   2. The glands adhere to the common carotid arteries just cranial to the bifurcation of the subclavian arteries.
   3. The arterial blood supply is via the cranial and caudal thyroid arteries, which originate from the common carotid artery.
   4. Venous circulation is via two to five thyroid veins, which empty directly into the jugular vein.
   5. Innervation may stem from the cervical sympathetic ganglion, as in mammals.

II. Physiology

A. General Functions
   1. Influence metabolism in general
   2. Regulate growth, reproductive organs, and heat
   3. Control the molting process

B. Hormone Metabolism
   1. The thyroid glands are composed of colloid-filled follicles.
      a. Colloid is a homogenous protein composed of iodinated protein thyroglobulin.
         (1) The thyroglobulin of birds is more iodinated than that of mammals.[12]

2. The thyroid glands of most birds are unique in that they lack calcitonin cells. However, calcitonin cells are found in the follicular epithelium of the thyroids of doves and pigeons.[12]
3. The size of the thyroids varies and is influenced by such factors as age, sex, environment, diet, activity level, and species.
   a. The weight of normal thyroid glands has been reported to be 0.02% of the bird's body weight.
4. Thyroglobulin contains monoiodotyrosine, diiodotyrosine, T_3, and T_4.[12]
   a. Most of the T_3 is produced by means of extrathyroidal monodeiodination of T_4 in the liver and kidneys.[12]
5. Unlike humans, birds do not have a T_4-binding globulin.[12]
   a. Serum prealbumin and albumin are the main protein carriers in avian species, although species variations exist.
6. The half-lives of T_3 and T_4 are similar to but shorter than that of mammals.
   a. The half-life varies from 3.3 to 8.3 hours, depending on the species;[13] which is three to 10 times shorter than in most mammals.
   b. The short half-life makes interpreting a single value for T_3 or T_4 difficult.
7. Excretion is through bile and urine.
8. Birds have lower plasma levels of T_4 than mammals, but their plasma levels of T_3 are similar or two to three times higher.
   a. The free T_4 level of chickens and pigeons has been reported to be higher than that of other warm-blooded vertebrates.[13]

III. Clinical Diseases

A. Goiter
1. Goiter develops as the result of an iodine deficiency, which causes diffuse thyroid hyperplasia.
2. Goiter has been reported in canaries, pigeons, budgerigars, and cockatiels.
   a. The condition has been reported primarily in white Carneaux pigeons.[14]
3. Iodine is lacking in most seed diets fed to birds.
   a. In pigeons, high soybean- and maize-based diets may potentiate goiter by increasing the need for iodine.[1]
4. Dietary deficiencies of iodine result in a decreased blood level of T_4 and T_3. Because of the lack of negative feedback, the anterior pituitary continues to produce TSH, resulting in proliferation of the thyroid follicular epithelium and accumulation of colloid, which causes severe thyroid enlargement.

B. Clinical Signs of Goiter
1. The clinical signs are related to enlargement of the glands:
   a. Inspiratory dyspnea and respiratory distress may be noted.

(1) A "squeaking" sound often is heard.
(a) May be caused by impingement on the trachea.
b. Vocalizations decrease.
c. Esophageal disease may be noted (e.g., difficulty swallowing, regurgitation, and delayed crop emptying).
d. Circulatory problems also may be seen.
e. The thyroids may or may not be palpable (the glands may enlarge as much as 100 to 200 times normal size).
2. Clinical signs in pigeons are different from those caused by diet-based goiters in psittacines.
a. Clinical signs include lethargy, obesity, and decreased fertility and hatchability.
b. The thyroid gland often is palpable.
c. Dyspnea may be present in severe cases.
d. Signs suggestive of hypothyroidism may be noted (e.g., myxedema and abnormal feather development).

C. Treatment of Goiter

1. Despite functional hypothyroidism, treatment consists of iodine supplementation.
2. Conversion from seed diets to formulated diets is recommended.
3. Initial iodine treatment may be provided by oral supplementation.
   a. Dilute Lugol's solution may be added to drinking water in a ratio of 1 drop per 250 ml of water daily; a decreasing dose schedule then is followed over the next few weeks.[15]
4. For life-threatening conditions, intramuscular injections of 20% iodine at a dosage of 0.3 ml/kg have been recommended.[15]
5. In most cases, clinical signs should improve within 5 days. The estimated daily requirement of iodine for a budgerigar is 20 µg per week.[1]

D. Hypothyroidism

1. Primary hypothyroidism has been recognized in poultry and is thought to be the cause of many clinical signs seen in pet birds, especially feather problems.
2. In psittacines, only one documented case confirmed by TSH stimulation has been reported.[16]

E. Clinical Signs of Hypothyroidism

1. Obesity
2. Nonpruritic feather loss
3. Feather picking
4. Feather discoloration
5. Mild leukocytosis, nonregenerative anemia, hypoalbuminemia, and hypercholesterolemia
6. Chronic bacterial or fungal infections
7. Delayed molt
8. In poultry:
   a. Delayed molt

b. Abnormal feather color, size, and structure
c. Delayed maturity and sexual development

F. **Diagnosis of Hypothyroidism**
1. A single basal T_3 or T_4 sample is not sufficient to document hypothyroidism because of the lower plasma circulating levels in birds; TSH or TRH stimulation tests are needed.
    a. T_3 and T_4 levels also are affected by handling, drugs, concurrent illness, environment, and temperature.
2. A response to thyroid hormone supplementation is not diagnostic of hypothyroidism and is contraindicated because of the possible side effects of oversupplementation.
3. The complete medical workup includes:
    a. CBC, panel, and electrolyte levels
    b. Whole body radiographs
    c. Skin and feather biopsy (especially with feather lesions)
4. TSH test
    a. The test is no longer commercially available.
        (1) The most recent protocol recommendation is:
            (a) 1 IU/kg of TSH is given intramuscularly, and a serum sample is collected 6 hours later.[17]
            (b) A T_4 level in the sample that is at least double the basal level supports the diagnosis of hypothyroidism.
5. TRH test
    a. This test is commercially available.
        (1) Protocol
            (a) The plasma level is determined, and then 50 µg/kg of TRH is given intravenously; a sample is collected 2 hours later.[18]
            (b) When TRH was given to Amazon parrots, the samples collected 2 hours later showed that the serum level of T_4 had decreased approximately 50%, whereas the T_3 level had increased 5% to 10%.
6. Determination of the free T_4 level
    a. This test currently is being used in small animal medicine; its use in avian medicine needs to be investigated.

G. **Treatment of Hypothyroidism**
1. Thyroid hormone replacement is the recommended treatment.
    a. Oral supplementation
        (1) 0.02 to 0.04 mg/kg of L-thyroxine is given orally every 12 to 24 hours.[15]
    b. Water supplementation
        (1) This method has more potential treatment failures because of variability in drinking habits, evaporation, and solubility.
            (a) 0.1 mg tablet of L-thyroxine is added to 4 to 12 ounces of water.[15]
            (b) The solution should be changed daily.
2. Treatment is monitored closely.

a. The postmedication serum thyroid level is determined.
   b. The response of clinical signs is noted.
3. Signs of oversupplementation:
   a. Tachypnea
   b. Tachycardia
   c. Weight loss
   d. Increased appetite
   e. Hyperactivity
   f. Possibly death

Parathyroid Glands

I. Anatomy

A. Description
1. The parathyroid glands are two pairs of small, yellowish glands located just caudal to each thyroid gland.
2. The two pairs may be fused in some species.
3. The cranial parathyroid gland on the right side often touches the thyroid gland; the left parathyroid gland usually does not touch the thyroid gland.
4. The glands consist of irregular groups and cords of chief cells encapsulated by connective tissue; oxyphil cells, which are found in human parathyroid glands, are absent.

II. Physiology

A. Hormone Metabolism
1. The parathyroid glands produce and secrete parathyroid hormone (PTH).
2. PTH appears to be the critical hormone for maintaining the plasma calcium level in birds.
   a. The normal range of avian blood calcium is 2 to 2.8 mmol/L (or 8 to 11.2 mg/dl); this value varies from species to species.
3. PTH is secreted in response to hypocalcemia; it increases plasma calcium levels by:
   a. Increasing calcium resorption from bone
   b. Decreasing renal excretion of calcium
   c. Increasing calcium absorption from the gastrointestinal tract
   d. Increasing renal synthesis of 1-25 dihydrocholecalciferol, the active form of vitamin D
4. During the egg-laying cycle, PTH increases resorption of medullary bone. It is estimated that 10% or more of total body calcium is needed for eggshell deposition. Most eggshell calcium comes from

medullary bone. Plasma phosphate decreases because of the increase in PTH caused by increased renal tubular secretion and decreased tubular resorption.
5. The response to PTH is more rapid in birds than in mammals.
   a. In chickens, a dose of 90 USP units of bovine PTH caused a maximal response 3 to 4 hours after injection.[19]

III. Clinical Diseases

A. Secondary Nutritional Hyperparathyroidism
1. This condition, which is common in birds, is the result of a calcium-deficient diet.
2. Diets that have a high seed, fruit, or raw meat composition are inadequate; they provide insufficient calcium and an excess of phosphorus, which results in parathyroid hyperplasia.

B. Clinical Signs of Secondary Nutritional Hyperparathyroidism
1. Hypocalcemia
2. Seizures
3. Pathologic fractures caused by excessive bone resorption
4. Osteodystrophy
5. Muscle cramps
6. Weakness

C. Treatment of Secondary Nutritional Hyperparathyroidism
1. Treatment involves parenteral or enteral calcium supplementation, depending on the severity of the signs.
2. If the condition is life-threatening, calcium is administered parenterally as follows:
   a. 50 to 100 mg/kg given intravenously slowly to effect[15]
   b. 50 to 100 mg/kg given intramuscularly (or subcutaneously in diluted form) twice a day[15]
3. When the condition has stabilized, calcium is administered enterally:
   a. 1 ml/kg of Neocalglucon given orally twice a day[15]
   b. 1 ml of Neocalglucon per 1 ounce of water (changed daily)[15]
4. Conversion to a balanced, formulated diet is recommended for management and for improved long-term prognosis.

D. Hypoparathyroidism
1. This condition has been suggested as the cause of hypocalcemia syndrome in African grey parrots.
   a. Although other psittacines suffer from hypocalcemia, African greys are the primary species seen with clinical signs.
   b. A study by Hochleithner et al.[20] in African greys showed:
      (1) Clinical signs of hypocalcemia (falling off perches, seizure activity, weakness, and abnormal behavior).

(2) Clinical signs of hypoparathyroidism in birds are similar to signs seen in dogs.
(3) Hypocalcemia, hyperphosphatemia, normal AP
(4) No bony changes
(5) Rapid response to treatment with calcium and vitamin D supplementation

c. A unique feature is that mature African greys often do not show bone demineralization radiographically, which would be expected with hypocalcemia.
d. The PTH level was undetectable in both affected and normal African greys; however, this may have been because the assay used was incapable of detecting the PTH level.
e. Altered PTH activity is thought to be the cause of hypocalcemic syndrome in African greys.
(1) This may be due to insufficient production of PTH, malfunction of target organs, or incompetent PTH receptors in target organs.

Ultimobranchial Glands

I. Anatomy

A. Description
1. The ultimobranchial glands are small, flat, pinkish, irregularly shaped glands located just caudal to the parathyroid glands.
2. The right ultimobranchial gland is more caudal than the left gland.
3. The glands consist of C-cells that are homologous to the C-cells of mammals.

II. Function

A. Production of Calcitonin from C-Cells
1. The biologic activity of the calcitonin produced in avian species is similar to that of purified salmon calcitonin.[19]
2. The avian plasma level of calcitonin is much higher than that found in mammals.[19]
3. The physiologic role of calcitonin in avian species is still unknown.
4. Neither ultimobranchialectomy nor exogenous calcitonin administration causes significant change in the plasma calcium concentration. There are two theories to explain this[19]:
   a. Because calcitonin circulates at a higher level, exogenous administration will have no further effect on calcium.
   b. Calcitonin is not a calcemic hormone in birds; its function is unknown.

Endocrinology of the Avian Reproductive System

I. Ovary

The ovary secretes estrogens, progesterone, and androgens. In most avian species, only the left ovary is present.

A. Anatomy
1. The ovary is just caudal to the adrenal gland, near the cranial pole of the kidneys. It is tightly enclosed in the left abdominal air sac.[21]
2. The size of the ovary increases as the FSH level increases.[22]
3. Right gonad
   a. The right ovary and oviduct are present in the embryo but usually regress and disappear by adulthood. Regression begins by day 10 in the chicken.[21]
   b. The right gonad normally is arrested at a testis-like stage of development.
      (1) If the left ovary is destroyed, the right gonad enlarges.
      (2) Spermatogenesis occurs only when the left ovary is removed within a month of hatch in the chicken.[21,23]
   c. A functional right ovary is present in a number of falconiformes and in the brown kiwi. About 5% of sparrows and pigeons have bilateral ovaries, although the right ovary may be smaller.[23,24]

B. Blood Supply
1. Ovarian artery: Branches off the left renolumbar artery or directly from the aorta, then divides into many branches, which supply a single follicular stalk.
2. Follicles are highly vascularized except for the stigma.
   a. The stigma is the region of the follicle that ruptures as a result of protease activity during ovulation.
   b. The stigma is present on the large follicles of most species.[21,23]

C. Nerve Supply
1. Cholinergic and adrenergic fibers supply the follicles.[21,23]

D. Follicles
1. The ovary of many adult, reproductively active birds consists of an arrangement of follicles.
   a. Unlike mammals, avian follicles do not mature in synchrony; rather, there is a hierarchy of rapidly growing follicles.
   b. A combination of follicle growth, atresia, and ovulation produces this hierarchy.
      (1) FSH regulates the pattern of follicle growth and maturation.
      (2) LH controls the discharge of the egg from the follicle (ovulation).

(3) The end of the breeding season is associated with atresia of all remaining follicles, and the ovary enters a quiescent phase.[21,23,25]

E. Preovulatory Follicles
1. Steroid hormones (estrogen, androgens, progesterone) are produced in the following cells:
   a. Thecal cells of the smaller follicles (estradiol and androgens)[21,26,27]
   b. Granulosa cells of the larger follicles (progesterone)[21,28]
2. Steroidogenesis is stimulated by LH.[29]

F. Postovulatory Follicles
1. Birds have no equivalent to the mammalian corpus luteum. The postovulatory follicle should be considered an alternative to the corpus luteum.[21,22,30]
2. Postovulatory follicles may produce steroid hormones.[21,23,31]

G. Atretic Follicles
1. Some follicles become enlarged and yolk filled but fail to reach the stage at which they are ovulated. The oocyte and yolk are reabsorbed.[23] Total resorption requires days in the chicken and months in the mallard.[31]
2. Normal physiologic causes of atresia include[23]:
   a. The transition from egg laying to incubation
   b. The onset of molt
3. Atresia or ovarian regression may also be induced by the administration of:
   a. Large amounts of exogenous prolactin[23]
   b. Gonadotropin-releasing hormone (GnRH) or GnRH agonists (e.g., leuprolide acetate)

H. Yolk
1. Protein and lipids formed in the liver are transported in the blood stream to the ovarian granulosa cells, which pass them to the oocyte, where they form yolk spheres. and fluid. The production of yolk (vitellogenesis) is regulated by gonadotropins and steroid hormones.[21,23]

II. Oviduct

A. Anatomy
1. The oviduct hypertrophies during egg laying to take up most of the left coelom secondary to elevated estrogen levels.[21,32]
2. It shrinks in length and width during the nonbreeding season. Therefore it is best to perform surgical removal during the nonbreeding season.
3. Right oviduct: The right oviduct is present in the embryo and regresses in most birds, although it is retained in several species, including raptors.[21]

B. Blood Supply
1. The cranial, middle, and caudal oviductal arteries are branches of the left cranial renal artery. These vessels are found in the dorsal mesentery and must be clamped and ligated when the oviduct is removed.
2. Veins draining the cranial part of the oviduct enter the caudal vena cava. Veins draining the caudal part of the oviduct enter the hepatic or renal portal systems.[21]

C. Infundibulum
1. The proximal funnel-shaped, thin-walled portion of the infundibulum engulfs the oocyte and yolk. Fertilization occurs in the distal tubular portion (chalaziferous region). In addition, the first layer of albumin (the chalaziferous layer) is laid down.

D. Magnum
1. The magnum is a highly coiled, highly glandular region; it is the largest part of the oviduct.
2. Most of the albumin (egg white), sodium, magnesium, and calcium are deposited in the magnum.

E. Isthmus
1. The isthmus is a short, muscular region that lays down the inner and outer shell membranes.
2. Calcification of the shell begins in the isthmus.

F. Uterus (Shell Gland)
1. The uterus is a muscular region that is highly vascularized during egg laying.
2. Plumping occurs before calcification.
3. Salts and water are taken into the albumin of the egg.

G. Calcification
1. Unbound calcium is deposited on the shell membranes.
2. In laying hens, about 100 to 150 mg of calcium is deposited per hour for 15 hours.[21]
3. Pigment also is deposited in the shell gland.

H. Vagina
1. The vagina is the thickest walled region of the oviduct.
2. This organ has no role in egg formation; it only helps the uterus expel the egg.

I. Endocrinologic Implications of Surgical Procedures on the Female Reproductive System
1. Ovariectomy: If the functional left ovary is removed experimentally in a chick less than 1 month of age, the vestige of the right ovary may develop into a testislike organ.[21,23]

2. Hysterectomy or salpingectomy: A hormonal feedback loop may occur between the uterus and ovary that inhibits follicular release of oocytes.
   a. Ovulation usually stops once the oviduct has been removed.
   b. Egg yolk peritonitis is a rare postoperative sequela, reported in one quail and one duck.[33,34]

III. Testis

The testes produce androgens and progesterone. Both testes should be functional in the adult bird.

A. Anatomy
1. The paired testes sit near the cranial pole of the kidneys, just caudal to the adrenal glands, near the abdominal air sacs.[30]
2. Size
   a. The size may increase 300- to 500-fold in seasonal breeders.
      (1) The seminiferous tubules increase in length and diameter.
      (2) Gonadotropins induce an increase in the number of Leydig's cells.[25]
   b. The testes of passerines and budgerigars are classic examples of testicular hypertrophy (which can be misinterpreted as neoplasia on radiographs).
   c. Testicular size does not change in some species, such as the zebra finch.[23,35,36]
3. Color
   a. In most breeds of domestic fowl and in many seasonal breeders, the color of the testes also changes with fluctuations in hormone levels.[21]
   b. An active testis often is white or yellow because of the collection of lipid in the interstitial cells, which are dispersed by the expanding seminiferous tubules. Gonadotropins induce this increase in lipid content.[21,25]
   c. In testes with a large number of melanocytes, the inactive testis is black, and the active testis is gray or white. Melanistic testes are found in the common starling, black leghorns, toucans, and mynah birds, as well as in some species of macaws, cockatoos, and conures.[30,35]

B. Blood Supply
1. The testicular artery is a branch of the cranial renal artery.
2. Birds have no pampiniform plexus; however, the vascular pattern becomes more prominent when the bird becomes sexually active.[23]

C. Seminiferous Tubules
1. The convoluted seminiferous tubules look like those found in mammals, but anastomoses are much more numerous. The lining consists of Sertoli cells and spermatogonia or germinal epithelial cells.[21,37]

2. Injections of mammalian FSH or androgen induce spermatogenesis in domestic fowl.[22]

D. Sertoli Cells (Sustentacular or Nurse Cells)
1. Serve as attachment sites for spermatids
2. Are primarily responsive to FSH
3. Produce inhibin and steroid hormones[37]

E. Leydig Cells (Interstitial Cells)
1. Are found between seminiferous tubules
2. Are responsible for the production of steroid hormones (primarily testosterone).[23,37]
3. Growth and maturation are stimulated by LH.[25]

F. Epididymis
1. Is located on the dorsomedial surface of the testicle
2. Consists of many efferent ductules that drain the rete testes or straight tubules
3. Is unlike the mammalian epididymis:
   a. Is not grossly visible
   b. Is not divided into a head, tail, and body
   c. Is not a sperm storage organ
4. Appendix epididymis: A segment of epididymal tissue that extends forward into the adrenal glands; it is present in many species.
   a. The efferent ductules of this tissue may secrete androgens after castration.[21]

G. Ductus Deferens
1. Runs parallel with the ureters just medial to the kidneys and enters the ventrolateral wall of the urodeum
2. Is the site of sperm fertilization and storage in most birds
3. Increases in size and becomes more convoluted during the breeding season

H. Seminal Sac or Glomus (Cloacal Promontory)
1. A ball of tissue at the distal segment of the ductus deferens that serves as an additional area for sperm storage
2. Is present in passerines and budgerigars
3. Enlarges during the breeding season, forming a prominent projection in the wall of the cloaca (ejaculatory papillae); passerines can be easily sexed at this time.[23,31]

I. Accessory Sex Glands
1. Are not present in birds
2. Exception: The cloacal gland or proctodeal gland, which is unique to the Japanese quail
   a. Androgen dependent
   b. Well developed in the male; rudimentary in the female
   c. Function unclear [23]

J. Puberty in the Cockerel
   1. Androstenedione levels are higher in the juvenile, whereas testosterone levels are higher in the adult. Androstenedione levels must decrease in the testes before testicular growth begins.
   2. After the testes begin to enlarge, spermatogenesis begins, the rate of comb growth increases, and LH and testosterone levels rise.

K. Testosterone in the Adult Bird
   1. The enzyme aromatase converts testosterone to its active metabolite, 5-alpha-dihydrotestosterone.[37]
   2. As in mammals, testosterone must be converted to estradiol in the central nervous system for expression of male sexual behavior.[37-39]
   3. The testosterone levels of captive males often are only a fraction of those seen in their wild counterparts.
   4. Testosterone levels are lower when males provide some care of young.[40]
   5. Testosterone levels increase when males arrive on breeding grounds and compete for territories and mates. The relationship between testosterone and male-male aggression is well established.
   6. High testosterone levels are "energetically expensive," and prolonged elevation may adversely affect survival.
   7. Testosterone levels are also correlated with vocalization in males and females.[35]

L. Surgical Procedures of the Testis
   1. Caponization/castration/orchidectomy
      a. Performed between 1 and 2 weeks of age in poultry
      b. Best performed in adults during the nonbreeding season
   2. Capons
      a. Grow slower and develop more fat, especially if the surgery is done after 5 months of age
      b. Have higher plasma FSH and LH levels because of a lack of endogenous feedback
      c. Are less aggressive than intact cocks.[23,41]
   3. Testicular tissue can persist after castration in birds with an appendix epididymis.[21]

M. Environmental Cues for Reproductive Activity (Box 16-1)
   1. Photoperiod
      a. Birds are highly photoperiodic.
      b. Of the psittacine species, cockatiels are especially sensitive to photostimulation.[25]
      c. Long days stimulate the secretion of GnRH.[35]
   2. Temperate versus equatorial climates
      a. In temperate climates, gonadal development and the onset of reproductive activity are strongly influenced by an increase in the photoperiod.[23]
      b. Photoperiod is less important in birds from equatorial regions. However, in captivity and in the absence of other cues, tropical and even equatorial species can be stimulated by day length.

> **Box 16-1**
>
> **Reproductive Strategies**
>
> *Continuous breeders:* Birds that are reproductively active throughout the year under optimal conditions (e.g., domestic hen, khaki Campbell duck[30]).
> *Noncontinuous or seasonal breeders:* Birds that have gonads that undergo periods of growth and regression (e.g., European starling, white-crowned sparrow[23]).
> *Opportunistic breeders:* Birds that have adapted to tropical or desert climates; breeding occurs when favorable conditions exist (e.g., mourning dove, cactus wren).
> *Determinate layers:* Birds that do not lay additional eggs if eggs are removed from the nest or destroyed (e.g., budgerigars, crows[21]).
> *Indeterminate layers:* Birds that lay more eggs if eggs are removed from the nest until they "recognize the correct number of eggs"[30] (e.g., domestic hen, most psittacines).

   c. Emperor penguins, which breed during the winter, are one of the few avian species that show gonadal growth during short days.[35]
3. Length of daylight
   a. Twelve to 14 hours of daylight has the strongest effect on the chicken; normal egg production can occur with 12 to 18 hours of light.[30]
   b. The rate of change in day length may also be important, with a gradual change being more natural.[35]
4. Photorefractoriness
   a. Long day length does not stimulate gonadotropin secretion; rather, the gonads regress dramatically.
   b. This period is highly species specific.[35]
5. Photosensitivity
   a. In most species, the gonads do not become photosensitive again until they are exposed to short day length.
   b. Some species need prolonged exposure to long days.[42] The photosensitive phase typically occurs 13 to 17 hours after the onset of dawn. Therefore a long day length does not need to be continuous as long as light occurs during this photosensitive phase.[35]
      (1) The circadian rhythm of photosensitivity is based primarily on dusk in the hen.
      (2) The photosensitive phase of the Japanese quail occurs 14 hours after dawn, therefore it is based on dawn.[22]
   c. Thyroid hormones
      (1) These hormones may affect the perception of day length.
      (2) T_4 treatment prevents recovery of photosensitivity in refractory birds.
6. Extraretinal photoreceptors
   a. An extraretinal photoreceptor must exist, because blind birds may also be photostimulated. This photoreceptor is believed to be in the hypothalamus.

b. Researchers once suspected that the pineal gland served as the extraretinal photoreceptor; however, gonadal activity continues in pinealectomized blind fowl.[43-45]
7. Although photoperiod exerts a strong effect on reproductive activity, other factors also are important, because reproductive activity may continue even when hens are placed in continuous darkness.[30,35]
   a. Rainfall stimulates reproductive behavior in many tropical and desert-dwelling species.
   b. Temperature extremes can be stressors, which diminish reproductive activity and semen production and lead to thinner eggshells.
   c. Male vocalization stimulates female cycling (photoperiod may have a direct effect on male vocalization).
   d. The presence of a mate (real or imagined) is needed to elicit nesting behavior in cockatiels.

N. Ovulation
1. Luteinizing hormone (LH)
   a. For most of the ovulation-oviposition cycle, a low level of tonic LH secretion is maintained.[22]
   b. LH and estrogen levels begin to increase about 3 weeks before the first ovulation.[42,43,46] Progesterone (P_4) levels increase about 2 week before ovulation.[43,47,53] LH and P_4 interact in a positive feedback system.[22,48]
      (1) P_4 stimulates the release of LH.
      (2) LH in turn stimulates a further increase in P_4 secretion.[37,48,54]
   c. Estrogen
      (1) Estrogen has not been found to stimulate LH release directly. However, the ability of P_4 to stimulate LH release depends on the hen's reaching critical blood levels of P_4 and estrogen.[22]
   d. LH surge: LH, P_4, testosterone, and estrogen levels peak 6 to 8 hours before ovulation.[23]
      (1) Crepuscular LH peak
         (a) A transient increase in LH occurs before the first ovulation.
         (b) This peak occurs at the onset of darkness in hens maintained under 14 to 16 hours of light and 8 to 10 hours of darkness.
         (c) This first, minor LH peak may play a role in setting the hen's "reproductive clock." The release of LH is believed to be very circadian in nature.[22]
2. Preovulatory uterine contractions
   a. These contractions are directly related to the preovulatory release of PGF_2-alpha by the largest follicle or the largest postovulatory follicle, or both.[43,49,50]
   b. The contractions occur 15 to 60 minutes before ovulation.[49]
3. Domestic fowl ovulate almost daily. Ovulation occurs within 1 hour of oviposition in hens kept under appropriate conditions.[23,51]

O. Expulsion of the Egg (Oviposition)

1. Abdominal and vaginal muscles and the uterovaginal sphincter relax, and the uterus contracts.[30]
2. Depending on the species, oviposition last seconds to hours.[30]
3. Different species lay their eggs at different times of the day.
4. Oviposition is stimulated by:
    a. Prostaglandins (PGs)
        (1) Prostaglandins are produced by the largest, preovulatory follicle.[43]
        (2) Prostaglandins are also produced in the uterus. However, little or no prostaglandin or its metabolites can be found in uterine venous blood. Therefore some researchers have concluded that the role of prostaglandins made in the uterus may be insignificant.[49,52]
        (3) Synthesis is stimulated by progesterone[53] and LH.[43,54]
        (4) Prostaglandin F_2-alpha (PGF_2-alpha)
            (a) PGF_2-alpha levels increase in oviductal myometrium and mucosa before oviposition, then decrease dramatically.[55,56]
            (b) The uterus has high-affinity receptors for PGF_2-alpha. Binding leads to intracellular calcium mobilization, which in turn leads to strong myometrial contractions.[57] Therefore prostaglandins stimulate uterine contraction when extracellular calcium is available.
        (5) Prostaglandin E (PGE)
            (a) PGE is produced mainly in the granulosa cells of the large, preovulatory follicles.[55,56,58]
            (b) The primary function of PGE is to relax the terminal portion of the oviduct, thereby allowing expulsion of the egg.[59]
            (c) PGE_1 and PGE_2 have dose-dependent effects in the vagina. PGE_1 is the more effective agonist.[59]
            (d) The vagina has high-affinity receptors for PGE; these may be necessary to competitively inhibit PGF_2-alpha.
            (e) PGE levels do not change significantly in the uterus during oviposition; however, PGE enhances the calcium mobilization effects of PGF_2-alpha on the uterus.[55-57]
            (f) PGE is much more potent than PGF_2-alpha in inducing oviposition, perhaps because oviposition is much easier with relaxation.[43]
    b. Arginine vasotocin (AVT)
        (1) The role of vasotocin in oviposition is unclear. It is known to stimulate uterine contraction in vitro.[43]
        (2) One study in poultry demonstrated that uterine muscle is most sensitive to vasotocin, then oxytocin, and finally mesotocin.[21,38]
            (a) Uterine sensitivity appears to increase as the time of oviposition nears.[55,56]
            (b) A transient, fivefold increase in the plasma vasotocin level has been associated with oviposition.

(3) In another study, vasotocin appeared to have no direct action on oviposition in domestic fowl.[61]
(4) It has been theorized that uterine contractions that occur secondary to the effects of prostaglandins may stimulate the release of vasotocin.[55,56] PGF_2-alpha and vasotocin probably work together to increase uterine contractions.[43]

5. Terminal oviposition (the last egg laid in a clutch)
   a. PGF_2-alpha levels fall as PGE_2 levels rise; this change in the ratio of prostaglandin concentrations may lead to completion of the clutch.[43,52]

P. Ovulation-Oviposition Cycle

1. The egg travels through the oviduct in about 25 to 26 hours in the chicken (the range is 24 to 28 hours)[22] with the exception of a pause day, on which no egg is laid. The laying interval is broken down as follows:

Infundibulum	15 to 30 minutes
Magnum	2 to 3 hours
Isthmus	1 to 2 hours
Uterus	20 to 26 hours
Vagina	Seconds [21,23]

2. The egg takes 40 to 44 hours to traverse the oviduct in the pigeon. The laying interval is about 48 hours in psittacines and 24 hours in passerines.[31]
3. The longer the ovulation-oviposition cycle, the more time the egg spends in the reproductive tract, especially in the uterus.[37]

Q. Calcium Metabolism

The eggshell is formed in the shell gland (uterus). The shell consists of about 95% crystallized calcium carbonate, or calcite.[32] Calcification is controlled by progesterone.[37]

1. The uterus mobilizes an enormous amount of calcium for eggshell formation.
   a. The uterus extracts up to 2.5 g of calcium from the blood stream over 15 hours in domestic fowl.[21,62-64]
   b. Calcium blood levels reach as high as 30 mg/dl during the laying process in psittacines.[23]
   c. All the calcium in the blood stream can be taken up by the uterus in 8 to 18 minutes.[23]
2. Sources of calcium
   a. Intestinal absorption from the diet
      (1) Reproductively active psittacines preferentially choose calcium-rich foods. The consumption of calcium increases daily during the prelaying period and during the early stages of shell calcification.[23,65]
      (2) A 1:1 or 2:1 calcium to phosphorus ratio is recommended.[30]
      (3) Changes in intestinal calcium absorption are regulated by vitamin D_3.[64]

b. Renal control of calcium levels
   (1) Normally, urinary calcium excretion is proportional to calcium intake; however, during shell formation calcium excretion decreases.[66,67]
   (2) Renal transport is important in calcium homeostasis in birds. Avian renal function adapts during reproduction to provide the calcium needed.[68]
      (a) Vitamin D_3 synthesis increases.
      (b) The number of PTH hormone receptors increases.
      (c) These changes appear to be estrogen dependent.
c. Mobilization of bone
   (1) Calcification of the medullary cavities of long bones occurs.
      (a) Polyostotic hyperostosis (osteomyelosclerosis) is the generalized medullary opacity of some or all bones.
      (b) This change occurs in the hen about 10 to 14 days before egg formation.
      (c) Calcification is stimulated by the synergistic action of androgens and estrogens.[65,69,70]
      (d) Calcium is laid down primarily in the femur and tibiotarsus of most birds. In budgerigars the humerus and femur are used.[30,70]
      (e) Medullary bone may replace hematopoietic tissue in as much as three fourths of the medullary cavity.
      (f) The skeleton may increase in weight by 25% during the prelaying period.[65]
   (2) Medullary bone is best suited for this role because the rate of calcium turnover has a half-life of only 2 days. The half-life is several months in cortical bone and several weeks in trabecular bone.[32]
   (3) Thirty percent to 40% of the calcium in eggshells is derived from medullary bone.[32,71,72] Mobilization of bone calcium stores occurs if:
      (a) The diet does not provide enough calcium
      (b) Eggs are laid during the early morning hours, when food (i.e., calcium-rich food) may not be available.
   (4) Polyostotic hyperostosis also may occur in nonlaying females and males. The pathogenesis of polyostotic hyperostosis in these birds is not known, but diseases such as estrogen-secreting tumors must be considered. Two theories for the process of polyostotic hyperostosis in these birds have been proposed:
      (a) Hyperestrogenism[30]
         (i) Some studies and case reports support this theory.[70,73]
         (ii) A study by Baumgartner et al.[74] did not support the theory; 30 of 35 affected budgerigars had normal ovaries with no evidence of estrogen-secreting tumors or other endocrine disease.

(b) Hepatic dysfunction[30]
       (i) The liver is responsible for inactivating estrogen.
3. Calcium deficiency
   a. Calcium mobilization eventually leads to a calcium deficiency.
   b. This deficiency leads to a decrease in FSH secretion, which halts egg laying. High-fat, low-calcium diets (e.g., seed diets) exacerbate calcium deficiency.[30]

R. Incubation
1. Incubation is associated with decreasing LH levels and increasing prolactin levels in galliformes.
   a. Serum prolactin levels increase slowly during incubation.[75]
      (1) Increasing prolactin levels are responsible for the shift from egg laying to incubation in the turkey.
      (2) Prolactin also induces ovarian regression or atresia.
      (3) Levels of the neuropeptide vasoactive intestinal peptide (VIP) increase dramatically during incubation. Studies in turkey hens suggest that VIP may be important in the regulation of prolactin.[37,76]
   b. The mechanism behind decreasing LH levels is unclear. An increase in prolactin, ovarian hormones, and neuronal stimulus associated with nesting may combine to inhibit LH secretion.
2. Brood patch (incubation patch)
   a. The brood patch is a region of increased vascularity, edema, and epidermal thickening.
   b. The patch probably develops secondary to the secretion of estrogen and prolactin.[35]

Workup for Reproductive Tract Disease

I. Signalment

A. Age
1. Puberty[21,30]

Species	Age when sexually active
Zebra finches, captive Japanese quail	2 months
Budgerigars, lovebirds, cockatiels	6 months to 1 year
Smaller passerines	First or second spring after hatch
Conures	1 to 2 years
Lories/lorikeets	2 to 3 years
Larger psittacines	3 to 6 years

Table 16-1
Predisposition to disease[21,33,36,78]

Species	Common disease processes
Budgerigars	Ovarian neoplasia, oviduct impaction, oviductal prolapse, chronic egg laying, egg binding
Canaries	Oviduct impaction, oviductal prolapse, chronic egg laying
Cockatiels	Ovarian cyst, oviduct impaction, chronic egg laying, egg binding
Ducks	Egg binding, egg yolk peritonitis, metritis
Finches	Egg binding, egg yolk peritonitis
Lovebirds	Chronic egg laying, egg binding
Gallinaceous birds	Ovarian neoplasia, salpingitis, internal laying, egg binding, oviduct impaction
Macaws	Egg yolk peritonitis

B. Gender
1. Determination
   a. Sexual dimorphism[30, 77]
      (1) Female raptors usually are about 30% larger than males.
      (2) Males often have a larger head and a wider and longer bill (e.g., this usually is obvious in toucans).
      (3) Feather, iris, and cere color also may suggest the gender.

II. History (Table 16-1)

A. Molt
1. Molt is associated with a complete regression of reproductive organs.
2. The molting cycle often is associated with a marked increase in LH, testosterone, estrogen, and thyroid hormone levels.

B. Diet
1. Determine the calorie, calcium, protein, and fat content.
2. Are any vitamin-mineral supplements provided?

C. Environmental Conditions
1. Photoperiod
2. Temperature
3. Location of cage
4. Recent environmental changes

D. Medical History
1. Potential exposure to toxic agents
   a. Mercury, cadmium and, to a lesser extent, copper, lead, zinc, aluminum, and nickel all inhibit adenosine triphosphate (ATP)-dependent calcium binding in the uterus, leading to thin-shelled eggs.[79]

b. With a calcium-deficient diet, zinc at lower concentrations (2800 ppm) may have a direct inhibitory effect on ovarian granulosa cell function.[80]
   c. Mercury is known to decrease spermatogenesis in some avian species.[30,36,81]
   d. Aflatoxins may cause follicular atresia.[30]
   e. Mycotoxins
   f. Copper fungicides in feed suppress spermatogenesis and lead to testicular atrophy.[30,36]
  2. Reproductive history
   a. Observed reproductive behavior
       (1) Regurgitation
       (2) Pair bonding
       (3) Copulation
       (4) Nest box inspection
       (5) Feeding
       (6) Mutual preening
       (7) Coordinated territorial defense in larger psittacines
           (a) Macaws lunge at the front of the cage.
           (b) Amazons fan the tail feathers and ruffle the nape feathers.
       (8) Paper shredding, nest building
       (9) Hiding under papers or seeking dark places
           (a) Has the bird laid eggs?
           (b) When was the last egg laid?
           (c) Were the eggs normal in size, shape, color, and shell thickness?
           (d) How often are eggs laid?
           (e) How many are in a clutch?
           (f) Was the bird hand raised or parent raised?
       (10) Mate selection
           (a) Some species can be force paired (i.e., quail, doves).
           (b) Force pairing may lead to aggression or decreased reproductive activity in some species (e.g., cockatiels[82]).

III. Physical Findings

A. Hands-Off Examination
  1. Nonspecific signs of illness include the following:
   a. Depresssion
   b. Ruffled feathers
  2. A change in secondary sex characteristics (e.g., cere color) may occur with ovarian tumors.[30]

B. Quick, Low-Stress Examination
  1. Assessment of body condition
   a. Obesity may be seen with salpingitis and infertility.
  2. Careful coelomic palpation
  3. Cloacal examination

IV. Diagnostics

A. Laboratory Tests
  1. Fecal and choanal Gram's stain
  2. Normal clinical pathologic changes associated with egg laying include[33]:
     a. A slight increase in the white blood cell count
     b. A slight increase in the packed cell volume and total protein level
     c. Possibly an increase in the alkaline phosphatase (ALP) level
     d. A threefold to 18-fold increase in phospholipids, fatty acids, and neutral fats
        (1) This increase is stimulated by ovarian hormones.
        (2) Plasma cholesterol does not increase.
     e. Possible doubling of blood glucose levels in some species
        (1) Blood glucose has been found to decrease significantly in turkey hens during incubation.
     f. Possible doubling of serum calcium levels
  3. *Chlamydia* screening
  4. Imaging
     a. Survey radiographs may not show a shell-less egg.
     b. Ultrasonography is especially helpful in larger birds such as ratites.
  5. Laparoscopy with biopsy, culture, and sensitivity testing and cytologic study of lesions
  6. Egg culture

V. Reproductive Disorders that May Involve Hormonal Therapy

A. Ovarian Cysts
  1. Single or multiple cystic ova have been reported in budgerigars, canaries, and pheasants. Ovarian cysts also are common in cockatiels.
  2. The cause of these cysts is unknown.
     a. A primary "endocrine disturbance" is suspected, because cystic ova frequently are associated with hyperostosis.
  3. The significance of the cysts also is unknown.
     a. Cystic ova may be an incidental finding during laparoscopy or at necropsy; however, in affected birds, future reproductive performance may suboptimal.[30,36,83]
  4. If the cysts are symptomatic, dyspnea and ascites are commonly seen.
  5. Ovarian cysts were successfully treated in two budgerigars with oral testosterone.[30,83] Human chorionic gonadotropin (HCG) also has been tried.

B. Chronic or Excessive Egg Laying
  1. Definition: Larger than normal clutch *or* repeated clutches regardless of the existence of a suitable mate or breeding season.

2. Etiology
   a. Imprinting is an important predisposing factor; this condition is especially common in hand-raised birds.
   b. Concurrent clinical disease.
3. Chronic egg laying may deplete nutritional stores, leading to:
   a. Egg binding
   b. Osteoporosis
   c. Malnutrition
   d. Weight loss (may occur secondary to constant regurgitation and depletion of nutrition stores)
   e. Feather loss, mild dermatitis around the cloaca as a result of masturbatory behavior
4. Therapy
   a. Behavioral modification should be introduced gradually. Techniques may include the following:
      (1) If one or more people in the household act as substitute mates, they should decrease time spent with the bird until egg laying stops.
      (2) Objects that stimulate masturbatory behavior or regurgitation should be removed (e.g., a favorite toy or a mirror).
      (3) Eggs should not be removed from the cage.
      (4) The bird may be moved to a location in another room.
      (5) Removal of the nest box is controversial. Some clinicians advocate initially leaving the nest box in an effort to get the bird to brood. The box then is removed.
   b. The photoperiod should be reduced to 8 hours of light per day.
   c. Nutritional status should be improved.
      (1) Healthy diet
      (2) Increased caloric intake
      (3) Adequate protein intake
      (4) Vitamin-mineral supplement
   d. Hormonal therapy
      (1) HCG
      (2) Leuprolide acetate
      (3) Testosterone (not commonly used)
      (4) Medroxyprogesterone (not currently recommended)
   e. Salpingohysterectomy

C. Dystocia/Egg Binding
1. Definition: Egg binding is the failure of the egg to pass through the oviduct at a normal rate; egg binding may or may not be associated with dystocia.
2. Most common locations of egg binding:
   a. Caudal uterus
   b. Vagina
   c. Vaginal-cloacal junction
3. The etiology usually is multifactorial and varies with the species.[30] Causes may include:

a. Genetic predisposition
b. Nutritional deficiencies
   (1) Calcium deficiency (extremely common; usually linked to egg binding through poor diet or chronic egg laying, or both)
   (2) Vitamin E or selenium deficiency (or both)
c. Bred out of season
d. Egg production in a virginal hen
e. Chronic or excessive egg production
f. Malformed eggs
g. Oviductal damage or disease (e.g., metritis)
h. Lack of exercise
i. Obesity
j. Concurrent stressors (temperature extremes, systemic disease)
k. Senility
l. Species predisposition
   (1) Ducks and psittacines are commonly affected.
   (2) Small species (e.g., budgerigars, lovebirds, finches, canaries, cockatiels) are most frequently affected.
4. Clinical disease may include:
   a. Acute onset of depression, straining, abdominal distention, dyspnea, lameness, bilateral paresis, or paralysis.
      (1) The bird may sit unsteadily on its perch with ruffled feathers and half-closed eyes.
      (2) Frequent tail bobbing or straining motions may be observed.
      (3) Canaries frequently droop their wings.[84]
      (4) Eventually, as the bird weakens, it moves to the bottom of the cage.
   b. Acute death may also occur, especially in small birds such as canaries and finches.
   c. Ratites rarely show signs other than persistent broodiness and cessation of egg laying.[36,85]
5. Diagnostics
   a. Quick, low-stress physical examination, stressing careful coelomic palpation and cloacal examination
   b. Radiographs
   c. Ultrasound scan (very helpful in ratites)
6. Supportive care: First, the patient's condition is stabilized. In general, egg-bound passerines should be treated more aggressively because death often occurs within hours. Additional treatment measures include:
   a. Fluid therapy
   b. Heat
   c. Humidity (incubator set at 29.4° to 35° C [85° to 95° F] with moist towels) or nebulization
   d. Address hypocalcemia and hypoglycemia
   e. Injectable vitamins A and D_3
   f. Consider systemic antibiotics
7. Medical techniques for removing the egg
   a. PGE

(1) PGE is applied topically to the uterovaginal sphincter.
(2) Hudelson and Hudelson[43] recommend using PGE_2 if dystocia continues for longer than 1 hour in a small bird or 3 to 5 hours in a large bird.
(3) If no adhesions are present, the egg usually is laid within 5 to 10 minutes.[43]
  b. PGF_2-alpha: *Not recommended*[43]
    (1) PGF_2-alpha does not relax the uterovaginal sphincter.
    (2) Administration may lead to uterine rupture or reverse peristalsis, or dilation of the sphincter may occur secondary to the constant pressure of the egg.
    (3) Systemic administration of PGF_2-alpha can cause serious adverse effects, such as hypertension, bronchoconstriction, and generalized stimulation of smooth muscle.
  c. Oxytocin or AVT
    (1) These two drugs appear to affect the uterus only; therefore fewer systemic reactions occur.[48]
    (2) Like PGF_2-alpha, oxytocin is unable to relax the uterovaginal sphincter and may lead to uterine rupture or reverse peristalsis. Dilation of the sphincter may occur from the constant pressure of the egg.
8. Manual techniques for removing the egg
  a. If medical therapy does not elicit contractions, the clinician must determine if the sphincter is dilated.
  b. Gentle massage and manual pressure may be applied with or without anesthesia.
  c. Cloacal or transabdominal ovocentesis may be required.
  d. Hysterectomy (salpingectomy) may be necessary if the egg is bound in the isthmus or if medical management is unsuccessful.
9. Complications
  a. An egg in the pelvis may compress blood vessels and the kidney, leading to shock.
  b. An egg may also interfere with passage of droppings, leading to ileus and renal dysfunction.
  c. Pressure necrosis, rupture of the oviduct, peritonitis, and abdominal hernias also are possible.

Adrenal Glands[119-121]

I. Anatomy

A. Description
1. The adrenal glands are small, yellow oval structures.
2. They are located just craniomedial to the kidneys and adjacent to the caudal vena cava.
3. These glands may be fused in the rhea and common loon.

4. The avian adrenal gland is not clearly divided into a cortex and medulla; rather, medullary and cortical tissue are intermingled.
   a. Cortical (interrenal) tissue
      (1) Cortical tissue makes up about 70% to 80% of the gland.
      (2) The tissue is arranged in cords made of double rows of cells.
      (3) The cords radiate from the center of the gland and loop against the inner surface of the connective capsule.
      (4) Anastomosing cords of granular vacuolated cells containing carotenoids give the gland its yellow color.
      (5) The arrangement of cell types along these cords creates structural zones:
         (a) The subcapsular zone produces aldosterone.
         (b) The larger inner zone produces corticosterone.
   b. Medullary tissue
      (1) Medullary tissue makes up about 25% of the adrenal gland tissue.
      (2) Chromaffin cells secrete norepinephrine and epinephrine.
5. Blood supply
   (1) Arterial supply: Cranial renal artery
   (2) Venous return: Caudal vena cava
   (3) Adrenal portal system
      (a) This system is present in some species, including the chicken.
      (b) Blood vessels extend between the adrenal glands and the muscles on the lateral surface of the body wall
6. Sympathetic nerves innervate the chromaffin cells of the medullary tissue.

B. Hypothalamic-Hypophyseal-Adrenal Axis
1. The hypothalamus releases CRF.
2. The adenohypophysis, or anterior pituitary gland, then releases ACTH.

C. Corticosterone
1. Corticosterone is the most common corticosteroid in avian species, whereas cortisol is the most common corticosteroid in mammals. Corticosterone has both glucocorticoid and mineralocorticoid activity, and its secretion is regulated by ACTH.
2. Corticosterone is an important mineral-regulating hormone in wild mallards in coastal estuaries and alkaline lake environments:
   a. It acts on the kidneys, nasal salt glands, and small intestine.
   b. ACTH release is stimulated by increases in:
      (1) Sodium
      (2) Chloride
      (3) Osmolality
   c. Glucocorticoids do not function as mineral-regulating hormones in birds that do not have nasal salt glands and cannot drink hyperosmotic water.

3. Stressors known to induce secretion of corticosterone include:
   a. Anesthesia
   b. Temperature extremes
   c. Food and water deprivation
   d. Frustration
   e. Handling
   f. Housing
   g. Hypovitaminosis A
   h. Immobilization
   i. Infection
   j. Noise
4. A nocturnal rise in plasma corticosterone has been found in chickens and pigeons.
   a. This suggests a nocturnal increase in ACTH secretion.
   b. Some researchers have suggested that glucocorticoids be administered during the morning hours to diurnal birds.

D. Corticosteroids in the Avian Embryo
1. Cortisone and cortisol are also made by the adrenal glands.
2. Levels of these corticosteroids fall around the time of hatch.
3. Levels reach zero in ducks and chickens over 2 weeks of age.

E. Aldosterone
1. Avian adrenals secrete much more corticosterone than aldosterone; the proportions vary with the species.
2. Unlike mammals, birds do not release aldosterone in response to hyperkalemia.
3. Hyponatremia or a fall in blood volume stimulates the release of renin.
4. Angiotensinogen > angiotensin I > angiotensin II acts directly on steroidogenic cells > aldosterone.
5. Carnivorous birds receive most of their water from the moisture in their food; they are much less sensitive to angiotensin II than granivorous species, such as quail.

F. Disease States Related to the Adrenal Gland
1. Stress marks (hunger traces)
   a. These are a common problem in growing feathers that represent a period of malnutrition or stress.
   b. Stress marks may also be caused by a single injection of steroids.
   c. Segmental dysplasia of the barbs and barbules occurs, which produces bilaterally symmetric lines perpendicular to the feather shaft.
2. No documented clinical diagnosis of spontaneous adrenal disease in a bird has ever been reported. However, adrenal lesions are frequently described at necropsy. In one study, 27% of psittacines had adrenal lesions at necropsy.
3. Hyperadrenocorticism
   a. ACTH stimulation test

(1) Dose and sampling times have been established for several avian species.
(2) Plasma corticosterone levels, not cortisol, should be measured.
b. Example: Pigeons
(1) 50 µg of ACTH gel is given and blood samples are drawn at time 0, 60, and 90 minutes; *or*
(2) 125 µg of ACTH gel is given and blood samples are drawn at 30, 60, 90, or 120 minutes.
(3) A normal bird has a 10- to 100-fold increase in corticosterone levels.
(4) Poststimulation levels range from 2.2 to 15 µg/dl.
4. There have been no confirmed reports of naturally occurring Addison's disease or hypoadrenocorticism in an avian species.
5. Amyloidosis
a. Amyloidosis is a common finding in waterfowl.
b. The adrenals may appear pale and enlarged.
6. In mallards that had eaten petroleum-contaminated food, the cells of the inner adrenal zone reportedly developed structural damage of the mitochondria, leading to decreased corticosterone levels.
7. Adrenal neoplasia
a. Adrenal tumors have been reported in a number of species, including:
(1) Bilateral adrenal adenoma in a budgerigar
(2) Unilateral adrenal carcinoma in a pigeon, three budgerigars, a ring-necked pheasant, and a mountain duck
(3) Unilateral adrenal adenoma in budgerigars, a chicken, two finches, a canary, and a white-crowned parrot
(4) Adrenal carcinoma with metastasis in a budgerigar
b. The overall incidence of adrenal tumors is not high.
(1) Adrenal adenomas are more common.
(2) Adrenal carcinomas are quite rare. In one budgerigar, the neoplastic cells locally infiltrated the testicular capsule, the cranial kidney, and the air sac. Distant metastasis was not seen.
c. Pheochromocytoma (chromaffinoma)
(1) This is a benign or malignant neoplasm of chromaffin tissue.
(2) The condition was reported in one juvenile broiler that died acutely. The affected left adrenal measured 15 mm in diameter.

G. Administration of Glucocorticoids to the Avian Patient
1. Glucocorticoids exert negative feedback action on hypothalamic and hypophyseal secretion, inhibiting basal and stress-induced corticosterone release.
a. Adrenocortical failure and shock may occur in a bird receiving steroids if the bird is exposed to stressors.
b. Sustained suppression of the hypothalamic-pituitary-adrenal (HPA) axis is common in humans given 30 mg of prednisone daily for longer than 1 week.
(1) High doses for a prolonged period may suppress the HPA axis for up to 1 year.

c. In a study that placed pigeons on a short course of high-dose glucocorticoid, HPA suppression was only temporary. However, the HPA axis was suppressed in these pigeons after only one oral administration or eye application of glucocorticoid.
2. High or even toxic blood levels may arise after topical administration of steroids.
   a. Alternate day therapy should be considered when steroids must be given for longer than 2 weeks.
      (1) The daily dose is doubled and given every other day.
      (2) Nonsteroidal antiinflammatory drugs (NSAIDs) may be used on "off" days.
3. The appropriate doses of glucocorticoids have not been fully established.

   Potency Table from H and H
Cortisone	25
Hydrocortisone	20
Prednisone	5
Methylprednisone	4
Triamcinolone	4
Dexamethasone	0.75

   a. Prednisone and Prednisolone
      (1) These are the agents of choice for antiinflammatory, antineoplastic, or immunosuppressive regimens.
      (2) Treatment of severe inflammatory disease is best divided into several doses throughout the day.
      (3) Once the desired effect has been achieved, the dosage is tapered.
   b. Dexamethasone
      (1) This is the agent of choice for spinal edema.
         (a) 2 mg/kg is given three times a day until improvement occurs.

Pancreas

I. Islets of Langerhans[122-124]

	Avian	Mammalian	Function
Alpha (A) cells	50%	20%	Secrete glucagon
Beta (B) cells	37%	70%	Secrete insulin
Delta (D) cells	13%	9%	Secrete somatostatin
Pancreatic polypeptide cells		1%	Unknown

A. Carbohydrate Metabolism in the Bird[122-125,127]
1. Glucagon: Glucagon levels are much higher in granivorous birds (e.g., domestic fowl, waterfowl) than in carnivorous birds, which are more like mammals in their islet cell composition.[125,126] Pancreatic

levels of glucagon in granivorous birds are two to five times higher than in mammals, and plasma levels are 10 to 50 times higher.
  a. The major regulator of glucose in granivorous birds is secreted by the A, or alpha, cells:
     (1) Secretion is stimulated by hypoglycemia, hyperaminoacidemia, and acetylcholine.
     (2) Secretion is inhibited by glucose, glucagon, insulin, and somatostatin.
  b. Glucagon stimulates:
     (1) Gluconeogenesis
     (2) Glycolysis
     (3) Lipolysis
  c. These actions increase free fatty acid and blood glucose levels.
  d. Diabetes mellitus
     (1) In avian diabetes, glucagon levels usually are elevated despite hyperglycemia.
     (2) Depending on the species involved, the cause of diabetes may vary. Avian diabetes mellitus in granivorous birds may be due to an excess of glucagon, not to too little insulin.
     (3) Glucagon intensifies ketogenesis and usually is elevated in diabetic ketoacidosis.[7]
2. Insulin
  a. A bolus of glucose or hyperaminoacidemia causes the level of insulin to double or triple.
  b. Vagal stimulation or administration of acetylcholine (Ach) does not affect insulin secretion in chickens.
3. Growth hormone
  a. The roles of GH in metabolism include:
     (1) Exerts short-term control of metabolism
     (2) Mobilizes lipid stores, thereby increasing the levels of free fatty acids
     (3) Decreases lipogenesis
     (4) Increases muscle glycogen stores
     (5) Decreases glucose utilization
  b. Insulin and glucagon have direct inhibitory effects on GH secretion in the duck.[129]

B. Diabetes Mellitus
1. Type I (insulin-dependent) diabetes[127,130]
  a. This condition results from the destruction of the beta cells.
  b. Insulin levels are below normal.
  c. The pathogenesis is unclear, although it may be related to diabetogenic hormones (i.e., GH, glucagon, glucocorticoids, and epinephrine).
  d. Treatment
     (1) Insulin administration; however, diagnosis in birds should be based on more than an elevated blood glucose level.
2. Type II (non-insulin-dependent diabetes)[7,9]

a. This condition is due to insulin antagonism; the cellular response to insulin is impaired.
   b. Underlying conditions include obesity and Cushing's disease.
   c. Insulin levels rise in response to glucose challenge.
   d. Treatment
      (1) Weight reduction
      (2) High-fiber diet
      (3) Oral hypoglycemic agents (in small animal medicine)
3. Transient diabetes mellitus[130]
   a. This disorder usually is a non-insulin-dependent condition.
   b. It is similar to a condition in cats with transient or induced diabetes. The animal is able to compensate for diabetes without developing clinical signs for prolonged periods, although clinical signs may develop during periods of severe stress or with underlying disease. Upon resolution of the underlying disorder, many cats spontaneously recover.
4. Signalment of diabetes mellitus
   a. The toco toucan is most commonly affected.[131,132]
   b. Of the pet bird species, diabetes mellitus is most commonly reported in budgerigars and cockatiels.[127,133]
   c. Diabetes also has been reported in the chicken, duck, goose, pigeon, Amazon parrot, African grey parrot, sulfur-crested toucan, and red-tailed hawk.[133,135-137]
5. History
   a. Diet
   b. Reproductive history
   c. Change in environment
   d. Recent medications (e.g., steroids)
   e. Vaccination status (e.g., paramyxovirus in pigeons)
6. Clinical disease[125,128,133-135,137-140]
   a. Polyuria, polydypsia, and polyphagia
   b. Diarrhea (possible)
   c. Nonspecific signs of illness (e.g., fluffed and ruffled feathers, weakness, weight loss)
7. Diagnostics
   a. Complete blood count (signs of secondary infection may be detected)
   b. Chemical profile
      (1) A persistently elevated fasting hyperglycemia (500 to 1800 mg/dl) may be present.[125]
         (a) Altman and Kirmayer[137] reported normal fasting blood glucose levels of 210 to 550 mg/dl.
         (b) Minick and Duke[141] reported normal levels of 196 to 318 mg/dl in normal and convalescing fasted bald eagles.
         (c) In normal fasted birds the range is 210 to 530 mg/dl.
         (d) Other conditions that must be ruled out are stress, egg yolk peritonitis, and acute pancreatitis.[138]

(2) Blood urea nitrogen, uric acid, AST, SBA, and calcium levels should be measured.
(3) Measuring the total protein level and protein electrophoresis may prove helpful.
(4) A HAI-titer for paramyxovirus should be done in pigeons.

c. Urinalysis[125,137]
   (1) Glucosuria
      (a) Avian urine normally contains no glucose. Because glucose can be completely filtered by the normal kidney, glucosuria indicates renal damage or marked hyperglycemia.
      (b) The renal threshold varies among species but is greater than 600 mg/dl in most birds.
   (2) Ketonuria
      (a) Ketones are not present in the normal bird's urine. The presence of ketonuria may be a poor prognostic indicator.
      (b) The only time energy metabolism may normally shift from metabolism of carbohydrates to fats is in the migrating bird. Somehow the migrating bird is able to prevent the accumulation of ketone bodies.
      (c) Fat serves as the primary energy source.

d. Insulin levels[124,125,142,143]
   (1) The measurement obtained may be affected by the facility and the test used, especially if testing on the species in question has not been performed before.
   (2) Normal reported values

Cockatiel	5.8 to 8.16 µU/ml
Amazon	7.7 µU/ml
Chicken	22 to 40 µU/ml
Duck	16 to 20 µU/ml
Bald eagle	1.42 to 5.44 µU/ml

      Insulin levels appeared to be higher in male bald eagles.

e. Glucagon levels
   (1) These levels in birds (1 to 4 ng/ml) normally are 10 to 50 times higher than in mammals[125,126]
   (2) Convalescing bald eagles had normal glucagon levels ranging from 229 to 1239 mg/dl; levels were higher in females.[141]
   (3) The level was markedly elevated (5255 pg/ml) in a diabetic toco toucan compared to four normal toucans.[131,132]
   (4) Glucagon levels were also elevated in an Amazon (14,222 pg/ml) compared to one normal Amazon.[127]
   (5) Not all diabetic birds in which glucagon levels were measured had elevated levels. For instance, a diabetic cockatiel had a glucagon level of 235 pg/ml compared to two normal cockatiels (780 to 964 pg/ml).
   (6) In another study, a diabetic macaw's plasma glucagon levels ranged from 684 to 2179 pg/dl; 18 normal macaws had levels ranging from 299 to 1190 pg/dl except for one macaw with a level of 1801 pg/dl.

 f. Glucose tolerance test
  (1) This test measures insulin levels after administration of an intravenous bolus of dextrose.
  (2) It is used to distinguish type I from type II diabetes in mammals.
  (3) One study in toco toucans revealed no change in insulin levels after a glucose challenge.[132]
8. Treatment of diabetes mellitus
 a. Supportive care is required.
  (1) Fluid therapy is especially important.
  (2) Supportive care may also include warmth, nutritional support, and antibiotics.
 b. The goal of long-term management is to alleviate the clinical signs and and induce weight gain.
 c. Insulin
  (1) Because some cases of avian diabetes mellitus may depend on glucagon, diabetic birds may show insulin resistance.
  (2) The apparent resistance of diabetic birds to insulin may be caused by the type of insulin used.[122]
   (a) The hypoglycemic effects of chicken insulin given to chickens was 10 times greater than the effects of mammalian insulin.[122]
   (b) Although mammalian insulin may fail to lower blood glucose levels effectively, it may prevent the severe weight loss commonly seen with uncontrolled diabetes mellitus
  (3) Although often reported to be ineffective in avian diabetes, isophane insulin (NPH), protamine zinc insulin (PZI; this has been discontinued), and ultralente insulin have been used in avian species.
   (a) Dosages vary considerably:
    (i) The initial dose suggested for a budgerigar is 0.002 U of isophane insulin. Most cases were regulated with single daily doses of 0.004 to 0.006 U.[137]
    (ii) The suggested initial dose for larger parrots is 0.01 U to 0.1 U.[143]
    (iii) 0.2 U/kg of NPH given twice daily successfully reduced blood glucose and glucagon levels in a hyperglycemic macaw.[144]
    (iv) 0.1 to 0.8 IU of PZI given twice daily and 1 to 2 IU of NPH given SID to toco toucans was successful.[131,132]
    (v) Smaller birds may need proportionally higher dosages.
 d. Somatostatin[139]
  (1) Clinical improvement was reported in a sulfur-breasted toucan on 3 µg/kg given subcutaneously twice a day. The synthetic analog, Sandostatin, was used.[139]

(2) Bonda[144] reported that administration of three times the mammalian dose failed to induce euglycemia in a diabetic macaw.
    e. Oral hypoglycemic agents have been used empirically in two diabetic cockatiels for short- and long-term management of clinical signs (Orosz and Pollock, personal communication).
 9. Monitoring
    a. Blood glucose
       (1) Ideally a glucose curve should be repeated several days after starting insulin therapy.
       (2) Glucose curves should be performed every 2 to 3 months to monitor and regulate patients.
    b. Urine glucose
       (1) This value is a useful way to monitor glucose levels in a nonstressful environment.
       (2) Urine glucose should be assessed two to three times a day.
       (3) Because urine glucose is a crude indicator of blood glucose levels, the goal should be mild glucosuria (slightly positive or a trace).
 10. Pathologic conditions of the liver are the most common sequelae reported in some diabetic birds. Conditions reported include:
    a. Hepatic lipidosis
    b. Hepatic fibrosis
 11. Diabetes mellitus has been associated with:
    a. Pancreatitis in an African grey parrot[136]
    b. Islet cell carcinoma in a parakeet[140]
    c. Renal tumors in budgerigars[125]
    d. Pancreatic islet atrophy and hyperplasia in toco toucans

C. Pancreatic Neoplasms
 1. Neoplasms reported include adenocarcinoma with carcinomatosis and islet cell (alpha cell) adenoma and carcinoma.[127,140]
 2. Most pancreatic carcinomas occur secondary to invasion from tumors of the female reproductive tract
 3. There are several reports of diabetes in association with renal tumors in budgerigars.

Pineal Gland (Epiphysis Cerebri)[145,146]

I. Anatomy

A. Description
 1. The pineal gland is located in the epithalamus.
 2. Morphologic types vary widely among species. In the chicken the pineal gland is a pink, cone-shaped structure lying between the cerebrum and the cerebellum.

3. Some cells resemble crude photoreceptors; others are secretory cells.

B. Influence of Light
   1. The pineal gland is strongly influenced by light from the following sources:
      a. Eyes
      b. Brain
      c. Cranial cervical ganglion
   2. The pineal gland still responds to light after the cranial cervical ganglion and eyes have been removed.

C. Melatonin Secretion
   1. The pineal gland secretes melatonin in phases:
      a. Secretion increases at night (scotophase).
      b. Secretion decreases during the day.
      c. Not all melatonin secretion is from the pineal gland in galliformes; a substantial amount of melatonin is secreted by the retina in the Japanese quail.
      d. The pineal gland plays a role in time-keeping.
         (1) The phasic secretion of melatonin transmits time-keeping information.
         (2) The pineal gland is thought to be the probable location of the biologic clock.

D. Locomotion
   1. The pineal gland also has a role in locomotion; the pineal is believed to be responsible for the circadian rhythm of locomotion.
   2. If the pineal gland is removed, house sparrows kept in constant darkness become atactic.

E. Reproduction
   1. The role of the pineal gland in reproductive activity is unclear.
   2. Surgical removal of the pineal gland does not appear to change the bird's response to environmental cues to reproduce.

F. Pineal Gland Neoplasia
   1. Pineoblastoma was reported in a cockatiel with polyuria and neurologic signs. The mass compressed adjacent brain parenchyma.

REFERENCES
   1. Lumeij JT: Endocrinology. In Ritchie BW, Harrison GJ, Harrison LR, editors: *Avian medicine: principles and applications,* Lake Worth, Fla, 1994, Wingers, pp 582-606.
   2. Lothrop CD, Loomis MR, Olsen JH: Thyrotropin stimulation test for evaluation of thyroid function in psittacine birds, *J Am Vet Med Assoc* 186:47-48, 1985.
   3. Hazelwood RL: Carbohydrate metabolism. In Sturkie PD, editor: *Avian physiology,* New York, 1986, Springer-Verlag, p 322.

4. Schlumberger HG: Neoplasia in the parakeet: spontaneous chromophobe pituitary tumors, *Cancer Res* 14:235-237, 1954.
5. Curtis-Velasco M: Pituitary adenoma in a cockatiel *(Nymphicus hollandicus)*, *J Assoc Avian Vet* 6(1):21-22, 1992.
6. Romagnano A et al: Pituitary adenoma in an Amazon parrot, *J Assoc Avian Vet* 9(4):263-270, 1995.
7. Slye M, Holmes HF, Wells HG: Intracranial neoplasms in lower animals: studies on the incidence and inheritability of spontaneous tumors in mice, *Am J Cancer* 15:1387-1400, 1931.
8. Reece RL: Observations on naturally occurring neoplasms in birds in the state of Victoria, Australia, *Avian Pathol* 21:3-32, 1992.
9. Lumeij JT: Endocrinology. In Ritchie BW, Harrison GJ, Harrison LR, editors: *Avian medicine: principles and applications,* Lake Worth, Fla, 1994, Wingers, p 586.
10. Alberts H et al: A water deprivation test for differentiation of polyuric disorders in birds, *Avian Pathol* 17:385-389, 1988.
11. Bennett RA: Neurology. In Ritchie BW, Harrison GJ, Harrison LR, editors: *Avian medicine: principles and applications,* Lake Worth, Fla, 1994, Wingers, p 736.
12. Wentworth BC: Thyroids. In Sturkie PK, editor: *Avian physiology,* New York, 1986, Springer-Verlag, pp 452-465.
13. Astier H: Thyroid gland in birds. In Epple A, Stetson MH, editors: *Avian endocrinology,* London, 1980, Academic Press, pp 167-189.
14. Hollander WF, Riddle O: Goiter in domestic pigeons, *Poultry Sci* 25:20-27, 1946.
15. Carpenter JW, Mashima TY, Rupiper DJ: *Exotic animal formulary,* Manhattan, Kan, 1996, Greystone, pp 91-179.
16. Oglesbee BL: Hypothyroidism in a scarlet macaw, *J Am Vet Med Assoc* 201:1699-1701, 1992.
17. Harms CA et al: Development of an experimental model of hypothyroidism in cockatiels *(Nymphicus hollandicus)*, *Am J Vet Res* 55:399-404, 1994.
18. Orosz SE, Oliver JW, Schroeder EC: TRH stimulation for the evaluation of thyroid function in Amazon parrots, Proc European Assoc Avian Vet, 1997, pp 41-45.
19. Kenny AD: Parathyroid and ultimobranchial glands. In Sturkie PK, editor: *Avian physiology,* New York, 1986, Springer-Verlag, pp 466-478.
20. Hochleithner M, Hochleithner C, Harrison GJ: Evidence of hypoparathyroidism in hypocalcemic African grey parrots, Avian Examiner, Delray Beach, Fla, 1996, HBD International, pp 1-4.
21. King AS, McLelland J: *Birds: their structure and function,* 1984, Orlando, Bailliere Tindall, pp 145-174.
22. Cunningham FJ: Ovulation in the hen: neuroendocrine control. In Clark JR, editor: *Oxford reviews of reproductive biology,* vol 9, 1987, Oxford, Clarendon Press, pp 96-136.
23. Johnson AL: Reproduction in the female and reproduction in the male. In Sturkie PK, editor: *Avian physiology,* New York, 1986, Springer-Verlag, pp 403-451.
24. Kinsky FC: The consistent presence of paired ovaries in the kiwi, J Orn, Lpz, 112:334-357, 1971.
25. Rosskopf WJ, Woerpel RW, editors: Avian reproductive endocrinology, *Vet Clin North Am* 21(6):1347-1359, 1991.
26. Robinson FE, Etches RJ: Ovarian steroidogenesis during follicular maturation in the domestic fowl, *Biol Reprod* 35:1096-1105, 1986.
27. Imai K, Nalbandov AV: Plasma and follicular steroid levels of laying hens after the administration of gonadotropins, *Biol Reprod* 19:779-784, 1978.
28. Huang ES, Nalbandov AV: Steroidogenesis of chicken granulosa and theca cells, *Biol Reprod* 20:454-461, 1979.

29. Marrone BL, Hertelendy F: Decreased androstenedione production with increased follicular maturation in theca cells from the domestic hen, *J Reprod Fertil* 74:543-550, 1985.
30. Joyner KL: Theriogenology. In Ritchie BW, Harrison GJ, Harrison LR, editors: *Avian medicine: principles and applications,* Lake Worth, Fla, 1994, Wingers, pp 748-786.
31. Lofts B, Murton RK: Reproduction in birds, vol III. In Farner DS, King JR, editors: *Avian biology,* New York, 1973, Academic Press, pp 1-107.
32. Hurwitz S: Calcium homeostasis in birds. In Aurbach GD, McCormick DB, editors: *Vitamins and hormones,* vol 45, New York, 1989, Academic Press, pp 173-221.
33. Rosskopf WJ, Woerpel RW: Avian obstetrical medicine, AAV Proceedings, pp 323-336, 1993.
34. McCluggage D: Hysterectomy: a review of select cases, Proceedings Association of Avian Veterinarians, pp 201-206, 1992.
35. Millam JR: Reproductive physiology. In Altman RB et al, editors: *Avian medicine and surgery,* Philadelphia, 1997, Saunders, pp 12-26.
36. Speer BL: Diseases of the urogenital system. In Altman RB et al, editors: *Avian medicine and surgery,* Philadelphia, 1997, Saunders, pp 633-643.
37. Ottinger MA, Bakst MR: Endocrinology of the avian reproductive system, *J Avian Med Surg* 9(4):242-250, 1995.
38. Watson JT, Adkins-Regan E: Testosterone implanted in the preoptic area of male Japanese quail must be aromatized to activate copulation, *Horm Behav* 23:432-447, 1989.
39. Balthazart J: Hormonal correlates of behavior. In Farmer DS, King JR, editors: *Avian biology,* vol 7, New York, 1983, Academic Press, pp 221- 265.
40. Kenton B, Millam JR: Photostimulation and serum steroids of orange-winged Amazon parrots, *Proc Assoc Avian Vet,* pp 435, 1994 (abstract).
41. Altman RB: Soft tissue surgical procedures. In Altman RB et al, editors: *Avian medicine and surgery,* Philadelphia, 1997, Saunders, pp 723-727.
42. Sharp PJ: Strategies in avian breeding cycles, *Animal Reproduction Science* 42:505-513, 1996.
43. Hudelson KS, Hudelson P: A brief review of the female avian reproductive cycle with special emphasis on the role of prostaglandins and clinical applications, *J Avian Med Surg* 10(2):67-74, 1996.
44. Yokohama K et al: The sites of encephalic photoreception in photoperiodic induction of the growth of the testes in the white-crowned sparrow, *Cell Tissue Res* 189:441-467, 1978.
45. Turek F: Diurnal rhythm and the seasonal reproductive cycle in birds. In Assenmacher I, Farner DS, editors: *Environmental endocrinology,* New York, 1978, Springer-Verlag, pp 144-152.
46. Senior BE: Oestradiol concentration in the peripheral plasma of the domestic hen from 7 weeks of age until the time of sexual maturity, *J Reprod Fertil* 41:107-112, 1974.
47. Williams JB, Sharp PJ: Control of the preovulatory surge of luteinizing hormone in the hen: the role of progesterone and the androgens, *J Endocrinol* 77:57-65, 1978.
48. Etches RJ et al: Follicular growth and maturation in the domestic hen, *J Reprod Fertil* 67:351-358, 1983.
49. Shimada K, Olson DM, Etches RJ: Follicular and uterine prostaglandin levels in relation to uterine contraction and the first ovulation of a sequence in the hen, *Biol Reprod* 31:76-82, 1984.
50. Tanaka K, Nakada T: Participation of the ovarian follicle in control of time of oviposition in the domestic fowl, *Poultry Sci* 53:2120-2125, 1974.

51. Van de Velde JP, Loveridge N, Vermeiden JPW: Parathyroid hormone responses to calcium stress during eggshell calcification, *Endocrinology* 115(5):1901-1904, 1984.
52. Olson DM, Shimada K, Etches RJ: Prostaglandin concentrations in peripheral plasma and ovarian and uterine plasma and tissue in relation to oviposition in hens, *Biol Reprod* 35:1140-1146, 1986.
53. Lundholm CE: Progesterone stimulates prostaglandin synthesis in eggshell gland mucosa of estrogen-primed chickens, *Comp Biochem Physiol* 103(1):217-220, 1992.
54. Etches RJ et al: Prostaglandin production by the largest preovulatory follicles in the domestic hen, *Biol Reprod* 43:378-384, 1990.
55. Saito N, Koike TI: Alterations in uterine contractility during the oviposition cycle in domestic hens, *Br Poult Sci* 33(3):671-676, 1992.
56. Saito N, Shimada K, Koike TI: Interrelationship between arginine vasotocin, prostaglandin, and uterine contractility in the control of oviposition in the hen, *Gen Comp Endocrinol* 67(3):352-347, 1987.
57. Molnar M, Asem EK, Hertelendy F: Differential effects of prostaglandin F_2-alpha and of prostaglandins E_1 and E_2 on cyclic 3'5'-monophosphate production and intracellular calcium mobilization in avian uterine smooth muscle cells, *Biol Reprod* 36(2):384-391, 1987.
58. Li J et al: Avian granulosa cell prostaglandin secretion is regulated by transforming growth factor alpha and beta and does not control plasminogen activator activity during follicular development, *Biol Reprod* 51(4):787-794, 1994.
59. Asem EK, Todd H, Hertelendy F: In vitro effect of prostaglandins on the accumulation of cyclic AMP in the avian oviduct, *Gen Comp Endocrinol* 66(2):244-247, 1987.
60. Murakami Y, Fujihari N, Koga O: Plasma levels of arginine vasotocin in the hen before and immediately after premature oviposition induced by prostaglandins and arachidonic acid, *Jpn Poult Sci* 27:346-355, 1990.
61. Nakada T et al: Studies on the role of arginine vasotocin in relation to ovipositon in laying hens, *Jpn Poult Sci* 31(5):358-362, 1994.
62. Hodges RD: pH and mineral ion levels in the the blood of the laying hen, *Comp Biochem Physiol Physiol* 28:1243-1257, 1969.
63. Simkiss K, Dacke CG: Ultimobranchial glands and calcitonin. In Bell DJ, Freeman BM, editors: *Physiology and biochemistry of the domestic fowl,* vol 1, New York, 1971, Academic Press, pp 481-488.
64. Eastin WC, Spaziani E: On the control of calcium secretion in the avian shell gland, *Biol Reprod* 19:493-504, 1978.
65. Nys Y: Progesterone and testosterone elicit increases in the duration of shell formation in domestic hens, *Br Poult Sci* 28:57-68, 1987.
66. Common RH: Observation on the mineral metabolism of pullets, *Journal of Agricultural Science* 23:555, 1933.
67. Common RH, Rutledge NA, Hale RW: Observations on mineral metabolism of pullets. VIII, The influence of gonadal hormones on the nutrition of calcium and phosphorus, *Journal of Agricultural Sciences* 36:64, 1948.
68. Elaroussi MA et al: Adaptation of the kidney during reproduction, *Poultry Sci* 72(8):1548-1556, 1993.
69. Dacke CG et al: Medullary bone and avian calcium regulation, *J Exp Biol* 184:63-88, 1993.
70. Schlumberger HG: Polyostotic hyperostosis in the female parkaeet, *Am J Pathol* 35(1):1-23, 1959.
71. Jowsey JR et al: Uptake of calcium by the laying hen and subsequent transfer from egg to chick, *Poultry Sci* 35:1234, 1956.

72. Mueller WJ, Schraer R, Schraer H: Calcium metabolism and skeletal dynamics of laying pullets, *J Nutr* 84:20, 1964.
73. Stauber E et al: Polyostotic hyperostosis associated with oviductal tumor in a cockatiel, *J Am Vet Med Assoc* 196(6):939-940, 1990.
74. Baumgartner R et al: Endocrinologic and pathologic findings in birds with polyostotic hyperostosis, *J Avian Med Surg* 9(4):251-254, 1995.
75. Millam JR, Zhang B, Halawani ME: Number of eggs in nest influences subsequent egg production and correlates inversely with plasma prolactin concentrations in cockatiels, Paper presented at the annual meeting of the American Society of Zoology, San Antonio Texas, 1990.
76. Mauro LJ et al: Effects of reproductive status, ovariectomy, and photoperiod on vasoactive intestinal peptide in the female tukey hypothalamus, *Gen Comp Endocrinol* 87:481-493, 1992.
77. Clubb SL: Nonsurgical means of sex determination in psittacine birds. In Bonagura JD, Kirk RW, editors: *Current veterinary therapy*. XII, Philadelphia, 1995, Saunders, pp 275-1278.
78. Morishita TY: Common reproductive problems in the backyard chicken, *Proceedings of the Association of Avian Veterinarians* :465-467, 1995.
79. Lundholm CE, Mathion K: Effect of some metal compounds on Ca^{++} binding and Ca-Mg-ATPase activity of eggshell gland mucosa homogenate from the domestic fowl, *Acta Pharmacolg Toxicol* 59(5):410-415, 1986.
80. Johnson AL, Brake J: Zinc-induced molt: evidence for a direct inhibitory effect of granulosa cell steroidogenesis, *Poultry Sci* 71(1):161-167, 1992.
81. Thaxton JP, Paarkhurst CR: Abnormal mating behavior and reproductive dysfunction caused by mercury in Japanese quail, *Proc Soc Exp Biol Med* 144:252-255, 1973.
82. Yamamoto JT: Reproductive activity of force-paired cockatiels, *Auk* 106:86-93, 1989.
83. Dorrestein GM: Physiology of the urogenital system. In Altman RB et al, editors: *Avian medicine and surgery*, Philadelphia, 1997, Saunders, pp 622-625.
84. Rupley AE: Aviculture and obstetrics. In *Manual of avian practice*, Philadelphia, 1997, WB Saunders, p 483.
85. Stonebreaker R: Ratites. In Altman RB et al, editors: *Avian medicine and surgery*, Philadelphia, 1997, Saunders, p 934.
86. Wissman MA: Preparation for the breeding season, *Proceedings of the Association of Avian Veterinarians* :387-389, 1992.
87. Speer B: A clinical approach to psittacine infertility, *Proceedings of the Association of Avian Veterinarians* :173-187, 1991.
88. Tilly JL, Kowalski KI, Johnson AL: Stage of ovarian follicular development associated with the initiation of steroidogenic competence in avian granulosa cells, *Biol Reprod* 44(2):305-314, 1991.
89. Foidart A, Balthazart J: Sexual differentiation of brain and behavior in quail and zebra finches: studies with a new aromatase inhibitor, R7613, *J Steroid Biochem Molec Biol* 53:267-275, 1995.
90. Harding CF: Hormonal modulation of neurotransmitter function and behavior in male songbirds, *Poult Sci Rev* 4:261-273, 1992.
91. Mohan J, Moudgal RP, Singh NB: Effects of cyproterone acetate and testosterone treatments on some physical parameters, angiotensin-converting enzyme activity and fertilizing ability of spermatozoa of domestic cocks, *J Vet Med* 37(7):499-505, 1990.
92. Takahashi T et al: Effect of estradiol-17 beta and progesterone on arginine vasotocin in nonlaying hens, *Poultry Sci* 73(3):468-471, 1994.

93. Liel Y et al: Evidence that estrogens modulate activity and increase the number of 1,25-dihydroxyvitamin D receptors in osteoblast-like cells, *Endocrinology* 130(5):2597-2601, 1992.
94. Johnson PA, Wang SY: The molecular biology and endocrinology of inhibin in the domestic hen. In Sharp PJ, editor: *Avian endocrinology*, Bristol, 1993, Society for Endocrinology, pp 297-308.
95. VanMontfort D et al: Source of immunoreactive inhibin in the chicken ovary, *Biol Reprod* 47:977-983, 1992.
96. Lewis WM et al: Evidence for inhibin in roosters, *Society for the Study of Reproduction* 40(suppl 1):109, 1989.
97. Rivier C et al: Inhibin: role and secretion in the rat, *Recent Prog Horm Res* 46:231-259, 1990.
98. Follet BK, Davies DT, Gledhill B: Photoperiodic control of reproduction in Japanese quail: changes in gonadotropin secretion on the first day of induction and their pharmacological blockade, *J Endocrinol* 74:449-460, 1977.
99. Miller SC: Rapid activation of the medullary bone osteoclast cell surface by parathyroid hormone, *J Cell Biol* 76:615, 1978.
100. Thiede MA et al: Expression of the parathyroid hormone–related protein gene in the avian oviduct, *Endocrinology* 129(4):1958-1966, 1991.
101. Lightfoot TL: Clinical use and preliminary data of chorionic gonadotropin administration in psittacines, *Proceedings of the Association of Avian Veterinarians*, pp 303-306, 1996.
102. Myers SA, Millam JR, Halawani ME: Plasma LH and prolactin levels during the reproductive cycle of the cockatiel, *Gen Comp Endocrinol* 73:85-91, 1989.
103. Bikle DD et al: Prolactin but not growth hormone stimulates 1,25-dihydrovitamin D3 production by chick renal preparations in vitro, *Endocrinology* 107(1):81-84, 1980.
104. Skwarlo-Sonta K: Prolactin as an immunoregulating hormone in mammals and birds, *Immunol Lett* 33(2):105-122, 1992.
105. Asboth G et al: PGE_2 binding synthesis and distribution in hen oviduct, *Am J Physiol* 248:380-388, 1985.
106. Wechsieng E, Claeys M, Houvenaghel A: Biotransformation of arachidonic acid in the hen oviduct, *Arch Int Pharmacodyn Ther* 244:348-350, 1980.
107. Mauro LJ et al: Alterations in hypothalamic VIP-like immunoreactivity are associated with reproduction and prolactin release in the female turkey, *Endocrinology* 125:1795-1804, 1989.
108. Mougdal RP, Jagmohan, Panda JN: Effect of prolactin, epinephrine and serotonin on follicular atresia and phosphatase levels in the ovarian follicles of hens, *Indian J Poult Sci* 25(1):37-43, 1990.
109. Johnson AL, Tilly JL: Effects of vasoactive intestinal peptide on steroid secretion and plasminogen activator activity in granulosa cells of the hen, *Biol Reprod* 38(2):296-303, 1988.
110. Norman AW: The hormone-like action of 1,25 (OH) 2-cholecalciferol in the intestine, *Vitam Horm* 32:325, 1974.
111. Halloran BP: Is 1,25-dihydroxyvitamin D required for reproduction? *Proceedings of the Society for Experimental Biology and Medicine* 191(3):227-232, 1989.
112. Castillo L et al: Production of 1,25-dihydroxyvitamin D3 and formation of medullary bone in the egg-laying hen, *Endocrinology* 104(6):1598-1600, 1979.
113. Carpenter JW, Mashima TY, Rupiper DJ: *Exotic animal formulary*, Manhattan, Kan, 1996, Greystone.
114. Millam JR, Finney H: Leuprolide acetate can reversibly prevent egg laying in cockatiels, *Proceedings of the Association of Avian Veterinarians*, p 46, 1993.
115. Harrison GJ: Progesterone implants in cases of chronic egg laying, *Proceedings of the Association of Avian Veterinarians*, pp 6-10, 1989.

116. University of California Davis Formulary.
117. Plumb DC: *Veterinary drug handbook,* ed 2, Ames, Iowa, 1995, Iowa State University Press.
118. Riley MR et al: *Drugs: facts and comparison,* Grand Rapids, Mich, Lippincott, 1999.
119. Latimer KS, Greenacre CB: Adrenal carcinoma in a budgerigar, *J Avian Med Surg* 9(2):141-143, 1995.
120. Rae M: Endocrine disease in pet birds, *Semin Avian Exotic Pet Med* 4(1):32-38, 1995.
121. Westerhof I, Pellicaan CHP: Effects of different application routes of glucocorticoids on the pituitary-adrenocorticol axis in pigeons, *J Avian Med Surg* 9(3):175-181, 1995.
122. Hazelwood RL: Carbohydrate metabolism. In Sturkie PD, editor: *Avian physiology,* New York, 1986, Springer-Verlag, pp 303-325, 494-501.
123. Hazelwood RL: Pancreas and pineal. In Sturkie PD, editor: *Avian physiology,* New York, 1986, Springer-Verlag, pp 494-501.
124. Hazelwood RL: Pancreatic hormones, insulin/glucagon molar rations and somatostatin as determinants of avain carbohydrate metabolism, *J Exp Zool* 232:647-652, 1984.
125. Lothrop C et al: Miscellaneous diseases. In Harrison GJ, Harrison LR, editors: *Clinical avian medicine and surgery,* Philadelphia, 1986, Saunders.
126. Sitbon G: Diabetes in birds, *Horm Metab Res* 12(1):1-9, 1980.
127. Lumeij JT: Endocrinology. In Ritchie BW, Harrison GJ, Harrison LR, editors: *Avian medicine: principles and applications,* Lake Worth, Fla, 1994, Wingers, pp 602-604.
128. Nelson RW: Disorders of the endocrine pancreas. In Ettinger SJ, editor: *Textbook of veterinary internal medicine,* ed 3, Philadelphia, 1989, Saunders, pp 1676-1720.
129. Foltzer C et al: Influence of insulin and glucagon on secretion of growth hormone in growing ducks, *J Endocrinol* 91(2):189-196, 1981.
130. Nelson RW: Disorders of the endocrine pancreas. In Nelson RW, Couto CG, editors: *Essentials of small animal internal medicine,* St Louis, 1992, Mosby, pp 561-585.
131. Murphy J: Diabetes in toucans, *Proceedings of the Association of Avian Veterinarians,* pp 165-170, 1992.
132. Douglas M: Diabetes mellitus in a Toco toucan, *Mod Vet Pract* 62:293-295, 1981.
133. Appleby RC: Diabetes mellitus in a budgerigar, *Vet Rec* 185:652-653, 1984.
134. Hochleithner M: Biochemistries. In Ritchie BW, Harrison GJ, Harrison LR, editors: *Avian medicine: principles and applications,* Lake Worth, Fla, 1994, Wingers, p 235.
135. Pendleton EA, Rogers D, Epple A: Diabetes mellitus in a red-tailed hawk, *Avian Pathol* 22:631-635, 1993.
136. Candeletta SC et al: Diabetes mellitus associated with chronic lymphocytic pancreatitis in an African grey parrot, *J Assoc Avian Vet* 7(1):39-43, 1993.
137. Altman RB, Kirmayer BA: Diabetes mellitus in the avian species, *J Am Anim Hosp Assoc* 12:531-532, 1976.
138. Rosskopf WJ, Woerpel RW: Avian obstetrical medicine, *Proceedings of the Association of Avian Veterinarians,* pp 323-336, 1993.
139. Kahler J: Sandostatin treatment for diabetes mellitus in a sulfur-breasted toucan, *Proceedings of the Association of Avian Veterinarians,* pp 269-273, 1994.
140. Ryan CP, Walder EJ, Howard EB: Diabetes mellitus and islet carcinoma in a parakeet, *J Am Anim Hosp Assoc* 18:139-142, 1982.
141. Minick MC, Duke GE: Simultaneous determinations of plasma insulin, glucose and glucagon in fasting, previously stressed but convalescing bald eagles, *Comp Biochem Physiol* 99(3):307-311, 1991.

142. Sitbon G et al: Endocrine factors in intermediary metabolism with special reference to pancreatic hormones. In *Avian endocrinology,* New York, 1989, Academic Press, pp 251-270.
143. Assenmacher I: The peripheral endocrine glands. In Farner DS, King JR, editors: *Avian biology,* vol 3, New York, 1973, Academic Press, pp 236-247.
144. Rae M: Endocrine disease in pet birds, *Semin Avian Exotic Pet Med* 4(1):36-37, 1995.
145. Bonda M: Plasma glucagon, serum insulin, and serum amylase levels in normal and a hyperglycemic macaw, *Proceedings of the Association of Avian Veterinarians,* pp 77-88, 1996.
146. King AS, McLelland J: *Birds: their structure and function,* Orlando, 1984, Bailliere Tindall, pp 210-211, 252-253.
147. Millam JR: Reproductive physiology. In Altman RB et al, editors: *Avian medicine and surgery,* Philadelphia, 1997, Saunders, pp 15-17.

17

Problems of the Bill and Oropharynx

Glenn H. Olsen

I. The Bill

A. Anatomy and Physiology
1. The bill, or beak, is described anatomically as the rostrum. The rostrum is composed of bone, keratin, dermis, vasculature tissues, and germinative tissues, similar to a horse's hoof. In this discussion, the terms *bill*, *beak*, and *rostrum* are used interchangeably.
2. The bones of the maxillary jaw and its associated horny covering make up the maxillary rostrum; the bones of the lower jaw form the mandibular rostrum.
3. The psittacine bill, or rhamphotheca (upper and lower), articulates with the skull through kinetic joints.
4. Psittacine and anseriform bills have well-developed mechanoreceptor nerve endings.[1] Taste is transmitted from the bill areas by cranial nerve VII, and sensation to the maxillary rhamphotheca is transmitted by the ophthalmic and maxillary divisions of cranial nerve V. The mandibular rhamphotheca is supplied by the mandibular division of cranial nerve V.
5. The growth rate of the bill varies, depending on the species.
6. Bills come in a variety of types, each serving a different purpose[2]:
    a. Stout, conical bills (finches, canaries) are used for seed cracking.
    b. Parrots have stout, hooked bills for cracking the seeds of fruits. Some have ridges on the bill edge, which provide additional help in seed cracking.
    c. Chisel-like bills (woodpeckers) are used to penetrate wood when searching for invertebrates.
    d. Long, thin bills (hummingbirds) are used to enter flowers to obtain nectar.
    e. Some insectivorous birds (e.g., warblers) have short, slender, pointed bills; others have wide mouths for catching insects on the wing.

f. Birds that probe for food (cranes, many shorebirds) have long, straight bills.
g. Some birds are filter feeders (some ducks, flamingos), which have a series of plates (lamellae) in their bills.
h. Raptors that eat flesh (hawks, eagles) have hooked bills for tearing.
i. Modifications for eating fish include spearlike bills (herons), serrated bills (mergansers), pouchlike bills for scooping up fish (pelicans), and elongated mandibles for skimming the water's surface (skimmers).
j. Many species use their bills for defense. The bills of large psittacines, herons, and cranes pose a danger to human handlers; in one case, a crane killed a man by driving its beak into his skull.

B. Beak Disorders
1. Bacterial diseases
   a. Any number of bacterial organisms can enter the vascular or germinative layers of the bill, causing infection and deformity.
   b. Chronic rhinorrhea from an upper respiratory infection can result in a groove from the nares down the beak on one or both sides (Fig. 17-1).
2. Mycotic diseases
   a. *Candida albicans,* a common pathogen of the bill, causes erosive lesions (Fig. 17-2).
   b. Trichothecene mycotoxins can cause necrotic lesions and beak deformity.[3]

Fig. 17-1 Chronic rhinorrhea has produced a groove in the beak of this budgerigar.

3. Viruses
   a. The psittacine beak and feather disease (PBFD) virus invades the germinative layers of the skin, including the bill, resulting in deformities (Fig. 17-3). Some birds develop only bill lesions and not the typical feather lesions of PBFD.
   b. Avian pox is a dry pox that can result in lesions of the bill. These lesions typically are crusty and are found in the area of the cere.
4. Parasites (see Chapter 20)
   a. *Knemidokoptes* mites can cause severe bill deformity in psittacines, especially budgerigars (Fig. 17-4). On close inspection, the bill and the area around it have a honeycombed appearance.
   b. *Oxyspirura* parasites cause bill deformities in cranes.
   c. Trichomoniasis
5. Trauma
   a. Punctures, lacerations, fractures, and avulsion of the bill are all possible outcomes of trauma.

Fig. 17-2 Erosive lesion (arrow) in a cockatoo caused by *Candida albicans*.

362 *Problems of the Bill and Oropharynx*

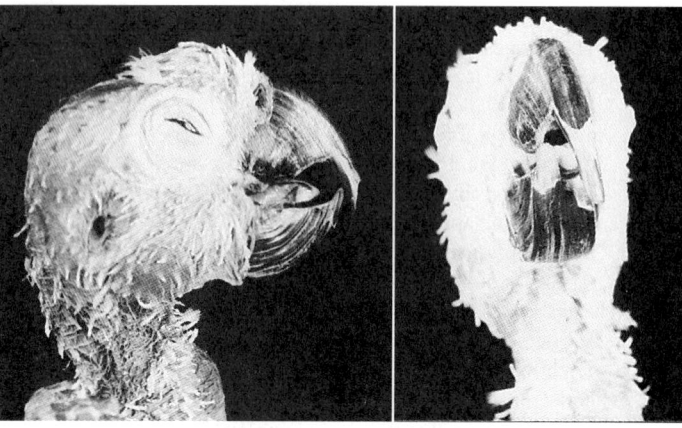

Fig. 17-3 Deformity of the beak in a cockatoo diagnosed with psittacine beak and feather disease.

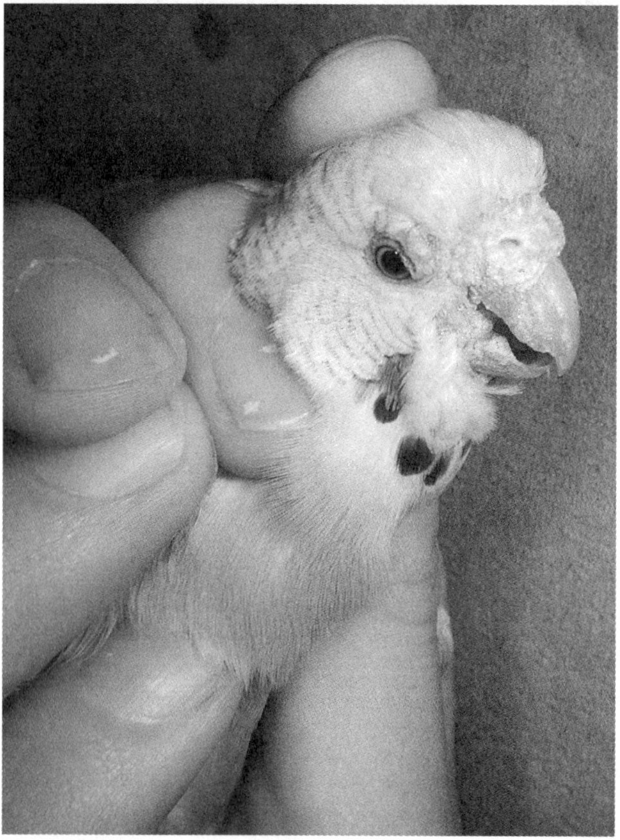

Fig. 17-4 A budgerigar with severe beak deformity caused by infestation with *Knemidokoptes* mites.

b. A traumatic wound can become secondarily infected, or the trauma may damage the germinative layers, causing the subsequent growth of a deformed bill.
  c. Traumatic fractures can occur if the bill is caught in an object or part of the cage, or from fighting with other birds. Bill trauma that occurs secondary to cage mate aggression is common in cockatoos.
  d. Bill trauma also may occur iatrogenically, as a result of zealous use of a mouth speculum or from trimming of the bill.
6. Nutrition
  a. Severe nutritional secondary hyperparathyroidism results in a pliable bill (rubber beak), which is seen frequently in pigeons, doves, and cockatiels.[4]
  b. In galliforms, embryonic deficiencies of certain B vitamins (i.e., biotin, folic acid, and pantothenic acid) can result in a deformed upper bill.
  c. Overgrowth of the upper beak in budgerigars has been associated with malnutrition and liver disease.[1,4]

C. Treatment of Bill Disorders

1. Bill injuries
  a. Initially, any hemorrhage is controlled, the bird is started on systemic antibiotics, and the patient's condition is stabilized. Hemorrhage can be controlled with ferric subsulfate, Nexaband, or electrocautery (using anesthesia), depending on the severity of the injury. *Silver nitrate is not used because it can cause esophageal burns.*
  b. Subsequently, under anesthesia, the wounds can be cleaned and debrided, and topical antibiotic therapy can be initiated. It is important to remember that the maxillary and mandibular bills are connected to the infraorbital sinus, and infections and hemorrhage can extend into the sinus.
2. Beak fracture fixation
  a. The initial therapy is as described above in "a" under Bill injuries.
  b. The goals of fracture repair are to stabilize and realign the beak. Various materials have been used, including hypodermic needles (small birds), wire, pins, screws, plates, and various cements.
  c. One common method of repairing fractured beaks involves the use of medical cyanoacrylics, such as the Cyno-Veneer kit (Ellman International Manufacturing, Inc., Hewlett, New York) or methoxycyanoacrylate (Sinomet 9000, 9010, and 9020, Henkel Adhesive Co., Kankakee, Illinois).[5,6]
    (1) The injury site is cleaned and flushed with dilute povidone-iodine (1%) or chlorhexidine (2%).
    (2) Additional cleaning of the fracture site may be done with a dental drill.
    (3) To give the acrylic better holding power on the bill, several small grooves are drilled partway into the bill above and below the fracture site.

(4) The two-part acrylic (after mixing) is applied to the entire bill area; it is important to fill any of the small, drilled grooves with the acrylic.
(5) After the beak is checked for proper alignment, a coat of quick-setting solution is added to harden the acrylic.
(6) Small plates, pins, or wire mesh may be placed over the initial acrylic layer and covered with a second layer of acrylic to further stabilize the fracture.
(7) Tints are available for the cyanoacrylic (Cyno-Veneer kit) to match bill color. Also, a retarder is available to slow the setting time when working with larger repairs.
(8) If the bird has associated soft tissue injuries, the skin should now be reapposed and held in place by sutures or tissue glue.
(9) In psittacines, fractures of the maxilla are harder to repair because of the kinetic joint between the upper bill and the skull.

3. Repair of bill deformities
   a. Bill deformities can result from any of the etiologies listed above.
   b. Overgrowth of the upper beak may require only frequent trimming.
   c. Bill deformities are thought to be related to liver disease. However, correcting the nutritional problems or treating the liver disease does not eliminate the need to continue trimming the upper bill.[4]

4. Scissors beak
   a. Scissors beak is a condition in which the lower bill is overgrown or deviated to one side, forcing the upper bill also to deviate.
   b. The condition usually is associated with damage to germinal beds in the embryo or neonate.[7]
   c. Scissors beak is not caused by feeding a neonate only from one side of the beak[8]; however, when hand-feeding, it is important to hold the bill properly. Equal pressure should be applied to each side of the bill, and the bill should not be held so firmly as to cause a trauma-induced problem. It also is important to clean the bill carefully after each feeding to reduce the chance of infection.

5. Correction of a severely deviated lower bill[7]
   a. The keratin layer is grooved with a motor tool (Dremel) without causing hemorrhage.
   b. The grooves are cleaned and disinfected; dilute povidone-iodine (1%) works well.
   c. A light coat of acrylic is applied.
   d. Nylon or stainless steel dental screen is put on over the acrylic, and a protrusion of the screen is created that will push against the bill tip to bring it into alignment.
   e. The screen is covered with more dental acrylic.
   f. The acrylic is shaped with the motor tool.

6. Correction of a severely deviated upper bill (bragnathism)
   a. This technique involves the placement of two KE wires or small pins; the first is placed at a right angle to the bill into the frontal bone just caudal to the maxilla joint and nares.

b. The second KE wire or pin is placed at a right angle to the bill at the point where the bill deviation is most severe. Small hooks can be formed on the protruding pin ends.
   c. Dental acrylic can be added around one or both pins for additional support.
   d. A dental rubber band is used to connect the two pins; the tension applied by the rubber band slowly pulls the beak into proper alignment.
   e. When the beak has returned to proper alignment, the rubber band is removed, but the pins are left in place until it is certain that a second application of tension will not be required.
   f. When the beak alignment is correct and has not shifted, the acrylic and pins are removed.

II. The Oropharynx

A. Anatomy
1. In birds, the oral and pharyngeal cavities are combined, forming an oropharynx.
2. The dorsal wall of the oropharynx consists only of a hard palate with a central opening, the choana. The choana is the distal extent of the nasal cavity, which is continuous with the infraorbital sinus and proximally begins at the nares. Birds do not have a soft palate or epiglottis.
3. Caudal to the choana lies the infundibular cleft, the common opening for the auditory tubes.
4. During swallowing, muscular action closes the walls of the choanal slit, which is followed by closure of the infundibular cleft and the glottis as the food bolus travels caudally.
5. Mucus-secreting salivary tissue lies along the walls of the oropharynx.
6. The avian tongue has a variety of shapes and functions, depending on the diet and ecologic niche of the species. Psittacine tongues are well muscled and used to help extract seed from a shell or husk. Fish-eating birds (herons, some cranes, and some waterfowl) have tongues with serrations that help hold the fish.
7. In some passerine species (e.g., crows), the oral mucosa in neonates is brightly colored; this acts as a feeding stimulus to the parent birds.

B. Diseases and Treatments
1. Oropharyngeal abscesses
   a. Oropharyngeal abscesses often develop from wounds or as a result of hypovitaminosis A (see Chapter 18). Hypovitaminosis A often is associated with sheets of basophilic cells seen on a choanal Gram's stain.
   b. The patient is pretreated with antibiotics and vitamin A (if a deficiency is suspected).
   c. The abscess material and the capsule surrounding it must be surgically removed. However, the hard palate is extremely vas-

cular, as are the other parts of the oropharynx. For this reason, hemorrhage is of utmost concern. Surgery may need to be performed in stages, combined with magnification and the use of microsurgical techniques. Radiosurgery is important to control hemorrhage.

   d. If the abscess is only lanced and drained, debris often remains attached to or infiltrating the tissue of the capsule; this remnant serves as the nidus of another abscess. Cytologic impression smears, cultures, and biopsy of the edges of the granuloma are important to diagnose the underlying cause.
2. Oropharyngeal papillomas
   a. Oropharyngeal papillomas are uncommon in birds except in macaws.
   b. Papillomas of the oral cavity can be safely removed using radiosurgery to control hemorrhage; however, they may recur.
   c. If the papillomas extend down into the crop or proventriculus, removal is not safely possible, and the prognosis for the patient is poor.
3. Granulomas
   a. Granulomas caused by *Mycobacterium*, other bacteria, or fungal pathogens are common.
   b. The diagnosis can be made by needle aspiration, cytologic studies (Gram's, Ziehl-Neelsen, or acid-fast stains), and cultures (aerobic, anaerobic, and fungal).
   c. Surgical excision coupled with systemic antibiotic therapy is the treatment of choice.
4. Poxvirus
   a. Poxvirus manifests as proliferative caseous lesions in the oropharynx that sometimes extend down the esophagus (diphtheritic form) or as nodular, crusty lesions, often ulcerated on the face, legs, and feet (dry or cutaneous form).
   b. Diagnosis is possible by cytologic examination, which will show large, eosinophilic, intracytoplasmic inclusion bodies (Bollinger's bodies, as seen microscopically with hematoxylin-eosin, Wright's, or Gimenez stains).
   c. Control methods include:
      (1) Screening outdoor aviaries to prevent mosquitoes from entering
      (2) Cleaning and disinfecting all hand-rearing feeding instruments, feed bowls, and other such equipment
      (3) Isolating infected birds from others, and feeding and handling the infected birds last
      (4) Avoiding situations that lead to wounds (poxvirus requires a break or alteration in the epithelium to enter the body)
   d. Treatment usually is supportive. Vitamin A is used in parrots, because hypovitaminosis A alters the epithelium so that the poxvirus can invade. Antibiotics and antifungals are indicated to treat secondary invaders. Birds that survive develop natural immunity and can develop lifelong protection.

e. Commercial vaccines are available for pigeons, doves, and canaries (Poximune-C, Biomune, Lenexa, Kansas). The poxviruses can be somewhat species specific, and the use of these vaccines in other avian species may not be efficacious.[9]
5. Pigeon herpesvirus (Smadels' disease)[1]
   a. Infection can result in diphtheritic membranes in the oropharynx and esophagus.
   b. Clinical signs include rhinitis, conjunctivitis, dyspnea, depression, and death. The disease is primarily a problem in squabs.
   c. The diagnosis is made by histologic identification of basophilic and eosinophilic intranuclear inclusion bodies.[10]
   d. No treatment or vaccine is available.
6. Parasites
   a. *Capillaria* spp. infest the tongue, oropharynx, and esophagus of raptors, psittacines, and gallinaceous birds. Lesions can appear as hemorrhagic inflammation at the beak commissures or as diphtheritic membranes of the oropharynx and tongue. The diagnosis is made by finding eggs in washings of the oropharynx, mouth, or esophagus.
   b. *Trichomonas* spp. form ulcerative, proliferative lesions in the oropharynx, esophagus, or crop. Raptors, pigeons, doves, psittacines, and passerines all are susceptible. The diagnosis is made by finding the organism in washes or wet mounts.
7. Vitamin A deficiency
   a. Hypovitaminosis A results in a metaplasia of all the salivary glands, including the submandibular and lingual salivary tissues. This can be observed as sheets of basophilic cells on a choanal Gram's stain.
   b. Characteristically, hypovitaminosis A is seen in psittacines fed only a seed diet.
   c. Treatment includes injection of vitamin A and an improved diet, such as a balanced, pelleted ration.
8. Fungal infections
   a. Fungal infections are caused by *Candida albicans* (or, more rarely, by *C. parapsilosis*) or *Aspergillus* spp.[11]
   b. Localized infections of the mouth or beak may appear as a white, caseous exudate.
   c. Infections respond to systemic antifungals (nystatin, fluconazole, itraconazole) plus debridement of lesions and topical application of 3% amphotericin B ointment, clotrimazole, or nystatin ointment. Care must be taken when using an azole antifungal agent or a flush preparation of amphotericin B with African gray parrots (likely toxic, may result in death).
9. Stomatitis
   a. General inflammation of the mouth is seen with burns caused by eating hot food, with chemical burns from chewing on a silver nitrate stick, and after ingestion of other caustic substances.
   b. Treatment is supportive: fluids, supplemental feedings if needed, and antibiotics for secondary infection. A pharyngos-

tomy tube may need to be placed temporarily for providing nutritional support.
10. Foreign bodies
   a. Thread or string wrapped around the tongue is sometimes seen, especially with small passerines (finches and canaries).
   b. Splinters from cages or toys may become embedded in the oropharynx and must be removed.
   c. Seed or seed hulls can lodge in the oropharyngeal mucosa and may cause a granuloma to form.

REFERENCES

1. Lumeij JT: Gastroenterology. In Ritchie BW, Harrison GJ, Harrison LR, editors: *Avian medicine: principles and applications,* Lake Worth, Fla, 1994, Wingers, pp 483-521.
2. King AS, McLelland J: *Birds: their structure and function,* London, 1984, Bailliere Tindall.
3. Olsen GH et al: Mycotoxin-induced disease in captive whooping cranes *(Grus americana)* and sandhill cranes *(Grus canadensis), J Zoo Wildlife Med* 26:569-576, 1995.
4. Harrison GJ: Disorders of the integument. In Harrison GJ, Harrison LR, editors: *Clinical avian medicine and surgery,* Philadelphia, 1986, Saunders, pp 509-524.
5. Harrison GJ: Surgical instrumentation and special techniques. In Harrison GJ, Harrison LR, editors: *Clinical avian medicine and surgery,* Philadelphia, 1986, Saunders, pp 560-567.
6. Clipsham R: Rhamphorthotics and surgical corrections of maxillofacial defects, *Semin Avian Exotic Pet Med* 3:92-99, 1994.
7. Martin HD, Ritchie BW: Orthopedic surgical techniques. In Ritchie BW, Harrison GJ, Harrison LR, editors: *Avian medicine: principles and applications,* Lake Worth, Fla, 1994, Wingers, pp 1137-1169.
8. Clipsham R: Surgical correction of beaks: a practical lab, *Proceedings of the Association of Avian Veterinarians,* pp 325-333, 1990.
9. Phalen DN: Viruses. In Altman RB et al, editors: *Avian medicine and surgery,* Philadelphia, 1997, Saunders, pp 281-322.
10. Hooimeiger J, Dorrestein GM: Pigeons and doves. In Altman RB et al, editors: *Avian medicine and surgery,* Philadelphia, 1997, Saunders, pp 886-909.
11. Goodman GJ, Widenmeyer JC: Systemic *Candida parapsilosis* in a 20-year-old blue-fronted Amazon, *Proceedings of the Association of Avian Veterinarians,* pp 105-119, 1986.

18

Avian Nutrition

Blake Hawley, Tracey Ritzman, and Thomas M. Edling

Nutritional disorders rarely present to the practitioner as a single deficiency or excess of a particular nutrient. More often, chronic malnutrition occurs. Birds frequently present with a wide array of clinical manifestations. To assess a bird's nutritional state, obtain a thorough dietary history and examine the dietary management practices of the owner.

Frequently, a close examination of actual foodstuffs and feeding practices, combined with a knowledge of the patient's food preferences, easily elucidate dietary deficiencies or excesses. Food preference and actual foodstuff consumption play a major role in assessing a bird's history and predisposition to nutritional disorders.

The age of the bird has some impact on its potential for nutritional disorders. Neonates may exhibit acute and severe manifestations of nutritional imbalances. This fact is related to the rapid growth rate of neonates, and higher demand for nutrients. Growing birds may gain 10% to 20% of their body weight per day. Once adult size and weight are achieved, however, the body's relative need for nutrients declines, and subtle chronic lesions are more commonly observed in adult and geriatric patients.

I. Nutrition of Young Birds (Neonate to 1 Year)

A. General Influences
1. Begin before fertilization of the egg
2. Quality and composition of the diet of the parent bird (especially if bird is reared by parents)
   a. Influence egg quality
   b. Influence hatchability
   c. Influence chick health

B. Causes of Nutritional Disorders
1. Deficiencies, excesses, or imbalances in nutrient components
2. Feeding techniques
3. Diet fed
4. Amount of food given

5. Frequency of feedings
6. Temperature of food

C. Diet History of Parents (see section on Adult Nutrition p. 375)
1. Seeds only
2. Seeds supplemented with fresh fruits and vegetables
3. Seeds supplemented with nutrient pellets
4. Table foods
5. Manufactured diets (as a total or percentage of total)

D. History of the Young Bird
1. Environment, which affects bioavailability of nutrients
2. Management
    a. Natural incubation
        (1) What are the parents' conditions?
        (2) What is the condition of the nest box?
        (3) How are the birds fed?
    b. Artificial incubation
        (1) Age when egg or young is pulled
    c. Hatching difficulties/Dead in shell
        (1) Deficiencies of most vitamins and minerals can lead to embryonic death if severe.
        (2) Riboflavin (vitamin B_2) deficiency in poultry has been described and may result in embryos dead in the shell, with edema and abnormal down.[1]
3. Nutrition of neonates
    a. Hatch to 3 days of age
        (1) The yolk sac provides the primary nutrition.[2]
        (2) Caloric/nutrient dilution of formula is expected.
        (3) Water and water-soluble nutrients are critical.
        (4) Electrolyte solutions should not be used.
        (5) Baby must develop adequate feeding response.
    b. 3 days to weaning
        (1) Examine the feeding charts and growth curves.
        (2) Babies should gain weight every day until the weaning process begins.
    c. Weaning
        (1) Weight loss often occurs at this time (5% to 10% body weight).
        (2) Bird is fully feathered, and at or above adult body weight initially.
        (3) Exploratory behavior begins—new food types are investigated by the bird.
    d. Adolescent to 1 year of age
        (1) Some growth still occurring
        (2) Ingredient selection by bird becomes a factor in its nutrition.
4. Diet
    a. Hand-fed vs. parent-raised
    b. Nutritional needs vary by species of bird

c. Evaluate by physical examination, weight gain, and condition of feathers
d. Commercial hand-feeding formula
   (1) Ensure diet is nutrient balanced.
   (2) Probiotics and enzymes aid digestion according to manufacturers.[3]
   (3) A lack of enzymatic activity may result in decreases in growth.[4]
   (4) Product quality control and consistency should be guaranteed by manufacturer.
   (5) Customer support or product information should be available via toll-free number.
e. Homemade hand-feeding formula
   (1) Proper nutrient balance and formula consistency difficult to achieve.[5,6]
   (2) Variation can occur between batches of formula prepared.
   (3) Nutrient stability is questionable without antioxidant or preservatives.[7]
   (4) Preparation time is high.
f. Weaning foods
   (1) Commercial products are available.
   (2) Birds should not be weaned onto seed-based diets.
   (3) Formulated foods can be mixed with fruit juices, mash/soft foods.
   (4) Fruits and vegetables can be added in limited quantities.
g. Water quality
   (1) Check source/filtration system.
   (2) Bacterial contamination, especially *Pseudomonas,* is common.[8]
   (3) Well systems may contain high mineral contents, as with some bottled/natural well water.
5. Feeding schedule
   a. Frequency of feedings: Crop should completely empty before next feeding
   b. Temperature of feed: Optimum is between 95° F and 100°F
   c. Amount—Percent body weight (10% to 20%) per feeding
6. Feeding utensils
   a. Syringe
      (1) Optimal method
      (2) Easy to measure amount fed
      (3) Safe
      (4) Sterilized between feedings or discarded
   b. Spoon: Can be used effectively, frequent spillage
   c. Feeding tube
      (1) High possibility of pharyngeal/esophageal trauma
      (2) Misplacement of feeding tube into trachea results in aspiration and possible death of bird.
7. Preventive health program and veterinary involvement[9]
   a. Physical examination
   b. Cultures—cloacal, choanal

c. Hematology/diagnostics
d. Vaccination
e. Gender determination

E. Determination of Nutritional Disorders
1. Based on physical examination and evaluation of organ systems
2. The considerations listed include only nutritional problems for developing a differential diagnosis list.
3. Skin/integument
    a. Normal: Pink, or pigmented (based on species); texture should be smooth and supple
    b. Red or dry, flaking skin result of inadequate water intake, formula fed too dry, dehydration, malnutrition[10]
    c. Pale, white skin result of anemia, low iron intake
    d. Choana normal: sharp papillae, projecting inward
    e. Blunted papillae result of vitamin A deficiency[11,12]
    f. Scales on plantar surface of feet—biotin deficiency
    g. Feathering: Color variation, abnormal loss, disturbed formation, odor, stress bars, molting pattern
        (1) The first feather coat of the neonate is natal down.
            (a) Different from adult feathers
            (b) Lack barbules on the central tips of the barbs
        (2) Natal down is eventually replaced by a set of downy or vaned feathers.
            (a) The length of time until adult plumage is reached varies by species.[2]
            (b) The rate at which adult feathers develop has been regulated, however, by improved nutrition in birds.[13]
        (3) Stress bars: Translucent lines that cross the feather shaft in 90-degree angles on the barbs and barbules.
            (a) If they extend through the shaft, it is believed they occur with the release of corticosteroids and are a result of dysfunctional growth at the epidermal collar of the developing feather.
            (b) Associated with nutritional deficiencies, especially methionine, but also hypovitaminosis D
            (c) Other etiologies involve illness and systemic disease.
            (d) Observing the presence of stress bars can give an indirect health and nutritional history of the bird during the last feather growth.
            (e) A few bars can be normal.
            (f) Disrupted feeding schedules may cause these.[9]
            (g) If they become chronic or numerous, then a thorough examination is warranted.[9]
            (h) When they extend only halfway through the shaft, it may signify aggressive neonatal preening.
        (4) Poor feather growth and breakage (along with dermatitis, perosis, pododermatitis)[1,10,14,15]

(a) Pantothenic acid deficiency in chicks also results in lack of contour feathers in growing cockatiels.[9]
(b) Biotin deficiency has similar gross effects to pantothenic acid[1,16] and folic acid.
(5) Curled wing tips result of arginine deficiency in chicks
(6) Feather achromatosis
(a) Choline and riboflavin deficiencies shown in young cockatiels, whereas a similar achromia can be caused by lysine deficiencies in chicken, turkey, and quail.[13]
(b) Iron, folic acid, and copper deficiencies are also implicated.
h. Nails: Evaluate shape, texture, strength
4. Musculoskeletal
a. Note size and proportion of chick in comparison to normal
b. Size and shape of head: Domed skull and head large in proportion to body indicate stunting because of poor or inadequate supply of nutrients.
c. Long bone formation, strength, symmetry: Rickets, folding fractures indicate inadequate calcium or poor calcium/available phosphorus ratio, inadequate vitamin D_3.[12,17,18]
d. Weakness, leg muscle atrophy, toes curled inward are associated with riboflavin deficiency
(1) On necropsy: Peripheral neuritis, Schwann cell proliferation, degeneration at neuromuscular endplates[19]
e. Splayed legs
(1) Nutrition (excess vitamin A, excess vitamin E, manganese deficiency, copper deficiency[20]) is only a component of multifactorial causes
(2) Other causes include management (e.g., chick maintained on slick surface)
(3) Treatment
(a) Evaluate and correct dietary imbalances
(b) Place chick in container that places legs underneath body, reducing the possibility of the legs slipping out from under the chick
5. Ophthalmic
a. Eyes: Development of palpebral fissures, time of opening should be noted on record
b. Infection susceptibility increases with hypovitaminosis A.[21]
6. Gastrointestinal: Digestive capability
a. Yolk sac
(1) Size at hatch
(2) External vs. internal
b. Umbilicus
(1) Open vs. closed at hatch
c. Mouth
(1) Oral paralysis in cockatiels occurs with vitamin E/selenium deficiencies, often misdiagnosed as "lockjaw" or *Clostridium* contamination of foodstuffs[22,23]

(2) White plaques in mouth or swelling of salivary glands—hypovitaminosis[12,24-26] or mycotoxicosis (T2 toxin)
  d. Crop: Size, shape, motility, wall thickness
      (1) Ingluvies (crop) stasis or malfunction, trauma, foreign body, regurgitation, aerophagia, change in wall thickness: multifactorial
      (2) Causes of crop impaction
          (a) Nutritional
              (i) High-fiber diets
              (ii) Foreign material ingestion
              (iii) Excess grit consumption
              (iv) Crop liths
          (b) Other
              (i) Cold food
              (ii) Cold environment
              (iii) Infrequent feedings of large amounts of food
  e. Proventriculus: Stasis or malfunction, enlargement, foreign body ingestion
      (1) Degeneration of ventricular musculature has been associated with vitamin E/selenium[20,27]
  f. Ventriculus: Malfunction, foreign body, grit impaction
      (1) Psittacine birds do not require grit in the diet
  g. Small intestine: maldigestion/malabsorption, enlargement
  h. Cloaca: Tenesmus, impaction, flatulence
  i. Droppings: Color and consistency, presence of whole seeds is abnormal, bacterial flora
7. Renal
  a. Abnormally colored urates (green or yellow)
      (1) Must determine if pathologic or dietary (B vitamin supplementation, food dyes, berries, fruits)
      (2) Most frequently indicate bacterial/viral infection
  b. Polydipsia, polyuria: Nutritional causes include the following:
      (1) Visceral gout as a result of oversupplementation with minerals,[28,29] hypervitaminosis D[30,31]
      (2) May occur in birds fed moist foods such as fruit, vegetables, or formula with too high water content
      (3) Excessive dietary sodium, magnesium
      (4) Calcium deficiency
      (5) High dietary fiber
  c. Black droppings: Lack of food, ulceration, or trauma to upper GI tract
  d. Red urine
      (1) Check with dipstick for occult blood.
      (2) Cytologic evaluation for red blood cells is important.
      (3) If no blood is found, consider idiopathic of caused by animal-based protein, which is not harmful and is usually transient.[32]
8. Hepatic
  a. Hepatomegaly, respiratory distress, hepatic encephalopathy

b. Hepatic lipidosis: Result of formula too high in fat, too frequent feedings, greater than 20% body weight per feeding (cockatoos are especially prone to this) and decreased biotin.
c. Hemochromatosis: Excessive levels of iron in the diet, combined with hereditary or physiologic processes not fully understood.

9. Central nervous system
    a. Seizures, ascending paralysis, opisthotonos, adrenal hypertrophy: Myelin degeneration of peripheral nerves and interruption of pathway within astrocytes, resulting from thiamine deficiency or ingestion of fish contaminated with thiaminase[78]
        (1) Treatment: Parenteral or oral B_1, response usually rapid, presumptive diagnosis based on response
    b. Ataxia, heat tilt, circling: Encephalomalacia caused by vitamin E/selenium deficiency[23,32,33]
        (1) Mechanism: Interrupts vascular integrity, destruction of erythrocytes, capillary hemorrhage, prominent edema, degeneration, and necrosis of neurons, cerebellar demyelination
    c. Cockatiel paralysis syndrome (weakness of legs, wings, and beak): Responds to vitamin E/selenium therapy, secondary to malabsorption
    d. Paralysis of neck muscles, wing tremors (accompanied by anemia and granulocytopenia)–folic acid deficiency[34]

II. Nutrition of Adult Birds (Older Than 1 Year)

To understand the nutritional status of an adult bird, evaluate the bird's diet. In addition to the overall diet evaluation, estimate the actual foods consumed. Different diets may work well for some situations, but not all, so consider how each diet can fit into the caretaker's lifestyle as well.

A. Diet History
1. Seeds only
    a. Considered the "natural diet for birds," but the commercially available seeds are not the common seeds eaten in the wild.
    b. Seed only diet is not adequate for pet birds
    c. Deficiencies
        (1) Vitamins: Vitamin A, vitamin D_3, riboflavin,[12,19] vitamin B_{12}, occasionally vitamin E[12,17,35,36]
        (2) Minerals: Calcium, sodium, iodine, iron, copper, zinc, manganese, and selenium[12,37]
        (3) Amino acids: Lysine and methionine, possibly arginine[12,17,38-40]
    d. Excesses
        (1) Fats[36]
    e. Imbalances
        (1) Amino acids[38-40]
        (2) Calcium/phosporus ratio[36]
        (3) Vitamin E, selenium

2. Seeds supplemented with fresh fruits and vegetables
   a. Primary method owners use to supplement seed-based diets
   b. Nutritional adequacy depends on types of fruits and vegetables fed
   c. Fruit products
      (1) Most fruits (fleshy portion) consist largely of water, sugar, and fiber.
      (2) Generally low in protein, fat, and minerals
      (3) Poor source of amino acids
      (4) Temperate climate fruit (apples, grapes, bananas, oranges, other citrus fruits) is a particularly poor source of nutrients
      (5) Certain tropical fruits (coconut, papaya, palm fruits) are high in fat and have low to moderate protein levels.
      (6) High fiber levels of fruits can decrease biotin availability[16]
   d. Vegetable origin products
      (1) Oil seeds: High protein (20% to 30%), high fat (30% to 50%); examples: sunflower, safflower, hemp, flax, and niger seed
      (2) High-carbohydrate seeds and cereal grains: Moderate protein (8% to 15%), low fat (2% to 5%); examples: millet, canary grass seed, corn, wheat, oats, barley, rice
      (3) High-fat legumes: Characterized by high protein (25% to 35%) and high fat (20% to 50%); examples: peanuts, soybeans
      (4) Low-fat legumes: High protein (20% to 25%), low fat (1% to 3%); examples: peas, beans
      (5) Pigmented, deeply colored vegetables are high in carotenoid pigments that can serve as precursors for vitamin A (namely, beta-carotene).
         (a) In general, the more pigmented the vegetable, the richer the source of total nutrients.
         (b) On a dry matter basis, some have moderate protein levels and may be good vitamin and mineral sources.
         (c) All are low in fat and high in fiber; examples of deeply pigmented vegetables: carrots, beets, sweet potatoes
      (6) Very light, pale vegetables are poor sources of nutrients, because they comprise mainly water and fiber; examples: lettuce, celery
   e. Deficiencies of these fruit- and vegetable-supplemented diets
      (1) Vitamins: Vitamin E (pale vegetables, fruits), vitamin D_3, riboflavin, vitamin B_{12}, occasionally vitamin E[12,35]
      (2) Minerals: Calcium, sodium, iodine, iron, copper, zinc, manganese, and selenium (depending on soils)[12,37]
      (3) Amino acids: Lysine and methionine, possibly arginine[38-40]
   f. Excesses of this diet: Sugars, fiber
   g. Imbalances: Calcium/phosphorus ratio[36]
3. Seeds supplemented with nutrients
   a. Type of supplementation is the most critical element of this diet type
      (1) Powder-coated seeds: Ineffective because powder falls to cage bottom when seed hull is discarded; nearly impossible to coat dehulled seeds effectively

- (2) Vitamin/mineral premixes for water: Ineffective because of dosing difficulties, and rapid breakdown of nutrients in aqueous environment
- (3) Nutrient pellet mixed into seed mixture: Most effective method of supplementing diet, because all missing nutrients can be placed into pellet, but requires that bird actually consume the pellet.
  b. Failure to consume supplement results in same problems as an all-seed diet.
4. Table foods
  a. Broad classification of foodstuffs makes analysis of group impossible.
  b. Many people do not routinely eat healthy diets, and many diets of industrialized nations are high in fat and sodium.
  c. AAV recommends an alternate diet (from formulated diets) consisting of 60% whole grains and grain products, 10% vegetables, less than 5% fruits, and up to 25% mature legumes.[41] Owner compliance is the greatest threat to this diet's success.[41]
  d. Animal-origin products—typically a rich source of nutrients. Good amino acid profile[39] for the animal proteins and certain "animal-origin" vitamins such as vitamins D_3 and B_{12}.
    (1) Egg is considered nature's most balanced protein source
      (a) Sole source of nourishment for the developing embryo
      (b) Cooked egg products may be used successfully in diet supplementation
    (2) Meats
      (a) Cooked meats, especially poultry and fish
      (b) May have rate-limiting amino acids, such as tryptophan and cysteine[38]
    (3) Dairy products
      (a) Hard cheese, yogurt
      (b) Limit milk because birds lack the enzyme lactase, which is required to break down the sugar in milk.
    (4) By-products
      (a) Poultry, wheat germ meal, etc.
      (b) Excellent sources of amino acids, minerals, and many vitamins.
      (c) High-quality products are an excellent dietary additive.
      (d) Poor quality products must be avoided, because they may contain a variety of compounds and/or contaminants that may create health concerns
      (e) Microbial products
        (i) Brewer's yeast and others: rich combination of amino acids, vitamins, and minerals, and can be an excellent vitamin B complex supplement
5. Manufactured diets (percentage of total)
  a. Fed as a complete diet, they remove the guesswork of nutrition and are 100% consumable pellets or extruded pieces

Table 18-1
Nutrient Profile Recommendations*

Nutrient		General psittacine profile Minimum level	General psittacine profile Maximum	General passerine profile Minimum level	General passerine profile Maximum
	Gross energy (kcal/kg)	3200	4200	3500	4500
	Total protein (%)	12		14	
	Linoleic acid (%)	1		1	
Amino acids	Lysine (%)	0.65		0.75	
	Methionine (%)	0.30		0.35	
	Methionine plus cystine (%)	0.50		0.58	
	Arginine (%)	0.65		0.75	
	Threonine (%)	0.40		0.46	
Vitamins, fat-soluble	Vitamin A activity (total) (IU/kg)	8000		8000	
	Vitamin D_3 (ICU/kg)	500	2000	1000	2500
	Vitamin E (ppm)	50		50	
	Vitamin K (ppm)	1.0		1.0	
Vitamins, water-soluble	Thiamine (ppm)	4.0		4.0	
	Riboflavin (ppm)	6.0		6.0	
	Niacin (ppm)	50.0		50.0	
	Pyridoxine (ppm)	6.0		6.0	
	Pantothenic acid (ppm)	20.0		20.0	
	Biotin (ppm)	0.25		0.25	
	Folic acid (ppm)	1.50		1.50	
	Vitamin B_{12} (ppm)	0.01		0.01	
	Choline (ppm)	1500		1500	

From Hawley SB: Year-end report of the nutrition and management committee, AAV Annual Meeting, Tampa, Fla, 1996.
*These nutrient profile recommendations were approved by the AAV Board of Directors in 1996. At the time of printing, they were published for public and industry comment by the Pet Food Committee of the American Association of Feed Control Officials.

    (1) An extruded diet combines raw ingredients such as eggs, corn, wheat, oats, amino acids, vitamins, and minerals into a mash, which is cooked at very high temperatures and pressures.
        (a) This cooking process destroys most bacteria (and some heat-labile vitamins, but these are easily measured and compensated for) and makes the diet highly digestible.
        (b) The extrusion, or cooking process, also makes the diet more palatable.
    (2) Pelleted diets are manufactured similarly, except at lower temperatures and lower pressure.
  b. Manufactured diets should meet the Nutrient Profile recommendations approved by the AAV Board of Directors in 1996[42] (Table 18-1).

Table 18-1—cont'd
Nutrient Profile Recommendations

		General psittacine profile		General passerine profile	
Nutrient		Minimum level	Maximum	Minimum level	Maximum
Minerals	Calcium (%)	0.30	1.20	0.50	1.20
	Phosphorus, total (%)	0.30		0.50	
	Calcium/total phosphorus	1:1	2:1	1:1	2:1
	Potassium (%)	0.40		0.40	
	Sodium (%)	0.12		0.12	
	Chlorine (%)	0.12		0.12	
	Magnesium (ppm)	600		600	
Trace minerals	Manganese (ppm)	65.0		65.0	
	Iron (ppm)	80.0		80.0	
	Zinc (ppm)	50.0		50.0	
	Copper (ppm)	8.0		8.0	
	Iodine (ppm)	0.40		0.40	
	Selenium (ppm)	0.10		0.10	

B. Life Stage and Activity Impact Nutrient Needs of the Adult
1. Young (fledged to sexual maturity)
   a. Dietary requirements of young adult birds in the wild would be greater than maintenance.
   b. During this stage of their life, they will be more active as they learn to fly, become efficient at food gathering and predator evasion, and establish territories.
   c. With companion animals these dietary requirements will not be as dramatic as their wild counterparts but will remain higher than maintenance.
   d. No specific dietary requirements during this stage other than a high-quality diet with adequate protein levels to aid in the development of adult musculature
2. Sexually mature adult
   a. Dietary requirements of mature, nonbreeding adults as companion animals are different from those in the wild.
      (1) In the native habitat, these birds would be actively seeking mates, which would increase dietary requirements because of courtship displays and nest activity.
      (2) Companion animals are frequently maintained in single habitats where mate selection is not available, or are paired by the owner.
   b. Activity levels of companion birds depend on their environment. Their dietary requirements can range from essentially maintenance to that of a young bird.

3. Breeding adult
   a. Dietary requirements of breeding birds are directly related to the number of eggs produced as well as their environment.
      (1) In species that lay a small number of eggs infrequently, the nutrients can most often be taken directly from body stores.
      (2) Hens sustaining a larger egg production over a prolonged time have dietary requirements that generally cannot be met through body stores.
   b. Supplements need to be introduced to the diet either through a breeding "formula" or direct supplementation of vitamins and minerals.
   c. If eggs are removed from the nest shortly after being laid, the dietary requirements are generally greater, because the hen might cycle again and produce another clutch.
   d. If adult birds are feeding their chicks, the diet fed the adult must be formulated to meet the requirements of the chicks.
4. Geriatric birds (age is 20% higher than average life span of species)
   a. Information of the needs of geriatric birds is extremely limited.
   b. Using other animals as models, assume that geriatric birds should be provided with a highly digestible (high-quality) diet that takes into account their reduced energy and protein needs while simultaneously slightly increasing some of their micronutrients and macronutrients to help compensate for a less efficient digestive system.

C. Macronutrients for Birds
1. Energy
   a. Not a specific nutrient, because it is not present in the diet in a specific chemical compound.
   b. Energy is required for maintaining the physiochemical environment and sustaining the electromechanical activities of the bird.
   c. The major body store of energy is adenosine triphosphate (ATP) and other high-energy phosphate bonds. How efficiently an animal is able to convert the potential energy available in foodstuffs into body energy stores is subject to individual variation and may explain the propensity toward or resistance to weight gain.[43]
   d. Energy intake is a highly variable component of energy balance. Energy is available in the diet through three major sources: fats, protein, and carbohydrates.
   e. Dietary energy to nutrient ratio
      (1) Diets are generally formulated by calculating the percentage or parts per million of a nutrient in the food; the true analysis by the nutritionist is of the dietary intake of each nutrient, not the relative quantities in the diet.
      (2) Birds typically eat at least the minimum amount of food required to satisfy their energy needs.

(3) In most species, optimum growth is seen in the range of 16% to 22% crude protein, depending on energy level of the diet. Feeding diets below 10% or above 30% crude protein are not advised.
2. Fatty acids are an essential portion of the avian diet.
   a. Used primarily as structural components of cell wall membranes
   b. The primary essential fatty acid for birds is linoleic acid. It cannot be synthesized in the bird's body.
   c. Extrapolating from other avian species, it can be predicted that linoleic acid requirements for most companion birds is 1.0% to 1.5% of the diet.[44] This is generally not a problem in most types of seed-based and processed diets.
   d. Deficiencies may lead to dry, flaky, or itching skin.
   e. Excesses can lead to obesity, fatty liver syndrome, locomotion problems, and infertility.
3. Proteins compose a combination of 22 different amino acids
   a. Ten amino acids cannot be manufactured in the bird's body and must be supplemented.
      (1) Essential amino acids include lysine, arginine, histidine, methionine, tryptophan, threonine, leucine, isoleucine, valine, phenylalanine.
      (2) Three other amino acids are formed through the modification of essential amino acids. These include lysine, hydroxylysine, tyrosine, formed from methionine, lysine, phenylalanine, respectively.
      (3) Nonessential amino acids (readily manufactured by the body) include alanine, aspartate, glutamate, glutamine, glycine, proline, serine, taurine, ornithine.
   b. Protein quality is determined by the balance of amino acids within the protein and the availability of the amino acids within the ingredients of the diet.
   c. The amino acid balance should closely profile the bird's body. This will provide the approximate proportion needed with no major deficiencies or excesses.
   d. The availability within the diet is affected by structural components or chemical compounds of the ingredients, which will increase or decrease the bioavailability of the amino acid. Because of these complex interactions, diets can meet the total protein requirements of an animal and still be deficient in an essential amino acid.
4. Carbohydrates are the most important form of energy for the body, because they are the only energy source readily available for use by the brain and generally efficiently metabolized.
   a. Energy is derived from the different carbohydrate groups of starches, monosaccharides, and disaccharides.
   b. Carbohydrates also form the fiber fraction or "indigestible" portion of the diet, which is primarily cellulose. This carbohydrate is essentially indigestible because birds lack the enzyme cellulase.

5. Water is essential to the bird's body.
    a. Provides for the homeostasis of the intracellular and extracellular fluids.
    b. Aids in digestion and absorption and is essential for cooling the body.
    c. Quality of water is critical to overall health. A poor or contaminated water supply can be a source of bacterial proliferation, which will be a detriment to the bird.
    d. Water intake will be influenced by the type of diet. Processed diets are lower in water content than an all-seed diet, and birds typically increase their water consumption on manufactured diets. This can sometimes result in more moist feces.

D. Nutritional Diseases of Adult Birds
1. Skin/integument/ epithelial tissues
    a. Itchy, dry skin: Linoleic acid deficiency, poor amino acid profile, food allergies
    b. Squamous metaplasia of the epithelial lining of the upper respiratory tract, gastrointestinal, urogenital tracts, uropygial gland: hypovitaminosis A[10,12,24-26]
        (1) Blunting or lack of choanal papillae. Hyperkeratosis of the epithelial surfaces can also occur. It is common on the metatarsal and digital pads of the feet, which can predispose a bird to developing pododermatitis.[10,14,15]
        (2) Clinical signs can also include nasal discharge, swelling around eyes, swollen sinuses, polyuria, polydypsia, dyspnea, or anorexia.
        (3) Masses (keratin cysts) may be present in oral cavity, white pustules or masses can occur in the oral cavity, crop, or nasal passages. Secondary bacterial infections commonly occur.
        (4) Treatment for vitamin A deficiency includes parenteral vitamin A weekly, then oral supplementation once improved, usually 2 to 3 weeks.[35] Treatment with antibiotics for secondary bacterial infections may be required.
    c. Other nutritionally based skin disorders include biotin, niacin, pantothenic acid deficiency, and mycotoxin ingestion.
2. Feathering
    a. Abnormal feather condition/feather: Brittle, frayed feathers are a common problem in many adult birds and may be the result of several nutritional inadequacies.
        (1) Diets deficient in minerals such as calcium, zinc, selenium, manganese, and magnesium may lead to a poor feathering condition.[45]
        (2) Feather breakage may also be due to hypovitaminosis B, which results in white streaks that weaken the feather.[45] Low niacin, folic acid, and total protein deficiencies are other contributors.[1]
    b. Retained feather sheaths can be a result of hyperkeratosis incited by nutritional inadequacies, along with infectious agents.[45]

When the sheaths are retained, there may appear to be an excess of pin feathers. Consider the nutritional state of a bird with such a condition.
   c. Feather picking: Malnutrition, along with many other causes such as organopathies, toxins, infections, boredom, and anxiety, has been implicated in feather picking.
      (1) Deficiencies in vitamin A, sulfur containing amino acids, arginine, niacin, pantothenic acid, biotin, folic acid, and salt have been shown to play a role in feather picking.[16,35,45]
      (2) A diet containing excess fat can also lead to self-mutilation.[46] Feather picking, for whatever reason, should prompt an evaluation of the nutritional needs of the bird because of an increased demand of amino acids such as methionine, lysine, and arginine to grow new feathers.
      (3) A bird with feather loss will suffer from increased heat loss and therefore will experience a higher metabolic rate (often as high as 60%). The energy needs in such a bird can approach up to 85% higher than in normal feathered birds at room temperature.[2] Therefore, the diet of these birds needs to be modified.
   d. Abnormal feather colors
      (1) Feather color is determined by two different processes: pigments deposited during feather development and structural features of the feather that alter the absorption of light.
         (a) Yellow, orange, and red colors are due to the fat-soluble carotenoids. Because these pigments cannot be synthesized, they must be provided for in the bird's feed.
            (i) Carotenoids can be obtained through natural foods such as yellow corn, carrots, nettle-leaf meal, and green, leafy vegetables or synthetically manufactured food additives.
            (ii) Xanthophyll must be present. Therefore, a diet that is high in carotenoids but still low in xanthophyll will result in a hypopigmentation of the feathers. Such diets often contain cod-liver oil, other fish oils, meat scraps, fish meal, soybean-oil meal, charcoal, or sulfur.[1]
         (b) Red, black, and brown colors are produced by melanin.
            (i) The pigment in melanin is derived from tyrosine and copper enzymatic reactions.
            (ii) Diets low in tyrosine, tyrosine-related amino acids, or copper will interfere with melanin production and cause hypopigmentation.[45]
         (c) Red patches on the normal green plumage—for example, peach-faced lovebirds—may be a result of diet change or a blood parasite. African greys also may show a color alteration manifesting as abnormal occurrence of red feathers as a result of excess dietary beta-carotene. An ad-

ditional cause of this color alteration in African greys is psittacine beak and feather disease virus.
- (d) Pigments that coat the feather and are secreted in lipid form from the glandular keratinocytes or the uropygial gland.
    - (i) Oil secretions help protect feathers and enhance color and sheet. Diets that minimize lipid secretion can actually cause a color change or promote wear of the feathers and resulting color alteration.[13]
    - (ii) Fatty diets, such as a prolonged feeding of bacon rind and bone, can cause oily feathers and, in the rose-breasted cockatoo, results in an increased pink color.[45] Such color changes occur without a molt.
- (e) Blue and white are known as structural colors, because they depend on light reflection off keratin in the feather.
    - (i) Essential amino acids in keratin that could affect this process are methionine, histidine, lysine, tryptophan, threonine, isoleucine, and valine.
    - (ii) Deficiencies in these amino acids cause structural alterations of keratin and resulting color changes.
    - (iii) Loss of structural blue color has been associated with such amino acid deficiencies and is seen often in psittaciformes; may be caused by a combination of amino acid deficiencies; a single amino acid deficiency of lysine can result in a green to yellow color change.[45]
- (f) Black
    - (i) Colors associated with altered keratin structure allowing melanin granules in the middle feather to absorb light will appear black.
    - (ii) Such keratin abnormalities are seen in malnourished or systemically ill birds and result in color changes from blue, green, or gray to black.[45]
- (g) Achromatosis—frequently noted on the primary feathers and tail feathers. Reported causes include lysine deficiency (quail), choline deficiencies (cockatiel), riboflavin deficiency (cockatiel).[47]
  e. Greasy appearance to feathers
    (1) Malfunction or damage to the uropygial gland because of hypovitaminosis A or general malnutrition.[10,15,48,49]
3. Musculoskeletal
  a. Obesity
    (1) Commonly seen in budgerigars, Amazon parrots, and rose-breasted cockatoos.
    (2) Can be nutritionally induced if fed an inadequate seed-based diet that is too high in fat.[50]
    (3) Can also be seen with simple overeating and a sedentary lifestyle.
    (4) Can occur concurrently with beak and nail problems when associated with fatty liver disease.[51,52]

b. Beak and nail overgrowth/malformation: Can be seen with malnutrition, and hypovitaminosis A.[53] Frequently associated with hepatic problems.
   c. Secondary nutritional hyperparathyroidism: Bone demineralization because of severe inverse calcium/phosphorus ratio.
   d. Osteomalacia
      (1) Under adequate natural lighting conditions birds do not need to have cholecalciferol supplemented in their diet.
      (2) Birds maintained under artificial light need to be supplemented with vitamin D_3. It is also advisable to supplement birds during egg laying and growth.
      (3) Vitamin D_3 affects the mineralization of bone in concert with calcium and phosphorus
   e. Muscular dystrophy: Characterized by light-colored streaks in muscle fibers, but also clinically related to nutritional myopathies, paresis/paralysis, and diminished play.
      (1) Selenium and vitamin E responsive disorder[23,32,54]
      (2) Vitamin A deficiency also reported[11,24,25]
   f. Lameness
      (1) Multifactorial interplay of vitamin D, calcium, and phosphorus
      (2) Obesity resulting from excessive intake of fat
      (3) Articular gout: Cause of gout is unknown, but it is commonly associated with renal-compromised birds on a diet that is high in protein.[30,31,55]
   g. Hypocalcemic tetany
      (1) Frequently seen in African grey parrots on a calcium-deficient diet such as seeds.
      (2) African greys do not require more calcium than other psittacine species, but exhibit signs of hypocalcemia at a lower threshold than other species.[56]
      (3) May also relate to decreased levels of PTH[57]
   h. Pododermatitis/bumble foot: Multifactorial origin with dietary components, including hypovitaminosis A, biotin, and excessive fat.[12,14,24-26]
4. Central nervous system/neurologic
   a. Seizures
      (1) Thiamine deficiency, zinc toxicity[28,29,36]
      (2) Calcium, phosphorus, and vitamin D imbalances[58,59] (see hypoglycemic tetany)
      (3) Mycotoxin-contaminated feedstuffs, especially aflatoxin, citreoviridin, tremorgens, fusariotoxins, vomitoxin, and ochratoxins[59-61]
      (4) Jerky, nervous movement progressing to frantic flapping and tonic-clonic seizures shown in pyridoxine (vitamin B_6) deficiency[61]
   b. Ataxia of head and limbs

(1) Neck paralysis, head ataxia: Folic acid deficiency[34]
(2) Head tilt, circling, torticollis: Encephalomalacia resulting from vitamin E deficiency[22,23,33,49]
(3) Opisthotonos and ataxia: Thiamine and pantothenic acid deficiencies[62,63]

c. Metabolic neuropathies
(1) Hepatic encephalopathy
   (a) Occurs secondary to severe liver disease, characterized by neurologic symptoms.
   (b) Caused by intoxication of the brain with the by-products of protein digestion and metabolism, which enter the portal system and are not detoxified by the liver.[64]
   (c) Neurologic signs usually occur after eating
   (d) Aromatic amino acids, methionine, and short-chain fatty acids[65]
   (e) Lactulose syrup, high-quality but low-protein, high-carbohydrate diet with vitamin supplementation can provide symptomatic relief if underlying liver disease is treated[66]
(2) Hypoglycemia: May result from starvation or general, chronic, and severe malnutrition[58]
(3) Botulism (limber neck): Uncommon in psittacine birds, common in waterfowl, results from ingestion of contaminated maggots, insects, or animal flesh[67]

5. Reproductive disorders
a. Egg binding/dystocias
(1) Multifactorial with nutritional components, including deficiency or excess of vitamin E, selenium, and vitamin D, calcium[27,45,68]
(2) Obesity[69]
b. Eggshell problems: Problems associated with egg shell structure such as thin walls, thick walls, etc., can be associated with an inadequate and/or improper calcium/phosphorus ratio in the diet, or hypovitaminosis D
c. Infertility: Many causes, poorly defined
(1) Malnutrition[27,45,69,70]
(2) Obesity[27,69]

6. Metabolic disorders
a. Goiter: Especially common in budgerigars on an all-seed diet. The disease is a result of an iodine deficiency resulting in diffuse thyroid hyperplasia.[37,71,72]
b. Hemochromatosis (iron storage disease): Commonly seen in fructivorous, insectivorous, and omnivorous birds. Commonly seen in toucans, mynahs, and birds of paradise. The etiology has not been fully determined, although it is probably both genetic and diet-related.
c. Blood clotting disorders: Associated with hypovitaminosis K and possibly calcium caused by the clotting mechanisms associated

with this vitamin.[73,74] This problem is commonly associated with birds on all-seed diets.[75]

d. Scurvy: Caused by the inability to maintain collagen and elastin in the blood vessel walls because of the lack of ascorbic acid (vitamin C) as a cofactor. Most birds can make their own ascorbic acid from glucose and a dietary supplement is not essential. Those birds that do need supplementation are passerines,[76,77] except for one galliformes, the willow ptarmigan.[78]

REFERENCES

1. Fowler ME: *Zoo and wild animal medicine,* Philadelphia, 1986, WB Saunders, pp 196-221.
2. Avian, Canine, and Feline Proceedings of the 1993 Practitioner's Symposium. Sponsored by the American Board of Avian Practitioners, *Avian Dermatology,* p 39.
3. Dustin LR: Hand feeding tips for the pediatric patient, *J Assoc Avian Vet* 4:23-25, 1990.
4. Adams C et al: Interaction between nutrition and *Eimeria acervuline* infection in broiler chickens: development of an experimental model, *Br J Nutr* 75:867-873, 1996.
5. Phalen DN et al: The avian urinary tract, form, function, diseases, *Proc Assoc Avian Vet,* 1990, pp 44-57.
6. Tekeshita K et al: Hypervitaminosis D in baby macaws, *Proc Assoc Avian Vet,* 1986, pp 341-345.
7. Engberg RM et al: Inclusion of oxidized vegetable oil in broiler diets: its influence on nutrient balance and on the antioxidative status of broilers, *Poultry Sci* 75:1003-1011, 1996.
8. Clubb SK et al: Water quality and aviculture, *Proc 4th Conf European Com AAV,* 1997, pp 147-153.
9. Bond MW: The veterinarian's role in nursery management, *Proc 4th Conf European Com AAV,* 1997, pp 147-153.
10. Bauck L et al: Dermatology. In Altman R et al, editors: *Avian medicine surgery,* Philadelphia, 1997, WB Saunders, pp 540-562.
11. Honour SM et al: Experimental vitamin A deficiency in mallards (Anas platyrhynchos): lesions and tissue vitamin A levels, *J Wildlife Dis* 31:277-288, 1995.
12. Hawley SB: What every bird owner should know about avian nutrition, Kaytee Technical Focus, Chilton, Wis, 1994, Kaytee products, Inc.
13. Lucas AM: *Avian anatomy integument,* Washington DC, 1972, US Government Printing Office, pp 229, 418.
14. Burgmann PM: Common psittacine dermatologic diseases, *Semin Avian Exotic Pet Med* 4:169-183, 1995.
15. Cooper JE, Harrison GJ: Dermatology. In Ritchie BW, Harrison GJ, Harrison LR, editors: *Avian medicine: principles and application,* Lake Worth, Fla, 1994, Wingers Publishing Inc, pp 607-639.
16. Oloyo RA: Studies on the biotin requirement of broilers fed sunflower seed meal based diets, *Arch Anim Nutr* 45:345-353, 1994.
17. Fudge AM: Clinical nutrition of companion birds, *Cal Vet* March/April pp 7-8, 1988.
18. Grone A et al: Hypophosphatemic rickets in rheas, *Vet Pathol* 32:324-327, 1995.
19. Wada Y et al: Peripheral neuropathy of dietary riboflavin deficiency in racing pigeons, *J Vet Med Sci* 58:161-163, 1996.

20. Dorrestein GM: Physiology of the musculoskeletal system. In Altman R et al, editors: *Avian medicine and surgery,* Philadelphia, 1997, WB Saunders, pp 529-539.
21. Elkan E, Zwart P: The ocular disease of young terrapins caused by vitamin A deficiency, *Path Vet* 4:201-222, 1967.
22. Ryan, T: Hypovitaminosis E in a cockatoo, *Comp Anim Prac* 19:31-32, 1989.
23. Campbell TW: Hypovitaminosis E: its effect on birds, *Proc 1st Internat Conf Zoological Avian Med,* 1987, pp 75-77.
24. Zwijnenberg RJG, Zwart: Squamous metaplasia in the salivary glands of canaries (a cased report), *Vet Quart* 16:60-61, 1994.
25. Austic RE, Scott ML: Nutritional diseases. In Calnec BW et al, editors: *Disease of poultry,* ed 9, London, 1991, Wolfe Publishing Ltd, pp 46-49.
26. Zwart P et al: Vitamin A deficiency in parrots, *Verhandl Ber 21 Int Symp Erkrank Zootiere,* Mulhouse 1979, pp 47-52.
27. Johnson AL: Reproduction in the female. In Sturkie PD, editor: *Avian physiology,* ed 4, New York, 1986, Springer-Verlag, pp 432-451.
28. Van Sant F: Zinc and clinical disease in parrots, *Proc Assoc Avian Vet,* 1997, pp 387-391.
29. Howard BR: Health risks of housing small psittacines in galvanized wire mesh cage, *J Am Vet Med Assoc* 200:1132-1136, 1992.
30. Siller WG: Renal pathology of the fowl: a review, *Avian Pathol* 10:187-262, 1981.
31. Leeson S et al: Gout and kidney urolithiasis. In Leeson S et al, editors: *Poultry metabolic disorders and mycotoxins,* Guelph, Ontario, 1995, University Books, pp 76-88.
32. Lumeij JT: Nephrology. In Ritchie BW, Harrison GJ, Harrison LR, editors: *Avian medicine: principles and application,* Lake Worth, Fla, 1994, Wingers Publishing Inc, pp 539-555.
33. Fuhrmann H, Sallman HP: The influence of dietary fatty acids and vitamin E on plasma prostanoids and live microsomal alkane production in broiler chickens with regard to nutritional encephalomalacia, *J Nutr Sci Vitaminol* 41:553-561, 1995.
34. Paul-Murphy J: Avian neurology, *Proc Assoc Avian Vet,* New Orleans, 1992, pp 420-432.
35. Ryan T: Vitamin A and its deficiency in birds, *Comp Anim Pract* 2:35-37, 1988/1989.
36. Cooper JE: Feeding exotic and pocket pets, *J Small Animal Pract* 31:482-488, 1990.
37. Bauck L: Nutritional problems in pet birds, *Semin Avian Exotic Pet Med* 4:3-8, 1995.
38. Wang X et al: Order of amino acid limitation in meat and bone meal, *Poultry Sci* 76:54-57, 1997.
39. Scwartz RW, Bray DJ: Limiting amino acids in 40:60 and 15:85 blends of corn:soybean protein for the chick, *Poultry Sci* 54:1814, 1975.
40. Edmonds et al: Limiting amino acids in low-protein corn-soybean meal diets fed to growing chicks, *Poultry Sci* 64:1519-1524, 1985.
41. Association of Avian Veterinarians: *Feeding,* Orlando, 1994, AAV Publication Office.
42. Hawley SB: *Year-end report of the nutrition and management committee,* AAV Annual Meeting, Tampa, Fla, 1996.
43. Devlin JT, Horton ES: Energy requirements. In *Present knowledge in nutrition,* International Life Sciences Institute Nutrition Foundation, Washington DC, 1990, pp 1-2.
44. Brue RN: Nutrition. In Ritchie BW, Harrison GJ, Harrison LR, editors: *Avian medicine: principles and application,* Lake Worth, Fla, 1994, Wingers Publishing, pp 63-95.

45. Macwhirter P: Malnutrition. In Ritchie BW, Harrison GJ, Harrison LR, editors: *Avian medicine: principles and application,* Lake Worth, Fla, 1994, Wingers Publishing, pp 842-861.
46. Hess L, Mauldin G, Rosenthal K: Nutrient contents of commonly fed companion bird diets, *Proc Assoc Avian Vet,* 1997, Reno, Nev, pp 229-232.
47. Roudybush TE, Grau CR: Food and water interrelations and the protein requirements for growth of an altricial bird, the cockatiel (*Nymphicus hollandicus*), *J Nutr* 116:552-559, 1986.
48. Wissman MA: Diseases of the uropygial gland, *Proc Assoc Avian Vet,* 1997, pp 317-320.
49. Lowenstein LJ: Avian nutrition. In Fowler M, editor: *Zoo and wild animal medicine,* Philadelphia, 1986, WB Saunders, pp 201-212.
50. Dabbert CB et al: Use of body measurements and serum metabolites to estimate the nutritional status of mallards wintering in the Mississippi Alluvial Valley, USA, *J Wild Dis* 33:57-63, 1997.
51. Branton SL et al: Fatty liver-hemorrhage syndrome observed in commercial layers fed diets containing chelated minerals, *Avian Dis* 39:631-635, 1995.
52. Fudge AM: Psittacine liver disease, an overview, *Proc 4th Conf European Com AAV,* 1997, pp 96-100.
53. Carpino MR, Phalen D: Beak deformities associated with malnutrition in hand-fed pediatric African grey parrots, *Proc Assoc Avian Vet,* 1997, pp 154-157.
54. Harrison GJ: Preliminary work with selenium/vitamin E responsive conditions in cockatiels and other psittacines, *Proc Assoc Avian Vet,* 1986, pp 257-262.
55. Enders F et al: Urolithiasis and unilateral renal atrophy in hand-reared juvenile psittacines, *Proc 4th Conf European Com AAV,* 1997, pp 46-49.
56. Hawley SB: Nutritional therapy of disease in the avian patient, *Proc 4th Conf European Com AAV,* 1997, pp 84-89.
57. Hochleithner M et al: Hypoparathyroidism in African grey parrots: clinical importance, *Proc 4th Conf European Com AAV,* 1997, pp 46-49.
58. Bennet A: Neurology. In Ritchie BW, Harrison GJ, Harrison LR, editors: *Avian medicine: principles and application,* Lake Worth, Fla, 1994, Wingers Publishing, pp 723-747.
59. Walsh MT: Seizuring in pet birds, *Proc Assoc Avian Vet,* 1985, pp 121-128.
60. Scheideler SE: A summary of mycotoxin (aflatoxin B_1 or vomitoxin) challenge in young ostrich chicks, *Proc Assoc Avian Vet,* 1997, pp 167-168.
61. Hasholot J, Petrak ML: Diseases of the nervous system. In Petrak ML, editor: *Disease of cage and aviary birds,* ed 2, Philadelphia, 1982, Lea & Febiger, pp 468-477.
62. Ward FP: Thiamine deficiency in a peregrine falcon, *J Am Vet Med Assoc* 159:599-601, 1971.
63. Gries CL, Scott ML: The pathology of thiamine, riboflavin, pantothenic acid and niacin deficiencies inn the chick, *J Nutr* 102:1269-1286, 1972.
64. Lumeij JT: Hepatology. In Ritchie BW, Harrison GJ, Harrison LR, editor: *Avian medicine: principles and application,* Lake Worth, Fla, 1994, Wingers Publishing, pp 522-537.
65. Center SA: Pathophysiology and laboratory diagnosis of liver disease. In Editing SD, editor: *Textbook of veterinary internal medicine,* Philadelphia, 1989, WB Saunders, pp 1462-1466.
66. Spalding MG et al: Hepatic encephalopathy associated with hemochromatosis in a toco toucan, *J Am Vet Med Assoc* 189:827-828, 1980.
67. Foreyt WJ, Abinanti FR: Maggot associated type-C botulism in game farm pheasants, *J Am Vet Med Assoc* 177:827-828, 1980.
68. Speer B: Diseases of the urogenital system, In Altman R et al, editors: *Avian medicine and surgery,* Philadelphia, 1997, WB Saunders, pp 625-644.

69. Roudybush TE, Grau CR: Calcium requirements for egg laying in cockatiels, *Am Fed Avic Watchbird* 18:10-13, 1991.
70. Joyner K: Theriogenology. In Ritchie BW, Harrison GJ, Harrison LR, editors: *Avian medicine: principles and application,* Lake Worth, Fla, 1994, Wingers Publishing, pp 748-804.
71. Blackmore DK: The pathology and incidence of thyroid dysplasia in budgerigars (*Milopsittacus undulatus*), *Vet Rec* 75:1068-1072, 1965.
72. Sasipreeyajan J, Newman JA: Goiter in a cockatiel (*Nymphicus hollandicus*), *Avian Dis* 32:169-172, 1988.
73. Rosskopf WJ, Woerpel RW: Psittacine conditions and syndromes, *Proc Assoc Avian Vet,* 1990, pp 432-459.
74. Rosskopf, WJ, Woerpel RW: Erythremic myelosis in conures: the hemorrhagic conure syndrome, *Proc Assoc Avian Vet,* 1984, pp 213-229.
75. Quesenberry KE, Hillyer EV: Supportive care and emergency therapy. In Ritchie BW, Harrison GJ, Harrison LR, editors: *Avian medicine: principles and application,* Lake Worth, Fla, 1994, Wingers Publishing, pp 382-416.
76. Roy RN, Guha BC: Species differences in regard to the biosynthesis of ascorbic acid, *Nature* 182:319, 1958.
77. Roy RN, Guha BC: Production of scurvy in a bird species, *Nature* 182:1689, 1958.
78. Hanssen I, Grav HJ, Steen JB, Lysnes H: Vitamin C deficiency in growing Willow Ptarmigan (*Lagopus lagopus lagopus*), *J Nutr* 109:2260, 1979.

19

Imaging Interpretation

April Romagnano and Nancy E. Love

Diagnostic imaging techniques are an integral part of patient evaluation in many avian species commonly seen in clinical practice. Although conventional radiography and contrast procedures remain the standard, alternative imaging modalities are now frequently used for pet and aviary birds. The diagnosis of many diseases in avian patients often is challenging without the use of conventional radiography. Optimizing imaging interpretation requires high-quality images and mastery of normal avian radiographic anatomy for both the species in question and the technique used. Imaging can often provide the same information as more invasive techniques, such as exploratory coelotomy or endoscopy.

I. Radiographic Technique

A. Technical Factors
  1. Isoflurane general anesthesia is recommended for avian radiography.[1]
  2. A standard small animal machine is used that has the following features:
     a. High milliampere (mA)
     b. High-detail film-screen combination
     c. Short exposure time ($1/60$ of a second or faster)
     d. No grid (<10 cm)
     e. Small focal spot
     f. Standardized focal film distance (40 inches)
  3. A technique chart that is adequate for several different species of birds should be compiled before radiographing compromised avian patients. A reliable chart helps ensure reproducible images, reduces the number of repeat examinations and the amount of personnel and patient exposure, and saves money. A variable kilovolt peak (kVp) chart is easy to use and common in small animal practice.
     a. When a technique chart is constructed, the lowest kVp, the highest mA, and the shortest exposure times possible are used.[1,2] The kVp settings will range from as low as 20 to as high as 70, depending on the film-screen combination and machine.

b. Birds typically are radiographed using a tabletop (nongrid) technique.
   (1) Either the bird is placed directly on the cassette, or the positioning device (Plexiglas avian restraint board) and the bird are placed on the cassette (Fig. 19-1).
   (2) The contact between the radiographic film and a properly positioned, motionless patient should be uniform.
   (3) The area of interest should be as close to the film as possible to reduce magnification and to optimize detail.[1]

B. Restraint
1. Anesthesia
   a. Most birds should be given general anesthesia before radiographic procedures to optimize positioning, to reduce patient stress, and to reduce the number of repeat radiographs.[1] Once anesthetized, birds can be positioned either on a Plexiglas board or directly on a cassette using other positioning aids such as masking tape, sandbags, foam blocks, and lead gloves (Fig. 19-1).
      (1) Masking tape is a safe, hygienic means of securing feathered appendages; it is easily placed and removed, leaving feathers intact.
      (2) Foam blocks are useful for alignment but must be disinfected or placed in a clean wrapping between patients; the same is true for sandbags, lead gloves, and other such aids.
      (3) In some cases general anesthesia may be contraindicated or simply not warranted, such as for a critically ill patient that is a poor anesthetic risk.[1]
   b. Parenteral administration of anesthesia generally is not recommended for psittacines (author, April Romagnano's, opinion).
2. Physical restraint
   a. Mechanical
      (1) Mechanical restraint is recommended for nonfractious avian species, for critically ill patients that are poor anesthetic risks, for screening of baby birds for metabolic bone disease, and for screening for heavy metal intoxication.
      (2) Nonfractious avian patients may be restrained on a Plexiglas avian restraint board, using extreme care and caution. This technique requires considerable skill and experience to be done safely.
      (3) Baby birds can be encouraged simply to sit or stand on the cassette; they then are radiographed with a vertical x-ray beam. They may also be radiographed with a horizontal beam while sitting or standing on the table.
      (4) Screening for heavy metal intoxication can be accomplished by placing the patient in a brown paper bag and then putting the bag directly on the cassette. This technique and the one described in (3) above compromise radiographic image quality but achieve a diagnosis without the need for direct patient restraint.

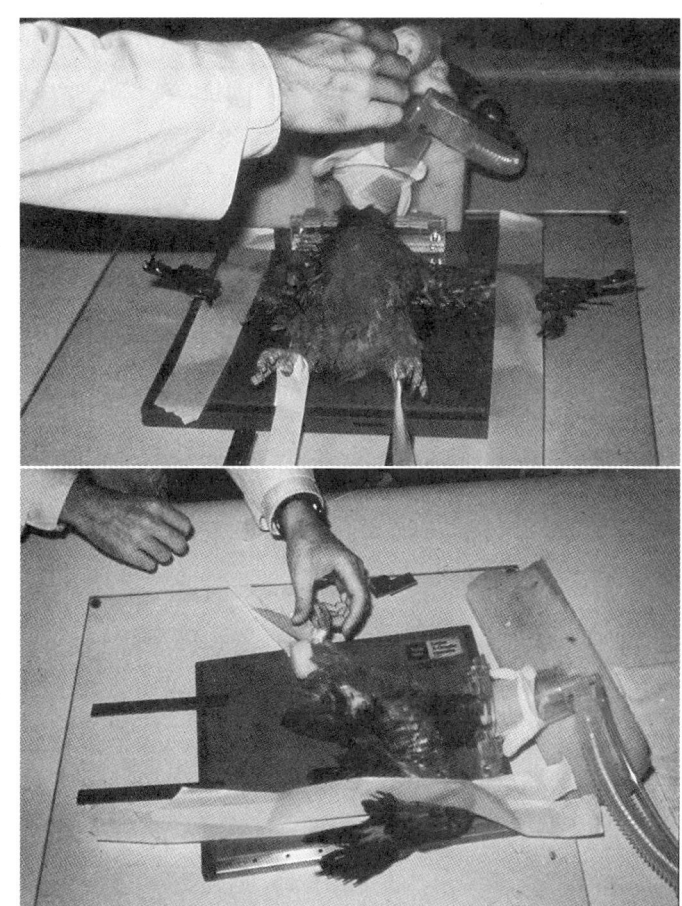

Fig. 19-1 A, Positioning for a ventrodorsal projection. For optimal positioning, the patient is placed under general anesthesia and restrained in dorsal recumbency with masking tape on a Plexiglas positioning board. The keel is directed (90 degrees) from the board, the wings are extended (90 degrees) from the body, and the legs are extended caudally. The film cassette is placed underneath the positioning board. **B,** Positioning for a left to right lateral projection. To optimize positioning, the patient is placed under general anesthesia and restrained on a Plexiglas positioning board. The bird is placed in right lateral recumbency with superimposition of the acetabula. The wings are extended caudodorsally and maintained in place with masking tape. Placing a small positioning sponge between the pelvic limbs can help prevent obliquity. The dependent limb is extended more cranially than the nondependent limb. (Courtesy Dr. Kevin Flammer)

b. Manual
      (1) Properly shielded personnel can restrain avian patients manually. However, we discourage this practice because of the risk of personnel exposure and because other techniques can be used successfully.

C. Positioning
   1. For routine, whole body evaluation in birds, a minimum of two radiographic views are made: the ventrodorsal (VD) view and the left to right lateral (LRL) view.[1]
      a. To ensure proper positioning, radiography of the skull requires general anesthesia.[1] To properly evaluate the skull, at least five views are recommended: the left to right lateral (LRL), the right to left lateral (RLL), the ventrodorsal (VD), the dorsoventral (DV), and the rostrocaudal (RCd, frontal sinus).[1]
      b. Collimation of the area of interest is particularly important in radiography of the skull and of the extremities. This technique helps reduce scatter and thus increase detail.
   2. VD positioning
      a. The patient is placed in dorsal recumbency with the keel directed 90 degrees from the plate.
      b. The wings are fully extended from the body (90-degree angle).
      c. The legs are extended caudally.
      d. If the patient is positioned correctly, the keel and spine are superimposed, and the femurs, scapulae, and acetabula are parallel.
   3. LRL positioning
      a. The patient is placed in LRL recumbency, and the acetabula are aligned on top of each other.
      b. The wings are pulled caudodorsally, with the right wing slightly cranial to the left.
      c. The legs are extended caudally, with the right leg positioned slightly cranially.
      d. If the patient is positioned correctly, the acetabula, ribs, coracoid, and kidneys are superimposed.

II. Skeletal System

A. Skull
   1. The cranium of birds is fused and contains numerous connections to the sinuses (Fig. 19-2).[3-5]
   2. The largest sinus is the infraorbital sinus, which is located rostral-ventral to the eyes.
   3. The upper bill forms a synovial joint with the frontal bone via the articular and quadrate bones (Fig. 19-2).
   4. The scleral ossicles form a bony ring that is visible radiographically in both eyes; psittacines have 12 ossicles per eye.[1,6]

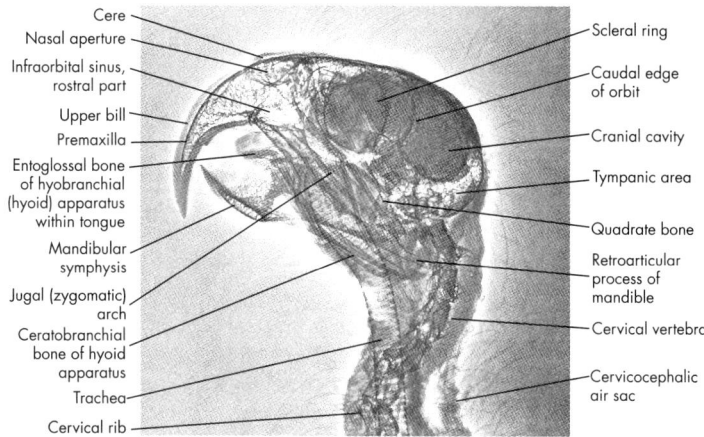

Fig. 19-2 Labeled right lateral xeroradiograph of an Amazon parrot's skull. (From Smith SA, Smith BJ: *Atlas of avian radiographic anatomy*, Philadelphia, 1992, Saunders.)

5. Significant species-specific soft tissue structures in the head region may include the cere, combs, wattles, or ear lobes.

B. Spine

1. The avian spine is divided into cervical, thoracic, synsacral (fused thoracic, lumbar, sacral, and caudal), free caudal, and fused caudal or pygostyle vertebrae (Figs. 19-3 and 19-4).[5]
2. Most parrots have 12 cervical vertebrae; budgerigars have 11 and swans 25.[1]
3. Incomplete ribs are present on the cervical vertebrae, and complete ribs (eight in parrots) are present on the thoracic vertebrae.
4. The uropygial gland, an integumental structure located above the distal spine of most species, may be visible on radiographs as an almond-shaped soft tissue mass.

C. Thoracic Girdle and Appendages

1. The thoracic girdle consists of the clavicle (which fuses distally to form the furcula), the coracoid, and the scapula.[4] These three bones meet, forming the triosseal canal, through which the tendon of the supracoracoideus muscle passes to abduct the wing; this muscle is large and visible radiographically.
2. The sternum contains a carina, or keel, which serves as the attachment point for the pectoral or flight muscles (Figs. 19-3 and 19-4).
3. Distal to the scapula the wing contains the humerus, the larger ulna, the radius, and the carpal bones. Growing flight feathers originate in the ulna and may be radiographically visible as soft tissue structures. Only the radial and the ulnar carpal bones are present in birds.[4] The distal

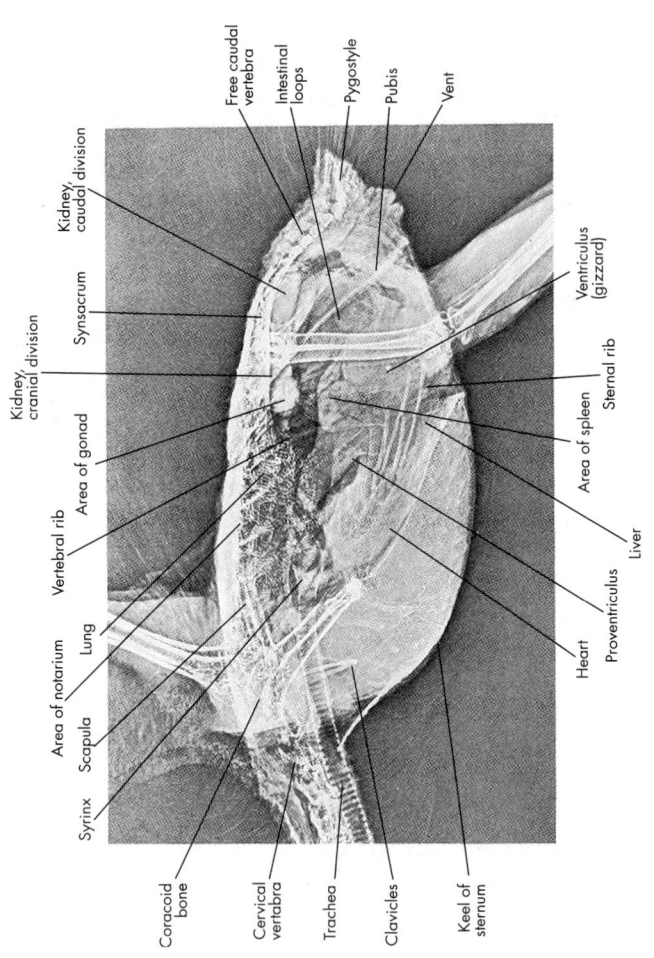

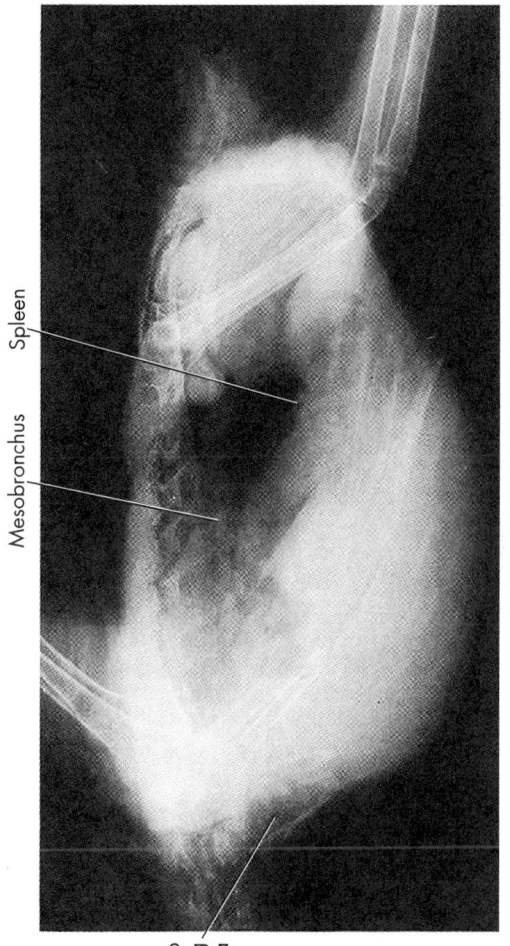

Fig. 19-3 Labeled right lateral xeroradiograph of an Amazon parrot's body. (From Smith SA, Smith BJ: *Atlas of avian radiographic anatomy*, Philadelphia, 1992, Saunders.)

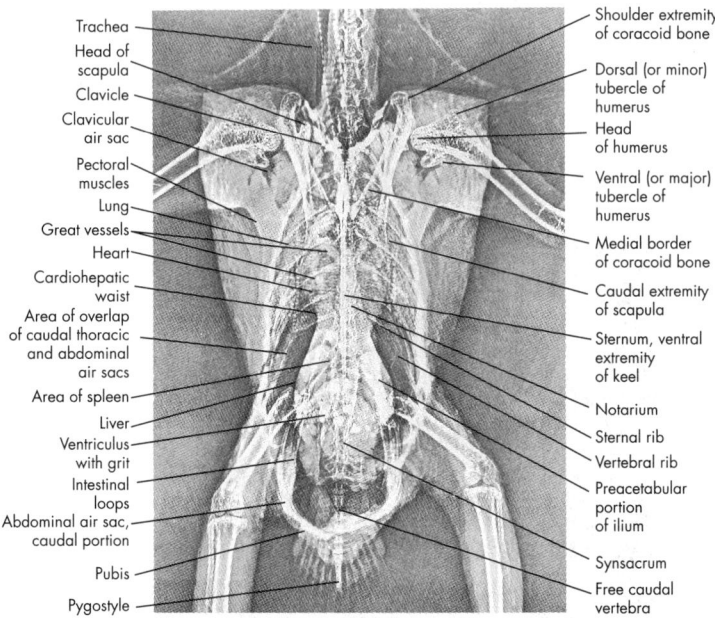

Fig. 19-4 Labeled ventrodorsal xeroradiograph of an Amazon parrot's body. (From Smith SA, Smith BJ: *Atlas of avian radiographic anatomy*, Philadelphia, 1992, Saunders.)

carpal bones are fused, forming the carpometacarpus. The manus has three digits: the alula (I), the major (II), and the minor (III) (Fig. 19-5).[5]

D. Pelvic Girdle and Appendages

1. The pelvic girdle consists of the fused ilium and ischium and the unfused pubis.
2. Distal to the pelvis lie the femur, the patella, the fibula, the tibiotarsus, the tarsometatarsus, and digits.
3. Most avian species have four digits on the pelvic limb. The digits are numbered one through four, starting with the most caudal medial toe. Each digit has one more phalanx then its digit number (i.e., one has two, two has three, and so on [Fig. 19-6]). The podotheca, or thickened foot skin, may be visible on radiographs.

E. Pathology

1. Avian skeletal trauma may result in fractures, sprains and, in rare cases, luxations.[1] Patients that may have a fracture should be radiographed as soon as possible to determine the type, location, and articular involvement and whether the fracture is simple or comminuted.[1]
2. In psittacines, metabolic bone disease and pathologic fractures are more common than traumatic injuries, infection or neoplasia.[1] Metabolic bone disease decreases bone opacity and causes cortical

thickening. The radiographic technique is important because overexposure can mimic decreased bone opacity. Hand-feeding or parental diets high in vitamin D_3 or calcium and phosphorus are notorious for causing metabolic bone disease and gout in neonates.[1] Subsequent osteomalacia (softening of the bones) and folding fractures (bending of the bones) typically are seen in the legs (valgus deformity, splay leg), spine (kyphosis, lordosis, and scoliosis), and sternum (compression).
3. In breeding season normal hens can develop hyperostosis or osteomyelosclerosis, which is depicted radiographically as an irregular increased medullary bone opacity.[8] This normal physiologic change can be exacerbated by hyperestrogenism and oviductal tumors in females, and by Sertoli cell tumors in males. The femurs, tibiotarsus, humerus, radius, and ulna are the bones affected by hyperostosis.[8]

III. Respiratory System

A. Trachea
1. The avian trachea is tubular and has radiographically visible complete tracheal rings (see Fig. 19-3). In most birds the trachea is seen coursing down the right side of the neck on the VD view; however, species vary.[7]
    a. In toucans and mynahs, the trachea deviates ventrally at the thoracic inlet.
    b. In curassows and spoonbills, the trachea is extremely elongated and coils between the skin and pectoral muscles.
    c. In swans and cranes, the elongated trachea coils within a sternal excavation.
    d. In male ducks of many species, the caudal part of the trachea forms a bulb, or bulla.
2. Luminal and extraluminal tracheal masses, strictures, or stenosis may be visible radiographically. The syrinx (voice box) is located at the end of the trachea at the base of the heart and may be visible radiographically.

B. Lungs
1. On lateral radiographs the normal avian lung has a honeycomb appearance because of visualization of "end-on" parabronchi (see Fig. 19-4).
2. The parabronchial infiltration caused by bacterial, granulomatous, or neoplastic lung disease enhances the parenchyma's honeycomb pattern.[1] Pneumonia generally causes similar changes at the caudal aspect of the avian lung. Granulomatous disease causes focal lesions, whereas pulmonary hemorrhage or edema may cause a diffuse parenchymal change.[1]
    a. Although these specific patterns have been described, it is our opinion that they are difficult to discriminate radiographically, except for large granulomatous focal lesions.

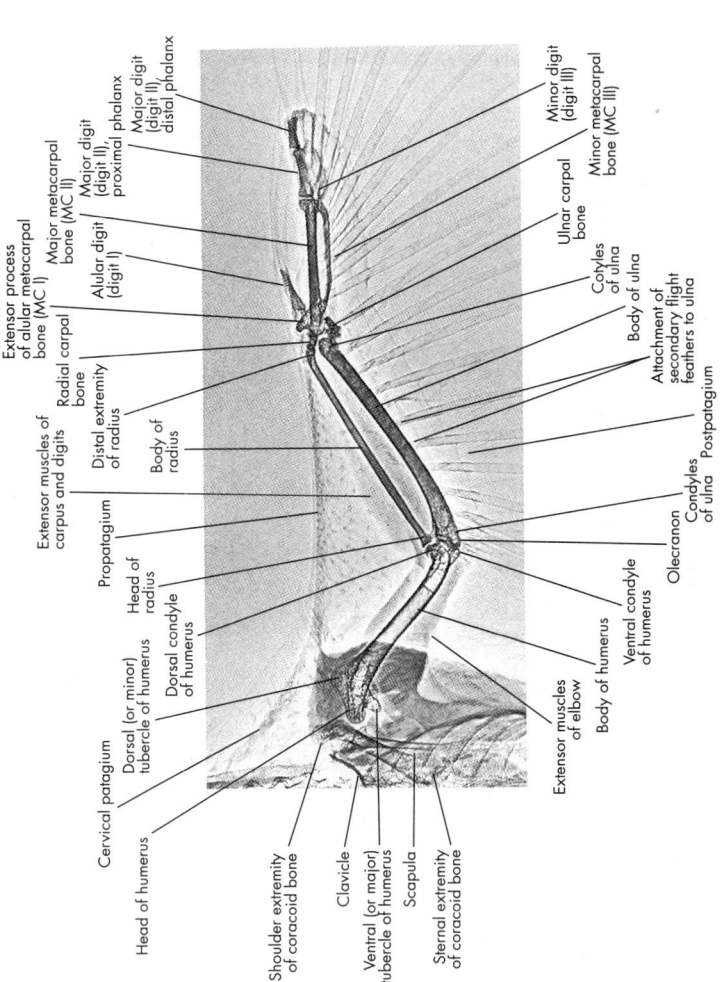

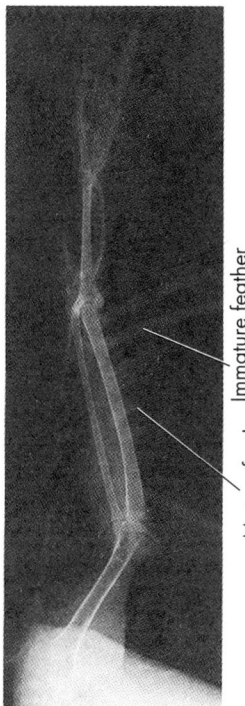

Fig. 19-5 Labeled ventrodorsal xeroradiograph of an Amazon parrot's wing. (From Smith SA, Smith BJ: *Atlas of avian radiographic anatomy*, Philadelphia, 1992, Saunders.)

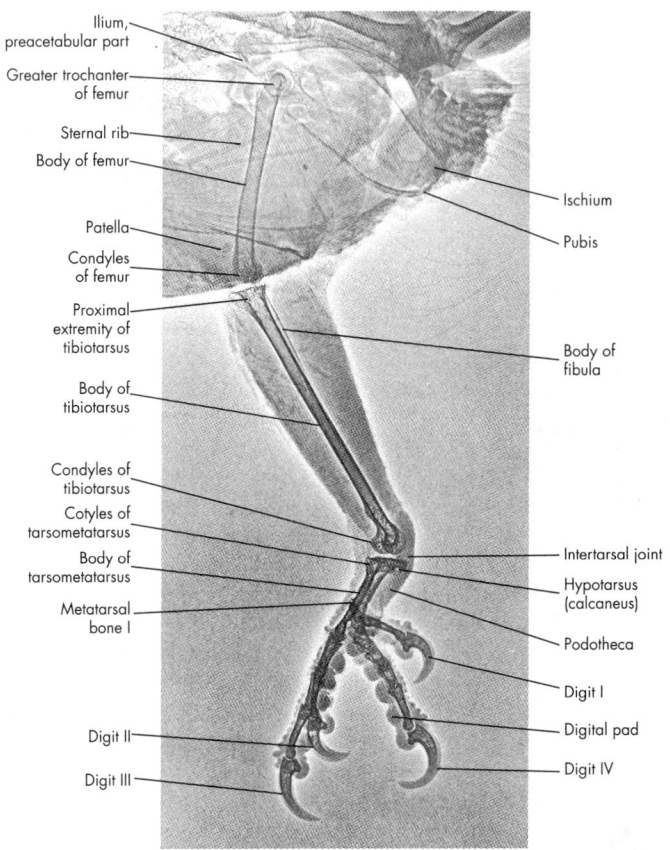

Fig. 19-6 Labeled mediolateral xeroradiograph of an Amazon parrot's leg. (From Smith SA, Smith BJ: *Atlas of avian radiographic anatomy*, Philadelphia, 1992, Saunders.)

C. Air Sacs
1. Birds have nine pulmonary air sacs: two cranial, two caudal thoracic, two abdominal, two cervical, and one clavicular. Parts of each are visible radiographically (see Figs. 19-3 and 19-4).[4,5]
2. In addition to the pulmonary air sacs, many psittacines have a cervicocephalic air sac, which originates from the infraorbital sinus. This air sac covers the head caudally and dorsally and may be partly visible radiographically (see Fig. 19-2).[5] In my experience (April Romagnano), the cervicocephalic air sac of Amazon parrots is most commonly hyperinflated after trauma or with respiratory disease.
3. Severe air sacculitis, with diffusely or focally thickened air sacs, can occur secondary to aspergillosis, mycobacteriosis, or severe bacterial or other fungal disease.

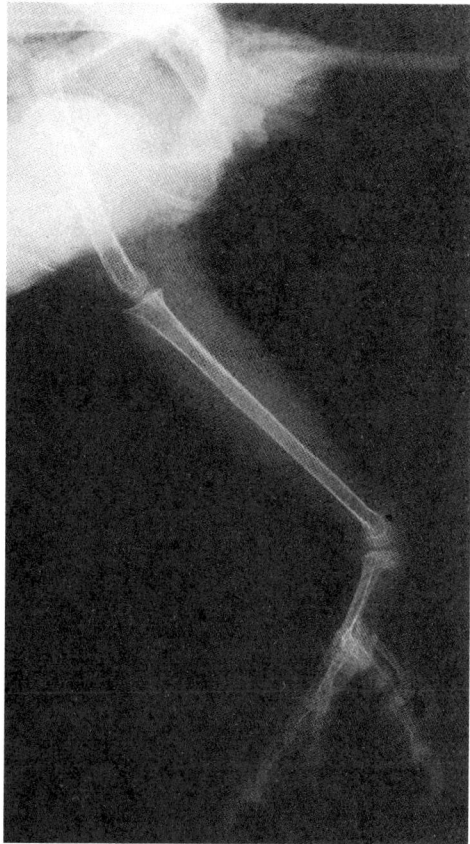

Fig. 19-6, cont'd For legend see opposite page.

    a. Air sacculitis that occurs secondary to aspergillosis may be complicated by bullae, which are visible radiographically because they hyperinflate and displace abdominal organs.
  4. In the case of blunt trauma, endoscopy, or rib or coracoid fractures, subcutaneous emphysema may occur in association with regionally affected air sacs.

D. Pneumatic Bones
  1. Several avian bones are classified as pneumatic because of their connection with the respiratory system via the air sac system.
    a. The cervical vertebrae connect to the cervical air sac.
    b. The thoracic vertebrae, ribs, and humerus connect to the interclavicular air sac.

      c. The synsacrum and femur connect to the abdominal air sacs.[1]
   2. Because of this connection, infection can be introduced into these bones directly from a localized respiratory infection.

IV. Heart

A. Radiographic Anatomy
   1. The avian cardiac silhouette lies between the second and sixth thoracic ribs. Although avian cardiac size and function are best assessed with echocardiography,[1] the width of the heart base (measured at the fifth thoracic vertebra in psittacines) should be about 50% of the total width of the coelomic cavity.[1]
   2. The entire heart and liver outline on the VD view creates an hourglass silhouette that is larger at the distal aspect. This classic cardiohepatic silhouette changes in size and shape with disease and organ displacement.
      a. Because macaws normally have a smaller liver compared to other psittacines, their cardiohepatic silhouette is more dramatic.
   3. Primary cardiac disease is quite rare in birds, except for congenital disease in baby birds.[1]
   4. Acquired cardiac disease, which is more common, falls into three radiographically visible categories: a heart that is enlarged, a heart that is diminished in size, and a heart that is affected by arteriosclerosis complicated by mineralization (Table 19-1).
      a. Arteriosclerosis complicated by mineralization is visible on radiographs because it causes enlargement of the great vessels and increased opacity of the caudal lung.

Table 19-1
Differential diagnoses for radiographic changes in the cardiac silhouette

Radiographic finding	Differential diagnoses
Cardiomegaly	Poxvirus Myxomatous valvular degeneration Endocarditis Hemochromatosis Chronic anemia Compression from extrinsic masses Pericardial effusion (secondary to chlamydiosis, eastern equine encephalitis, sarcocystosis, polyomavirus, neoplasia, or tuberculosis)
Microcardia	Critically dehydrated or hypovolemic patients
Enlarged great vessels (arteriosclerosis)	Older birds fed high-fat diets Very young birds

V. Digestive System

A. Radiographic Anatomy

1. The digestive system consists of the cervical esophagus, the crop, the thoracic esophagus, the proventriculus, the ventriculus, the intestines, and the cloaca (see Figs. 19-3 and 19-4).
2. The crop lies to the right of the midline on the VD view.
3. The proventriculus lies to the left of the crop on the VD view and is dorsal to the liver on the lateral view.[1,4]
4. On conventional radiographs the ventriculus is identifiable as a large, round organ located caudoventral to the proventriculus on the lateral view; the ventriculus frequently contains radiopaque material, which facilitates its identification (see Fig. 19-4).
5. The intestines occupy the caudodorsal portion of the abdominal cavity (see Fig. 19-4).[1] They typically are unremarkable unless dilated by fluid or air.
6. The pancreas is located in the right cranioventral abdomen within the duodenal loop and is not visible radiographically if of normal size. Pancreatic enlargement is unusual.[1]
7. The intestines terminate in the cloaca, which may be visible on radiographs if it contains appreciable air, fluid, feces, and urates (see Figs. 19-3 and 19-4).
8. Dilation of any part of the gastrointestinal (GI) tract is considered abnormal except in neonatal and juvenile birds before weaning.[1]

B. Pathology

1. Distention of the intestinal tract with fluid is abnormal in most birds, except toucans and mynahs.[1]
2. A nonspecific finding, fluid-filled intestines may result from various disease states, including intraluminal infection or inflammation or luminal obstruction.
3. Dilation of the intestines with gas is abnormal in many avian species and should be further investigated with other imaging techniques, such as GI contrast studies, fluoroscopy, and ultrasonography.[1]

C. Neonatal GI Radiology

1. In parent- and syringe-fed baby birds, the GI tract, especially the proventriculus, normally is dilated and predominately fluid filled until the bird is weaned.
2. Dilation of the proventriculus in weaned juvenile or adult birds is an indicator of disease. Typically, such a proventriculus is filled with gas or food and is evident on radiographs.
   a. Differential diagnoses for dilation of the proventriculus (Table 19-2) include the following:
      (1) Bacterial, fungal, or parasitic proventriculitis
      (2) Proventricular dilation disease

Table 19-2
Differential diagnoses for radiographically evident dilation of the gastrointestinal tract

Area of dilation	Differential diagnoses
Proventriculus	Proventriculitis (parasitic, bacterial, ot fungal) Proventricular dilation disease (PDD) Other neuromuscular disease Heavy metal toxicity Impaction Foreign body Normal baby bird
Ventriculus	Ventriculitis (parasitic, bacterial, or fungal) PDD or other neuromuscular disease Impaction Foreign body Neoplasia
Intestines	Gastroenteritis (parasitic, bacterial, or fungal) PDD or other neuromuscular disease Neoplasia Luminal or extraluminal mass obstruction Foreign body
Cloaca	Cloacitis Papillomas PDD or other neuromuscular disease Cloacolith Retained soft-shelled egg Traumatic dilation Idiopathic dilation

    (3) Impaction
    (4) Presence of a foreign body
    (5) Heavy metal toxicity (e.g., lead poisoning)
        (a) It should be noted that lead toxicity may be seen in patients that may or may not have a metallic foreign body; therefore the serum lead level should be determined in patients with clinical signs.

VI. Spleen

A. Radiographic Anatomy
1. The spleen of psittacine birds is best depicted on a lateral whole body radiograph (see Fig. 19-4).[5]
2. The spleen often is difficult to see because it silhouettes with the organs of the GI tract.
3. Normally the spleen is slightly dorsal to the proventriculus, and its diameter is less than two times the width of the femur. Hence, splenomegaly usually can be identified easily on lateral radiographs as a larger then normal, circumscribed, soft tissue opacity dorsal to the proventriculus.
4. On VD radiographs an enlarged spleen is seen midline and immediately cranial to the pelvis.

Table 19-3
Differential diagnoses for splenomegaly

Splenic changes	Differential diagnoses
Uniform splenomegaly	*Infectious diseases:* Chlamydiosis; viral, bacterial, and mycobacterial disease *Parasitic disease:* Sarcocystosis *Neoplastic disease:* Lymphoma *Metabolic diseases:* Lipidosis and hemochromatosis
Irregular splenomegaly	*Neoplastic diseases:* Hemangiosarcoma, fibrosarcoma, and leiomyosarcoma

5. Differential diagnoses for splenomegaly are listed in Table 19-3.

VII. Liver

A. Radiographic Anatomy and Pathology

1. The liver of psittacine birds does not normally extend beyond a line drawn from the coracoid to the acetabulum on the VD view (see Fig. 19-3).[1]
2. Loss of the classic hourglass cardiohepatic waist may indicate liver, heart, or proventricular enlargement.
3. If the liver is enlarged, the abdominal air sacs may be compressed, the heart may be slightly displaced cranially, the proventriculus may be displaced dorsally, and the ventriculus may be displaced caudodorsally.
4. On the lateral view the liver should not extend beyond the sternum (see Fig. 19-4).
5. Hepatomegaly may be differentiated from an enlarged proventriculus by using the VD view. However, both the lateral and VD views are necessary for a proper comparison. Definitive differentiation of hepatomegaly from an enlarged proventriculus is best achieved with positive or negative contrast techniques.
6. Differential diagnoses for hepatomegaly are presented in Table 19-4.

VIII. Kidneys

A. Radiographic Anatomy

1. The avian kidneys lie deep within the renal fossa. They are securely attached to the synsacrum and are flattened dorsoventrally.[1]
2. Each kidney has three divisions (cranial, middle, and caudal).
3. The ureter empties directly into the cloaca, because birds do not have a urinary bladder.
4. Kidneys are best viewed on lateral radiographs, in which their silhouettes are superimposed (see Fig. 19-4).
5. In the VD view the cranial pole of the cranial kidney is seen at the cranial edge of the pelvis.

Table 19-4
Differential diagnoses for hepatomegaly

Liver changes	Differential diagnoses
Hepatomegaly	*Infectious diseases:* Chlamydiosis; viral, bacterial, mycobacterial, and fungal disease *Primary neoplasia:* Biliary adenocarcinoma, hepatocellular carcinoma, fibrosacoma, hemangiosarcoma, hepatoma, and lymphoma *Metastatic neoplasia:* Adenocarcinoma, fibrosarcoma, and melanoma *Parasitic diseases:* Toxoplasmosis, sarcocystosis, flukes, and Plasmodium infection *Metabolic diseases:* Lipidosis, hemochromatosis, gout, and pansteatitis
Microhepatica	Hepatic cirrhosis; also, may occur normally in macaws

Table 19-5
Differential diagnoses for nephromegaly

Kidney changes	Differential diagnoses
Uniform nephromegaly	*Infectious diseases:* Bacteria, *Chlamydia* organisms *Neoplastic diseases:* Adenocarcinoma, lymphoma, embryonal nephroma *Metabolic diseases:* Lipidosis, gout, and dehydration (gout and dehydration increase kidney radiopacity; gout causes renal mineralization, which is evident radiographically) Toxicity: Heavy metal intoxication
Irregular nephromegaly	*Neoplastic diseases:* Adenocarcinoma, embryonal nephroma *Abscesses:* Bacterial, fungal organisms *Cystic kidney disease:* Idiopathic
Enlargement of the cranial pole of the cranial kidney	*Kidney enlargement:* See above *Adrenal enlargement:* Infection, inflammation, or neoplasia *Gonadal enlargement:* Normal cyclic change (testicular enlargement), infection, inflammation, or neoplasia

B. Pathology
1. Renal enlargement causes a caudodorsal soft tissue projection that advances in a ventral direction. To determine if nephromegaly is present, the radiograph should be compared with those of clinically normal patients of the same species and sex and of similar size, age, and weight.
2. The standard kidney length of a large psittacine is thought to be approximately 3 cm. Avian kidneys normally are surrounded by air[1]; hence, kidneys longer than 3 cm or loss of the air shadow in a medium to large psittacine suggests renal enlargement.[1]
3. Table 19-5 shows the differential diagnoses for renal enlargement.

IX. Gonads and Associated Structures

A. Radiographic Anatomy
1. The avian gonads are located ventral to the cranial pole of the cranial division of the kidneys, directly caudal to the adrenal gland.
2. Gonadal enlargement during the breeding season is normal in adults and may be visible on radiographs. Enlarged testes resemble a uni-

Table 19-6
Differential diagnoses for gonadal and associated organ enlargement

Organ changes	Differential diagnoses
Testicular enlargement	*Neoplastic diseases:* Seminoma, Sertoli cell and interstitial cell neoplasia, and lymphoma *Infectious diseases:* Bacterial or fungal orchitis
Ovarian enlargement	*Neoplastic diseases:* Adenocarcinoma, adenoma, granulosa cell tumor, lipoma, fibrosarcoma, and carcinomatosis *Retained yolks:* Ectopic or misovulated eggs located within coelomic cavity
Oviductal enlargement	*Neoplastic diseases:* Adenomatous hyperplasia, adenocarcinoma, adenoma, and carcinomatosis *Infectious diseases:* Salpingitis, egg yolk peritonitis *Eggs:* Retained eggs (soft shelled or hard) *Cystic oviducts:* Idiopathic

form soft tissue mass in the caudodorsal abdomen; an active ovary may resemble an irregular cluster of grapes.
3. The testes are paired in males.
4. Females typically have only a left ovary.

B. Pathology

1. Testicular or ovarian enlargement and adrenal enlargement easily can be confused with nephromegaly in the caudodorsal abdomen. Short of surgery or endoscopy, nephromegaly can be differentiated from gonadal or adrenal enlargement by intravenous excretory urography or ultrasonography (see section XII, part A, below).
2. Ovarian neoplasia is more common than oviductal neoplasia in psittacines. Ovarian tumors can enlarge to one-third body weight, causing massive organ displacement in the cranioventral direction.
3. Abdominal herniation, ascites, and cysts are common sequelae to neoplasia of the ovary and oviduct.
4. The most significant clinical signs that occur secondary to abdominal enlargement are dyspnea and unilateral or bilateral paresis or paralysis.
5. The differential diagnoses for gonadal and associated organ enlargement are presented in Table 19-6.

X. Radiographic Illustration

See Figs. 19-7 through 19-14.[1]

Text continued on p 418

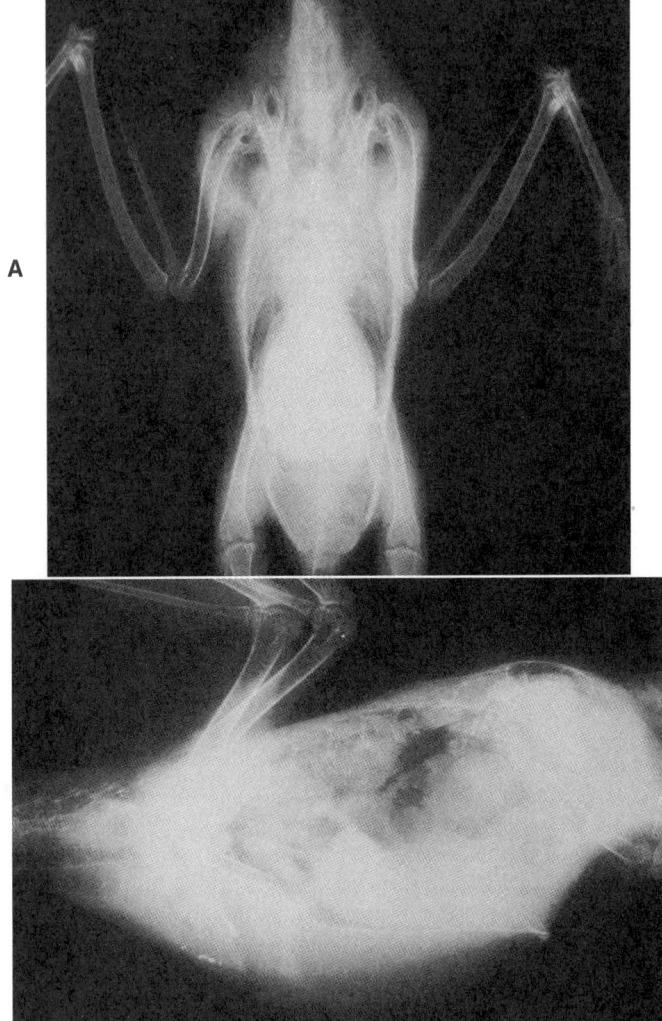

Fig. 19-7 Proventricular dilation. **A,** Ventrodorsal radiograph of an adult female African grey parrot. The proventriculus is grossly dilated with gas and fluid. **B,** On the right lateral projection, the spleen and small intestines can be seen "through" the distended organ. The margin of the proventriculus can be seen extending laterally beyond the margin of the cardiac silhouette on the ventrodorsal view.

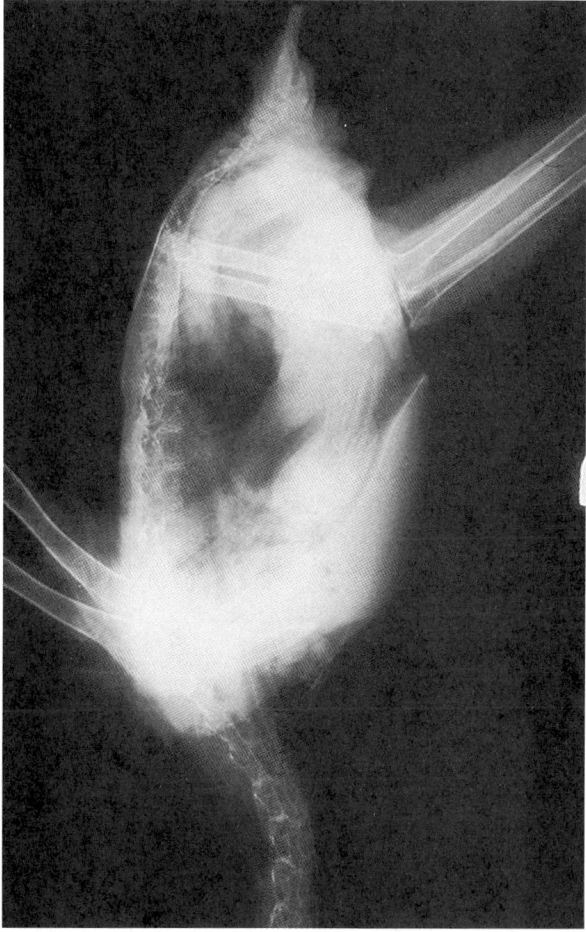

Fig. 19-8 Splenomegaly. Right lateral radiographic view of a Moluccan cockatoo with splenomegaly. The enlarged spleen is seen as a circumscribed, soft tissue opacity dorsal to the caudal aspect of the proventriculus and to the cranial aspect of the ventriculus. The spleen usually is best seen on the lateral view.

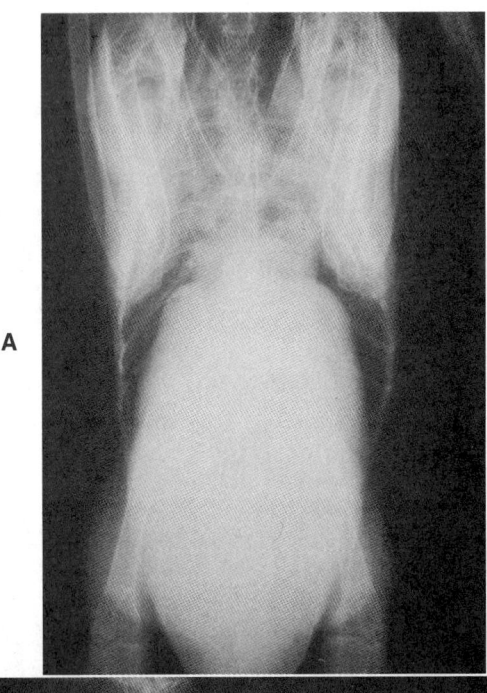
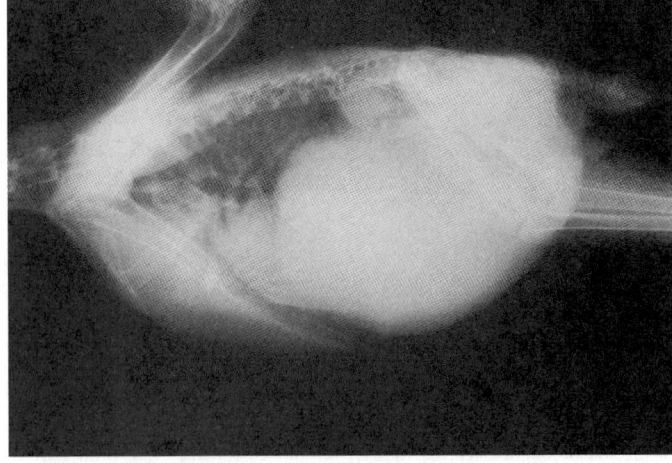

Fig. 19-9 Hepatomegaly. Ventrodorsal **(A)** and right lateral **(B)** radiographs of an adult toucan. The liver is grossly enlarged, causing loss of the cardiohepatic "waist" and cranial displacement of the cardiac silhouette.

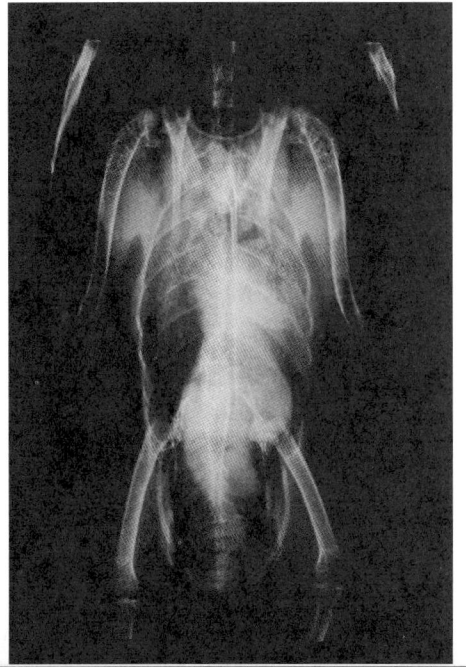

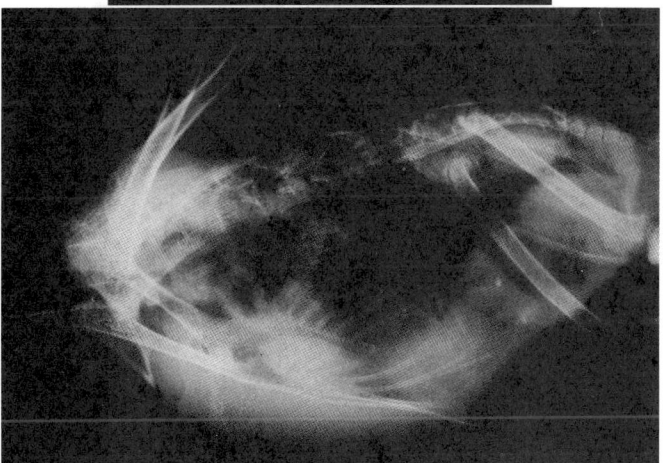

Fig. 19-10 Pulmonary disease. Ventrodorsal **(A)** and right lateral **(B)** radiographs of an adult female Amazon parrot. Ill-defined soft tissue opacity can be seen in the left lung, with increased parabronchial infiltrate. The ill-defined, increased soft tissue opacities in the left thoracic air sac are best seen on the ventrodorsal view. Infectious or inflammatory disease involving the lung and air sac should be considered in the diagnosis.

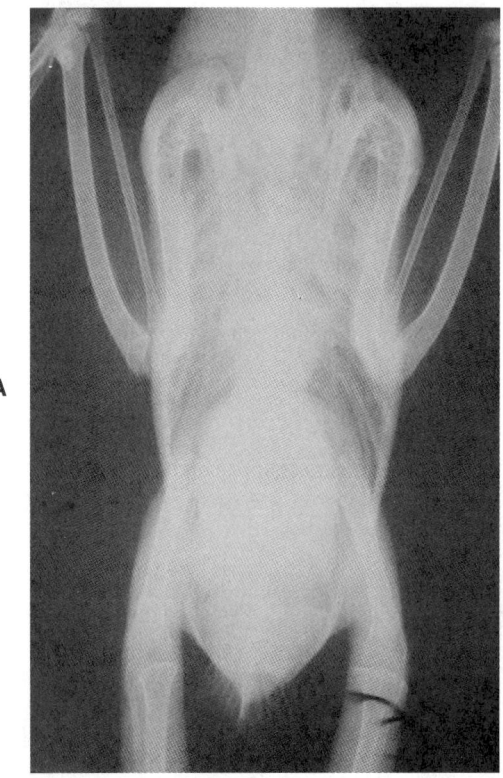

Fig. 19-11 Renomegaly. Ventrodorsal **(A)** and right lateral **(B)** radiographs of an adult male Amazon parrot. Note the enlarged cranial pole region of the kidneys on the lateral view. The kidney margins can be seen extending cranial and lateral to the pelvis on the ventrodorsal view. This enlargement is consistent with renomegaly; other differential diagnosis considerations would include an active gonad or, less likely, an enlarged adrenal gland.

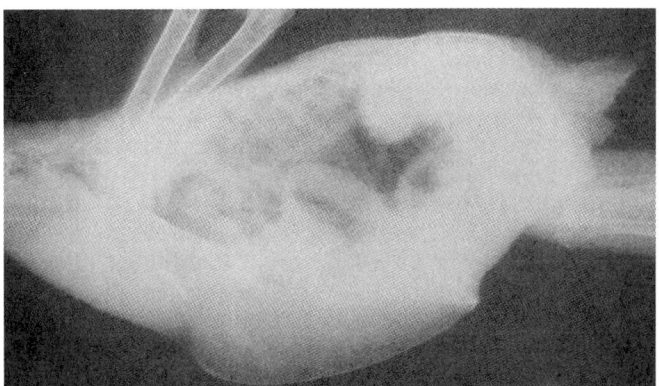

Fig. 19-11, cont'd For legend see opposite page.

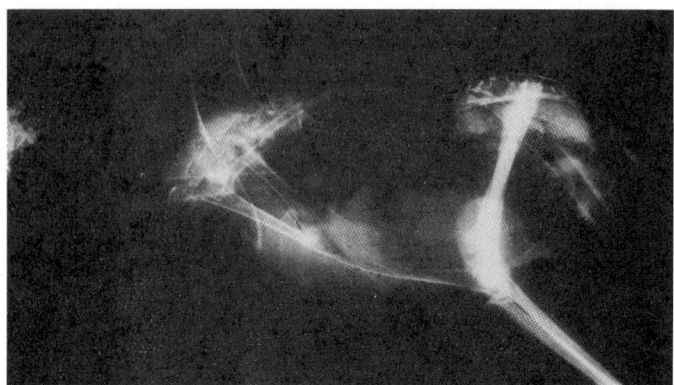

Fig. 19-12 Renal mineralization. Right lateral radiograph of an adult Amazon parrot showing generalized, increased mineral opacity involving the pelvic long bones. The kidneys are more opaque than the other coelomic soft tissue structures and are the same opacity as the pelvic limbs.

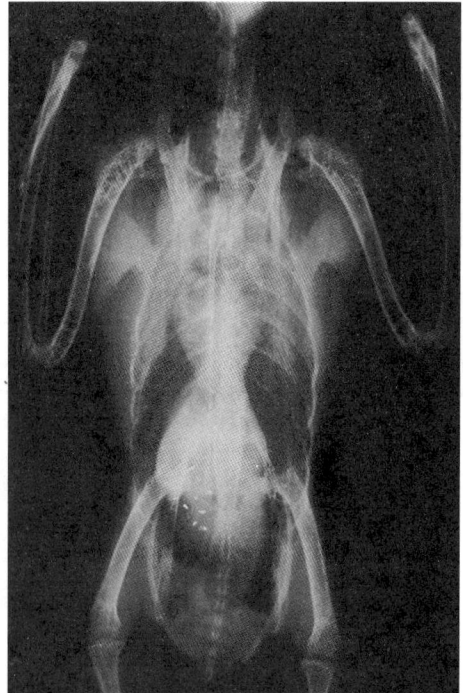

Fig. 19-13 Lead toxicity. Ventrodorsal radiographs of an adult female Amazon parrot. Multiple metallic opacities can be seen in the ventriculus. Note that these opacities are more opaque than grit, which normally may be found in the ventriculus.

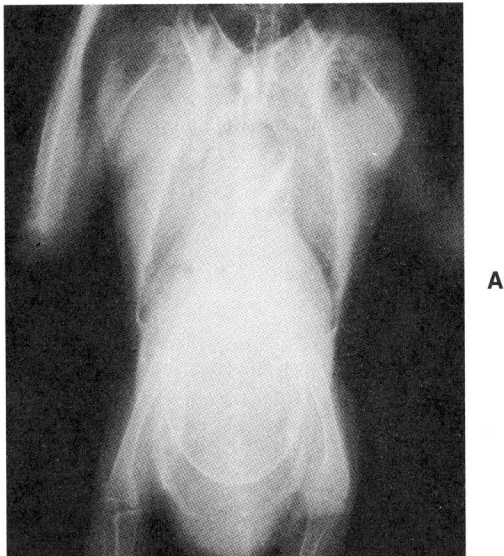
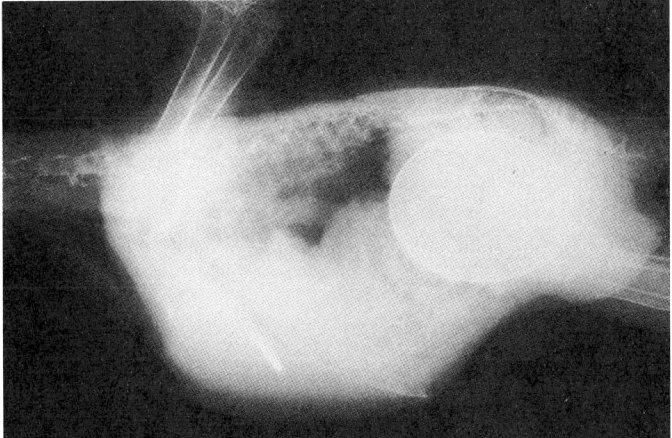

Fig. 19-14 Retained egg. Ventrodorsal **(A)** and right lateral **(B)** radiographs of an adult female *Eclectus* parrot. The large, ovoid, mineral opacity mass in the caudal coelomic cavity is a retained egg. The egg is causing cranioventral displacement of gastrointestinal tract structures. On the ventrodorsal view, note the artifactual loss of the cardiohepatic "waist" as a result of the mass effect of the egg. Incidental findings include an identification chip in the keel musculature and a motion artifact (wing movement) on the ventrodorsal view.

XI. Contrast Techniques

A. GI Studies
1. Selected radiopaque contrast agents are administered by gavage for GI tract studies. The following agents can be used for this purpose:
   a. Barium sulfate suspension: 10 to 15 ml/kg
   b. Iohexol (Omnipaque): 1 ml/100 g (a nonionic, low-osmolar, water-soluble, iodinated contrast medium made by Nycomed, Princeton, NJ)
2. Diatrizoate meglumine and diatrizoate sodium injection (Renografin-76) (an ionic, high-osmolar, water-soluble, iodine contrast medium made by Solvay Veterinary, Mendota Heights, Minn, or Hypaque-76, Nycomed, Princeton, NJ) also can be used to evaluate the GI tract.
   a. Renografin-76 is contraindicated in critically ill patients because it is hypertonic and draws fluids into the GI tract, exacerbating debilitation and dehydration.
   b. If this contrast agent is aspirated, it can cause severe pulmonary edema.
3. Survey radiographs should be made without contrast before all contrast studies are done to establish a baseline and to aid in the assessment of the patient's general health and condition.
4. Mechanical restraint or general anesthesia may be used with extreme care. If the bird is anesthetized, it must be allowed to recover fully after the survey radiographs before contrast material can be given safely by gavage.
5. Care must be taken not to give too much contrast medium, especially barium. Surprisingly small amounts of barium (10 to 15 ml/kg) are needed to achieve adequate and diagnostic coating of the GI tract.
6. If the bird is anesthetized, during recovery it should be held in an upright position and the throat swabbed clean (author April Romagnano's experience).

B. Barium Sulfate
1. Barium sulfate suspension is an excellent, reasonably priced, positive contrast agent. However, it may be contraindicated in cases of possible GI tract perforation, regurgitation, or vomiting because it can cause extensive tissue reactions if leaked or aspirated.
2. Water-soluble contrast medium (iohexol) is absorbed if introduced into the coelomic cavity, but barium sulfate causes granulomas that become "walled off" in companion animals.
3. During a barium study, sequential images in the VD and LRL positions should be taken 30 minutes apart until the contrast reaches the area of interest.[1]
4. Use of barium sulfate suspension precludes subsequent ultrasound scans (or surgery in some cases) until the barium sulfate passes from the GI tract.

C. Iohexol
1. Iohexol is a water-soluble, nonionic, iodinated, low-osmolar contrast medium commonly used for companion animal myelography.[9] It may be used as a GI contrast agent.
2. If aspirated, iohexol is absorbed from the coelomic cavity with minimal tissue reaction.
3. Although adequate, in the author's experience (April Romagnano) iohexol studies often produce uneven luminal coverage and bubbles.
4. Sequential images in the VD and LRL positions should be taken 15 minutes apart until the contrast reaches the area of interest.[1]
   a. Iohexol moves approximately twice as fast as barium through the avian GI tract (author April Romagnano's experience).
5. Iohexol does not preclude subsequent ultrasound scans, endoscopy, or surgery.

D. Fluoroscopy
1. In birds, fluoroscopic evaluation of the GI tract allows the study of organ movement in real time, which may be the best way to monitor GI motility.
2. Imaging can be done in the following manner: The bird is gavaged with barium sulfate suspension (10 to 15 ml/kg) and placed unanesthetized in a small box. A horizontal x-ray beam technique then is used to make the radiographs.
3. Fluoroscopy recently has gained importance in avian diagnostics for screening and evaluating of proventricular dilation disease.[10]

E. Urography
1. Intravenous excretory urography, a rather uncommon radiographic procedure, requires the use of intravenous, water-soluble contrast medium (Renografin-76.)
2. The indications for this study include defining mass lesions associated with the urinary tract and delineating the shape and size of the kidneys.
3. All studies should include survey films and sequential VD and LRL images at 1-, 2-, 5-, 10-, and 20-minute intervals.[1]

XII. Alternative Imaging Modalities

Although conventional radiography currently is the most accessible imaging modality in avian diagnostics, evaluation of caudal coelomic structures often is hindered by poor detail of the coelomic cavity. GI contrast studies or alternative imaging modalities such as ultrasonography (US), computed tomography (CT), and magnetic resonance imaging (MRI) may allow noninvasive evaluation of the coelomic cavity.

A. Uses

1. Avian cardiac size and function are best assessed with echocardiography, a specific ultrasonographic examination.
2. Radiographic evaluation of the avian brain and spinal cord is rarely done because cerebral angiography and myelography are difficult, potentially dangerous, and currently impractical clinically in birds.[11] For these reasons, little published information is available on avian nervous system imaging.
3. Nuclear scintigraphy currently has limited application in birds, although a few studies have demonstrated its usefulness in evaluation of the avian thyroid gland and spine.[12-14]

B. Ultrasonography

1. In a sonographic evaluation, tissues interact with high-frequency sound waves to generate an image. This is a noninvasive technique that may not require anesthesia if a tractable patient can be restrained manually.
2. Ultrasonography is particularly useful for delineating caudal coelomic masses, disease in the liver, and diseases of the female reproductive tract.[1,15]
3. This technique also can be used to differentiate soft-shelled eggs from egg yolk peritonitis and to perform ultrasonographically guided liver biopsies and echocardiography.[1,15]
4. Focal peripheral pulmonary lesions sometimes can be seen.
5. As with radiographs, the spleen often is difficult to see on ultrasound scans.
6. Less common uses for ultrasonography in birds include ocular and bone contour studies.[1]
7. Compared with studies in mammals, the use of ultrasonography in birds may be limited by the patient's size and by the extensive air sac systems, which limit sonographic windows.
8. Because most avian patients are small, a high-frequency transducer is recommended. The 7.5 and 10 MHz transducers are good choices; however, a 5 MHz transducer may be used for larger patients or with a "stand off" pad for smaller birds.

C. Computed Tomography

1. Computed tomography is good for diminishing anatomic complexity by evaluating regions in multiple planes. The technique uses digital cross-sectional images that may be reconstructed in a three-dimensional format; further sagittal and dorsal planes also can be evaluated.
2. Computed tomography is useful for evaluating cranial, ocular, otic, respiratory system, and spinal lesions in birds, because these areas are poorly evaluated by conventional radiography and contrast techniques (see above).
3. Computed tomography is minimally invasive but uses ionizing radiation.

4. The use of intravenous contrast (iohexol) is optional.
5. General or parenteral anesthesia (depending on the species) is required, because it is imperative that the patient not move.
6. Avian anatomic imaging studies using computed tomography have described the lungs and air sacs of psittacines[16] and the tomographic anatomy of the golden eagle[17] and African grey parrot.[18]

D. Magnetic Resonance Imaging
1. Magnetic resonance imaging is one of the most significant advances in medical imaging this century. In birds it is best performed using a 1.5 Tesla magnet and a human knee coil. The bird is anesthetized with isoflurane, and the head and coelomic cavity can then be imaged in the dorsal, sagittal, and transverse planes to produce T_1-weighted, T_2-weighted, and contrast medium–enhanced T_1-weighted images (gadolinium-DTPA is the most commonly used intravenous contrast medium).
2. In avian patients, magnetic resonance imaging offers many advantages over conventional radiography and computed tomography, including multiplanar imaging, lack of ionizing radiation, and greatly improved soft tissue contrast resolution.
3. Magnetic resonance imaging has been used to describe the anatomy of the normal eye and orbit of a euthanized screech owl.[19] Recently it was shown that magnetic resonance imaging of the avian brain and coelomic cavity allowed accurate identification of anatomic structures in anesthetized pigeons.[20]
4. The human knee surface coil was an appropriate size for imaging of the pigeon coelomic cavity and brain. This coil also is appropriate for mid- to large-sized psittacines (unpublished observations, April Romagnano).

E. Nuclear Scintigraphy
1. Nuclear scintigraphy requires the administration of radioactive isotopes "tagged" with organ-specific probes.[21] A gamma camera is used to image the radioactivity emitted from the patient.[21]
2. Before the patient is released to its owner, an overnight (or longer) hospital stay may be required to ensure that the bird is no longer radioactive.
3. To prevent the patient from moving, mechanical or chemical restraint, or general anesthesia is required.
4. Nuclear scintigraphy is not commonly used in birds,[12] but small animal applications include evaluation of bone, the thyroid gland,[13,14] the glomerular filtration rate, transcolonic portal vein hepatic perfusion, lung perfusion, and occult bleeding.
5. Bone scintigraphy ("bone scans") for the evaluation of active lesions probably is the most useful application of nuclear scintigraphy in birds.[12] This technique can aid in the assessment of areas difficult to evaluate on conventional radiographs, such as the avian spine, skull, and distal extremities.[12]

a. An increase in radiotracer activity (hot spots) may be seen with fractures, infection, inflammation, or neoplasia.
b. Regions of decreased radiotracer uptake (cold spots) may indicate sequestra, nonunions, or infarcts.[21]

XIII. Conclusion

Imaging is an integral part of the complete diagnostic evaluation of an avian patient. Time spent learning anatomy and the indications for specific imaging studies and modalities, as well as the formulation of a reliable technique chart, will make imaging safer, more rewarding, and more effective.

Alternative imaging is in its infancy in avian medicine. However, we predict that this area will develop considerably in the near future, increasing the information available and facilitating accessibility.

REFERENCES

1. McMillian MC: Imaging techniques. In Ritchie BW, Harrison GJ, Harrison LR, editors: *Avian medicine: principles and applications,* Lake Worth, Fla, 1994, Wingers, pp 246-326.
2. Morgan JP: Technique charts. In Morgan JP, editor: *Techniques of veterinary radiography,* ed 5, Ames, Iowa, 1993, Iowa State University Press, pp 78-85.
3. Paul-Murphy JR et al: Psittacine skull radiography: anatomy, radiographic technique, and patient application. *Vet Radiol* 31:125-131, 1990.
4. Smith SA, Smith BJ: Radiographic evaluation of general avian anatomy. In Smith SA, Smith BJ, editors: *Atlas of avian radiographic anatomy,* Philadelphia, 1992, Saunders, pp 5-18.
5. Smith SA, Smith BJ: Orange-winged Amazon. In Smith SA, Smith BJ, editors: *Atlas of avian radiographic anatomy,* Philadelphia, 1992, Saunders, pp 19-44.
6. Walsh MT: Radiology. In Harrison GJ, Harrison LR, editors: *Clinical avian medicine and surgery,* Philadelphia, 1986, Saunders, pp 201-233.
7. King AS, McLelland J: Respiratory system. In King AS, McLelland J, editors: *Birds: their structure and function,* Philadelphia, 1984, Bailliere Tindall, pp 110-144.
8. Quesenberry KE, Hillyer EV: Supportive care and emergency therapy. In Ritchie BW, Harrison GJ, Harrison LR, editors: *Avian medicine: principles and applications,* Lake Worth, Fla, 1994, Wingers, pp 382-416.
9. Williams J et al: Use of iohexol as a gastrointestinal contrast agent in three dogs, five cats, and one bird, *J Am Vet Med Assoc* 202:624-627, 1993.
10. Taylor M: Fluoroscopic examination of the motility of the psittacine proventriculus and ventriculus: a new method to aid in screening for PDD? *Proceedings of the International Avicultural Society,* 1996.
11. Harr KE et al: A myelographic technique for avian species, *Veterinary Radiology and Ultrasound* 38:187-192, 1997.
12. Lung NP, Ackerman N: Scintigraphy as a tool in avian orthopedic diagnosis, *Proceedings of the American Association of Zoo Veterinarians, Annual Meeting,* p 45, 1993.
13. Harmes CA et al: Development of an experimental model of hypothyroidism in cockatiels *(Nymphicus hollandicus), Am J Vet Res* 55:399-404, 1994.
14. Harmes CA et al: Technetium-99m and iodine-131 thyroid scintigraphy in normal and radiothyroidectomized cockatiels *(Nymphicus hollandicus), J Vet Radiol Ultrasound* 35:473-478, 1994.

15. Krautwald-Junghanns ME, Ender SF: Ultrasonography in birds, *Semin Avian Exotic Pet Med* 3:140-146, 1994.
16. Krautwald-Junghanns ME, Schumacher F, Tellhelm B: Evaluations of the lower respiratory tract in psittacines using radiology and computed tomography, *J Vet Radiol Ultrasound* 34:382-390, 1993.
17. Orosz SE, Toal RL: Tomographic anatomy of the golden eagle *(Aquila chrysaetos), J Zoo Wildlife Med* 23:39-46, 1992.
18. Love N, Flammer K, Spaulding K: The normal computed tomographic (CT) anatomy of the African grey parrot *(Psittacus erithacus)*: a pilot study, *Proceedings of the American College of Veterinary Radiology*, 1993.
19. Morgan RV, Donnell RL, Daniel GB: Magnetic resonance imaging of the normal eye and orbit of a screech owl *(Otus asio), J Vet Radiol Ultrasound* 35:362-367, 1994.
20. Romagnano A et al: Magnetic resonance imaging of the brain and coelomic cavity of the domestic pigeon *(Columba livia domestica), J Vet Radiol Ultrasound* 37:431-440, 1996.
21. Berry CR, Daniels GB: *Handbook of veterinary nuclear medicine,* Raleigh, 1996, North Carolina State University Press.

20

Parasitism of Caged Birds

Victoria L. Clyde and Sharon Patton

Parasitism is not a common problem in captive-born caged birds compared with other groups of animals, and frequently this diagnosis is overlooked. Some parasites cause little to no harm in their bird hosts, but others may cause moderate to severe disease. The significance of many parasitic infections is unknown. It is important that the avian veterinarian include parasitism as a diagnostic differential when examining birds.

I. Clinical Signs of Parasitism[1-5]

A. Clinical Signs Potentially Indicative of Parasitism
1. Stunting and poor growth of neonates
2. Increased neonate mortality
3. Diarrhea—chronic, intermittent, watery, mucoid, or bloody
4. Malodorous droppings
5. Dyschezia, tenesmus, straining to defecate
6. Gastrointestinal obstruction
7. Oral or esophageal plaques
8. Regurgitation
9. Debilitation, weight loss, hypoproteinemia, anemia
10. Voice changes, coughing, sneezing
11. Dyspnea, open mouth breathing, "gaping"
12. Restlessness, irritability, ataxia, neurologic signs
13. Subcutaneous nodules
14. Proliferative, honey-combed masses of nonfeathered skin
15. Dry skin, pruritus, feather loss, feather picking, self-mutilation
16. Sudden death

Box 20-1
Parasitic Differential Diagnoses for Clinical Signs in Birds

Gastrointestinal (Anorexia, Regurgitation, Diarrhea, or Straining)
Giardia spp.
Hexamita spp.
Trichomonads
Eimeria spp.
Isospora spp.
Cryptosporidium spp.
Atoxoplasma spp.
Ascarids
Capillarids
Gastrointestinal spirurids
Trematodes
Cestodes (many)

Oral Lesions
Trichomonas spp.
Capillarids

Dyspnea, Gaping, or Respiratory Changes
Syngamus trachea
Sternostomid respiratory mites
Tracheal mites
Trichomoniasis (severe)
Sarcocystis falcatula
Atoxoplasma spp.
Toxoplasma gondii

Neurologic Signs
Cerebral nematodiasis
Toxoplasmosis

Illthrift
Giardia spp.
Ascarids
Capillarids
Trematodes
Cestodes

Acute to Peracute Death
Sarcocystis falcatula
Atoxoplasma spp.
Toxoplasma gondii
Tracheal mites

Poor Growth or Juvenile Mortality
Giardia spp.
Trichomonas spp.
Isospora spp.
Cryptosporidium spp.
Eimeria spp.
Atoxoplasma spp.

Dermatologic Abnormalities
Giardia spp.
Knemidokoptic mites
Pelecticus spp.

II. Diagnostic Workup Plan for Parasitism

(Box 20-1)

A. Birds at Risk for Parasitism
1. When maintained in raised-floor cages, captive-born caged birds have little access to many parasites. Therefore, annual testing for parasites, as done in companion mammals, is not usually performed in these birds.
2. Examinations for parasites should be performed routinely on wild caught birds, birds housed on dirt-floor aviaries, or whenever clinical signs of parasitism are present.

B. Routine Testing of Wild Caught Birds, or Birds Kept in Dirt-Floored Enclosures
1. Fecal flotation (Sheather's sugar and zinc sulfate)
2. Direct saline smear of feces
3. Gram's stain of feces
4. Consider empirical treatment

C. Gastrointestinal, Respiratory, or Neurologic Signs
1. Fecal flotation (Sheather's sugar and zinc sulfate)
2. Direct saline smear of feces or sputum
3. Gram's stain of feces

D. Stunting, Poor Growth, Increased Juvenile Mortality
1. Direct saline smear of feces, oropharyngeal and crop swabs
2. Fecal flotation (Sheather's sugar and zinc sulfate)
3. Diagnostic necropsy and histologic evaluation

E. Subcutaneous Nodules
1. Aspiration of nodules and cytologic evaluation of contents
2. Surgical exploration and removal of adult worms if present
3. Empirical ivermectin therapy for microfilariae

F. Proliferative, Honey-Combed Masses of Nonfeathered Skin
1. Empirical ivermectin therapy
2. Skin scrape if necessary

G. Dry Skin, Pruritus, Feather Loss, Feather Picking, Self-Mutilation
1. Skin scrape
2. Tape prep and cytology of affected skin
3. Skin biopsy
4. Direct saline smear of feces
5. Fecal flotation (Sheather's sugar and zinc sulfate)
6. Gram's stain of feces

III. Laboratory Techniques

(Box 20-2)

A. Considerations for Testing
1. Laboratory tests may not detect reproductive products (eggs, larvae, oocysts, cysts) in infected animals if the parasite is immature, produces few reproductive products or produces these products only periodically.
2. The number of eggs, oocysts, or other parasite products recovered does not predict the actual number of parasites present.
3. The appropriate test must be used.
   a. A zinc sulfate flotation recovers cysts of *Giardia* spp. with much less distortion than a sugar flotation.
   b. The most common laboratory techniques for antemortem detection of internal parasites involve examination of feces and blood.[2,4-8]

B. Fecal Examination
1. Fresh samples should be examined if possible. Feces can be collected on waxed paper placed on the floor of the cage.

> **Box 20-2**
>
> **Laboratory Tests for Endoparasites in Birds**
>
> **Direct Smear with Saline**
> *Giardia* spp. trophozoites
> Trichomonads
> Motile protozoa and larvae
> **Direct Smear/No Saline**
> *Hexamita* spp.
> *Cochlosoma* spp.
> **Zinc Sulfate Flotation**
> *Giardia* spp. cysts
> Most nematode eggs
> Nematode larvae
> Coccidian oocysts
> Tapeworm eggs (if present in feces)
> Some trematode eggs
> Few Acanthocephalan eggs
> **Sugar Flotation**
> *Cryptosporidium* spp. oocysts
> Most nematode eggs
> Coccidian oocysts
> Tapeworm eggs (if present in feces)
> Some trematode eggs (usually distorted)
> Few Acanthocephalan eggs
> **Sodium Nitrate Flotation**
> Many nematode eggs
> Coccidian oocysts
> Tapeworm eggs (if present in feces)
> **Sedimentation**
> Most any parasite reproductive stage but the entire sediment must be examined
> **Blood Smears**
> Microfilariae
> Trypanosomes
> Hemosporidia
> **Special Stains**
> *Giardia* (Gram's stain)
> *Cryptosporidium* (modified acid fast)
> **Immunoassay**
> *Toxoplasma gondii* (serum- MAT)
> *Giardia* (fecal- ELISA, IFA)
> *Cryptosporidium* (fecal-ELISA,IFA)
> **Histopathologic Examination of Tissues**
> *Toxoplasma*
> *Atoxoplasma*
> *Sarcocystis*

2. If fresh samples can not be examined, or if it is necessary to send the specimen to another laboratory for analysis, the sample should be mixed with an equal amount of 10% formalin.
3. Always treat feces with respect! Remember they may contain parasite products, bacteria, or viruses that can be zoonotic.
4. Individual samples are best; however, useful information can be obtained from pooled samples from aviaries and flocks. The possibilities of false-negative results must be considered.
5. Always observe the feces for abnormal consistency and color, melena, mucus, parasites (tapeworm segments or nematodes), or foreign objects. These may indicate parasitic infection, malabsorption, dietary indiscretion, or other problems.
6. Remove as much urate as possible before performing fecal examinations. Washing the fecal sample with water before examination may reduce urate crystals.[9]

C. Direct Smear Technique

1. To detect motile forms of protozoa and helminth larvae that may be distorted or killed by flotation and other concentration procedures.
2. Place several drops of saline on a large (2-inch × 3-inch) glass slide. If not available, 1-inch × 3-inch microscope slides are adequate. Certain protozoa such as *Cochlosoma* spp. and *Hexamita* spp. may be easier to find if no saline is added[4] (Box 20-2).

3. With an applicator stick, mix a small amount of feces in the saline on the slide. The preparation should be transparent enough to read newsprint through the smear.
4. Place a coverslip, preferably an extra wide (22 mm × 40 mm) coverslip, over the suspension.
5. Examine the entire area of the coverslip with the 10× objective of the microscope. Use the 40× objective to scrutinize any moving or otherwise interesting objects. Also note any helminth eggs, mites, or oocysts present.
6. After the wet film has been examined for motile forms, a drop or two of iodine solution can be placed at the edge of the coverslip, or a new slide can be prepared and a drop of iodine added before the coverslip is placed on the preparation. This will kill and stain any protozoan trophozoites and stain their cysts.

D. Flotation Solutions

1. No single ideal flotation fluid exists.
2. Sheather's sugar solution (specific gravity, 1.275)[7-9] is an inexpensive and easily prepared solution suitable for the recovery of a wide variety of coccidian oocysts and helminth eggs. This solution does not form crystals or distort most eggs. It may distort some delicate larvae and protozoa. The quantities listed below are sufficient to prepare approximately 1000 ml of sugar solution.
    a. Heat 355 ml of water in a double boiler. The mixture is easily burned, and is best prepared in a double boiler to avoid caramelization.
    b. With constant stirring, slowly add 454 g (1 lb) cane or beet sugar, obtained from the local grocery store.
    c. Continue heating until all of the sugar is dissolved.
    d. Add 6 ml of 10% formalin solution as preservative to prevent the growth of molds.
    e. Allow solution to cool, and check specific gravity with a hydrometer if possible.
    f. Transfer to wash bottles and label appropriately.
3. Zinc sulfate solution (specific gravity, 1.180)[8] is an easily prepared solution suitable for the recovery of protozoan cysts (such as *Giardia* spp.) without distortion. Zinc sulfate also floats a variety of helminth eggs and larvae, coccidian oocysts, and some trematode eggs from fecal samples without distortion. This solution will form crystals, and slides should be examined immediately after preparation.
    a. Slowly mix 350 g of reagent grade granular zinc sulfate, obtained from a veterinary or biologic supply house, with 1000 ml distilled water.
    b. Stir until dissolved and check with a hydrometer, if possible.
    c. This solution can be purchased with commercially available diagnostic test kits.
4. Sodium nitrate solution (specific gravity, 1.20) is an easily prepared solution adequate for floating many nematode eggs.[11] This solution

will form crystals and distort eggs if allowed to sit longer than 20 minutes. Slides should be examined immediately.
   a. Dissolve 378 g of reagent grade granular sodium nitrate obtained from a veterinary or biologic supply house in 1000 ml of water.
   b. Check specific gravity with a hydrometer, if possible.
   c. This solution can be purchased with commercially available diagnostic test kits.

E. Direct Flotation Procedure
1. To concentrate and recover helminth eggs, protozoan cysts, and coccidian oocysts from fecal samples without centrifugation
2. Relies on gravitational force to assist in separation of the fecal debris from the parasite products levitated by the flotation solution
3. All of the flotation media described previously can be used with this technique. If urates[9] are a problem, see step j.
4. Steps
   a. Thoroughly mix a small amount of fresh feces (approximately 2 g or the size of the end of your thumb if it is possible to collect that amount of feces) with approximately 12 ml of a flotation solution in a 3 oz paper cup or other suitable container.
   b. Place two layers of cheesecloth over a second cup or similar container. Pour the suspension through it.
   c. Rinse the cup with a small amount of flotation solution and pour this through the cheesecloth; gently squeeze the remaining solution from the cheesecloth by pressing the cheesecloth containing the fecal debris between a tongue depressor and the side of the paper cup.
   d. Discard the cheesecloth containing the feces, the tongue depressor, and the dirty cup into a properly designated biohazard bag.
   e. Pinch the rim of the cup to form a pouring spout and pour the filtrate into a disposable centrifuge tube, pill vial, or other suitable container. Rinse out all filtrate remaining in the cup with additional flotation solution and pour into tube; if necessary, add more flotation solution until tube is nearly full.
   f. Add flotation solution to the tube until a slightly convex meniscus is obtained; apply a 22×22 mm coverslip to the meniscus.
   g. Allow 10 to 20 minutes for the eggs, etc., to float to the surface. If sugar is used as the flotation solution, 15 to 20 minutes may be necessary; zinc sulfate and sodium nitrate require 10 minutes.
   h. Remove the coverslip carefully by picking it straight up and placing it on a microscope slide.
   i. Examine the entire coverslip with the $10\times$ objective of the microscope. Confirm identifications with higher-power objectives.
   j. To reduce urate crystals, wash sample with water before addition of flotation solution
      (1) Follow steps a to e above, but substitute water for the flotation solution.

(2) Allow fecal/water suspension to stand for 15 to 20 minutes.
(3) Carefully decant the water leaving the sediment and approximately 1 ml of water in the tube.
(4) Stir water and sediment; add flotation solution to the tube and mix well with an applicator stick; discard stick and add additional solution until a slightly convex meniscus is obtained; apply a 22 × 22 mm coverslip to the meniscus.
(5) Continue with steps g to i.

F. Centrifugal Flotation Procedure

1. To concentrate and recover helminth eggs, protozoan cysts, and coccidian oocysts from fecal samples by use of centrifugal force.
2. All of the flotation media described previously can be used with this technique. If urates[9] are a problem, see step k.
3. Steps
   a. Thoroughly mix a small amount of fresh feces (approximately 2 g or the size of the end of your thumb if it is possible to collect that amount of feces) with approximately 12 ml of a flotation solution in a 3 oz paper cup or other suitable container.
   b. Place two layers of cheesecloth over a second cup or similar container and pour the suspension through it; rinse the cup with a small amount of flotation solution and pour this through the cheesecloth.
   c. Gently squeeze the remaining solution from the cheesecloth by pressing the cheesecloth containing the fecal debris between a tongue depressor and the side of the paper cup.
   d. Discard the cheesecloth containing the feces, the tongue depressor, and the dirty cup into a properly designated biohazard bag.
   e. Pinch the rim of the cup to form a pouring spout and pour the filtrate into a 15 ml disposable centrifuge tube.
   f. Rinse out all filtrate remaining in the cup with additional flotation solution and pour into tube; if necessary, add more flotation solution until tube is nearly full.
   g. Place centrifuge tube in the centrifuge, and balance with additional tubes if necessary. Add additional flotation solution to the tube until a slightly convex meniscus is obtained.
   h. Apply a 22 × 22 mm coverslip to the meniscus. Close centrifuge and spin at 1500 RPM for 5 minutes.
   i. Remove the coverslip carefully by picking it straight up and placing it on a microscope slide.
   j. Examine the entire coverslip with the 10× objective of the microscope. Confirm identifications with higher power objectives.
   k. To reduce urate crystals, wash sample with water before addition of flotation solution
      (1) Follow steps a through f above, but substitute water for the flotation solution.
      (2) Place centrifuge tube in the centrifuge, and balance with additional tubes if necessary. Close centrifuge and spin at 1500 RPM for 5 minutes.

(3) Remove tube from centrifuge and carefully decant the water, leaving the sediment and approximately 1 ml of water in the tube.
(4) Stir water and sediment; add flotation solution to the tube and mix well with an applicator stick; discard stick and add more flotation solution until tube is nearly full.
(5) Continue with steps g through j.

G. Formalin Ethyl Acetate Sedimentation[8]

1. To concentrate and recover a wide variety of helminth eggs and larvae, protozoan cysts, and coccidian oocysts from fecal samples without distortion.
2. Kits to perform this test are commercially available.
3. Steps
   a. Thoroughly mix a small amount of fresh feces (approximately 2 g or the size of the end of your thumb if it is possible to collect that amount of feces) with approximately 10 ml of water in a 3 oz paper cup or other suitable container.
   b. Place two layers of cheesecloth over a second cup or similar container and pour the suspension through it; rinse the cup with a small amount of water and pour this through the cheesecloth.
   c. Gently squeeze the remaining solution from the cheesecloth by pressing the cheesecloth containing the fecal debris between a tongue depressor and the side of the paper cup.
   d. Discard the cheesecloth containing the feces, the tongue depressor, and the dirty cup into a properly designated biohazard bag.
   e. Pinch the rim of the cup to form a pouring spout and pour the filtrate into a 15 ml disposable polypropylene-screw cap centrifuge tube.
   f. Rinse out all filtrate remaining in the cup with additional water and pour into tube. Tube should be nearly full.
   g. Place cap on tube and place tube in centrifuge, and balance with additional tubes if necessary. Close centrifuge and spin at 1500 RPM for 5 minutes.
   h. Remove tube from centrifuge, and remove top. Carefully decant supernatant, leaving the surface layer of the sediment pellet at the bottom of the centrifuge tube.
   i. Add 9 ml of 10% buffered neutral formalin (BNF) to the centrifuge tube, stir with wooden applicator stick until the sediment is resuspended. Add 3 ml of ethyl acetate to the BNF-fecal suspension. Cap the tube and shake it vigorously.
   j. Remove the cap from the tube carefully because pressure may build up during shaking. Replace cap and place tube in the centrifuge; balance with additional tubes if necessary and recentrifuge at 1500 RPM for 5 minutes.
   k. Remove the tube from the centrifuge after spinning is completed, and remove the cap carefully.

l. Rim (ring) the debris layer with a wooden applicator stick to dislodge, and carefully decant the debris and supernatant fluid leaving the sediment in the tube. If possible, swab tube with a cotton-tipped applicator stick to remove any remaining ethyl acetate and debris from the sides of the tube. A small amount of formalin may remain with the sediment.

m. With a capillary pipette, transfer the sediment pellet in bottom of the centrifuge tube to a clean glass microscope slide and cover the preparation with a glass coverslip. For better visibility, it may be necessary to add a few drops of formalin to the sediment before the coverslip is applied. The preparation should be transparent enough to read newsprint through the slide. (The larger glass slides, 75 mm × 50 mm, and coverslips, 22 mm × 40 mm, are preferable for these preparations.)

n. Examine the preparation by systematically scanning the entire coverslip for parasite eggs, larvae, cysts, and oocysts with the 10× objective of the microscope. Confirm identifications with higher power objectives.

H. Detergent Sedimentation

1. To concentrate and recover a wide variety of helminth eggs and larvae, protozoan cysts, and coccidian oocysts from fecal samples without distortion.
2. Steps
   a. Thoroughly mix fresh feces about the size of the end of your thumb (2 to 5 g) with approximately 30 ml of a detergent solution in a 3 oz paper cup or other suitable container. (Prepare detergent solution by mixing 5 ml of liquid detergent with 995 ml of water. Add 8 drops of alum (i.e. aluminum potassium sulfate). Use a detergent that is inexpensive and contains as few additives as possible; "fancy" detergents contain compounds that may suspend the parasites in solution and prevent them from sedimenting.
   b. Place two layers of cheesecloth over a second cup or similar container and pour the suspension through it; rinse the cup with a small amount of detergent solution and pour this through the cheesecloth.
   c. Gently squeeze the remaining solution from the cheesecloth by pressing the cheesecloth containing the fecal debris between a tongue depressor and the side of the paper cup.
   d. Discard the cheesecloth containing the feces, the tongue depressor, and the dirty cup into a properly designated biohazard bag.
   e. Pinch the rim of the cup to form a pouring spout and pour the filtrate into a 50 ml disposable centrifuge tube.
   f. Rinse out all filtrate remaining in the cup with additional detergent solution and pour into tube. Tube should be nearly full.
   g. Allow solution to stand for 15 to 20 minutes.

h. Carefully decant supernatant fluid. Refill the tube with detergent solution; mix well, and allow to sediment for another 15 to 20 minutes.
i. Carefully decant supernatant fluid without disturbing the sediment layer.
j. With a capillary pipette, transfer a small amount of the sediment from the bottom of the centrifuge tube to a clean glass microscope slide and cover the preparation with a glass coverslip. For better visibility, it may be necessary to add a few drops of water to the sediment before the coverslip is applied. The preparation should be transparent enough to read newsprint through the slide. (The larger glass slides, 75 mm × 50 mm, and coverslips, 22 mm × 40 mm, are preferable for these preparations.)
k. Examine the preparation by systematically scanning the entire coverslip for parasite eggs, larvae, cysts, and oocysts with the 10× objective of the microscope. Confirm identifications with higher-power objectives.
l. Alternate procedure[6] – Smaller amounts of feces may be mixed with 12 ml detergent solution, strained and sedimented in 15 ml centrifuge tubes. These tubes can sediment as described above or be centrifuged at 1500 RPM for 3 minutes.

H. Immunoassays
1. To detect antigen produced by a parasite or specific antibodies produced against the parasite by the bird.
2. Coproimmunoassays detect antigen in the stool.[12,13]
    a. ELISA (enzyme-linked immunosorbent assay) tests
        (1) Most of the commercially available immunodiagnostic tests are ELISA tests.
        (2) These tests detect parasitic infection by capturing antigens with a parasite-specific antibody that is coated on a membrane, plastic wand, or a plastic microwell.
        (3) At this time, these tests are not specifically designed for use in birds and results should be interpreted carefully.
        (4) Tests are currently available for *Giardia* and *Cryptosporidium* spp. diagnoses.
    b. Immunofluorescent antibody tests
        (1) Require a fluorescent microscope
        (2) A specifically labeled antibody reacts with the cyst wall of the organism causing it to "glow" when viewed with the fluorescent scope.
        (3) Tests are available for *Giardia* and *Cryptosporidium* spp. diagnoses.
3. Serologic tests
    a. Most of the available tests detect antibodies in the blood produced in response to infection.

b. A positive antibody titer indicates that the bird has at some time been infected with the organism, but it does not prove that the organism is responsible for clinical disease.
c. Types
   (1) *Toxoplasma gondii*
      (a) Several tests are available that measure the concentration of *T. gondii* specific antibodies in the blood.[14]
      (b) The modified agglutination test (MAT) works particularly well for detection of *T. gondii* antibodies in bird serum.[15]
      (c) Latex agglutination, hemagglutination, and IFA tests are also available.
      (d) Remember that the presence of IgG antibodies indicates infection with *T. gondii* at some time, perhaps in the past. *T. gondii* IgG titers may remain elevated for years.
      (e) A *T. gondii* IgM titer indicates recent infection.
      (f) Infection, even recent infection, is not synonymous with disease.

IV. Clinical Treatment of Parasitism in Caged Birds

A. General Considerations
1. Few antiparasitics have been evaluated for efficacy and safety in caged birds, and none is labeled for use in these species.
2. A valid veterinarian-client-patient relationship and effective communication of potential hazards is necessary prior to prescription of these drugs.
3. Information regarding the sensitivity of avian parasites to antiparasitics is scant, and efficacy should be evaluated by appropriate retesting after initial treatment.
4. Because reinfection can mimic resistance, appropriate hygiene and control measures must be incorporated into all treatment regimens.
5. Direct dosing of individual birds is preferred over the use of medicated feed or water, which often results in nontherapeutic concentrations of drug.
6. Commonly used antiparasitics in cage birds include the following:[2,4,5,16-18]
   a. Antiprotozoals
      (1) Amprolium
         (a) 2 to 4 ml of a 9.6% solution per gallon of drinking water 5 days per month
         (b) For coccidiosis
      (2) Carnidazole
         (a) 20 mg/kg PO
         (b) For trichomoniasis, labeled for use in domestic pigeons
      (3) Fenbendazole
         (a) 50 mg/kg PO every 24 hours for 3 days

(b) For giardiasis, efficacy unproved in birds
(c) Toxic reactions, including death, have occurred in lories, pigeons, and doves dosed at 100 mg/kg.[49]

(4) Metronidazole
   (a) 25 mg/kg PO every 12 hours for 5 to 10 days
   (b) Or 50 mg/kg PO every 24 hours for 5 to 10 days
   (c) For giardiasis, trichomoniasis; may see resistance
   (d) 250 mg tablets may be crushed and mixed with juice, yogurt, or applesauce[18]

(5) Primaquine
   (a) 0.03 mg/kg PO every 24 hours for 3 days
   (b) For malaria or tissue forms of atoxoplasmosis

(6) Pyrimethamine
   (a) 0.5 mg/kg PO every 12 hours for 14 to 28 days
   (b) Or 40 to 80 mg/L drinking water
   (c) Or 100 mg/kg food
   (d) For toxoplasmosis and sarcocystosis, use in conjunction with trimethoprim-sulfa drug
   (e) Supplement with folate and B vitamins

(7) Trimethoprim-sulfamethoxazole or trimethoprim-sulfadiazine
   (a) 30 mg/kg PO every 12 hours for 5 to 10 days
   (b) For coccidiosis, or toxoplasmosis in conjunction with pyrimethamine

(8) Paromomycin sulfate[13]
   (a) For *Cryptosporidium*: 100 mg/kg mixed in egg food daily for 5 days
   (b) Regimen reduced morbidity and mortality in aviary finches[13]

b. Anthelmintics
  (1) Chlorsulon
    (a) 20 mg/kg PO every 14 days for three treatments
    (b) For trematodiasis

  (2) Fenbendazole
    (a) For ascaridiasis: 25 mg/kg PO repeat in 14 days
    (b) For capillariasis: 50 mg/kg PO every 24 hours for 5 days
    (c) For cestodiasis: 50 mg/kg PO every 24 hours for 3 days
    (d) Toxic reactions, including death, have occurred in lories, pigeons, and doves dosed at 100 mg/kg.[49]

  (3) Ivermectin
    (a) For nematodiasis: 0.2 to 0.4 mg/kg PO, SQ, IM, topical in small species
    (b) May dilute with propylene glycol or water, depending on formulation of product used (read label)
    (c) Ivermectin degrades in light

  (4) Levamisole
    (a) 15 mg/kg PO
    (b) Or 5 mg/kg SQ
    (c) Use only if less toxic anthelmintics are not effective.

(d) Toxic reactions include regurgitation, ataxia, and death.
- (5) Pyrantel pamoate
  - (a) 7 mg/kg PO, repeat in 14 days
  - (b) For nematodiasis
- (6) Praziquantel
  - (a) PO
  - (b) For cestodiasis, some trematodes
- c. Acaricides
  - (1) Ivermectin
    - (a) 0.2 to 0.4 mg/kg PO, SQ, IM, topical in small species
    - (b) For tracheal or knemidokoptic mites
    - (c) May dilute with propylene glycol or water, depending on formulation of product used (read label)
    - (d) Ivermectin degrades in light
  - (2) Pyrethrin powder
    - (a) Light dusting to affected birds

V. Specific Parasites of Concern

(Box 20-3)

A. Flagellated Protozoa
1. *Giardia psittaci*[2,5,9,19-21]
   - a. Life cycle
     - (1) Direct
     - (2) Infective cyst shed in feces
     - (3) Environmental contamination occurs because of prolonged survival of cyst in organic matter and water.
     - (4) Transmission by ingestion
     - (5) Parasite excysts, and lives in small intestine where it reproduces by binary fission
   - b. Clinical syndrome
     - (1) Asymptomatic carriers are common.
     - (2) Small birds such as budgerigars, lovebirds, and cockatiels are frequently infected.
     - (3) Heavy infections may result in intestinal malabsorption of fats and nutrients.
     - (4) Clinical signs include chronic to intermittent diarrhea with loose, malodorous, mucoid stools; lethargy, anorexia, and mortality especially in juveniles; dry skin, pruritus, and feather picking.
   - c. Diagnosis
     - (1) Cysts and trophozoites are shed irregularly in feces, so repeated fecal examination of fresh fecal samples (less than 10 minutes old) may be required.
     - (2) A single negative sample does not rule out infection.

Box 20-3

Host Avian Species and Associated Parasites

Finches and Canaries
  Sternostomid respiratory mites
  Trichomonads
  Tapeworms
  Knemidokoptic mites
  Syngamus trachea*
  Spirurid*
  Toxoplasma gondii
  Atoxoplasma spp.
  Isospora spp.
  Cryptosporidium spp.
  Cochlosoma spp.
Budgerigars
  Trichomonads
  Giardia spp.
  Knemidokoptic mites
Australian Parakeets
  Proventricular spirurids*
  Nematodes*
Cockatiels
  Giardia spp.
  Cryptosporidium spp.
  Hexamita spp.
  Ascarids*
  Knemidokoptic mites
Lorikeets
  Coccidia
  Ascarids*
Macaws
  Capillaria spp.*
  Ascarids*
  Pelecitus spp.
  Proventricular spirurids*

Toucans
  Giardia spp.
  Coccidia
  Ascarids*
  Capillaria spp.*
Parrots
  Giardia spp.
  Cryptosporidium
  Trichomonads
  Spirurids*
  Ascarids
  Pelecitus spp.
Columbiformes
  Trichomonas spp.
  Giardia spp.
Galliformes
  Histomonas spp.*
  Cestodes*
  Capillaria spp.*
  Heterakis spp.*
  Ascarids*
  Spirurids*
  Syngamus trachea*
  External parasites
Ratites
  Cerebral nematodiasis* (*Baylisascaris* spp./*Chandlerella quiscali*)
  Cestodes* (e.g., *Houttuynia struthionus* – ostrich)
  Syngamus trachea*
  Proventricular spirurids*
  Trichostrongylids* (e.g. ventricular wireworm)
  Feather lice

*In birds with access to dirt, intermediate, or paratenic hosts.

(3) On direct saline smear evaluation, motile trophozoites will have a gliding motion, and two "eyespots" or nuclei can be seen.

(4) Visualization of trophozoites can be assisted by iodine staining of fresh feces or trichrome staining of feces stored in polyvinyl alcohol.

(5) Occasionally, trophozites can be identified in routine Gram's stain of feces. Zinc sulfate flotation concentrates cysts.

(6) An enzyme-linked immunosorbent assay test that detects *Giardia*-specific antigen in aqueous extracts of human feces is available, but its use in diagnosis of avian giardiasis remains controversial.

d. Treatment
   (1) Intensive treatment of all exposed birds with metronidazole[9,18] or fenbendazole.[22]
   (2) Administration via drinking water may not be effective.[21]
   (3) Recurrence is common and may be due to reinfection.
e. Environment
   (1) Reinfection resulting from environmental contamination is common and is frequently misinterpreted as resistance, because cysts can be passed in feces as soon as 5 days after reinfection.
   (2) Thorough cage cleaning to remove organic debris must precede attempts at disinfection.
   (3) Quaternary ammonium compounds and 10% Clorox solutions are effective in inactivating cysts.[23]
   (4) Prevent exposure to feces by insect control and the use of adequately sized cages with a low stocking density, elevated food and water bowls, and a floor grate.
f. Zoonosis
   (1) Zoonotic potential is unclear.
   (2) *G. psittaci* may be host restricted, but a presumptive case of human infection has been reported.[21]

2. *Trichomonas gallinae* and related species[9,24-26]
   a. Life cycle
      (1) Direct, without formation of cyst.
      (2) Poor environmental stability of trophozoite requires close contact between birds for transmission.
   b. Clinical syndrome
      (1) Most common in pigeons, doves, and raptors; but occurs sporadically in psittacines and other small caged birds.
      (2) White to yellow plaques or necrotic masses form in the oral cavity and esophagus, causing regurgitation, anorexia, dyspnea, and debilitation.
      (3) Neonates are infected through parental feeding and present with poor growth and acute death.
      (4) Asymptomatic shedders are common.
   c. Diagnosis
      (1) Direct saline smear of oral lesions.
      (2) Motile, pyriform trophozoites have one nucleus or "eyespot," a polar flagella, and an undulating membrane.
      (3) Trichomonads have a characteristic jerky motion which is easily discriminated from the gliding movements of *Giardia* spp.
   d. Treatment
      (1) Oral metronidazole.[18] Relapses after cessation of treatment occur more commonly in psittacines than columbiformes.
      (2) Carnidazole tablets (Spartrix; Wildlife Pharmaceuticals, Inc., Fort Collins, CO) are manufactured for treatment of pigeons.
      (3) In pigeon colonies, routine treatment is advisable to decrease the number of carrier birds and prevent parental contamination of neonates.

(4) Amprolium (Corid; Merial, Iselin, NJ) is often used in gallinaceous birds.
  e. Environment
    (1) Protect water sources from contamination by wild birds and avoid overcrowding.

B. Apicomplexan Protozoa: The Coccidia
  1. *Isospora* spp. and *Eimeria* spp.[2,4,5,9,27]
    a. Life cycle
      (1) Direct
      (2) Oocyst passed intermittently in feces must sporulate in the environment to become infective.
      (3) Once produced, the oocysts may remain infective in the environment for months.
      (4) A few species of *Eimeria* have an extraintestinal life cycle and can parasitize the liver, kidney, and lungs.
    b. Clinical syndrome
      (1) Life cycle makes infection of caged birds less likely, but infection can be seen in aviary-housed species, especially canaries, finches, mynahs, toucans, lories, quail, and pheasant.
      (2) Subclinical infection is common, but young birds that are stressed or overcrowded may develop a mucoid or bloody diarrhea with resultant dehydration.
      (3) In severe cases, destruction of the intestinal epithelium may induce malabsorption, anemia, and hypoproteinemia.
    c. Diagnosis
      (1) Oocysts can be identified on routine fecal flotation. Asexual stages can be identified in intestinal mucosa on histopathologic examination.
    d. Treatment
      (1) Trimethoprim-sulfa medication
      (2) Break fecal-oral cycle by elevating off cage floor, reducing number of birds per cage, and improving hygiene
      (3) Amprolium may be used for treatment or prevention, but resistance may develop
    e. Environment
      (1) Protect food and water sources from contamination by wild birds and insects, and avoid overcrowding.
  2. *Cryptosporidium* spp.[13,27,28]
    a. Life cycle
      (1) Direct, with infective oocysts passed in stool.
      (2) Oocysts are infective when passed in feces.
      (3) Transmission occurs by ingestion of the oocyst. Endogenous sporulation of the oocyst in the gut can result in autoinfection.
      (4) Unlike other coccidians, *Cryptosporidium* spp. develop intracellularly but extracytoplasmically. Cryptosporidiosis is usually limited to the upper gastrointestinal tract of psittacines but may also involve the respiratory and urinary epithelial surfaces.

b. Clinical syndrome
  (1) Cryptosporidial infections are usually self-limiting.
  (2) Prolonged or severe clinical disease is seen most commonly in juvenile cockatiels and may be secondary to immunosuppression induced by agents such as polyoma virus or psittacine beak and feather disease.
  (3) Regurgitation and weight loss are secondary to gastroenteritis.
c. Diagnosis
  (1) The oocyst is very small (4 to 6 μm) and floats in a higher plane than other parasite products. Thus, the oocysts are commonly overlooked on routine fecal examinations. The oocysts are readily concentrated by sugar flotation and take on a "pink glow" when observed under the microscope.
d. Treatment
  (1) No effective treatment is currently available, but Paromomycin has been used successfully.[13]
  (2) Control measures include elimination of immunosuppressive diseases, environmental improvements, and the removal of persistently infected birds from the flock.
e. Environment
  (1) Hygiene measures may help to break the fecal-oral cycle, but the oocysts are infective when shed and are resistant to disinfectants.
f. Zoonosis potential
  (1) Species of *Cryptosporidium* that infect birds do not appear to be zoonotic to humans.

3. *Atoxoplasma* spp.[2,29,30]
  a. Life cycle
    (1) Direct
    (2) Asexual reproduction of *Atoxoplasma* spp. in vital organs such as lung, liver, spleen, and intestine incites a mononuclear inflammatory response.
    (3) Sexual stages in the intestinal tract result in passage of oocysts that sporulate in the environment.
  b. Clinical syndrome
    (1) Canaries and other passeriformes such as finches and mynahs are most frequently infected.
    (2) Clinical signs and mortality are greatest in juveniles, which can show diarrhea, anorexia, weight loss, ill-thrift, ruffled feathers, abdominal distention caused by hepatomegaly and acute death.
    (3) Asymptomatic shedders are common.
  c. Diagnosis
    (1) Diagnosis of clinically affected animals is rarely made antemortem because mortality usually occurs before the production of oocysts.
    (2) Asexual stages may be identified in buffy coat smears, liver biopsies, or postmortem tissue samples.

(3) The small oocysts (20 μm) may be present in fecal flotation of asymptomatic shedders, but are difficult to distinguish from many *Isospora* spp. of birds.
  d. Treatment
     (1) No consistently effective treatment is readily available. Primaquine may suppress the tissue forms.
4. *Toxoplasma gondii*[14,15,31-34]
  a. Life cycle
     (1) Both direct and indirect life cycles.
     (2) Oocysts are produced only in the enteroepithelial cycle of felids.
     (3) Birds become infected by the ingestion of infective oocysts from the environment or tissue cysts in raw or undercooked meat.
     (4) The parasites multiply asexually in cells as tachyzoites before becoming quiescent in tissue cysts as bradyzoites.
  b. Clinical syndrome
     (1) *T. gondii* is an uncommon parasite of caged birds.
     (2) Acute onset of "sick bird syndrome" with possible neurologic signs or acute death is the common presentation.
     (3) Specific signs are variable because of the wide variety of body systems that can be affected.
  c. Diagnosis
     (1) Serologic testing or histologic examination of tissues
     (2) Immunoperoxidase staining is helpful in the identification of tissue stages.
     (3) Birds are an intermediate host, so oocysts are not produced.
  d. Treatment
     (1) Combined trimethoprim-sulfamethoxazole with pyrimethamine.
  e. Environment
     (1) Eliminate sources of infection: oocysts (cat feces) and cysts (undercooked or raw meat).
     (2) Insect control to avoid transfer of oocysts through the environment
  f. Zoonosis
     (1) Birds do not produce infective oocysts; therefore, pet birds are not contagious to people. Sources of contamination for birds are also potential sources of contamination for people.
5. *Sarcocystis falcatula*[2,35-38]
  a. Life cycle
     (1) Indirect, with psittacine birds acting as an intermediate host.
     (2) *Sarcocystis falcatula* is a coccidian parasite that undergoes sexual multiplication in the intestine of the opossum (*Didelphis virginiana*) in North America.
     (3) Infective sporocysts are shed in opossum feces, which may be deposited into aviaries directly or transported by cockroaches.

b. Clinical syndrome
      (1) Asexual reproduction of the parasite in the psittacine results in severe vasculitis and pneumonitis.
      (2) Old world psittacines are very susceptible to clinical disease.
      (3) Adult new world psittacines have more resistance, but neonates can be fatally infected.
      (4) Peracute death is common.
      (5) Marked dyspnea and yellow urates may be observed before death.
      (6) Pulmonary edema and hemorrhage are consistent findings at necropsy.
   c. Diagnosis
      (1) Histopathologic examination of affected tissues is required for the identification of schizonts or merozoites in pulmonary tissue.
      (2) These tissue stages may be overlooked by inexperienced pathologists.
   d. Treatment
      (1) Treatment can be attempted with trimethoprim-sulfamethoxazole and pyrimethamine.[38]
   e. Environment
      (1) Control disease by limiting access of opossums to aviaries, removing food from cages overnight, and cockroach control.

C. Hemoparasites[1-4,39,40]
  1. *Hemoproteus* spp.
  2. *Leukocytozoon* spp.
  3. *Plasmodium* spp.
  4. Flagellated protozoans
     a. These parasites may be observed in wild caught birds, but are often nonpathogenic. Avian hemoparasites are rare in captive-born caged birds.

D. Helminths[4,9]
  1. Frequency
     a. Infection of caged birds with helminths is less common than with protozoal parasites, because these birds have little access to the intermediate hosts many of the helminths require to complete their life cycle.
     b. Parasitism with helminths is common, however, in birds kept in dirt-floored enclosures, and in wild or free-ranging birds.
  2. Cestodes[4,9]
     a. Cestodiasis occurs occasionally in parrots, especially wild-caught African gray parrots, cockatoos, or in birds kept in dirt-floored enclosures.
     b. Fecal flotations are usually negative for eggs, which are trapped in proglottids and not distributed throughout the feces.
     c. Proglottids are shed infrequently, but may be observed by owners.

d. Straining or unthriftiness may also be noted.
e. Tapeworms can be treated effectively with praziquantel.
3. Trematodes[4,9]
   a. Infection of birds with digenetic trematodes requires the consumption of an intermediate host and thus is rarely seen in companion birds.
   b. Cases usually involve imported, wild-caught old world psittacines.
   c. Treatment can be attempted with praziquantel or chlorsulon.
4. Nematodes[4,9]
   a. Nematodes are occasionally seen in companion and aviary birds.
   b. Because most infections require access to larvated eggs, intermediate hosts, or feces of other species, nematode problems are diagnosed most frequently in birds kept in dirt-floored enclosures.
   c. When present, infections usually respond to treatment with anthelmintics such as pyrantel pamoate, fenbendazole, or ivermectin.
   d. Ascarids[2-5,9,41,42]
      (1) Life cycle
         (a) Direct
         (b) Eggs are not infective until they larvate 2 to 3 weeks after passage in feces, but they can persist in moist environments for prolonged periods.
         (c) Ascaridiasis is most commonly seen in psittacines with access to the ground.
         (d) Infection with nonpsittacine species of ascarids can result in aberrant larval migration.
      (2) Clinical syndrome
         (a) Intestinal ascaridiasis can cause anorexia and diarrhea with resultant malabsorption, weight loss, and stunting.
         (b) Heavy infections can cause gastrointestinal obstruction and death.
         (c) Cerebrospinal nematodiasis can result in ataxia, torticollis, and death.
      (3) Diagnosis
         (a) Routine fecal flotation will confirm intestinal ascaridiasis.
         (b) No eggs are passed in cerebrospinal nematodiasis; therefore, diagnosis is based on clinical signs with confirmation by postmortem examination of neural tissue.
      (4) Treatment
         (a) Pyrantel pamoate, fenbendazole, or ivermectin is useful in the treatment of intestinal ascaridiasis.
         (b) Intestinal blockage and death may occur after treatment of heavily parasitized birds.
         (c) Cerebrospinal nematodiasis is untreatable.
      (5) Environment
         (a) Limit access of birds to feces and insects; keep birds in raised-floor cages when feasible
         (b) Provide a dry environment; and prevent access of raccoons, other mammals, and wild birds to the aviary.

e. *Capillaria* spp.[2,9,43]
   (1) Life cycle
      (a) Direct for most *Capillaria* spp. of caged birds, although a few species utilize earthworms as an intermediate host.
      (b) Eggs larvate in 2 weeks and remain infective in the environment for prolonged periods.
   (2) Clinical syndrome
      (a) Adults burrow into the mucosa of the intestinal tract, causing anorexia, regurgitation, diarrhea, and weight loss.
      (b) Heavy infections can result in ulceration, anemia, and death.
      (c) Esophageal infections can cause gaping and difficulty in swallowing.
   (3) Diagnosis
      (a) Bipolar eggs are seen on routine fecal flotation
   (4) Treatment
      (a) Ivermectin
      (b) Recheck fecal flotations 1 week after treatment as resistance can occur.
   (5) Environment
      (a) Prevent access to ground.
f. *Spiruroidea*[2-4,9,44]
   (1) Life cycle
      (a) Indirect, usually requiring an intermediate host
   (2) Clinical syndrome
      (a) Several syndromes reflecting spirurid infection can be observed in psittacines.
      (b) Proventricular and ventricular (gizzard) worms such as *Spiroptera* spp., *Dispharynx* spp., and *Tetrameres* spp. infect the upper gastrointestinal tract inducing epithelial hyperplasia.
      (c) Penetration of the mucosa can lead to coelomitis.
      (d) Conjunctival nematodiasis results from infections with adult spirurids including *Oxyspirura* spp., *Thelazia* spp., *Caratospira* spp., and *Annulospira* spp.
   (3) Diagnosis
      (a) Examination of a direct saline smear of a proventricular wash or conjunctival flush; fecal flotation (difficult to identify ova).
   (4) Treatment
      (a) Ivermectin
      (b) Manual removal of killed parasites in conjunctival sac requires anesthesia and conjunctival flushes.
   (5) Environment
      (a) Control intermediate host, or limit bird's access to intermediate host.
g. *Syngamus trachea*[1,45]
   (1) Life cycle

(a) Can be either direct or indirect utilizing invertebrate transport hosts.
(b) Eggs are coughed up and passed out in the feces where the infective larvae develop inside the egg.
(c) After migrating through the lung parenchyma, adult worms develop in the bronchial tree and trachea.

(2) Clinical syndrome
(a) *Syngamus trachea* (gapeworm) infections are rare in psittacine birds but are common in wild birds and reported in ratites.
(b) When present, young birds are most often affected and can display voice changes, coughing, marked dyspnea, and bloody tracheal secretions.

(3) Diagnosis
(a) Transtracheal illumination can demonstrate large, bright red worms that appear Y-shaped because of permanent coitus of the small male with the larger female. Direct evaluation of tracheal secretions or flotation of feces will show characteristic double operculated eggs.

(4) Treatment
(a) Fenbendazole or ivermectin

(5) Environment
(a) Prevent access to ground

h. *Filariidea*[2,4,46,47]
(1) Life cycle
(a) The filarid nematodes have a migrating microfilarial stage with adults located subcutaneously or in the body cavity, ocular chambers, heart, or air sacs.
(b) Indirect life cycle requires arthropod intermediate host.

(2) Diagnosis
(a) Antemortem diagnosis is usually limited to subcutaneous filariasis by *Pelecitus* spp. in which aspiration or surgical exploration of nodules on the feet and legs of birds shows microfilariae or adult filarids.

(3) Treatment
(a) Treatment requires surgical removal of adults and ivermectin to kill microfilarial stages.

E. Arthropods
1. Tracheal mites – *Sternostoma tracheacolum*[1,2,9,45]
   a. Life cycle
      (1) Complete entire life cycle on the bird, so direct or close contact between birds is required for transmission.
   b. Clinical syndrome
      (1) Canaries, finches, parakeets, and cockatiels can harbor tracheal mites.
      (2) Lady Gouldian finches are commonly affected.
      (3) Signs are most severe in juveniles, and consist of dyspnea, coughing, and sneezing. Mortality may occur.

c. Diagnosis
      (1) Tiny dark mites moving in the trachea may be visualized by transillumination of the trachea.
      (2) Eggs can sometimes be identified in feces or sputum.
   d. Treatment
      (1) Ivermectin is an effective treatment.
      (2) All incontact birds should be treated simultaneously.
2. Knemidokoptic mites[1-4,48]
   a. Life cycle
      (1) Completes entire life cycle on the bird, so direct or close contact between birds is required for transmission.
   b. Clinical syndrome
      (1) The "scaly leg and face" mite (*Knemidokoptes* spp.) seen in budgerigars, canaries, and other small birds induces proliferative, honey-combed masses of the nonfeathered skin, especially around the beak and on the legs.
   c. Diagnosis
      (1) The clinical appearance is pathognomonic.
      (2) Knemidokoptic mites in other species of birds may incite only pruritis and feather loss around the head and neck without the characteristic proliferative lesions. In these cases, diagnosis can also be made by visualization of mites in skin scrapings, by biopsy, or by clinical response to treatment.
   d. Treatment
      (1) Treat all exposed birds with ivermectin.
      (2) Two to three treatments at 10- to 14-day intervals may be required for eradication.

REFERENCES

1. Barnes HJ: Parasites. In Harrison GJ, Harrison LR, editors:. *Clinical avian medicine and surgery*, Philadelphia, 1986, WB Saunders, pp 472-485.
2. Greiner EC, Ritchie BW: Parasites. In Ritchie BW, Harrison GJ, Harrison LR, editors: *Avian medicine: principles and application*, Lake Worth, Fla, 1994, Wingers, pp 1107-1029.
3. Keymer IF: Parasitic diseases. In Petrak ML, editor: *Diseases of cage and aviary birds*, ed 2, Philadelphia, 1982, Lea and Febiger, pp 535-598.
4. Greiner EC: Parasitology. In Altman RB, Clubb SL, Dorrestein GM, Quesenberry K, editors: *Avian medicine and surgery*, Philadelphia, 1997, WB Saunders, pp 332-349.
5. Clyde VL, Patton S: Diagnosis, treatment, and control of common parasites in companion and aviary birds, *Semin Avian Exot Pet Med* 5:75-84, 1996.
6. Hendrix CM: Internal parasites. In Pratt PW, editor: *Laboratory procedures for veterinary technicians*, St Louis, 1997, Mosby, pp 307-384.
7. Sloss MW, Kemp RL: *Veterinary clinical parasitology*, ed 5, Ames, Ia, 1978, Iowa State University Press, pp 1-23.
8. Zajac AM: Fecal examination in the diagnosis of parasitism. In Sloss MW, Kemp RL, Zajac AM, editors: *Veterinary clinical parasitology*, Ames, Ia, 1994, Iowa State University Press, pp 3-16.

9. Greve JH: Gastrointestinal parasites. In Rosskopf WJ, Woerpel RW, editors: *Diseases of cage and aviary birds,* ed 3, Baltimore, 1996, Williams and Wilkins, pp 613-619.
10. Sheather AL: Detection of worm eggs in the faeces of animals and some experiences in the treatment of parasitic gastritis in cattle, *J Comp Pathol Ther* 36:71-90, 1923.
11. O'Grady MR, Slocombe JOD: An investigation of variables in a fecal flotation technique, *Can J Comp Med* 44:148-154, 1980.
12. Mohan R: Evaluation of immunofluorescent and ELISA tests to detect *Giardia* and *Cryptosporidium* in birds, *Proc Annu Conf Assoc Avian Vet* 62-64, 1993.
13. Clubb S: What is your diagnosis (cryptosporidium infection in gouldian finches [*chloebia gouldiae*])? *J Avian Med Surg* 11:41-42, 1997.
14. Patton S: Diagnosis of *Toxoplasma gondii* in birds, *Proc Annu Conf Assoc Avian Vet* 75-78, 1995.
15. Orosz SE, Mullins JD, Patton S: *Toxoplasma gondii* in two ratites, *J Assoc Avian Vet* 6:219-222, 1992.
16. Marshall R: Avian anthelmintics and antiprotozoals, *Semin Avian Exot Pet Med* 2:33-41, 1993.
17. Sikarskie JG: The use of ivermectin in birds, reptiles, and small mammals. In Kirk RW, editor: *Current veterinary therapy IX – small animal practice,* Philadelphia, 1986, WB Saunders, pp 743-745.
18. Rosskopf WJ, Woerpel RW: Practical avian therapeutics with dosages of commonly used drugs. In Rosskopf WJ, Woerpel RW, editors: *Diseases of cage and aviary birds,* ed 3, Baltimore, 1996, Williams and Wilkins, pp 255-259.
19. Erlandsen LS, Bemrick WJ: SEM evidence for a new species, *Giardia psittaci, J Parasitol* 73:623-629, 1987.
20. Fudge AM, McEntee L: Avian giardiasis: syndromes, diagnosis, and therapy, *Proc Annu Conf Assoc Avian Vet* 155-164, 1986.
21. Scholtens RG, New JC, Johnson S: The nature and treatment of giardiasis in parakeets, *J Amer Vet Med Assoc* 180:170-173, 1982.
22. Barr SC, Bowman DD, Heller RL: Efficacy of fenbendazole against giardiasis in dogs, *Am J Vet Res* 55:988-990, 1994.
23. Zimmer JF, Miller JJ, Lindmark DG: Evaluation of the efficacy of selected commercial disinfectants in inactivating *Giardia muris* cysts, *J Am Anim Hosp Assoc* 24:379-385, 1988.
24. Garner MM, Sturtevant FC: Trichomoniasis in a blue-fronted Amazon parrot (*Amazona aestiva*), *J Assoc Avian Vet* 6:17-20, 1992.
25. Murphy J: Psittacine trichomoniasis, *Proc Annu Conf Assoc Avian Vet* 21-24, 1992.
26. Ramsay EC, Drew ML, Johnson B: Trichomoniasis in a flock of budgerigars, *Proc Annu Conf Assoc Avian Vet* 309-311, 1990.
27. Patton S: An overview of avian coccidia, *Proc Annu Conf Assoc Avian Vet* 47-51, 1993.
28. Ley DH: Avian cryptosporidiosis, *Proceedings of the First International Conference on Zoological and Avian Medicine,* 1987, pp 299-303.
29. Flammer K: Clinical aspects of atoxoplasmosis in canaries, *Proceedings of the First International Conference on Zoological and Avian Medicine,* 1987, 33-35.
30. Dorrestine GM: Passerines. In Altman RB, Clubb SL, Dorrestein GM, Quesenberry K, editors: *Avian medicine and surgery,* Philadelphia, 1997, WB Saunders, pp 867-885.
31. Dhillon AS, Thacker HL, Winterfield RW: Toxoplasmosis in mynahs, *Avian Dis* 26:445-449, 1982.
32. Howerth EW et al: Fatal toxoplasmosis in a red lory (*Eos bornea*), *Avian Dis* 35:642-646, 1991.

33. Parenti E et al: Spontaneous toxoplasmosis in canaries (*Serinus canaria*) and other small passerine cage birds, *Avian Pathol* 15:183-197, 1986.
34. Vickers MC et al: Blindness associated with toxoplasmosis in canaries, *J Am Vet Med Assoc* 200:1723-1725, 1992.
35. Clubb SL et al: An acute fatal illness in old world psittacine birds associated with *Sarcocystis falcatula* of opossums, *Proc Annu Conf Assoc Avian Vet*, 1986, pp 139-149.
36. Hillyer EV et al: An outbreak of sarcocystis in a collection of psittacines, *J Zoo Wildl Med* 22:434-445, 1991.
37. Page CD et al: Antemortem diagnosis and treatment of sarcocystosis in two species of psittacines, *J Zoo Wildl Med* 23:77-85, 1992.
38. Quesenberry K: Disorders of the musculoskeletal system. In Altman RB, Clubb SL, Dorrestein GM, Quesenberry K, editors: *Avian medicine and surgery*, Philadelphia, 1997, WB Saunders pp 523-539.
39. Van Der Heyden N: Identification, pathogenicity and treatment of avian hematozoa, *Proc Annu Conf Assoc Avian Vet*, 1985, pp 163-174.
40. Van Der Heyden N: Hemoparasites. In Rosskopf WJ, Woerpel RW, editors: *Diseases of cage and aviary birds*, ed 3, Baltimore, 1996, Williams and Wilkins, pp 627-629.
41. Armstrong DL et al: Cerebrospinal nematodiasis in blue and gold macaws and scarlet macaws associated with *Baylisascaris procyonis*, *Proceedings of the First International Conference on Zoological and Avian Medicine*, 1987, pp 489-490.
42. Myers RK, Monroe WE, Greve JH: Cerebrospinal nematodiasis in a cockatiel, *J Am Vet Med Assoc* 183:1089-1090, 1983.
43. Helmboldt CF et al: The pathology of capillariasis in the blue jay, *J Wildl Dis* 7:157-161, 1971.
44. Brooks DE, Greiner EC, Walsh MT: Conjunctivitis caused by *Thelazia* sp. in a Senegal parrot, *J Am Vet Med Assoc* 183:1305-1306, 1983.
45. Rosskopf WJ, Woerpel RW: Respiratory parasites. In Rosskopf WJ, Woerpel RW, editors: *Diseases of cage and aviary birds*, ed 3, Baltimore, 1996, Williams and Wilkins, pp 620-622.
46. Allen JL et al: Subcutaneous filariasis (*Pelecitus* sp.) in a yellow-collared macaw (*Ara auricollis*), *Avian Dis* 29:891-894, 1985.
47. Greenacre CB et al: Adult filarioid nematodes (*Chandlerella* sp.) from the right atrium and major veins of a Ducorps' cockatoo, *J Assoc Avian Vet* 7:135-137, 1993.
48. Greve JH: Parasites of the skin. In Rosskopf WJ, Woerpel RW, editors: *Diseases of cage and aviary birds*, ed 3, Baltimore, 1996, Williams and Wilkins, pp 623-626.
49. Papendick R et al: Suspected fenbendazole toxicity in birds, *Proc Annu Conf Am Assoc Zoo Vet* 144-146, 1998.

21

Endoscopic Diagnosis

Michael Taylor

The unique anatomy of the avian respiratory system makes endoscopic visualization of organs relatively easy. Focal, directed illumination with magnification (FDIM) is especially useful in smaller avian patients because it causes minimal trauma.

I. Fiberoptic Endoscopy

A. General Principles
1. The transmission of light through glass fibers is based on the principle of internal reflection.
2. Different refractive indices
3. The degree of curvature is directly proportional to the amount of light lost; that is, a straight light cable gives maximum transmission.

II. Equipment

A. Rigid Endoscopes (Fig. 21-1)
1. Rigid endoscopes are available in a wide variety of diameters and lengths.
2. The fine diameters are most suitable for avian patients that weigh less than 1 kg. A 1.9-mm diameter rigid endoscope is available commercially (Karl Storz Veterinary Endoscopy, Goleta, Calif).
   a. This is the finest diameter currently available using high-quality optics (rod lenses).
   b. Less light is transmitted and a smaller image size is obtained with this endoscope; therefore it is best suited for avian patients weighing 20 to 200 g or for use in small spaces (e.g., infraorbital sinuses, external ear canal).

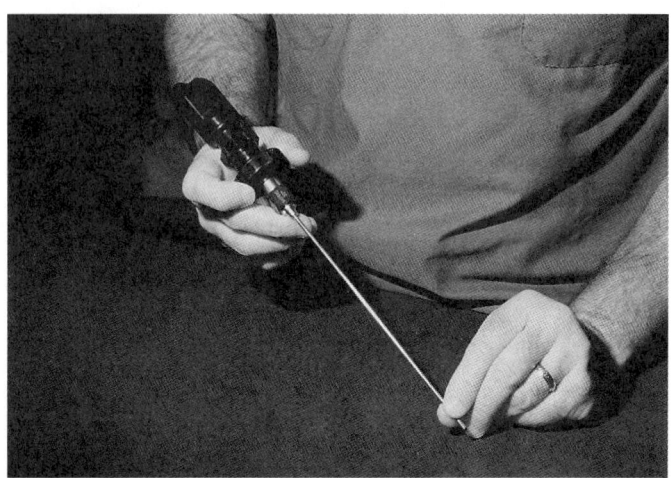

Fig. 21-1 Correct positioning of the hands to support a rigid endoscope with attached video camera. The operator is right-handed. Note that the thumb and index finger of the left hand are used to control and support the tip of the endoscope.

B. Flexible Endoscopes
1. In small animal medicine, 5 mm and 10 mm flexible endoscopes are often used.
2. A long endoscope with a distal tip that is deflectable in four planes is very useful in large birds (e.g., anseriforms, gruiforms).
3. A flexible endoscope is awkward to use in birds that weigh less than 1000 g.
4. Smaller diameter flexible endoscopes have poor image quality compared with similarly sized rigid systems.
5. The avian crop, esophagus, proventriculus, and ventriculus are linear in arrangement, and most portions can be viewed with a small-diameter, rigid scope with a 30-degree distal viewing lens.

C. Biopsy Forceps
1. Biopsy forceps traditionally are used with flexible endoscopes.
2. Biopsy forceps recently were introduced with the new rigid avian diagnostic system developed.[1]
    a. Consists of a 2.7 mm telescope with a no. 67065 C sheath, (Karl Storz Veterinary Endoscopy, Goleta, Calif.).
    b. The forceps can be easily and atraumatically delivered to the distal lens surface and accurately guided to the target.
    c. Several cup sizes and shapes are available:
        (1) 3 French (Fr), 5 Fr elliptical cups (no. 67071 ZJ and no. 67161 Z, respectively, Karl Storz Veterinary Endoscopy): Routinely used in small and medium-sized birds.

(2) 5 Fr round cups (no. 27071 Z, Karl Storz Veterinary Endoscopy): Causes less tissue penetration because of the cup shape.

D. Grasping Forceps
1. Rigid grasping forceps
   a. "Alligator" forceps and other types have been endoscopically guided as secondary instruments.
   b. Rigid forceps pose problems similar to those encountered with rigid biopsy instruments.
   c. Two types of forceps have been introduced with the rigid diagnostic set:
      (1) Fine (3 Fr): For retrieval of small objects and blunt dissection.
      (2) Coarse (5 Fr): For retrieval of larger foreign bodies.
2. Flexible grasping forceps
   a. Flexible grasping forceps are routinely used in flexible endoscopic applications.
   b. A variety of shapes and jaw types are available.
   c. These forceps are excellent for foreign body retrieval.

E. Scissors
1. Semiflexible 3 Fr scissors are available for the Karl Storz Veterinary Avian Diagnostic Set (item no. 11501 EK).
2. These scissors are superb for incising air sacs and peritoneal cavities.

F. Infusion Needle
1. A long, 22 gauge infusion and aspiration needle has been introduced as part of the rigid avian diagnostic set (Karl Storz Veterinary Avian Diagnostic Set).
2. A Teflon guide eases passage of the needle in the sheath and can be used to control the depth of needle penetration.
3. The tip of the infusion needle may also be used as a sharp instrument for the incision of an air sac or a pleura, or it can be energized with an electrosurgical unit.

G. Catheters
1. Sterile 3.5 and 5 Fr feeding catheters (Sovereign, Sherwood Medical, St. Louis, Mo) may be used for aspiration or infusion of liquids under the precise guidance of the endoscope.

III. Anesthesia

See Chapter 24 for a more detailed discussion of anesthetics.

A. Isoflurane
1. Isoflurane is the anesthetic of choice for most avian patients for endoscopic procedures.

IV. Applied Anatomy

A. Visualization
1. The following characteristics of lesions are described:
   a. Location
   b. Color
   c. Size
   d. Shape
   e. Consistency

B. Approaches
1. External acoustic meatus
   a. The external acoustic meatus is a highly variable structure among the different orders of birds.
      (1) A 1.9 mm endoscope must be used to examine Amazons and smaller psittacines.
      (2) A 2.7 mm endoscope is useful in *Ara* and *Cacatua* species and in most raptors.
      (3) A 4 mm endoscope may be used in some strigiforms because of their large size.
   b. Structures seen with the endoscope in the external acoustic meatus include:
      (1) Epithelium of the ear canal
      (2) Tympanum (any outward projections should be noted)
      (3) Columella
      (4) Extracolumellar cartilage
2. Oropharynx
   a. Although the oropharynx is easily accessed through the open bill, great care must be taken while examining some species, such as psittacines, that have strong bills and can damage equipment and personnel.
   b. Anesthesia often is required for a thorough examination.
   c. FDIM is especially useful in small patients, in which visualization of oral lesions may be very difficult with conventional equipment.
   d. Structures seen through the endoscope include:
      (1) Papillae
      (2) Choana
      (3) Infundibular cleft
      (4) Nasal septum
      (5) Conchae
      (6) Salivary glands
      (7) Anterior trachea (Fig. 21-2)
3. Ingluvies
   a. The crop may be examined with either rigid or flexible equipment.
   b. The best results are achieved in birds weighing less than 800 g using a rigid 2.7 or 4 mm endoscope of the appropriate length.

Fig. 21-2 Normal tracheal mucosa in an Amazon parrot.

    c. Either a rigid or a flexible endoscope with a forceps may be used to retrieve foreign objects.
    d. FDIM greatly aids the diagnosis of crop disease, including visual inspection of lesions and debris.
4. Proventriculus and ventriculus
    a. Either a flexible or a rigid endoscope may be used, depending on the size of the bird.
    b. Passing a flexible endoscope across the crop and into the thoracic esophagus may prove difficult in small to medium-sized birds.
    c. The added length of the typical flexible endoscope is most useful in large birds, especially if a 10 mm charged-coupled device (CCD) endoscope can be used to improve image quality.
    d. The best visualization in smaller patients is achieved with a rigid endoscope introduced through an ingluviotomy incision into the thoracic esophagus.
    e. The Storz system sheath has infusion ports for saline flushing or air infusion, as well as an instrument port for a flexible grasping forceps; it has been used for foreign body retrieval in birds ranging from 300 to 1200 g.
5. Air sacs
    a. Clavicular air sac
        (1) Approach
            (a) The patient is fasted to empty the crop and proventriculus.
            (b) The patient then is placed in the dorsal recumbent position.
            (c) The skin is incised over the caudoventral border of the thoracic inlet, with care taken to avoid the ingluvies.
            (d) The ingluvies is gently displaced to the right side.

(e) A small incision is made on the midline through the glistening clavicular air sac membrane.
    (2) Structures seen with the endoscope in the clavicular air sac include:
        (a) Trachea
        (b) Syrinx
        (c) Bifurcation and primary bronchi
        (d) Esophagus
        (e) Base of the heart
        (f) Brachycephalic trunk
        (g) Carotid arteries
        (h) Thyroids
        (i) Parathyroids
 b. Cranial thoracic air sacs
    (1) Approach
        (a) The patient is placed in the lateral recumbent position.
        (b) The incision is made on the ventrolateral thoracic wall, caudal to the last sternal rib in the region of the lateral notch.
    (2) Structures seen with the endoscope in a cranial thoracic air sac include (Fig. 21-3):
        (a) Lungs
        (b) Pulmonary arteries
        (c) Heart (pericardial sac)

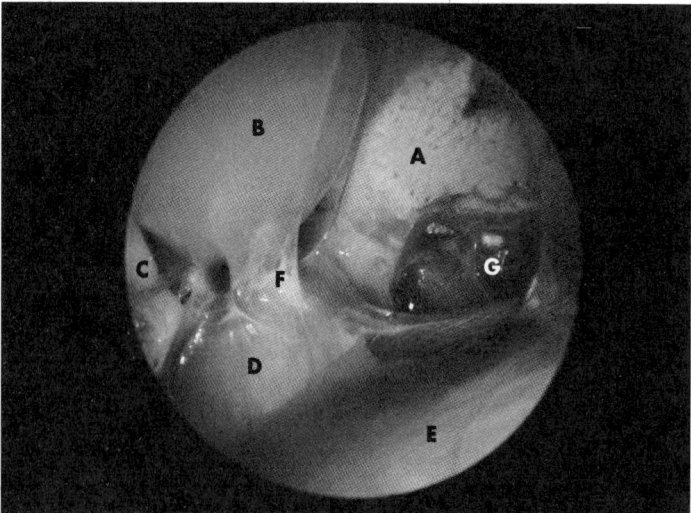

Fig. 21-3 Cranial thoracic air sac of an orange-winged Amazon (*Amazona amazonica*). The heart is visible in the pericardial sac. Structures seen in this view include (**A**) lung, (**B**) medial intercostal muscle, (**C**) rib, (**D**) heart, (**E**) proventriculus, (**F**) pericardial sac attachment point, (**G**) cranial thoracic air sac ostium.

(d) Liver
(e) Ribs
(f) Confluent membranes of the cranial and caudal thoracic air sacs
c. Caudal thoracic air sacs
   (1) This is *the major insertion point* for routine avian endoscopic examinations (Fig. 21-4).[2]
   (2) Approach cranial to the femur
      (a) The patient is placed in the lateral recumbent position with the wings extended dorsally.
      (b) Entry on the left side allows examination of the greatest number of structures, including the ovary and reproductive tract.
      (c) The upper leg is extended and held caudally.
      (d) The insertion point is located by visualizing the center of a triangle formed by the cranial muscle mass of the femur, the last rib, and the ventral border of the synsacrum.
      (e) This entry frequently places the endoscope between the seventh and eighth ribs.
   (3) Approach caudal to the femur
      (a) Positioning is the same as for the approach cranial to the femur except that the upper leg is extended cranially.
      (b) The entry site is located where the semimembranosus muscle (muscle flexor cruris medialis) crosses the last rib.[2]
      (c) The semimembranosus is reflected dorsally, and blunt entry is made through the body wall just caudal to the last rib.
      (d) The view from this point is similar to the view from the cranial insertion; however, the cranial insertion positions the tip of the endoscope in the cranial to middle portion

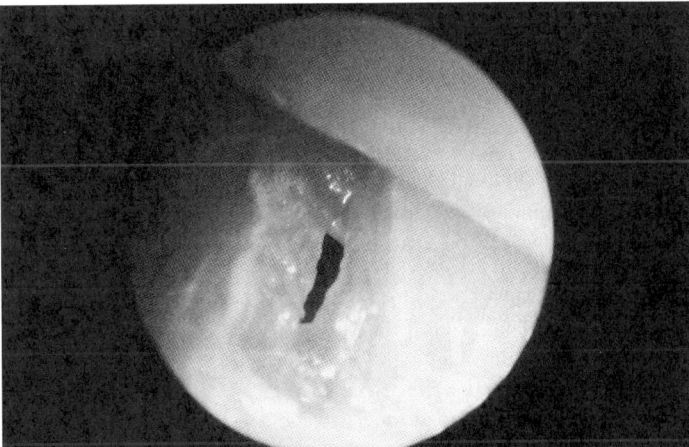

Fig. 21-4 Entrance puncture in the lateral body wall just ventral to the edge of the semitendinosus muscle. The site is magnified 10 to 12 times.

456 *Endoscopic Diagnosis*

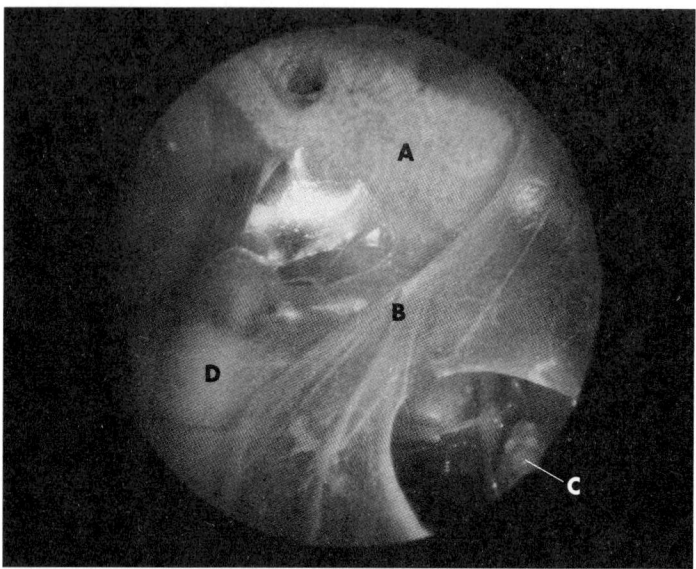

Fig. 21-5 Caudal thoracic air sac of the red-tailed hawk (*Buteo jamaicensis*) with incision of the confluent thoracic-abdominal air sac walls. The operator can see into the abdominal air sac. Structures that are visible include (**A**) lung, (**B**) air sac wall, (**C**) ovary, and (**D**) proventriculus.

of the air sac, whereas the caudal insertion enters the air sac at its caudal border.
- (e) This approach offers potential access to both the cranial thoracic air sac and the abdominal air sac (Fig. 21-5).
- (4) Structures seen with the endoscope in a caudal thoracic air sac include:
  - (a) Lung with a large ostium
  - (b) Proventriculus
  - (c) Liver
  - (d) Confluent walls of the cranial and caudal thoracic air sacs
  - (e) Confluent walls of the caudal thoracic air sac and abdominal air sac
- d. Abdominal air sacs
  - (1) Approach via a caudal thoracic air sac
    - (a) The previously described approach to a caudal thoracic air sac is used.
    - (b) The operator then passes through the confluent air sac walls to enter the abdominal air sac.
  - (2) Approach via the prepubic area
    - (a) Entry is made directly ventral to the acetabulum, approximately midway between the last rib and the pubis and just dorsal to the ventral border of the flexor cruris medialis.
    - (b) This approach enters directly into the abdominal air sac.

(3) Approach via the postischial area[2]
  (a) Entry is made dorsal to the pubic bone, caudal to the ischium, and dorsolateral to the vent.
  (b) The endoscope first enters the caudal (and only the superficial) portion of the intestinal peritoneal cavity; it must be pushed through the confluent intestinal peritoneal cavity-abdominal air sac membranes to enter the abdominal air sac.
(4) Structures seen with the endoscope in the abdominal air sac include:
  (a) Proventriculus
  (b) Ventriculus
  (c) Spleen
  (d) Adrenal glands
  (e) Gonad (one or more)
  (f) Kidneys
  (g) Intestines

6. Peritoneal cavities
  a. Ventral hepatic peritoneal cavity
    (1) The ventral hepatic peritoneal cavity surrounds the entire ventral surface of the liver.
    (2) The cavity is divided on the midline by the ventral mesentery into the left and right ventral hepatic peritoneal cavities.
    (3) Approach via the ventral midline
      (a) Entry is made on or just caudal to the side of the midline, just caudal to the border of the sternum.
      (b) A large fat pad often is present in the caudal ventral hepatic cavity.
      (c) The endoscope will enter the left or right ventral hepatic cavity; the opposite side may be entered by passing the endoscope through the ventral mesentery.
      (d) The approach via the ventral midline provides the best access to both the right and left lobes of the liver.
    (4) Approach via the caudal thoracic air sac
      (a) The liver appears tantalizingly approachable from the caudal thoracic air sacs, but the confluent air sac-peritoneal cavity membranes *must* be incised to allow better visualization and handling of the organ.
      (b) This approach is convenient when liver lesions are noted during a traditional examination of the abdomen.
      (c) This approach is contraindicated when ascites is present because the fluid will spill from the ventral hepatic peritoneal cavity into the air sac and possibly the lung.
  b. Intestinal peritoneal cavity
    (1) This cavity is approached using the postischial insertion points as described for the abdominal air sac (above).
    (2) There is a potential space that can be insufflated.
    (3) The intestinal peritoneal cavity straddles the midline, running from the region of the lungs caudal to the cloaca.

7. Pleural cavity
   a. Approach via the intercostal spaces
      (1) The intercostal space just caudal to the end of the scapula is the best location in most species.
      (2) Entry is made just dorsal to the level of the scapula.
      (3) The intercostal muscles are very thin and must be dissected carefully.
      (4) The visceral pleura is very thin; pleural space may be almost nonexistent.
      (5) In many species the pleural space is large enough to be entered.

V. Endoscopic Biopsy and Specimen Collection

A. Kidneys
1. Approaches
   a. Entry into the caudal thoracic air sac is used to approach the abdominal air sac, or entry is made directly into the abdominal air sac.
2. Techniques
   a. The operator must decide where the biopsy should be taken (i.e., the cranial, middle, or caudal division of the kidney).
   b. A cup or spoon-shaped biopsy forceps should be used.
3. Indications
   a. Polyuria, polydipsia
   b. Persistently elevated uric acid levels
   c. Abnormal gross appearance of renal parenchyma (Fig. 21-6)

B. Liver
1. Approaches (see the description for entering the ventral hepatic peritoneal cavities, above)
2. Techniques
   a. The liver's surface is examined thoroughly to determine if lesions are focal or diffuse.
   b. The caudal or lateral borders of the liver are easily grasped for representative samples of diffuse hepatic disease.
   c. Focal lesions may be more difficult to harvest.
   d. Hemorrhage is rare in patients with adequate coagulation parameters because the crush and cut of the biopsy forceps release tissue thromboplastin.
3. Indications
   a. Evidence of hepatic disease that does not respond to treatment.
      (1) The evidence may be clinical, radiographic, or biochemical: aspartate aminotransferase, gamma glutamyl transferase, lactate dehydrogenase, bile acids beyond normal range.
   b. Bile acids elevated more than twice the reference range for longer than 14 days.

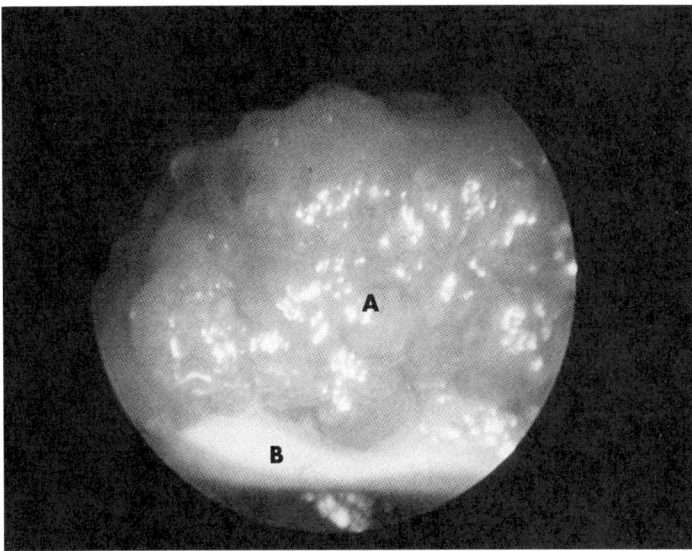

Fig. 21-6 Severe chronic kidney disease in a Tucuman Amazon (*Amazona tucumensis*). Note the rough, irregular surface of the middle division of the kidney. Structures that are visible are (**A**) kidney and (**B**) ureter.

  c. In my experience, clinicians frequently wait too long to perform hepatic endoscopy and biopsy because of fear of the procedure, leading to nonspecific "biopsy of chronicity."

C. Air Sac
1. Approaches
   a. Cranial or caudal air sac approach, as described in detail above, is used.
   b. Air sacculitis often is discovered during a routine caudal air sac examination (e.g., for gender determination).
2. Techniques
   a. Cup biopsy or grasping forceps may be used to harvest air sac debris for cytologic and microbiologic studies.
   b. Biopsies are best collected from the edge of the air sac puncture site (e.g., between the caudal thoracic and abdominal air sacs).
3. Indications
   a. Radiographic irregularities in air sacs
   b. Assessment of findings from a routine examination (Fig. 21-7)

D. Lungs
1. Approaches
   a. The caudal lung areas may be approached from the caudal or cranial thoracic air sacs (entry into these is described above).

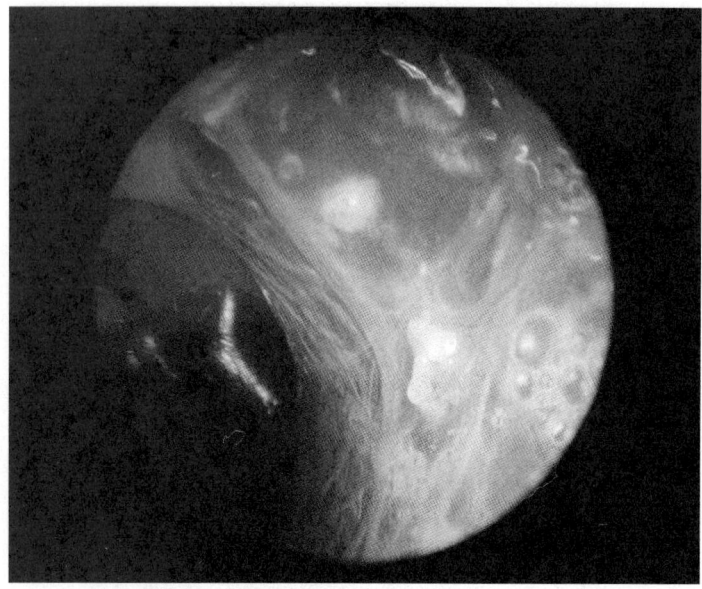

Fig. 21-7 Granulomatous air sacculitis of the left caudal thoracic air sac in a green-winged macaw (*Ara chloroptera*).

      b. The dorsolateral surface is best accessed by the intercostal approach (described above).
  2. Techniques
      a. From the caudal approaches, the combined air sac and pleura is thick and may need to be incised (e.g., using the fine infusion needle or scissors [Fig. 21-8]).
      b. Deep penetration of the parenchyma is avoided to help prevent trauma to major blood vessels.
      c. FDIM enhances the accuracy of biopsy collection, especially for small lesions.
  3. Indications
      a. Diffuse pulmonary disease (Fig. 21-9)
      b. Focal granulomatous disease
      c. Radiographic, clinical, or auscultative abnormal findings

E. **Ventriculus**
  1. Serosal and muscularis biopsies of the ventriculus can be diagnostic for proventricular dilation disease (also called neuropathic gastric dilation or wasting disease).
  2. At least two sites near blood vessels are chosen to ensure harvest of nerve.
  3. Entry is made via the caudal portion of the left caudal thoracic air sac (left caudal border of the ventriculus or left of the midline, just caudal to the sternum).

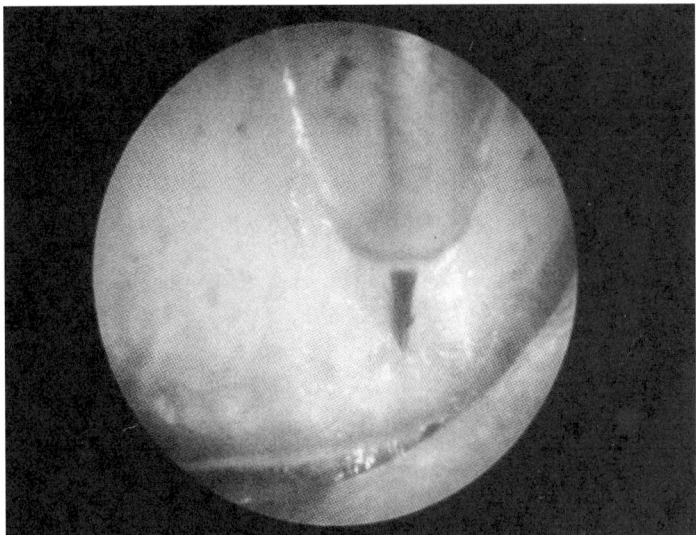

Fig. 21-8 View of the fine needle and Teflon sheath inserted into the sheath and used to approach the air sac and pleura of the lung in the caudal thoracic air sac.

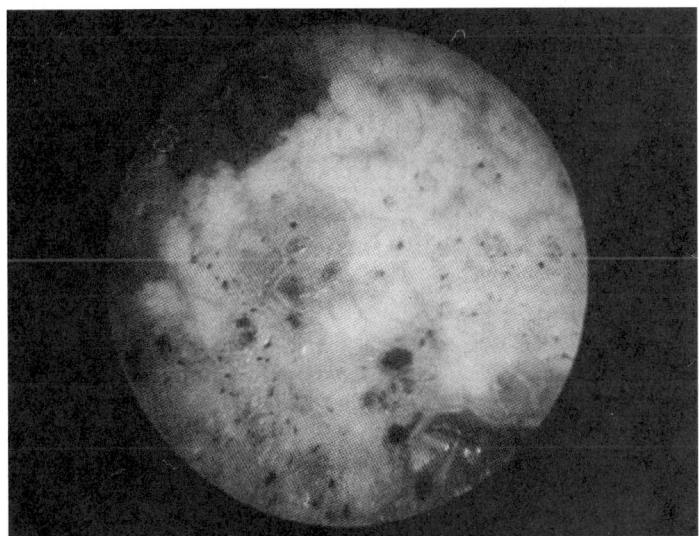

Fig. 21-9 Pneumoconiosis of the lung viewed from the left caudal thoracic air sac in a yellow-naped Amazon (*Amazona ochracephala*).

4. This procedure may require heavier biopsy forceps (e.g., 7 or 9 Fr).

VI. Endoscopic Surgery
A. General Considerations
1. Most avian patients weigh less than 1 kg and have a large, platelike sternum, making large, open surgical approaches difficult.
2. FDIM combined with appropriate hand instrumentation shows great promise for reducing patient trauma while allowing precise surgical therapy in smaller avian patients.
3. Avian clinicians must learn to use the magnification and reach afforded by endoscopy to scale their efforts to the patient's size.

VII. Endodocumentation
A. Still Photography
1. Newer, high-speed films have allowed acceptable imaging with less light required.
   a. The higher the ISO value of the film, the more sensitive it is to light, but the larger the grain structure present in the film.
      (1) This may lead to unsuitable graininess if the end use requires enlargement of the original image.
2. The 35 mm format is used for still endophotography because of its comparable size in relation to the normal ocular image and because of the ease of handling modern single lens reflex cameras.
3. As in all other types of photography, the key limiting factor, besides equipment considerations, is light.
   a. The quality and amount of light that can be delivered to the site is limited by the design of the endoscope and the type of light source used.
   b. It soon becomes evident that a specialized, high-intensity light source is required to obtain good quality photographic images.
   c. Various manufacturers have developed systems to allow photographic imaging using endoscopic equipment. The most successful of these use a flash generator that is fired in synchronization with the camera's shutter and is controlled by a through-the-lens (TTL) metering system.
4. For the past 6 years I have achieved excellent results using a powerful flash generator with TTL control (Model 600, Karl Storz Endoscopy, Tuttlingen, Germany). The high output of this unit has allowed the use of fine-grain transparency films in the ISO range of 50 to 100. The strobe effect of the flash has the added benefit of stopping motion.
5. Uses
   a. Case documentation (for the medical record or to share with the client or colleagues)
   b. Publication

B. Videotaping
1. The greatest practical advances in endodocumentation have occurred in the field of video imaging.
   a. Improvements in the CCD chips allow greater sensitivity to low light levels combined with higher resolution when recording on formats such as S VHS and Hi8.
2. Specialized endovideo cameras consist of a soakable hand piece that contains the CCD chip, a focusable lens, and a quick connector.
   a. The unit is attached to the controller by a sealed cable.
   b. The controller contains all the electronic circuitry for the camera and is placed out of the surgical field.
3. The sterilized camera may be used for real-time visualization of procedures; some clinicians also prefer it as an aid to performing certain manipulations.
   a. The procedure is observed on a monitor without the need for the surgeon to view through the ocular piece, improving ergonomics and reducing fatigue.
   b. Parts or all of the examination can be recorded for later review; this recording can be used as a research tool, allowing comparison of many different examinations.
4. Still images can be captured from recorded material using a video printer. Such prints are adequate for the medical record but are seldom suitable for publication.
5. Uses
   a. Case documentation (particularly useful in showing relationships in space and time)
   b. Teaching demonstrations (for clients or colleagues)

REFERENCES
1. Taylor M: Diagnostic application of a new endoscopic system for birds, *Proceedings of the European Conference of Avian Medicine and Surgery*, pp 127-131, 1993.
2. Taylor M: Endoscopic examination and biopsy techniques. In Ritchie BW, Harrison GJ, Harrison LR, editors: *Avian medicine: principles and applications*, Lake Worth, Fla, 1994, Wingers, pp 327-354.

22

Avian Anesthesia

Darryl J. Heard

I. General Anesthesia Considerations

A. Ability of the Veterinarian
1. The ability to safely restrain and anesthetize a wide range of birds is an essential skill for an avian veterinarian.
2. Prolonged complex procedures in critical patients increases the level of skill required; in some instances anesthesia is the major factor limiting a procedure.

B. Knowledge of the Veterinarian
1. Understanding of anatomy and physiology
2. Knowledge of available regimens, preparation, perioperative stabilization and support, appropriate equipment, attentive monitoring, and practice

II. Preanesthetic Preparation

A. History and Physical Examination
1. Obtain a comprehensive history and perform a physical examination before induction.
2. Chemical restraint is necessary for a complete examination of some birds.
3. In some debilitated patients, anesthesia is used to avoid restraint-induced stress and struggling.
4. Preoperative assessment is primarily directed at evaluating cardiopulmonary function and reserve.
   a. Observe the patient in a quiet environment for tachypnea, abnormal breathing patterns (e.g., open-mouth), and tail bobbing.
   b. Assess exercise tolerance by how fast respiration normalizes after brief restraint.

c. Examine the nares for obstruction that will interfere with mask induction.
d. Auscultate the heart for abnormal rate and rhythm, and murmurs.
e. Evaluate hydration by examining the mouth for dryness and peripheral veins for turgidity, and by rolling the skin between fingers to assess turgor.
 5. Palpate the crop and abdomen for distention.

B. Diagnostics
 1. A minimal clinicopathologic database includes the following:
    a. Packed cell volume (PCV)
    b. Total protein
    c. Blood glucose
 2. Further hematologic tests, clinical chemistries, and radiographs are performed as indicated.
 3. Many common avian infectious and toxic diseases impair hepatic and renal function, the primary excretion routes for parenteral anesthetics.

C. Fasting
 1. Indications
    a. Decrease regurgitation and passive reflux.
    b. Decrease proventricular and ventricular distention that does the following:
       (1) Interferes with normal respiratory airflow
       (2) Increases the chances for perforation during laparoscopy
 2. Duration related to the following:
    a. Size
       (1) Large birds (heavier than 500 g) are fasted at least 12 hr.
       (2) In small birds (e.g., budgerigar, canary) the period is proportionately less (6 to 12 hr).
       (3) One author recommends overnight fasting for all birds regardless of size.[1]
    b. Clinical condition
       (1) Ileus and/or gastrointestinal obstruction—fast, and administer fluids, electrolytes, and glucose either intravenously (IV) or intraosseously (IO).
       (2) Empty fluid-distended crops by gavage.
    c. Species
       (1) Vultures actively regurgitate when restrained and may passively reflux large volumes of fluid under anesthesia
       (2) I have also observed perianesthetic regurgitation in other raptors, seabirds, crows, and mynahs.
    d. Diet—sunflower kernels may be in the proventriculus of parrots even after 24 hr of fasting.
 3. Remove grit, toys, paper, and wood chips from the cage to prevent ingestion.
 4. Offer water at least 2 to 3 hr before induction.

5. Deep anesthesia promotes passive gastrointestinal reflux by decreasing smooth muscle tone.

D. Stabilization
1. Ideally, stabilization occurs before anesthesia—correct severe anemia, fluid, and metabolic disorders.
2. Attain vascular access preoperatively
   a. In the critical patient
   b. When extensive hemorrhage is anticipated
3. Cardiopulmonary compromised patients
   a. Preoxygenate (oxygen cage or mask)
   b. Avoid struggling, which increases oxygen consumption and cardiac work

---

III. Premedication

A. Reasons for Not Using Premedication
1. Prolongs recovery
2. Variable effects

B. Phenothiazine Tranquilizers
1. Examples
   a. Acepromazine[2,3]
   b. Chlorpromazine[2,3]
2. Effects
   a. Long-acting
   b. Ineffective
   c. Peripheral vasodilation and hypotension

C. α_2-Adrenoceptor Agonists (Xylazine)[4]
1. Prolonged induction and recovery
2. Excitement and occasional convulsions
3. Inadequate surgical analgesia
4. Hypoventilation[5]
5. Ratites, produces good sedation in healthy animals

D. Benzodiazepines (diazepam, midazolam, zolazepam)
1. Physiologic effects[6,7]:
   a. Dose-dependent sedative-hypnosis
   b. Muscle relaxation
   c. Minimal cardiopulmonary depression
   d. Anticonvulsant
   e. Amnesia
   f. Not analgesic
2. Indications
   a. Anxiolysis in wild, caught, or aviary birds
   b. To facilitate handling during induction and minor nonpainful procedures

c. Improve muscle relaxation and duration of effect of dissociative anesthetics
  3. Disadvantages
     a. Unpredictable effect
     b. Dysphoria
     c. Struggling
     d. Incoordination and ataxia
  4. Diazepam
     a. Solubilized in propylene glycol[8]
        (1) Hypotension and cardiac collapse when given rapidly IV
        (2) Effect exacerbated by dehydration and hypovolemia
     b. Chickens (2.5 mg/kg IV)[6]
        (1) Mild sedation
        (2) Hypotension
        (3) Bradycardia
        (4) Bradypnea
        (5) Hypothermia
  5. Midazolam[7]
     a. Water soluble
     b. Short-acting (15 to 30 min)
     c. Slightly more potent than diazepam
     d. Canada geese (2 mg/kg IM)[9]
        (1) Moderate sedation adequate for radiographic positioning in 15 and lasting no more than 20 minutes
        (2) Moderate tachypnea
  6. Zolazepam
     a. Potent
     b. Long-acting
     c. Combined with tiletamine in the commercial preparation, Telazol

E. **Opioid Analgesics**
  1. Indications
     a. Decreased anesthetic requirement
     b. Analgesia
     c. Sedation
  2. Disadvantages
     a. Limited knowledge of effects in birds
     b. Respiratory depression
     c. Bradycardia, hypotension
  3. Morphine (chicken 0.1, 1.0, and 3.0 mg/kg IV)[10]
     a. Pure µ-opioid agonist
     b. Dose-dependent reduction in isoflurane MAC (approximately 15% to 50%)
     c. Heart rate and mean arterial blood pressure not significantly affected

F. **Parasympatholytics (Atropine, Glycopyrrolate)**
  1. Not routinely used
  2. Physiologic effects

a. Inhibition of respiratory and salivary secretions
   b. Inhibition of gastrointestinal motility (including the crop)
   c. Prevention of vagal bradyarrhythmias[11]
3. Glycopyrrolate (0.01 to 0.03 mg/kg IM or IV)[8,12]
   a. Potent
   b. Relatively long-acting
   c. More selective antisecretory agent than atropine
4. Atropine (0.04 to 0.1 mg/kg IM or IV)[8,12]
   a. Relatively short duration of effect (30 min)
   b. Fast onset, so it is indicated in cardiac emergencies

IV. Local Anesthesia

A. Reasons for Infrequent Use
1. Potential adverse drug effects
2. Stress associated with physical restraint

B. Most Useful in Tame or Sedated Large Birds
1. Examples: pigeons, poultry, ratites, waterfowl

C. Technique
1. Small volume (0.5 to 1.0 ml) syringes and small gauge (25 to 27) needles.
2. Use lowest available concentration.
3. Avoid vessels or highly vascular areas.
4. Calculate volumes to avoid overdosage.
5. *Do not* use epinephrine solutions.

D. Adverse Effects
1. Usually the result of overdosage
   a. For example—if 0.2 ml 2% lidocaine is used in a 200 g bird, the dose = 20 mg/kg, which is greater than the toxic dose of 4 to 10 mg/kg reported in mammals[8]
   b. In budgerigars subcutaneous LD_{50} of procaine as low as 200 mg/kg[13]
2. Clinical signs of toxicity and treatment[11]
   a. Dose-dependent
   b. Initial excitement and seizures—controlled with diazepam (0.1 to 1 mg/kg IV)
   c. Depression
   d. Respiratory arrest—intubate and ventilate
   e. Cardiovascular collapse—fluids (10 to 20 ml/kg IV), calcium gluconate (100 mg/kg IV)

E. Longer-Acting Local Anesthetics
1. Example: bupivicaine[11]
2. More potent than lidocaine
3. Affect the cardiovascular system at lower dosages

V. Parenteral Anesthesia

1. Usually reserved for short procedures to avoid prolonged recovery and mortality
2. An accurate weight and appropriate sized syringe are essential for small volume injection.
3. A means of ventilation and a source of oxygen should always be available.

A. Routes of Administration
1. Intramuscular (IM)
   a. Pectoral or thigh muscles
   b. The renal portal system[14] does not clinically appear to affect either duration or quality of anesthesia produced by hindleg injection.
2. Intravenous (IV)
   a. Basilic (ulnar) vein—usually avoided because it is difficult to provide good hemostasis in a struggling bird
   b. Medial saphenous vein
   c. Right jugular vein—slow injection to avoid marked myocardial depression and retroperfusion of the brain with high drug concentrations
   d. Dosages are 50% to 70% of IM
3. Intraosseous (IO)
   a. Indicated when IV access is impossible or too time-consuming[15]
   b. Distal end of the ulna[16]—note some birds (e.g., pelicans) have air-filled ulnas.
   c. Proximal tibiotarsus

B. Parenteral Drugs
1. Metabolic drug scaling
   a. Metabolic scaling of parenteral drug doses between species and diverse body sizes has been evaluated and described by Sedgwick[17]
   b. In general, parenteral drug dosage required to produce a given anesthetic level varies nonlinearly with size; the smaller the bird, the greater the dose/unit of body weight
   c. Sedgwick, however, has demonstrated that many drug dosages are uniform when related to the metabolic size of an animal[17]
      (1) To calculate such a dosage the daily minimum energy cost (MEC) is first determined using the equation $MEC = K \times M_b^{0.75}$ where M_b = body weight (kg) and K = 129 and 78 for passerines and nonpasserines, respectively. The total dose of a drug is then calculated and divided by the MEC value to give a MEC dosage (mg/kcal).[17]
      (2) This MEC dosage can then be used to calculate drug dose in larger or smaller birds by first calculating MEC for that bird, and multiplying by the MEC dosage.

(3) For example, the MEC for a 30 g budgerigar is $78 \times 0.03^{0.75}$ = 5.6 kcal/day.
(4) If the MEC dose of ketamine is 0.2 mg/kcal, this bird would require 1.1 mg or 36 mg/kg.

2. Propofol
   a. Ultrashort acting
   b. Noncumulative
   c. Requires intravenous administration
   d. Physiologic effects (chicken,[18] duck,[19] pigeon[20])
      (1) Dose- and rate-dependent
      (2) Smooth and rapid induction
      (3) Good muscle relaxation of short duration
      (4) Marked cardiopulmonary depression similar to thiopental[11]
      (5) Arrhythmias—single, or runs, of ventricular premature contractions and ventricular tachycardia (chicken)[18]
   e. *Very narrow safety margin* in spontaneously breathing birds[18,20]
   f. Doses
      (1) Pigeon: 14 mg/kg[20]
      (2) Chicken: 6.8 mg/kg, infusion 1.0 mg/kg/min[18]
      (3) Duck: 10 mg/kg[19]

3. Ketamine
   a. Dissociative anesthetic—cateleptoid state[11,21]
      (1) Open eyes
      (2) Occasional purposeful skeletal movements
      (3) Hypertonus independent of stimulation
   b. Water-soluble—IM or IV administration
   c. IM[4]
      (1) Initially, incoordination and opisthotony
      (2) Relaxation within 1 to 3 min
      (3) Large birds—manual restraint may be necessary to avoid neck and leg trauma.
      (4) Duration of maximal effect is dose-dependent
   d. Use a moderate IV dose to achieve a short duration of effect.
   e. Prolongation of anesthesia—give 30% to 50% of the original dose to effect.
   f. Recovery
      (1) Characterized by incoordination, excitement, head shaking, and wing flapping[4]
      (2) Caused by redistribution followed by hepatic biotransformation and/or renal excretion[8]
      (3) Prolonged by repetitive dosing, renal disease, and severe dehydration
   g. Physiologic effects
      (1) Blood pressure and cardiac output are generally maintained because of increased sympathetic activity and catecholamine release

(2) However, directly depresses myocardial contractility, and may cause cardiac collapse in birds with marginal cardiovascular reserve[11]

(3) In ducks (20 mg/kg IV)[5] and chickens (30 to 120 mg/kg IM)[21] it did not significantly affect respiration, whereas in red-tailed hawks (30 mg/kg IM)[22] it produced mild hyperventilation.

(4) In budgerigars,[23] high dose produced respiratory arrest, followed by cardiac arrest.

(5) In penguin, gallinule, water rail, golden pheasant, turaco, and hornbill, it produced only light sedation and excited recovery.[4]

h. Dosages

(1) In chickens,[24] the median effective IV dosage for at least 15 min anesthesia was 14 mg/kg, median lethal dosage was 67.9 mg/kg, and surgical analgesia was not apparent at dosages less than 60 mg/kg.

(2) In budgerigars,[23] IM lethal dosage was approximately 500 mg/kg, whereas an adequate anesthetic dosage was estimated to be within 50 to 100 mg/kg IM.

4. Diazepam-ketamine

a. Diazepam improves muscle relaxation, anesthetic duration, and recovery.

b. Surgical anesthesia is not as good as xylazine-ketamine.

c. Cardiopulmonary depression is less.

d. Chicken (2.5 mg/kg IV + 75 mg/kg IM)[6]

(1) Rapid tranquilization and loss of the righting reflex

(2) Recovery in 90 to 100 min

(3) Opisthotonus common

(4) Short-term myotonic limb contractions present in all birds

(5) Pain reflexes elicited at all times

(6) Although bradycardia may be observed, blood pressure, respiration rate, and body temperature remain stable.

e. Raptors (1 to 1.5 mg/kg + 30 to 40 mg/kg IV)[25]

(1) Minor to major surgical procedures

(2) Rapid injection–prolonged apnea, cardiac arrhythmias, and increased risk of death–administer divided doses at intervals of 2 to 3 min

(3) Overdosage observed in fat birds

5. Xylazine-ketamine

a. Xylazine improves muscle relaxation, duration of surgical analgesia, and quality of recovery.

b. Xylazine prolongs induction and recovery and increases the likelihood of adverse cardiopulmonary effects.[4,5]

c. Physiologic effects

(1) Eyes remain open and palpebral reflex is preserved.[4]

(2) Respiratory depression, acidemia, and hypoxemia (duck,[5] red-tailed hawk[26])

(3) Moderate hyperthermia (duck)[5]

(4) Bradycardia (red-tailed hawk,[26] pigeon,[27] goshawk[27])
    d. Duck (1 mg/kg + 20 mg/kg IV)[5]
        (1) Respiratory depression, acidemia, and hypoxemia
        (2) Moderate hyperthermia
    e. Great horned owl (0.15 mg/kg + 15 mg/kg IM)[28]
        (1) Smooth, rapid induction
        (2) Transient apnea and hypoventilation
        (3) Bradycardia
    f. Turkey vulture (1 mg/kg + 10 mg/kg IM)[29] —rapid induction of a consistent level of anesthesia (induction time 5.4 plus or minus 1 min and duration 109.8 + 25.4 min)
    g. For convenience equal volumes of xylazine (20 mg/ml) and ketamine (100 mg/ml) are combined (final concentration 10 mg/ml to 50 mg/ml) and a dosage based on ml/kg is calculated.[30,31]
6. α_2-adrenoceptor antagonists
    a. Reverse effects of α_2-adrenoceptor agonists.
    b. *Do not* reverse the effects of ketamine.
    c. Yohimbine
        (1) Budgerigar (0.275 mg/kg IV)—shortened recovery when administered 40 to 45 minutes after xylazine-ketamine[32]
        (2) Guineafowl (1 mg/kg IV)[33]
        (3) Red-tailed hawk (0.1 mg/kg IV)—no profound cardiopulmonary changes[26]
        (4) Ostrich (0.125 mg/kg IV)[34]
        (5) Excitement and mortality occasionally occur in mammals at dosages greater than 1 mg/kg.
    d. Tolazoline: turkey vulture (15 mg/kg IV)[29]
7. Zolazepam-tiletamine
    a. The dissociative anesthetic tiletamine is combined with the potent benzodiazepine zolazepam in the commercial combination Telazol.
    b. Tiletamine[8]
        (1) Approximately three times as potent as ketamine
        (2) Long duration of effect–prolonged recovery
    c. Duck (13 mg/kg IM)[35]
        (1) Adequate anesthesia for liver biopsy in 15 minutes
        (2) Recovery in 3 to 5 hours
        (3) Occasional additional doses (3 mg/kg) required
    d. Great horned and screech owls (5 and 10 mg/kg IM)[36]
        (1) Smooth induction and recovery
        (2) Brief period of bradypnea
        (3) Moderate tachycardia at higher dose
        (4) Prolonged recovery = 4 to 5 hours
    e. Red-tailed hawk (10, 15, 20, 40 mg/kg IM)[36]
        (1) No loss of consciousness
        (2) Bradycardia and bradypnea
        (3) Prolonged recoveries characterized by catalepsy, opisthotonus, and ataxia

VI. Inhalation Anesthesia

A. Minimum Anesthetic Concentration (MAC) or Dose (MAD)
1. Measure of inhalant anesthetic potency
2. Anesthetic concentration (vol%) that produces immobility in 50% of an anesthetized population subjected to a noxious stimulus[37-39]
3. The more potent an inhalation anesthetic, the lower its MAC value
4. MAC is similar within and across species and is decreased by hypothermia.[40]
5. Maintenance surgical anesthesia vaporizer settings are approximately 25% higher than MAC (e.g., halothane MAC = 0.85 " 0.09% in chickens, maintenance = 1% to 1.5%)[11,37]
6. In spontaneously breathing large birds, discrepancies between vaporizer setting and anesthetic effect are overcome by controlled ventilation and increased flow rate.

B. Inhalation Anesthetic Agents
1. Avian respiratory system the most efficient gas exchange system among all air-breathing vertebrates.[14]
   a. Uptake and excretion of inhalation anesthetic agents are more rapid in birds than mammals.
   b. This efficiency, in combination with large volume air sacs, makes birds more susceptible than mammals to inhalant anesthetic overdosage.
2. Methoxyflurane
   a. *Not recommended* for use in avian practice
   b. Advantages
      (1) Potent
      (2) Very low saturated vapor pressure = decreased potential for lethal concentrations[11]
      (3) Inexpensive
      (4) Analgesic
      (5) Muscle relaxant
      (6) No myocardial sensitization to catecholamines[11]
   c. Disadvantages
      (1) High tissue solubilities result in prolonged induction and recovery[11]
      (2) Dose-dependent cardiopulmonary depression[11,12]
   d. Induction vaporizer settings greater than 3%, maintenance 0.2% to 0.5%
2. Halothane
   a. *Not recommended* for use in avian practice
   b. Advantages[11]
      (1) Potent
      (2) Relatively low tissue solubilities
      (3) Inexpensive
      (4) Moderate muscle relaxation

c. Disadvantages[11,37]
      (1) Dose-dependent cardiopulmonary depression
      (2) Myocardial sensitization to catecholamine-induced arrhythmias
      (3) Potential hepatotoxicity
   d. Duck (2% to 2.5%)[41]
      (1) Moderate tachycardia
      (2) Mild hypotension
      (3) Moderate to severe bradypnea
      (4) Arrhythmias (ventricular fibrillation, ventricular bigeminy, and multifocal ventricular rhythms)
   e. Induction vaporizer setting should not exceed 2%, maintenance equal to 1% to 1.5%
3. Isoflurane
   a. Preferred inhalant anesthetic
   b. Advantages[11,12]
      (1) Rapid induction and recovery
      (2) Less depression of cardiac output than other agents
      (3) No myocardial sensitization to catecholamines
      (4) Lower incidence of hepatotoxicity
      (5) Moderate muscle relaxation
   c. Disadvantages
      (1) Dose-dependent cardiopulmonary depression
      (2) Hypotension
      (3) Relatively expensive
   d. Duck (2.0% to 3.0%)[39,41]
      (1) Mild tachycardia (20% to 30% increased)
      (2) Hypotension (10% decreased)
      (3) Bradypnea (50% decreased)
      (4) No arrhythmias
      (5) All effects less then those of halothane (2% to 2.5%)
      (6) Significant hypoventilation
      (7) Neither respiration rate nor tidal volume good determinant of adequacy of ventilation
   e. Sandhill crane (1 and 1.5 MAC)[38]
      (1) Minimal effects on cardiovascular function
      (2) Moderate to marked hypoventilation
   f. Chicken (mechanically ventilated, end-tidal 2.1%)[42]: significantly lower threshold for electrical fibrillation of the heart when compared with halothane (end-tidal 1.2%) or pentobarbital (30 mg/kg IV)
   g. Induction vaporizer setting less than or equal to 3%, maintenance 1.5% to 2%
4. New inhalants—desflurane and sevoflurane
   a. Advantages: low blood gas partition coefficient—more rapid induction and recovery than isoflurane, precise control of alveolar anesthetic concentrations
   b. Disadvantages

(1) Desflurane requires expensive vaporizer and is relatively expensive.
(2) Sevoflurane requires agent-specific vaporizer.
(3) Otherwise, same as isoflurane
5. Nitrous oxide (N_2O)
    a. Use controversial mainly because of misunderstanding of avian respiratory anatomy
        (1) Although it will expand closed gas-filled spaces within the body, it will not cause air sac expansion because they communicate with the outside[11]
        (2) Some diving birds (e.g., pelicans, gannets) have subcutaneous air pockets that do not communicate with the respiratory system, and therefore use can result in subcutaneous emphysema.[43]
    b. Advantages[11,12]
        (1) Odorless
        (2) Very low tissue solubilities—rapid uptake as well as excretion
        (3) Decreased anesthetic requirement of concurrently administered anesthetic agents because of its analgesic effect
        (4) Facilitates uptake of other inhalation anesthetics through the second gas effect
        (5) Minimal cardiovascular depression
    c. Disadvantages
        (1) Very low potency
        (2) Decreases inspired oxygen concentration—do not use in birds with marginal respiratory reserve

C. Inhalation Anesthetic Equipment[44]
1. Anesthesia machine
    a. Requisite parts
        (1) Precision vaporizer
        (2) Flowmeter
    b. Inappropriate and hazardous to use inhalant poured on cotton wool
2. Breathing system
    a. Nonrebreathing systems are used in birds at least 10 kg to minimize respiratory resistance and mechanical dead space.
    b. Most commonly use modified Jackson Rees and its modifications[44] (Fig. 22-1)
        (1) Flow rate approximately 3 × minute ventilation (respiration rate × tidal volume)
        (2) Flows at least 200 ml/min impair accurate vaporizer function, and slow equilibration of vaporizer settings
        (3) Connect to the fresh gas outlet that runs from the vaporizer to the circle breathing system on most conventional machines
    c. In large birds heavier than 10 kg (e.g., ratites), a circle breathing system is used with appropriate size rebreathing bag and absorber canister for the tidal volume of the bird.

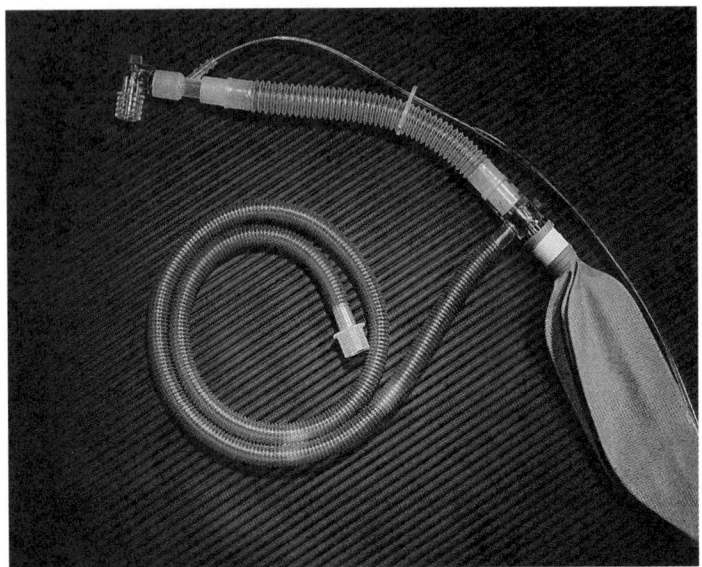

Fig. 22-1 Modified Jackson Rees circuit. (Courtesy SurgicVet/Anesco, Inc.)

3. Masks
   a. Medium sized birds—commercial dog and cat masks
   b. Small birds—can be made from syringe cases
   c. Cranes and other long-billed birds—an elongated mask can be crafted from a 60-ml syringe case
4. Endotracheal tubes
   a. Three major types
      (1) Cuffed
         (a) Not used in small birds because the cuff increases the outside tube diameter
         (b) Cuff not routinely inflated because it can cause ischemic necrosis of tracheal mucosa
      (2) Uncuffed—either purchased or fashioned from urinary and over-the-needle catheters
      (3) Cole
         (a) Uncuffed
         (b) A narrow and thick portion—thin portion is introduced into the trachea and the thick portion seals against the glottis
         (c) Originally designed to decrease respiratory resistance in small patients[44]
5. Scavenge system
   a. Essential when using inhalation anesthesia to remove waste gases to the outside[44]

b. Anesthetic gases represent a potential health risk to personnel[12,44]
c. Nonrebreathing systems
    (1) Difficult to scavenge, particularly when a mask is used for maintenance
    (2) A scavenging device for mask in birds has been described.[45]
    (3) Other systems have been described for laboratory animals.[46,47]

D. Induction
1. Techniques
    a. Injectable anesthetic
    b. Low dose injectable anesthetic plus inhalant anesthesia
    c. Inhalant anesthetic
2. Mask induction
    a. Gently place the mask over the beak to include the nares.
    b. Allow the bird to breathe oxygen.
    c. Once respiration is regular, do one of the following:
        (1) Add the inhalant in incremental steps at 30 second intervals—used to avoid breath holding
        (2) Begin with maximum recommended inhalant anesthetic concentration—usually less excitement
    d. Once the bird relaxes, rapidly decrease the concentration back to maintenance, and be alert for indications of excessive depression in the initial postinduction period.
3. Awake intubation
    a. An alternative to mask induction in some species of birds (e.g., pigeons, small raptors, wading birds)
    b. The same induction protocol as for mask induction
    c. Do not ventilate the bird to increase the rate of anesthetic uptake—may prove fatal.
4. Aquarium induction
    a. Easy
    b. Offers the potential for injury as the bird struggles, and variable anesthetic concentration
5. Injectable anesthesia
    a. Ketamine IV or IM
    b. Propofol IV

E. Intubation
1. Advantages
    a. Allows assisted/controlled ventilation
    b. Protects against aspiration
    c. Allows lower respiratory tract suction
    d. Minimizes fresh gas flow and dead space
2. Tracheal mucosal trauma
    a. Caused by inflated cuffs and rough intubation
    b. Promoted by complete tracheal rings that prevent expansion
    c. Severe trauma will result in tracheal stenosis.

3. Technique
   a. Select the largest diameter tube possible to provide an adequate seal and reduce airway resistance.
   b. Limit tube length extending from the patient to reduce mechanical dead space.
   c. Identify the glottis at the base of the tongue.
   d. Visualization is facilitated by grasping the tongue with gauze or a cotton-tipped applicator and pulling it forward.
   e. Open the mouth, with gauze placed around the upper and lower beak, to expose the glottis.
   f. Turn the tube bevel sideways to pass through into the trachea.
   g. Some species (e.g., flamingo) do not allow good visualization of the glottis. Pass the tube blindly by palpating the trachea externally with the head and neck in extension.
   h. Unlike mammals, birds can still vocalize when intubated because of the syrinx at the tracheal bifurcation. Furthermore, vocalization can occur on both inhalation and exhalation, and be induced by mechanical ventilation.
   i. Light sources to illuminate the glottis include the following:
      (1) Penlights
      (2) Laryngoscopes with small straight blades
      (3) Head lamps
      (4) Goose neck lamps
      (5) Examination lights
   j. In awake birds 1% lidocaine is topically applied to the glottis before the lubricated tube is placed.
   k. A roll of gauze squares is taped in the mouth to prevent the endotracheal tube being severed.
   l. Tape the tube in place and lubricate the eyes.
   m. Protect the down eye because the avian cornea protrudes from the head.

F. Muscle Relaxants
1. Indicated during inhalation anesthesia to provide the following:
   a. Good muscle relaxation
   b. Prevent movement during delicate (e.g., ophthalmic) surgery
   c. Reduce anesthetic requirement
2. Disadvantages
   a. Require ventilatory support
   b. Absence of analgesia
3. Atracurium
   a. Ultrashort-acting nondepolarizing muscle relaxant.
   b. Chicken (adjunct to isoflurane)[48]
      (1) Muscle relaxation monitored with a nerve stimulator
      (2) Effective dosage producing 95% twitch depression in 50% of birds was 0.25 mg/kg IV
      (3) Duration of effect 30 to 40 minutes, increasing to 45 to 55 minutes when the dosage increased to 0.45 mg/kg

(4) Minimal tachycardia and hypertension
(5) Edrophonium (0.5 mg/kg IV) hastened return of normal muscle strength.

VII. Perianesthetic Monitoring and Support

A. **Major Determinant of Anesthetic Success, Particularly in Critical Patients**

B. **Level Determined by Patient Status and the Procedure to Be Performed**

C. **Regular Assessment of Depth and Cardiopulmonary Function**

D. **Depth**
   1. Most useful guides
      a. Response to painful stimuli
      b. Muscle tone: Jaw tone is used to subjectively assess muscle tone.
      c. Palpebral reflex: Usually present until light surgical anesthesia, whereas the corneal response persists to deeper levels
      d. Respiratory rate and depth—at medium anesthetic planes is regular and tidal volume normal.
   2. Cardiac arrest—piloerection and pupillary dilation
   3. Anesthetic level should be appropriate to the procedure; some procedures are likely to be more painful than others and depth is adjusted accordingly.
   4. Very painful stimuli include the following:
      a. Plucking feathers
      b. Manipulation of bone
      c. Visceral traction

E. **Allometric Scaling of Physiologic Variables**
   1. Physiologic variables vary with animal size.
      a. This relationship is not linear for most physiologic variables monitored during anesthesia.
      b. An allometric equation of the general form $y = a \times x^b$ can be used to describe many of these relationships.[49,50]
   2. Sedgwick[50] described the use of allometric equations to determine expected variable reference ranges during anesthesia and serve as guidelines for the anesthetized avian patient.
   3. Changes of 20% above or below the calculated value are considered abnormal.[50]

F. **Cardiovascular Monitoring**
   1. Resting heart rate (beats/min) is estimated from the allometric equation $f_h = 155.8\ M_b^{-0.23}$ (M_b = body weight in kg).[49]
   2. Auscultation

a. Used to determine rate and rhythm
b. Cardiac sounds are muffled by the sternum and pectoral musculature, whereas the greatest intensity sounds are heard below the sternum and at the thoracic inlet.
c. Although continuous auscultation is impracticable during a surgical procedure, a good quality stethoscope should always be available.
d. Esophageal stethoscopes are used in some species but are difficult to position in birds with crops.
3. Electrocardiography (ECG)
   a. Continuous heart rate and rhythm monitoring during long and complex procedures
   b. An ECG machine should be able to do the following:
      (1) Monitor high heart rates
      (2) Freeze the ECG for interpretation
      (3) Provide a printout
   c. ECG quality is maximized using electrode paste, needles, or stainless steel suture to improve lead contact.
   d. Leads I or II are usually monitored.
4. Ultrasonic Doppler flow apparatus
   a. Provides the following:
      (1) An audible signal of arterial flow
      (2) Continuous monitoring of heart rate and rhythm
      (3) Detection of sudden changes in pressure
   b. A pediatric probe is placed over either a basilic or metatarsal artery.
   c. The pencil probe attachment is used to assess cardiac and peripheral blood flow in emergencies.
5. Arterial blood pressure
   a. Indirect measurement[14]
      (1) Determined in large birds in the same manner as mammals
      (2) The appropriate sized cuff (width = 70% to 80% of circumference) must be used or the values will be too high or too low.
   b. Direct measurement
      (1) Indicated in large birds (e.g., ratites) for prolonged procedures and/or where hypotension/hypovolemia is anticipated.
      (2) Place over-the-needle catheters (18 to 24 gauge) in either the basilic or metatarsal arteries and connect through heparinized saline filled pressure tubing to a pressure transducer placed at the level of the heart.
      (3) Secure the catheter with either sutures or glue.
      (4) The arterial line also allows blood sampling for blood gas measurement.
   c. Normal[14]
      (1) Varies among species, sexes, age, and breed
      (2) In many birds higher than expected for mammals
      (3) Apparent hypertension has been described in an ostrich.[51]
6. Orthostatic hypotension
   a. Results from a pooling of blood in the extremities

b. Exacerbated by the following[11]:
      (1) Dehydration
      (2) Hypovolemia
      (3) Anesthesia
   c. Prevented during the perianesthetic period by avoiding rapid changes in body position
   d. Severe orthostatic hypotension will decrease venous return and induce cardiac arrest.
   e. A common time for this to occur is at the end of a surgical procedure when the drapes are removed and the recovering bird is rapidly picked up before returning to recovery.

G. Cardiovascular Support
   1. Intravenous catheterization
      a. Place in either the basilic (ulnar) or jugular veins.
      b. The medial saphenous vein is used in some large birds (e.g., cranes, ratites).
      c. Small-gauge butterfly catheters (25 or 27 gauge) are temporarily placed in very small birds.
      d. Use over-the-needle catheters in larger birds.
      e. The basilic vein is most easily entered proximal to the elbow joint, where it straightens out along the humerus.
      f. Secure the catheter with either sutures or glue.
   2. Intraosseous catheterization
      a. Indications (see above)
      b. Technique[16]
         (1) At least 500 g: 18- to 22-gauge spinal needle
         (2) Small birds and neonates: 25- to 30-gauge hypodermic needles
         (3) Use cannula long enough to penetrate to the ulna midpoint.
         (4) Remove feathers from lateral surface of distal ulna and aseptically prepare.
         (5) Palpate the lateral notch on the distal ulna, and centrally position the needle parallel to the median plane of the bone.
         (6) Advance the needle with a slight rotating movement into the supported ulna.
         (7) The needle should pass without resistance into the marrow cavity.
         (8) Aspirate the needle to check for bone marrow, and flush with heparinized saline to prevent occlusion and check for free flow.
         (9) If the needle is correctly placed the fluid should flow easily, and the basilic vein should clear as the fluids pass out of the bone.
         (10) Flush the cannula twice daily with heparinized saline.
      c. Pigeon, lactated Ringer's solution infused into an IO catheter reached the systemic circulation in 30 seconds, and an infusion rate of 20 ml/kg/hr was achieved.[52]

3. Fluids
   a. Establish vascular access before major surgical procedures and provide slow infusion (5 to 10 ml/kg/hr) of balanced electrolyte solution.
   b. Add dextrose as indicated.
   c. Use commercial small volume syringe infusion pumps for accurate fluid administration.
   d. Guidelines for rate of fluid infusion have not been established and are extrapolated from those for mammals.
4. Blood transfusion
   a. Excessive hemorrhage is a common cause of mortality because of the small size of most patients.
   b. Birds are better able than mammals to tolerate severe blood loss.[14]
   c. Land and nonflying birds (chicken, pheasant) are less able to tolerate severe blood loss than flying species (pigeon, duck).[14]
   d. Total blood volume ranges from 6% to 11% of body weight.[14]
   e. Estimated hemorrhage of 5% to 10% of blood volume is treated with three times the volume in balanced electrolyte solution.
   f. Indications:
      (1) Hemorrhage at least 20% to 30% of circulating blood volume
      (2) Hemoglobin concentration at least 8 g/dl and PCV at least 15%
   g. Other factors to consider:
      (1) Hemoglobin saturation
      (2) Adequacy of cardiac output
   h. Sources (decreasing desirability)
      (1) Siblings
      (2) Same species
      (3) Same order
      (4) Do not transfuse across orders (e.g., pigeon into parrot) because of the short RBC half-life and the subsequent breakdown products.
   i. Technique
      (1) Approximately 10% of a donor's blood is collected, necessitating two donors be available.
      (2) Collect blood into a heparinized syringe using a 23- or 25-gauge butterfly catheter.
      (3) Gently mix during collection and immediately transfuse.
      (4) Avoid the anticoagulants citrate phosphate dextrose and lithium EDTA, which can produce a calcium responsive tetany.
      (5) Lithium EDTA will produce hemolysis in some species (e.g., ostriches).

H. Respiratory Monitoring
1. Respiration rate
   a. Resting rate (breaths/min) is estimated from the allometric equations $f_{resp} = 17.2\ M_b^{-0.31}$ (M_b = body weight in kg).[49]
   b. Technique
      (1) Observation of thoracic movement

(2) Observation of expiratory condensation in a clear endotracheal tube
(3) Commercial respiratory monitor attached to endotracheal tube
2. Ventilation
   a. Tidal volume (ml) and minute ventilation (ml/min) are estimated from the allometric equations $V_t = 13.2\ M_b^{1.08}$ and $V = 284\ M_b^{0.77}$, respectively.[49]
   b. Cannot assess visually
   c. Commercial flowmeters
      (1) Not accurate in small patients
      (2) Increased mechanical dead space and resistance
   d. Most accurate method of assessment is determination of arterial partial pressure of carbon dioxide ($PaCO_2$).
3. Blood gas analysis
   a. For evaluation of ventilation, oxygen transport and delivery, and metabolic status
   b. Correct to body temperature for interpretation[53]
   c. Values are similar to mammalian and interpretation is the same.[14]
4. Pulse oximetry
   a. Measurement of hemoglobin saturation
   b. Presently available commercial systems have not been validated for use in birds and appear to be inaccurate.

I. Respiratory Support
1. Routinely assist ventilation during anesthesia for the following reasons:
   a. Birds have a slower rate of breathing (one third) and increased tidal volume (4 times) than a mammal of the same size[14]–small changes in respiration rate will therefore have a greater effect on overall minute ventilation than in a mammal.
   b. Birds very susceptible to the respiratory depressant effects of anesthetics
   c. Potential for hypoventilation is further exacerbated by hindrance of normal thoracic movements (e.g., heavy surgical drapes or leaning hands on the bird), and pathologic conditions such as obesity and intraabdominal masses.
   d. Cannot visually assess adequacy of ventilation
2. Atelectasis
   a. Although the lungs are fixed to the thoracic wall, atelectasis may still be a problem in anesthetized birds.
   b. Several dorsally recumbent patients die under anesthesia, and subsequent histologic examination has shown collapse and obstruction of the small airways (parabronchi and air capillaries).
   c. Assisted ventilation may help.
3. Other causes of ventilatory difficulties include the following:
   a. Accumulation of blood and fluids in the air sacs with subsequent migration to the lungs.
      (1) If this is suspected, the bird is held vertical until it recovers.
      (2) Hemostasis should be meticulous during surgery.

(3) Do not flush the coelomic cavities.
  b. Dorsal recumbency
     (1) In chickens results in a 40% to 50% decrease in tidal volume and 20% to 50% increase in respiratory rate, with an overall 10% to 60% decrease in minute ventilation.[54]
     (2) Adverse effects are thought to be due to visceral compression of the air sacs.[54]
  c. High inspired oxygen fraction (ducks anesthetized with 1.4% isoflurane)[55]
     (1) Greater than 40% inspired oxygen fraction–decreased tidal volume, decreased respiration rate, increased $PaCO_2$
     (2) Possible mechanism–decreased activity of oxygen-sensitive chemoreceptors

J. Temperature Monitoring and Support
1. Normal body temperature[14]
   a. Important for normal metabolic rate
   b. Avian deep-body temperatures are generally higher than mammalian.
   c. Large flightless and diving birds and species that are capable of torpor tend to have low body temperatures.
   d. Fluctuates during the course of a day, the smaller the bird the greater the fluctuation
   e. Decreased by fasting
   f. Dehydrated birds (e.g., ostrich and turkey) are not able to prevent hyperthermia as well because of decreased evaporative cooling.
   g. Respiratory evaporative cooling is extremely important.
2. As core body temperature decreases:
   a. Peripheral vasoconstriction
   b. Bradycardia
   c. Hypotension
   d. Ventricular fibrillation
3. As body temperature returns to normal:
   a. Peripheral vasodilation = hypotension?
   b. Increased oxygen requirement = hypoxemia?
   c. Increased glucose requirement = hypoglycemia?
4. Monitor temperature continuously or intermittently.
   a. Esophageal probes, although optimal for core body temperature measurement, are difficult to place in birds with crops.
5. Hypothermia is reduced by the following:
   a. Decreasing anesthetic time
   b. Minimal use of surgical preparation solutions
   c. Using warming fluids
   d. Wrapping the bird in bubble-wrap
   e. Using circulating water blankets
   f. Using heat lamps
6. Electric heating pads and heat lamps will produce severe burns if placed too close to the patient.

K. Glucose Monitoring and Support
1. Most patients are small and consequently their glucose and glycogen reserves are small. Hence, blood glucose is assessed preoperatively in debilitated animals and monitored intermittently during prolonged procedures and recovery.
2. Under anesthesia, hypoglycemia may only manifest as the following:
   a. Nonresponsive bradycardia
   b. Hypotension
   c. Pupillary dilation
3. Administer glucose when blood levels fall below 200 mg/dl.
4. Although a bolus of 50% dextrose (1 to 2 ml/kg IV) is given in emergencies (glucose ≤60 mg/dl), it is better to use an infusion to prevent rebound hypoglycemia.
5. Hypertonic solutions of SC dextrose produce dehydration.

VIII. Resuscitation

A. General Considerations
1. Follow the mammalian protocol (airway, breathing, circulation, and drugs).[56]
2. Time between detection and initiation of resuscitation is critical; most avian patients are small, and consequently tissue oxygen reserves are rapidly exhausted during cardiopulmonary arrest.

B. Prevention
1. Perioperative stabilization
2. Attentive monitoring

C. Facilitation of Successful Resuscitation
1. Intubation
   a. If it is not possible to intubate a bird, flapping of the wings appears to provide some air movement and may also facilitate venous return to the heart from the pectoral musculature.
2. Ventilatory support
3. Preplacement of vascular access lines
4. If necessary all the drugs required during resuscitation can be administered IO.[15]
   The author has successfully resuscitated several small birds using this technique.

IX. Recovery

A. Critical Period
1. Carefully restrain in a warm, quiet environment.
2. Monitor.

B. Primary Determinants of Duration and Quality
1. Anesthetic agents
2. Procedure duration
3. Magnitude of physiologic dysfunction incurred

C. Causes of Prolonged Recovery
1. Hypothermia
2. Hypoglycemia
3. Anesthetic overdose

D. Possible Cause of Unexplained Deaths
1. Rewarming a bird that is hypovolemic and/or hypoglycemic may result in vasodilation and increased metabolic demand for glucose and oxygen–these phenomena may explain some unexplained deaths that occur a few hours into recovery.

X. Anesthesia for Specific Problems
A. Neonatal and Pediatric
1. Two types of avian neonate
   a. Altricial
   b. Precocial
2. Altricial neonates hatch at an earlier stage of development than precocial neonates.
3. Regardless of neonate type, it is assumed that hepatic and renal function is immature, and parenteral drug excretion is therefore likely to be prolonged.
4. Glycogen reserves are small and hypoglycemia may develop perioperatively.
5. All neonates develop hypothermia rapidly because of their large surface area to volume ratio. Altricial neonates have very poor thermoregulatory ability.[14]
6. Important interventions
   a. Provide adequate warmth for neonates during the perianesthetic period.
   b. Minimize anesthetic period.
   c. Aim for rapid return to normal function.
7. Neonates are likely to become dehydrated during prolonged procedures.
   a. High water requirements
   b. Water loss from the respiratory system when using inhalant anesthesia
   c. Placement of an IO catheter will be lifesaving in some procedures
8. Regurgitation is common.
   a. Remove food from full crops.
   b. Fasting is recommended to reduce the crop volume.
   c. Sufficient food may still remain in the proventriculus and ventriculus for reflux.

9. Air sac volume is smaller in neonates because of high gastrointestinal volume.
10. For anesthesia mask induce with isoflurane, intubates, and ventilates.

B. Ophthalmic
1. Surgery requires a dilated, centrally fixed eye.
   a. Do not achieve this by increasing the depth of anesthesia—endangers the patient.
   b. Use a muscle relaxant (see above) to provide relaxation without deep anesthesia.
2. The oculocardiac reflex (vagally induced bradycardia caused by manipulation of the globe) is not a major problem in birds.
3. Enucleation will often result in much hemorrhage. Be prepared for a transfusion.

C. Radiography
1. Requirements
   a. Rapid recovery of outpatients
   b. Good relaxation for diagnostic quality
2. Ketamine alone is inadequate relaxation and prolonged recovery.
3. Ketamine plus either midazolam or diazepam results in decreased ketamine dose required and better relaxation.
4. The author prefers an inhalation agent, usually isoflurane, for good relaxation and short duration of effect.

D. Respiratory (Upper Airway Obstruction)
1. Preoxygenate
2. Isoflurane for restraint often preferable to physical restraint alone
3. Unique respiratory system allows oxygen and inhalation anesthetics to be delivered through the air sacs.[57]
   a. In an emergency, oxygen is insufflated directly into the air sacs through a line connected to a large-gauge needle.
   b. If the airway obstruction is complete, a large tube is rapidly placed.
   c. Place the air sac tube either from behind the back leg, or through the paralumbar fossa as for surgical sexing.
      (1) Make a skin incision with a scalpel, then use a small hemostat to penetrate the muscle and peritoneum.
      (2) Place the largest tube possible—for medium to large birds sterile endotracheal tubes, whereas small catheters are used in smaller birds.
      (3) Inflation of a cuffed tube decreases the likelihood of its displacement.
   d. A technique for placement of a tube in the clavicular air sac has been described and evaluated in phenobarbital sedated ducks.[58]
      (1) Cannulation of the clavicular air sac resulted in a significant increase in tidal volume and minute ventilation.
      (2) PaO_2, $PaCO_2$, heart rate and mean arterial blood pressure remained unchanged from control values.

e. Air sac cannulas can be used for administration of inhalant anesthetics in spontaneously breathing birds.
f. Periodically check the cannula for occlusion with secretions.

E. Surgical Sexing
1. Always use some form of analgesia.
2. Isoflurane is excellent for this procedure, because it allows rapid induction and recovery.
3. Alternatively, low dose IV ketamine in the medial metatarsal vein will give approximately 10 minutes of surgical analgesia and restraint, with a shorter recovery than with the use of a combination.
4. In restrained large birds, some veterinarians infiltrate the laparoscopy site with local anesthetic—avoid overdosage (see above).

XI. Anesthesia for Select Avian Groups

A. Parrots (Psittaciformes), Pigeons (Columbiformes)
1. Mask induce with an inhalation agent, intubate, and maintain on a nonrebreathing system.
2. Administer ketamine (IV medial saphenous vein or IM) in difficult-to-restrain birds.
3. A xylazine-ketamine combination can also be used.[30]

B. Poultry (Galliformes)
1. Same as for parrots and pigeons, and has been reviewed by Hartsfield and McGrath[59]

C. Raptors (Falconiformes and Strigiformes)
1. Require firm and quiet restraint, with control of the head and talons to avoid injury.
2. A towel wrapped around the body will prevent wing flapping.
3. Use a falconry hood to decrease struggling.
4. Large raptors often require chemical restraint (ketamine, diazepam-ketamine) before induction of inhalation anesthesia.
5. In African vultures ketamine alone (18 to 42 mg/kg IM) produces marked salivation, excitation and convulsions, inadequate surgical anesthesia, and a prolonged recovery.[4]
6. Strigiformes (owls) are more sensitive than falconiformes (hawks, eagles, etc.) to ketamine.[4]
7. Barred, long-eared and short-eared owls require only half the ketamine dosage of great horned, screech, and saw-whet owls.[25]

D. Ratites (Struthioniformes, Rheiformes, Casuariiformes)
1. Physical restraint
   a. Reviewed by Jensen[60]
   b. Struggling is common, may be very violent, and can cause trauma to the animal and veterinarian.

c. All ratites kick forward, and placing a cover over the eyes will facilitate handling in some animals, as will darkness.
   d. Keeping the head lowered will discourage an animal from kicking and jumping.
   e. Do not sit on a restrained ratite, because it will interfere with respiration.
2. Anesthesia
   a. Hospitalize large ratities in padded stalls.
   b. For examination and drug administration darken the stall, place a hood over the head, and compress the bird against a wall using a covered foam pad.
   c. Give intramuscular and intravenous injections in the thigh and basilic vein, respectively.
   d. Inhalant anesthesia alone, except in young or moribund ratites, is hazardous.
   e. During anesthesia secure the legs and wrap the feet to prevent injury if the animal should lighten during a procedure.
      (1) Avoid prolonged struggling with the legs restrained, which can lead to exertional rhabdomyolysis (capture myopathy).
      (2) Pad the head and neck and place in extension, avoid abnormal body positions.
   f. For direct blood pressure measurement and arterial blood gas analysis during major or prolonged procedures, identify and cannulate arteries on the lateral and dorsal tarsometatarsus and ulnar.
   g. Recovery is a critical period.
      (1) Place in sternal recumbency, and support the head and neck until the bird is able to hold it up itself.
      (2) Prevent the bird from rising until it is fully awake. Use a heavy foam pad tented over the top of it.
      (3) Provide a quiet, darkened environment to prevent unnecessary struggling.
3. Emu
   a. Ketamine (25 mg/kg IM, plus 5 mg/kg IV every 10 min)[61]
      (1) Adequate restraint for minor procedures
      (2) Prolonged and difficult recovery
      (3) Large induction volume
   b. Ketamine plus either xylazine or diazepam
      (1) Improved induction and recovery
      (2) Decreased ketamine dosage
4. Ostrich
   a. Xylazine/ketamine/isoflurane/nitrous oxide[62, 63]
      (1) Xylazine 2.0 mg/kg IM
      (2) Wait 15 minutes for sedation and wing drooping.
      (3) Give ketamine 4 mg/kg IV to effect.
      (4) Intubate and maintain with isoflurane in oxygen and N_2O.
   b. Zolazepam-tiletamine (4 to 12 mg/kg IM, 2 to 8 mg/kg IV)[62-65]
      (1) Rapid, smooth induction
      (2) Small injection volume

(3) Prolonged excitable recovery
(4) Occasional apnea
c. Carfentanil and etorphine[34,62]
(1) Excitement during induction
(2) Occasional apnea
(3) Rapid reversal with diprenorphine
(4) Small injection volume
(5) Use acepromazine and xylazine to improve immobilization.

REFERENCES

1. Harrison GJ: Pre-anesthetic fasting recommended. (In My Experience), *J Am Assoc Avian Vet* 5:126, 1991.
2. Grono LR: Anesthesia of budgerigars, *Aust Vet J* 37:463, 1961.
3. Jordan FTW, Sanford J, Wright A: Anesthesia in the fowl, *J Comp Pathol Ther* 70:437-448, 1960.
4. Samour JH et al: Comparative studies of the use of some injectable anaesthetic agents in birds, *Vet Rec* 115:6-11, 1984.
5. Ludders JW: Effects of ketamine, xylazine, and a combination of ketamine and xylazine in Pekin ducks, *Am J Vet Res* 50:245-249, 1989.
6. Christensen J et al: Comparison of various anesthetic regimens in the domestic fowl, *Am J Vet Res* 48:1649-1657, 1987.
7. Harvey SC: Hypnotics and sedatives. In Gilman AG et al, editors: *The pharmacological basis of therapeutics,* New York, 1985, Macmillan, pp 339-371.
8. Booth NH, McDonald LE, editors: *Veterinary pharmacology and therapeutics,* ed 6, Ames, Iowa, 1988, Iowa State University Press.
9. Valverde A et al: Determination of a sedative dose and influence of midazolam on cardiopulmonary function in Canada geese, *Am J Vet Res* 51:1071-1074, 1990.
10. Concannon KT, Dodam JR, Hellyer PW: Influence of a mu- and kappa-opioid agonist on isoflurane minimal anesthetic concentration in chickens, *Am J Vet Res* 56:806-811, 1995.
11. Stoelting RK: *Pharmacology and physiology in anesthetic practice,* Philadelphia, 1987, JB Lippincott.
12. Paddleford RR, editor: *Manual of small animal anesthesia,* New York, 1988, Churchill Livingstone.
13. Klide AM: Avian anesthesia, *Vet Clin North Am* 3:175-186, 1973.
14. Sturkie PD, editor: *Avian physiology,* New York, 1986, Springer-Verlag.
15. Otto CM, McCall Kaufman G, Crowe DT: Intraosseous infusion of fluids of fluids and therapeutics, *Comp Cont Educ* 11:421-431, 1989.
16. Ritchie BW, Otto CM, Latimer KS, Crowe DT: A technique of intraosseous cannulation for intravenous therapy in birds, *Comp Cont Educ* 12:55-58, 1990.
17. Sedgwick C, Pokras M, Kaufman G: Metabolic scaling: using estimated energy costs to extrapolate drug doses between different species and different individuals of diverse body sizes, *Proc Am Assoc Zoo Vet* 249-254, 1990.
18. Lukasik VM et al: The cardiopulmonary and respiratory effects of propofol anesthesia in chickens, *Vet Surg* 25:525, 1996.
19. Machin KL, Caulkett NA: The cardiopulmonary effects of propofol in mallard ducks, *Proc Am Assoc Zoo Vet* 149-154, 1996.
20. Fitzgerald G, Cooper JE: Preliminary studies on the use of propofol in the domestic pigeon (*Columbia livia*), *Res Vet Sci* 49:334-338, 1990.
21. Salerno A, Van Tienhoven A: The effect of ketamine on heart rate, respiratory rate and EEG of white leghorn hens, *Comp Biochem Physiol* 55:69-75, 1976.

22. Kollias GV Jr, McLeish I: Effects of ketamine hydrochloride in red-tailed hawks (*Buteo jamaicensis*). I. Arterial blood gas and acid base, *Comp Biochem Physiol* 60:57-59, 1978.
23. Mandelker L: A toxicity study of ketamine HCl in parakeets, *Vet Med/ Small Anim Clin* 68:487-488, 1973.
24. McGrath CJ, Lee JC, Campbell VL: Dose-response anesthetic effects of ketamine in the chicken, *Am J Vet Res* 45:531-534, 1984.
25. Redig PT, Duke GE: Intravenously administered ketamine HCl and diazepam for anesthesia of raptors, *J Am Vet Med Assoc* 169:886-888, 1976.
26. Degernes LA et al: Ketamine-xylazine anesthesia in red-tailed hawks with antagonism by yohimbine, *J Wildl Dis* 24:322-326, 1988.
27. Lumeij JT: Effects of ketamine-xylazine anesthesia on adrenal function and cardiac conduction in goshawks and pigeons. In Redig PT, Cooper JE, Remple JD, Hunter DB, editors: *Raptor biomedicine,* Minneapolis, 1993, Minnesota Press, pp 145-149.
28. Raffe MR et al: Cardioprespiratory effects of ketamine-xylazine in the great horned owl. In Redig PT, Cooper JE, Remple JD, Hunter DB, editors: *Raptor biomedicine,* Minneapolis, 1993, Minnesota Press, pp 150-153.
29. Allen JL, Oosterhuis JE: Effect of tolazoline on xylazine-ketamine-induced anesthesia in turkey vultures, *J Am Vet Med Assoc* 189:1011-1012, 1986.
30. Harrison GJ, Harrison LR, editors: *Clinical avian medicine and surgery including aviculture,* Philadelphia, 1986, WB Saunders.
31. Kaufman E, Pokras M, Sedgwick C: Anesthesia in waterfowl, *AAV Today* 2:98, 1988.
32. Heaton JT, Brauth SE: Effects of yohimbine as a reversing agent for ketamine-xylazine anesthesia in budgerigars, *Lab Anim Sci* 42:54-56, 1992.
33. Teare JA: Antagonism of xylazine hydrochloride-ketamine hydrochloride immobilization in guineafowl (*Numidia meleagris*) by yohimbine hydrochloride, *J Wildl Dis* 23:301-305, 1987.
34. Raath JP, Quandt SKF, Malan JH: Ostrich (*Struthio camelus*) immobilisation using carfentanil and xylazine and reversal with yohimbine and naltrexone, *J S Afr Vet Assoc* 63:138-140, 1992.
35. Carp NZ et al: A technique for liver biopsy performed in Pekin ducks using anesthesia with Telazol^R, *Lab Anim Sci* 41:474-475, 1991.
36. Kreeger TJ et al: Immobilization of raptors with tiletamine and zolazepam (Telazol). In Redig PT, Cooper JE, Remple JD, Hunter DB, editors: *Raptor biomedicine,* Minneapolis, 1993, Minnesota Press, pp 141-144.
37. Ludders JW, Mitchell GS, Schaefer SL: Minimum anesthetic dose and cardiopulmonary dose response for halothane in chickens, *Am J Vet Res* 49:929-932, 1988.
38. Ludders JW, Rode J, Mitchell GS: Isoflurane anesthesia in sandhill cranes (*Grus canadensis*): minimal anesthetic concentration and cardiopulmonary dose-response during spontaneous and controlled breathing, *Anesth Anal* 68:511-516, 1989.
39. Ludders JW, Mitchell GS, Rode J: Minimal anesthetic concentration and cardiopulmonary dose response of isoflurane in ducks, *Vet Surg* 19:304-307, 1990.
40. Quasha AL, Eger EI, Tinker JH: Determination and application of MAC, *Anesthesiology* 53:315-334, 1980.
41. Goelz MF, Hahn AW, Kelley ST: Effects of halothane and isoflurane on mean arterial blood pressure, heart rate, and respiratory rate in adult Pekin ducks, *Am J Vet Res* 51:458-460, 1990.
42. Greenlees KJ et al: Effect of halothane, isoflurane, and pentobarbital anesthesia on myocardial irritability in chickens, *Am J Vet Res* 51:757-758, 1990.

43. Reynold WT: Unusual anesthetic complication in a pelican, *Vet Rec* 113:204, 1983.
44. Dorsch JA, Dorsch SE: *Understanding anesthesia equipment: construction, care and complications,* ed 3, Baltimore, 1994, Williams & Wilkins.
45. Altman RB: A method for reducing exposure of operating room personnel to anesthetic gas, *J Am Assoc Avian Vet* 6:99-101, 1992.
46. Franz DR, Dixon RS: A mask system for halothane anesthesia of guinea pigs, *Lab Anim Sci* 38:743-744, 1988.
47. Mauderly JL: An anesthetic system for small laboratory animals, *Lab Anim Sci* 25:331-333, 1975.
48. Nicholson A, Ilkiw JE: Neuromuscular and cardiovascular effects of atracurium in isoflurane-anesthetized chickens, *Am J Vet Res* 53:2337-2342, 1992.
49. Schmidt-Nielsen K: *Scaling. Why is animal size so important?* New York, 1984, Cambridge University Press.
50. Sedgwick C: Allometrically scaling the data base for vital sign assessment used in general anesthesia of zoological species, *Proc Ann Meet Am Assoc Zoo Vet* 360-369, 1991.
51. Matthews NS, Burba DJ, Cornick JL: Premature ventricular contractions and apparent hypertension during anesthesia in an ostrich, *J Am Vet Med Assoc* 198:1959-1961, 1991.
52. Lamberski N: Fluid dynamics of intraosseous fluid administration in birds, *J Zoo Wildl Med* 23:47-54, 1992.
53. Powell FL, Scheid P: Physiology of gas exchange in the avian respiratory system. In King AS, McLelland J, editors: *Form and function in birds,* vol 4, New York, 1989, Academic Press, pp 393-437.
54. King AS, Payne DC: Normal breathing and the effects of posture in *Gallus domesticus, J Physiol* 174:340-347, 1964.
55. Seaman GC, Ludders JW, Hollis NE, Gleed RD: Effects of low and high fractions of inspired oxygen on ventilation in ducks anesthetized with isoflurane, *Am J Vet Res* 55:395-398, 1994.
56. Robello CD, Crowe Jr DT: Cardiopulmonary resuscitation: current recommendations, *Vet Clin North Am Small Anim Pract* 19:1127-1149, 1989.
57. Rosskopf WJ, Woerpel RW: Abdominal air sac breathing tube placement in psittacine birds and raptors, its use as an emergency airway in cases of tracheal obstruction, *Proc Am Assoc Avian Vet* 215-217, 1990.
58. Rode JA, Bartholow S, Ludders JW: Ventilation through an air sac cannula during tracheal obstruction in ducks, *J Am Assoc Avian Vet* 4:98-101, 1990.
59. Hartsfield SM, McGrath CJ: Anesthetic techniques in poultry, *Vet Clin N Am Sm Anim Pract* 2:711-730, 1986.
60. Jensen JM: Ratite restraint and handling. In Fowler ME, editor: *Zoo and wild animal medicine - current therapy 3,* Philadelphia, 1993, WB Saunders, pp 198-200.
61. Grubb B: Use of ketamine to restrain and anesthetize emus, *Vet Med Sm Anim Clin* 78:247-248, 1983.
62. Cornick JL, Jensen J: Anesthetic management of ostriches, *J Am Vet Med Assoc* 200:1661-1666, 1992.
63. Jensen JM, Johnson JH, Weiner ST: *Husbandry and medical management of ostriches, emus and rheas,* College Station, Tex, 1992, Wildlife and Exotic Animal Teleconsultants.
64. Stewart JS: Husbandry, medical and surgical management of ratites: Part II, *Proc Am Assoc Zoo Vet* 119-122, 1989.
65. Van Heerden J, Keffen RH: A preliminary investigation into the immobilising potential of tiletamine/zolazepam mixture, metomidate, a metomidate and azaperone combination and medetomidine in ostriches (*Struthio camelus*), *J S Afr Vet Assoc* 62:114-117, 1991.

23

Limb Dysfunction

Glenn H. Olsen, Patrick T. Redig, and Susan E. Orosz

I. Orthopedic Principles

A. General Considerations[1]
1. The cortices of bone are strong but brittle; they offer less holding power for orthopedic devices, and they shatter easily.
2. The limited amount of cancellous bone makes finding enough bone for grafts difficult.
3. At the proximomedial end of the humerus is a foramen, where the extrathoracic portion of the clavicular air sac extends into the medullary space of the humerus. This connection with the respiratory system must be kept in mind with humeral lavage and extension of infection. Extensions into the bones of the thoracic girdle, sternum, and ribs also are present.
4. A limited amount of soft tissue is present over the bones of the thoracic and pelvic limbs, making bone and soft tissue healing difficult to achieve.
5. Avian skin is thin and lacerates easily, which often leads to open fractures; this factor also further complicates the problem of healing.

II. Anatomy of the Thoracic Girdle

The bones of the thoracic girdle include the coracoid, clavicle, and the scapula.

A. Sternum
1. Although not part of the thoracic girdle, the sternum is associated with the coracoid bones. The sternum has a ventral projection, or keel, which increases the surface area available for the attachment of the powerful pectoral muscles.
2. The keel is deeper in birds with a powerful, flapping flight. Its size varies in relation to the type of flight, and it is absent in ratites.
3. The sternum is longer in diving birds.

4. In cranes and swans, the cranial portion of the sternum is excavated and contains one or more loops of the trachea.[2]
5. The sternum protects and supports the viscera as the bird flies through the air. Therefore the space between the distal end of the sternum and the pelvic bones is much reduced compared to that in mammals.
6. Extensions of the clavicular air sac into the sternum are found in some species.

B. Ribs
1. The ribs are not truly part of the forelimb, but their function or the pathologic processes affecting the ribs can affect forelimb function.
   a. The ribs are composed of two elements with an articulation in the middle. The two parts are the dorsal element (vertebral rib), which projects caudally, and the ventral element (sternal rib), which projects cranially.
2. The sternal ribs in birds articulate with the sternum and provide strength while supporting the viscera. They are similar to the costal cartilages of mammals. Ribs also resist dorsoventral compression with the downstroke.
3. The vertebral ribs articulate with the vertebral bodies via transverse processes.
4. In diving birds, the ribs are flattened and lengthened to reduce resistance as they move through the water.
5. Some of the vertebral ribs have a caudally directed uncinate process for attachment of the trunk musculature, which helps to strengthen the thoracic cage against compressive forces.

C. Scapula
1. The scapula is attached to the ribs by ligaments.
2. The scapula is longer in strong flying birds compared to short-burst flyers.
3. In ratites and frigate birds, the scapula is fused to the coracoid.
4. The scapula articulates with the coracoid and clavicle cranially.[3] The scapula and coracoid form part of the glenoid cavity; its lateral arrangement is important for abduction and adduction. The junction of the scapula, coracoid, and clavicle forms the foramen triosseum, through which the tendon of the supracoracoideus muscle passes as a wing elevator.

D. Clavicle
1. The pectoral muscles originate in part from the clavicle.
2. The widest angle between the fused portion of the clavicles (the furcula) is found in birds that are strong flyers.
3. The clavicles act as transverse spacer bars to keep the wings spread, particularly with the power exerted during the downstroke.
4. In parrots and owls, the clavicles contain fibrous tissue; clavicles may be absent in some parrots and ratites.

E. Coracoid
  1. The coracoid bones articulate with the sternum distally and with the clavicle, scapula, and humerus proximally; this proximal attachment of bones forms the foramen triosseum, or triosseal canal. The supracoracoideus muscle traverses this foramen; its directional change is important in providing the upstroke of the wing.
  2. With its articulation to the sternum, the coracoid acts as a strut to hold the wing to the sternum and also transmits the power to the thoracic cage. This short but broad bone prevents the thoracic wall from collapsing during flight.
  3. When a coracoid has been fractured or subluxated, the bird is unable to move its wing above the horizontal. It is important to attempt palpation of this bone when performing an orthopedic examination. It may be difficult to palpate in species with thickened pectoral muscles.

F. Pectoral Musculature
  1. Superficial pectoral muscle
     a. Origin: Sternum, clavicle, and keel
     b. Insertion: Ventral pectoral crest and the proximal humerus (a small slip may insert on the propatagial complex).
     c. Innervation: Pectoral nerves from the ventral fascicle
     d. Contraction of the superficial pectoral muscle causes a downstroke and depression of the leading edge of the wing to reduce turbulence.
  2. Deep pectoral muscle
     a. Origin: Keel, clavicle, and sternum
     b. Insertion: Tendinous onto the pectoral crest
     c. Innervation: Pectoral nerves from the ventral fascicle
     d. The deep pectoral muscle pulls the wing forward immediately before the downstroke; it also is important in gliding flight and bringing the wing back to its normal position when the wing is blown out of its gliding position.
     e. This muscle is poorly developed in owls, psittacines, and birds that flap rapidly to sustain flight.

G. Supracoracoideus Muscle
  1. Origin: Sternum and coracoid
  2. Insertion: Dorsal tubercle of the humerus; the tendon passes through the foramen triosseum
  3. Innervation: Supracoracoid nerve branch from the ventral fascicle
  4. Contraction of the supracoracoideus muscle creates the upstroke of the wing as it changes direction in the triosseal canal.
  5. The supracoracoideus is large in penguins, in birds that take off at a steep angle, and in birds that hover.

H. Coracobrachialis Caudalis Muscle
  1. Origin: Coracoid and sternum
  2. Insertion: Ventral tubercle of the humerus

3. Innervation: Coracobrachialis caudalis nerve branch from ventral fascicle
4. Contraction of the coracobrachialis caudalis muscle depresses the wing, but the true action is not well understood.

III. Anatomy of the Proximal Thoracic Limb (Brachium)

A. Humerus (Fig. 23-1)
1. The humerus is broad at the proximal end and has some curvature; the amount of curvature is based on the type of flight.
2. The proximal one third to one half lies against the body wall; here the body wall acts as a natural splint for fractures if the wing is bound to the body.
3. The pectoral crest is a broad, laterally projecting blade that is the insertion point of both the superficial and deep pectoral muscles.
4. The head of the humerus articulates with the glenoid cavity of the scapula.
5. The clavicular air sac extends through a large medial pneumatic foramen into the humerus.

B. Muscles of the Humerus (Figs. 23-2 and 23-3)
1. Tensor propatagialis muscle
   a. Origin: Coracoid or clavicle, or both
   b. Two major insertions:
      (1) The pars longus tendon inserts onto the extensor process of the carpometacarpus.
      (2) The pars brevis tendon inserts on the proximal radius and ulna by attaching to the origin of the secondary remiges, forming the interremigial ligament.

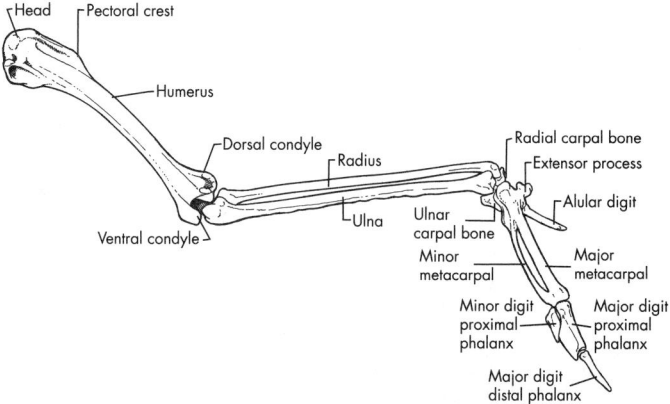

Fig. 23-1 Skeletal anatomy of the avian pectoral limb.

c. Innervation: Axillary nerve
d. The propatagialis complex of muscles is large in psittacines; it extends the metacarpus and digits while flexing the elbow.
e. As its name implies, the tensor propatagialis muscle tenses the propatagium, which is important for controlling the angle of the wing for lift[3]; tensing the propatagium prevents it from vibrating or causing turbulence as the air moves across its surface. The propatagium serves the same function as the wing on an airplane.
2. Deltoideus major muscle
   a. Origin: Scapula and clavicle
   b. Insertion: Dorsolateral humerus from the pectoral crest distally
   c. Innervation: Axillary nerve
   d. The deltoideus major muscle pulls the wing caudally and dorsally, which is important in flapping flight.[3,4]
3. Deltoideus minor muscle
   a. Origin: Scapula and clavicle
   b. Insertion: Proximal dorsolateral humerus
   c. Innervation: Axillary nerve

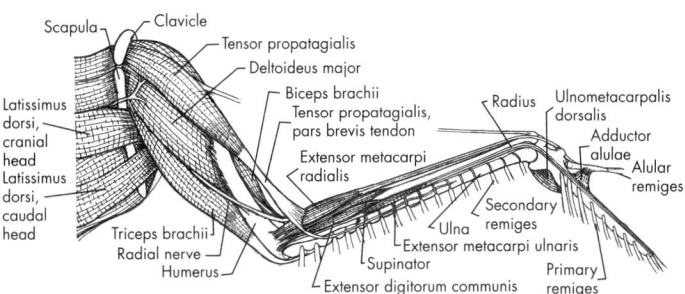

Fig. 23-2 Orientation of the muscles of the dorsal wing.

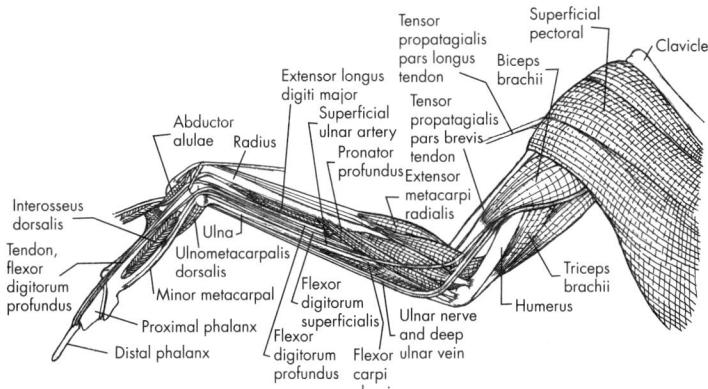

Fig. 23-3 Orientation of the muscles of the ventral wing.

d. The deltoideus minor muscle pulls the humerus cranially; this occurs as part of the upstroke.
4. Triceps brachii muscle
    a. Origin from as many as three heads: The long head takes origin from the scapula; the other two heads arise from the humerus[3]
    b. Insertion: Elbow joint onto the proximal ulna
    c. Innervation: Radial nerve
    d. The major action of the triceps brachii muscle is to extend the elbow; however, the long head draws the humerus caudally.[1,3]
5. Biceps brachii muscle
    a. Origin: Cranial border of the proximal humerus
    b. Insertion: On the proximodorsal radius and ulna
    c. Innervation: Bicipital nerve (a branch of the medianoulnar nerve)
    d. The biceps brachii muscle acts as the major flexor of the forearm.
6. Latissimus dorsi muscle
    a. Two independent heads of origin:
        (1) Cranial head: Attaches to the caudal cervical and cranial thoracic spinous processes.
        (2) Caudal head: Originates from the caudal thoracic vertebrae and, in some species, from the synsacrum.
    b. Insertion:
        (1) Cranial portion: Inserts between the two parts of the triceps.
        (2) Caudal portion: Has variable insertion points on the humerus, depending on the species; absent in pigeons and finches.[3]
    c. Innervation: Nerve to the latissimus dorsi muscle, branch from the dorsal fascicle.
    d. The latissimus dorsi muscle retracts and folds the wing against the body.

Table 23-1 summarizes the muscles of the pectoral limb and their associated actions.

Table 23-1
Muscles of the pectoral limb and their associated actions

Muscles	Action
Supracoracoideus	Elevation of the humerus
Pectoralis (both)	Depression of the humerus
Coracobrachialis dorsalis, deltoideus minor	Protraction of the humerus
Latissimus dorsi, deltoideus major, triceps (partial)	Retraction of the humerus (folding the wing against the body)
Biceps brachii, brachialis	Flexion of the elbow joint
Triceps brachii	Extension of the elbow joint
Supinator	Dorsal rotation of the elbow
Pronator	Ventral rotation of the elbow
Extensor metacarpi radialis	Extension of the carpal joint
Extensor metacarpi ulnaris	Flexion of the carpal joint

IV. Anatomy of the Forearm (Antebrachium)
A. Radius and Ulna (see Fig. 23-1)
  1. The radius and ulna often are slightly bowed (they are more bowed in birds with a flapping flight than in birds with a long, gliding flight); the distal ends of both bones curve caudally as they articulate with the carpal bones.
  2. The ulna is larger than the radius and is located caudal to it.
  3. The secondary remiges, or secondaries, insert directly into the bone on the caudal surface of the ulna. These feathers have a ligament that attaches to adjacent feathers, forming the interremigial ligament, which helps to keep the feathers moving in a coordinated manner and keeps them spaced equally.
  4. Reciprocal apparatus
     a. The radius and ulna act together when extending or flexing the manus and elbow; when the elbow flexes, the carpus flexes and vice versa.
     b. Extension of the proximal wing moves the ulna distally, thereby pushing the carpometacarpus into extension.

B. Major Muscles of the Dorsal Antebrachium (see Fig. 23-2)
  1. Extensor metacarpi radialis muscle
     a. Origin: Dorsal epicondyle of the humerus
     b. Insertion: Often two tendons onto the extensor process of the first metacarpal bone
     c. Innervation: Radial nerve
     d. The extensor metacarpi radialis muscle flexes the elbow and extends the carpus and metacarpus.
  2. Common digital extensor muscle
     a. Origin: Dorsal epicondyle
     b. Insertion: By two tendons:
        (1) Short tendon: Inserts on the alular digit
        (2) Long tendon: Inserts on the base of the proximal end of the major digit
     c. Innervation: Radial nerve
     d. The common digital extensor muscle extends the carpal joint and flattens the alula against the carpometacarpus to reduce lift.[3-5]
  3. Supinator muscle
     a. Origin: Dorsal epicondyle
     b. Insertion: Proximocranial margin of the radius
     c. Innervation: Radial nerve
     d. The supinator muscle elevates the cranial margin of the wing and flexes the elbow[3,4]; this action is important for lift and for stalling.
  4. Extensor metacarpi ulnaris muscle (ulnaris lateralis)
     a. Origin: Dorsal epicondyle of the humerus
     b. Insertion: Ulna and secondary remiges; major and minor metacarpal bones

c. Innervation: Radial nerve
d. The extensor metacarpi ulnaris muscle flexes the antebrachium and manus (the name of this muscle is a misnomer, because its major action is flexion, not extension).

C. Major Muscles of the Ventral Antebrachium (see Fig. 23-3)
1. Pronator superficialis and profundus muscles
   a. Origin: Ventral epicondyle of the humerus (both muscles)
   b. Insertion: Ventral caudal edge of the radius, with the insertion of the superficialis (pronator brevis) more proximal than that of the profundus (pronator longus)
   c. Innervation: Median nerve
   d. The pronator superficialis and profundus muscles depress the propatagium, causing depression of the cranial margin of the wing; this action results in descending flight.
2. Flexor digitorum superficialis muscle
   a. Origin: Ventral epicondyle of the humerus (this muscle may be composed of several muscle bellies; it is located more caudally than the flexor digitorum profundus)
   b. Insertion: Proximal end of the distal phalanx of the major digit (the tendon of insertion crosses over that of the profundus to insert along the cranial border of the major digit)
   c. Innervation: Median nerve
   d. The flexor digitorum superficialis muscle flexes the distal end of the wing ventrally or it may extend the wing.
3. Flexor digitorum profundus muscle
   a. Origin: Palmar surface of the ulna (this muscle and its tendon are located cranial to the flexor digitorum superficialis)
   b. Insertion: Distal phalanx of the major digit (it is crossed by the superficial flexor tendon and inserts deep to it)
   c. Innervation: Median nerve
   d. The flexor digitorum profundus muscle flexes the carpal joint and depresses and extends the major digit[1]; these actions are important for flex gliding.
4. Flexor carpi ulnaris muscle
   a. Origin: Ulna and secondary remiges, in which it has two or three muscle bellies
   b. Insertion: Ulnar carpal bone
   c. Innervation: Ulnar nerve
   d. The flexor carpi ulnaris muscle extends the secondary remiges.
5. Extensor longus digiti majorus muscle
   a. Origin: Ventral radius along its caudal margin (it moves from a ventral location to a dorsal one by slipping over the radial carpal bone)
   b. Insertion: Crosses the carpus to insert onto the dorsal end of the distal phalanx of the major digit
   c. Innervation: Radial nerve
   d. The extensor longus digiti majorus muscle extends the major digit and carpus by its dorsal insertion.

V. Anatomy of the Distal Wing (Manus) (see Fig. 23-1)
A. Carpus
1. The carpus is composed of only radial and ulnar carpal bones.
2. The carpus represents the proximal row of carpal bones, with the distal row absent.
3. The ulnar carpal bone is absent in ratites.

B. Carpometacarpus
1. The carpometacarpus consists of fused major metacarpal and minor metacarpal bones; this fusion of bones and the architecture of a flat plane is important for the attachment of the primary remiges (primaries).
2. A large extensor process on the proximal carpometacarpus represents metacarpal I.
3. The primary remiges insert on the dorsal surface of the carpometacarpus.

C. Digits
1. Alular digit
   a. The alular digit sometimes is called "the bastard wing"; it represents the pollex, or thumb.
   b. This digit is a single bone with a large degree of mobility: flexion, extension, adduction, and abduction.
   c. The alular digit prevents stalling in slow flight when raised or abducted.
2. Major digit
   a. The major digit has a proximal and distal phalanges, which is flattened dorsoventrally.
   b. This digit provides an important surface for the insertion of the distal primary remiges.
3. Minor digit
   a. Only the proximal phalanx is present.

D. Muscles of the Dorsal Manus (see Fig. 23-2)
1. Interosseous dorsalis muscle (located in the intermetacarpal space)
   a. Origin: Major and minor metacarpal bones
   b. Insertion: Cranial surface of the phalanges of the major digit
   c. Innervation: Deep ramus of radial nerve
   d. The interosseous dorsalis muscle extends the major digit.
2. Ulnometacarpalis dorsalis muscle (flexor metacarpi caudalis)
   a. Origin: Distal ulna, caudal surface
   b. Insertion: Minor metacarpal bone and each of the insertions of the primaries
   c. Innervation: Deep ramus of radial nerve
   d. The ulnometacarpalis dorsalis muscle pulls the primary remiges medially or proximally to prevent a gap between the primaries and the secondaries.

3. Flexor digiti minoris muscle (found distal to the ulnometacarpidorsalis muscle, running along the caudal border of the carpometacarpus)
   a. Origin: Metacarpals
   b. Insertion: Onto the minor digit
   c. Innervation: Ulnar nerve
   d. The flexor digiti minoris muscle flexes the minor digit

E. Muscles of the Ventral Manus (see Fig. 23-3)
   1. Abductor alulae muscle (a small muscle on the cranial surface of the ventral side of the carpometacarpus)
      a. Origin: Tendon of the extensor metacarpi radialis muscle
      b. Insertion: Onto the proximoventral alula
      c. Innervation: Median nerve
      d. The abductor alulae muscle pulls the alula toward the flat plane of the carpometacarpus.
   2. Flexor alulae muscle lies adjacent to the abductor alulae muscle on the ventral side of the carpometacarpus.
      a. Origin: Major metacarpal bone
      b. Insertion: Just distal to the insertion of the abductor
      c. Innervation: Deep ramus of median nerve
      d. The flexor alulae muscle keeps the alula close to the carpometacarpus and flexes it.
   3. Interosseous ventralis muscle (located in the intermetacarpal space ventrally)
      a. Origin: Major and minor metacarpal bones
      b. Insertion: Caudally on the phalanges of the major digit
      c. Innervation: Ulnar nerve
      d. The interosseous ventralis muscle flexes the major digit.

VI. Fractures of the Thoracic Limb

A. Coracoid Fractures
1. General aspects
   a. These fractures are a common problem, but they often are misdiagnosed.
   b. The history often indicates that the bird flew into something or was hit by a moving object.[6]
   c. The bird will be able to fly only short distances, if at all, and will be unable to lift its wing above the horizontal.
   d. Proper positioning is necessary for a true ventrodorsal radiograph (see Chapter 19).[7]
2. Fixation options
   a. Minimal fracture displacement and fractures in small birds (<300 g) require a figure-eight wrap to immobilize the wing and a body wrap to hold the wing to the body, followed by cage rest.
   b. Fractures in larger birds or displaced fractures in small birds may have a better prognosis with internal fixation.[6,7]

(1) The skin incision is made over the clavicle and extends one half the length of the keel.
(2) The cranial portions of the pectoral and supracoracoid muscles are dissected from the keel and clavicle and reflected laterally.
(3) The clavicular artery is within the pectoral muscle near the middle of the clavicle.
(4) The intramedullary pin is passed retrograde out the shoulder joint area, the fracture ends are aligned, and the pin then is driven retrograde into the proximal section of the coracoid.[6]
(5) The muscles are sutured back to insertion ends at the sternum and clavicle with polydioxanone suture (Ethicon, Inc., Somerville, NJ). The skin is closed in a routine manner.
(6) The patient is kept in a figure-eight wrap with body wrap to hold the wing immobilized for 7 to 10 days.[6]
(7) When radiographs show sufficient healing, the pin is removed from the shoulder.

3. Complications
  a. If the intramedullary pin is pushed too far through the coracoid, it may enter the sternum or even the heart.
  b. If the pin is not removed after the bone has healed or if the pin causes severe damage to the shoulder, ankylosis of the shoulder joint may result.
  c. Large calluses may partly occlude the esophagus.[1]

B. Fractures of the Scapula or Clavicle
1. General aspects
   a. Fractures of the scapula or clavicle are more unusual than coracoid fractures.
   b. Scapular and clavicular fractures usually are diagnosed by radiographs.
2. Fixation options
   a. For small birds (<300 g), the treatment is simply cage rest with no other intervention.
   b. For larger birds, the affected wing is bandaged to the body (figure-eight wrap of wing followed by body wrap) and cage rest for 3 weeks.

C. Fractures of the Humerus
1. General aspects
   a. The diagnosis usually is made by palpation; radiographs are done to determine the extent of the fracture.
   b. Humeral fractures are typically open, spiral, and unstable, with great displacement of fragments.
   c. The pectorals pull the proximal fragment dorsally.
   d. The triceps pulls the distal fragment ventrally and foreshortens the humerus.
   e. The longer the time between fracture and repair, the greater the complications from the actions of the muscles.

2. Repair of fractures of the proximal humerus (Fig. 23-4)
    a. Proximal humeral fractures sometimes are not greatly displaced and may respond to the treatment of placing the affected wing in a figure-eight wrap and using a body wrap for immobilization.
    b. Another immobilization technique involves wrapping the wings to the body with two bandages that encircle the body, one at the shoulder and proximal humerus and the other at the elbows.[6] A figure-eight bandage then is wrapped around the body, crossing over the back.
    c. Complications with fracture reduction by bandaging include rotation of the wing, malunion, or diminished ability to fly.
3. Repair of fractures of the diaphyseal humerus (midshaft)
    a. Humeral fractures distal to the deltoid crest usually are severely displaced because the pectoral, supracoracoid, and triceps muscles pull on the fractured end.
    b. External skeletal fixation (ESF)
        (1) Type I external skeletal fixation uses pins that penetrate both cortices. (Note: It is of no concern if they penetrate skin—most of the time they do.) (Fig. 23-5)
            (a) To obtain better stabilization in some cases, two type I external skeletal fixators can be used at a 90-degree angle to each other.
            (b) Type I external skeletal fixation is most useful when the fracture site is open and possibly contaminated.
        (2) Type II and type III external skeletal fixation (see Fig. 23-5) cannot be used on most humeral fractures because the skin over the medial humerus is close to or convergent with the thoracic body wall.

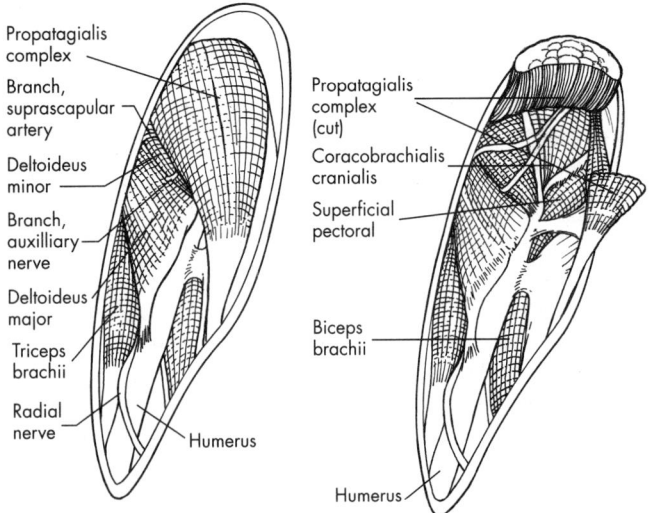

Fig. 23-4 Dorsal approach to the proximal and midshaft humerus.

c. Intramedullary (IM) pins
   (1) An IM pin is placed retrograde through the pectoral crest on the proximal section of the fractured humerus.[6]
   (2) It is possible but more difficult to insert an IM pin normograde into the humerus at the pectoral crest.
       (a) Complications, primarily periarticular fibrosis or ankylosis of the elbow, are a common sequela if the IM pin penetrates the elbow joint.
d. ESF-IM tie-in (Fig. 23-6)
   (1) A combination of IM pins and an ESF device gives the best results.[10]
   (2) An IM pin 50% to 60% of the diameter of the marrow cavity is positioned at a right angle to the distal end of the distal fragment and away from the dorsal edge of the triceps tendon.
   (3) After the initial hole is made in the humerus, the pin is inserted normograde.
   (4) The fracture is reduced, and the clinician continues to drive the pin into the proximal fragment.
   (5) Two positive-profile ESF pins are placed, one in the proximal humerus and one in the distal humeral condyles.
   (6) If needed, two additional ESF pins can be inserted on either side of the fracture.
   (7) The end of the IM pin is bent at a right angle to the shaft of the humerus. The ESF pins are bent at a right angle to be parallel with the shaft of the humerus
   (8) All pins are connected by an acrylic bar (Fig. 23-6).

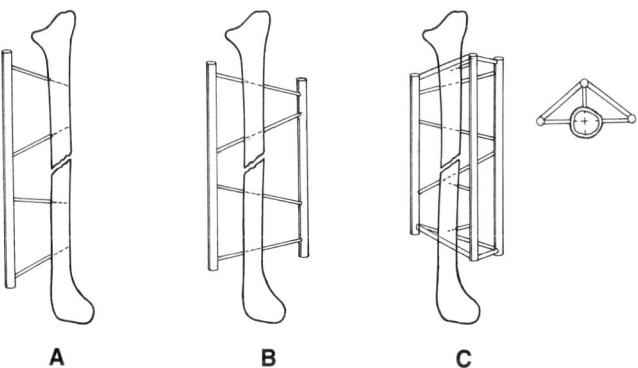

Fig. 23-5 Types of external skeletal fixation used for birds. **A,** Type I penetrates the skin and both cortices of the bone but does not penetrate the skin on the other side. **B,** Type II penetrates the skin, both cortices, and the skin on the opposite side; there are connecting bars on both sides. **C,** Type III combines types I and II: a type I device is placed at a 90-degree angle to a type II device, and additional connecting rods are placed between the fixators near their ends.

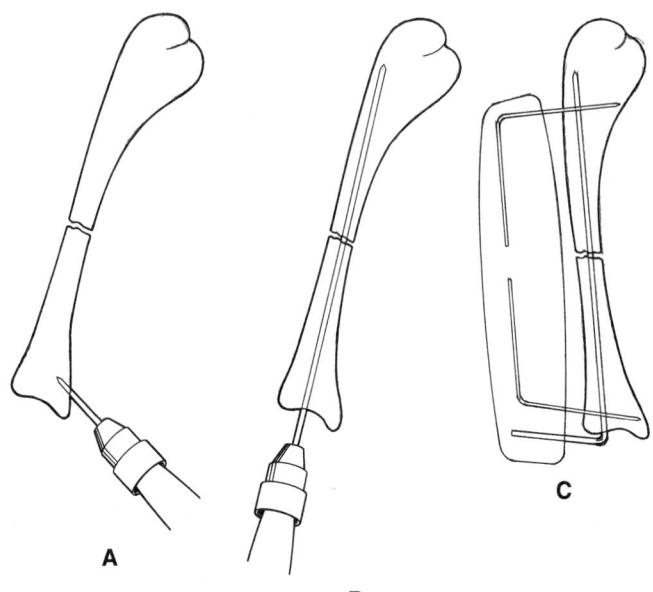

Fig. 23-6 An IM-ESF tie-in for a diaphyseal fracture of the humerus. **A,** A trocar-tipped IM pin is placed at a 70- to 90-degree angle to the long axis of the humerus and lateral to the triceps tendon. **B,** As the IM pin is passed through the cortex, the pin angle is changed until the pin is in the marrow cavity (the pin diameter should be 50% to 60% of the diameter of the marrow cavity). **C,** Two positive-threaded ESF pins are placed in the humerus. The ends of the IM pin and the ESF pins are bent, and all three exposed pins are incorporated into a thermoplastic stabilization bar.

4. Repair of fractures of the distal humerus
   a. Fractures of the distal humerus can be both difficult to repair and prone to complications, especially if the fracture or repair involves damaging the elbow joint (Fig. 23-7).
   b. For uncomplicated fractures, the dry, clean, medullary cavity is filled with bone cement (polymethyl methacrylate, Surgical Simplex-P, Howmedica, Inc., Rutherford, NJ or LVC Bone Cement, Zimmer, Warsaw, Ind) and a shuttled, threaded stainless steel IM pin is inserted.
   c. Transarticular external skeletal fixators are recommended for fractures close to or involving the elbow joint.[7,8]
      (1) Five threaded external skeletal fixator pins are required for stabilization: two in the proximal humeral fragment, one in the distal humeral fragment, and two in the ulna.
   d. Complications, including ankylosis, periarticular fibrosis, and loss of flight capabilities, are common sequela of distal humeral fractures.

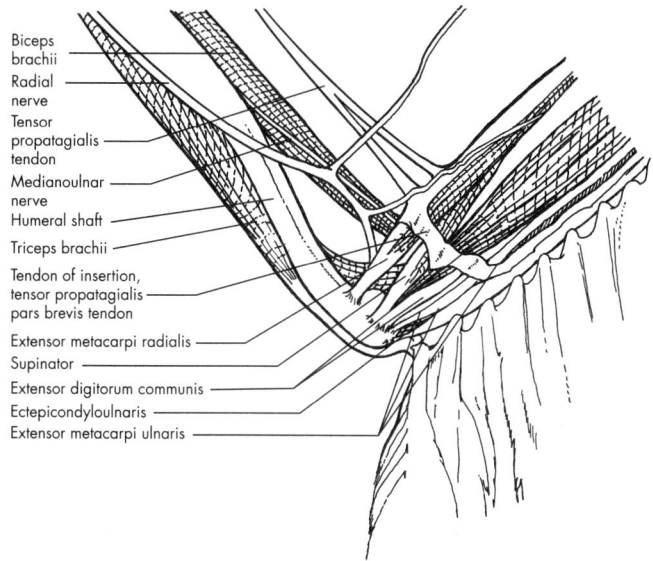

Fig. 23-7 Surgery around the distal humerus and proximal radius and ulna is difficult because of the number of anatomic structures that cross the elbow joint.

D. Fractures of the Radius and Ulna
1. General aspects
   a. The diagnosis is made by palpation and with radiographs.
   b. The prognosis often is good, especially if only one of the bones is fractured.
2. External coaptation splinting
   a. This technique is extremely useful if only one of the two bones is fractured, especially in small caged birds.
   b. A figure-eight wrap is the most commonly used form of external coaptation splinting for avian patients (Fig. 23-8).
   c. In falconry, a braille sling is used.[9] The sling is made from a long, thin piece of cloth or, more often, leather. A slit is made in the material, and the carpus is placed through the slit. One end then goes under the metacarpals and the radius and ulna, and the other end goes over the humerus. The sling is then tied near the elbow; next, one end is run over the elbow and digit areas, the other end is run under them, and the two ends are tied where they come together.
   d. External coaptation bandaging or splinting is contraindicated if the bone or bones are not in good alignment at the fracture site, because synostosis between the radius and ulna can occur and eradicate or diminish normal flight capabilities.

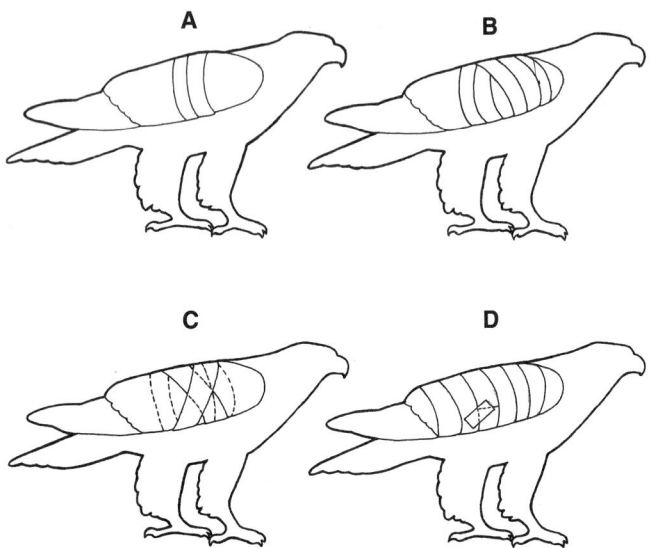

Fig. 23-8 Forming a figure-eight wrap. **A,** Using a nonadhesive bandage material (e.g., Vetrap), the carpal area is circled to seat the bandage. **B** and **C,** The bandage then is wrapped around the elbow caudally and back around the carpus in a figure-eight pattern. **D,** The exposed wing is covered evenly with a final layer of Vetrap, and the end is secured with a piece of adhesive tape.

3. Internal fixation of the radius or ulna (or both)
    a. These fractures are repaired with IM pins, external skeletal fixation, or a combination of one or both and coaptation.
    b. The radius commonly is stabilized by internal fixation for partial restoration of load sharing. Failure to stabilize the radius adequately can lead to delayed healing or nonunion.
    c. If only the ulna is fractured and displacement is minimal, a figure-eight bandage may suffice for immobilization.
    d. If both the radius and ulna are fractured, optimal stabilization and healing can be achieved by IM pinning of the radius and use of a type I ESF device in the ulna.[10]
4. Placement of IM pins in the radius and ulna
    a. Placement of the IM pins is critical to avoid prolonged loss of function.[10]
        (1) Radius: The pin is positioned retrograde with the exit location distally adjacent to the carpus joint (Fig. 23-9).
        (2) Ulna: The ulna is pinned normograde, with the pin inserted into proximal fragment first. The location of insertion is just below the triceps tendon insertion.
            (a) The spot between the second and third to last secondary feathers is located.

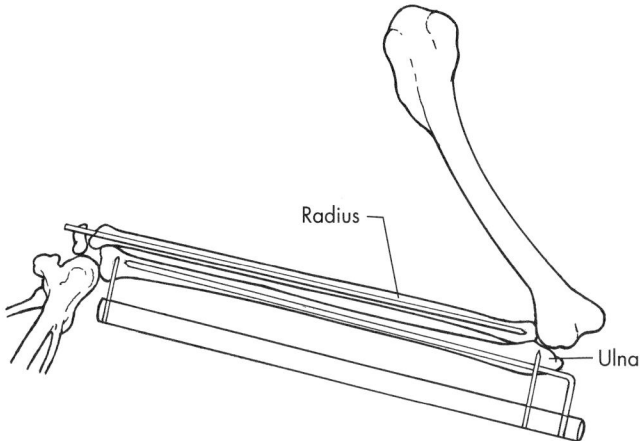

Fig. 23-9 An IM-ESF tie-in for fractures of the radius and ulna. The IM pin is placed retrograde into the radius, exiting distally adjacent to the carpal joint. An IM pin in the ulna exits below the elbow. Two positive-profile type I ESF pins are placed near the ends of the ulna for additional stabilization. Acrylic-filled latex tubing can be used for the stabilization bar connecting the ESF pins and the bent IM pin in the ulna. In some cases a conventional type I four pin ESF is used on the ulna.

      (b) A pilot hole is drilled with a sharp trocar point of a pin.
      (c) The pin is replaced with the next larger size pin with the trocar point removed from the threaded end.
      (d) The blunt-ended pin is run down through the proximal fragment and then the distal fragment. The blunt end prevents accidental penetration into the carpal joint.
  b. A combination using an IM pin in the radius and type I ESF in the ulna is best for cases in which both bones are fractured or in some cases of displacement when just the ulna is fractured (Fig. 23-9).[10]
  c. An IM-ESF combination is useful for stabilizing ulnar fractures, especially fractures near the metacarpus, or if an unstable radial fracture is present.
  d. External skeletal fixators that cross the elbow joint are useful for stabilizing very proximal fractures of the ulna or very distal fractures of the humerus.[7]
  e. Synostosis of the radius and ulna is a frequent complication of fracture healing. Synostosis can prevent full flight. However, the synostosis can be repaired surgically.
      (1) The osseous material bridging the two bones is removed from subcutaneous areas and abdominal fat is placed in the defect

510 *Limb Dysfunction*

between the radius and ulna to prevent the synostosis from reforming.

E. Fractures of the Metacarpals
1. General aspects
   a. Metacarpal fractures typically are high-energy fractures, caused by flying into an object (e.g., wire, powerline, or fence) or by a projectile.
   b. Little soft tissue covers the bone, and the soft tissue often is damaged severely. The many anatomic structures in the metacarpal area complicate repair (Fig. 23-10).
   c. These fractures usually are open and comminuted.
2. Treatment options
   a. Coaptation splinting
      (1) This technique is suitable for transverse, reducible fractures, especially if only one metacarpal bone is fractured.

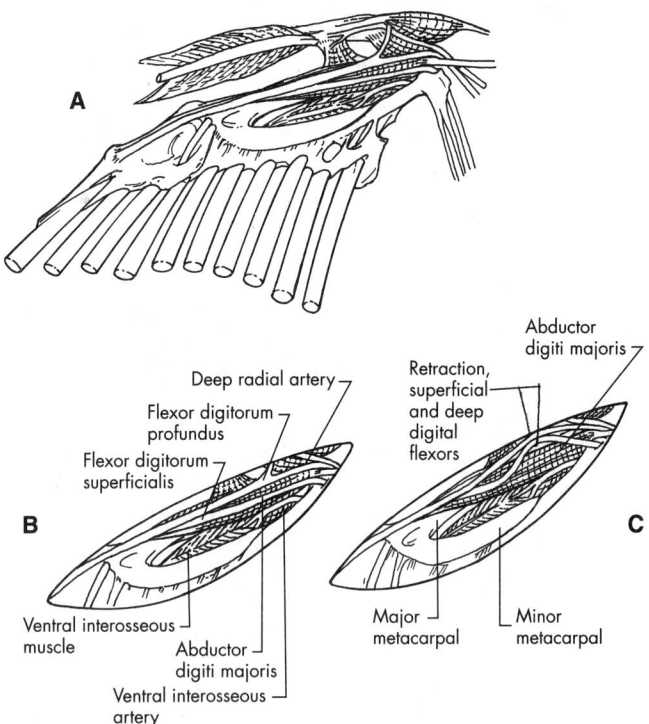

Fig. 23-10 Overview of the complicated anatomy in the manus. **A,** View of the ventral metacarpals and associated structures. **B,** View from a ventral incision made between the two metacarpal bones. **C,** View showing tendons of the deep and superficial digital flexors retracted to allow exposure of the proximal end of the major metacarpal.

(2) Splints made from moldable veterinary thermoplastic splint material work best.
(3) Two pieces are used, one on each side of the manus, extending from the carpal joint to beyond the last digit. Adhesive tape is wrapped around the splints and manus, and the wing then is placed in a figure-eight wrap.
(4) Possible complications include failure of bones to heal (nonunion) and loss of function in adjacent joints related to fracture or long-term (weeks) immobilization of the wing, or both.
  b. External skeletal fixation
    (1) This technique is useful in highly comminuted fractures or in cases with extensive soft tissue injury.
    (2) An IM-ESF tie-in gives the best results as long as the soft tissue injury is not too severe.[10]
    (3) IM pinning alone is not recommended, because the stability it affords is not sufficient for fracture healing.[10]

VII. Luxations of the Pectoral Limb

Luxations occur primarily at four locations: the proximal coracoid, the scapulohumeral joint, the elbow, and the carpus.

A. Proximal Coracoid Luxation
1. This type of luxation is associated with chest injuries.
   a. On radiographs the base of the coracoid is below the edge of the sternum.[10] The injury is reparable if the coracoid is outside the sternum; a lateral radiograph is taken to verify this.
2. A proximal coracoid luxation is repaired surgically[10]:
   a. An incision is made over the area, and the pectoral muscle is elevated from its attachment to the keel.
   b. Blunt dissection is used to locate the cranial edge of the sternum and coracoid bone.
   c. An osteotome is used to lever the coracoid into place. It usually is held in place in the small depression by the forces of the musculoskeletal system.
   d. The pectoral muscle is reattached to the keel using sutures, and the skin closed.
   e. After surgery, the wing is kept bandaged to the body (figure-eight wrap and body wrap) for at least 3 weeks.

B. Scapulohumeral (Shoulder) Luxation
1. Radiographically this type of luxations appears as an asymmetry between the humeral heads and lateral processes of the scapular bones when comparing the normal and abnormally functioning sides.
2. Soft tissue damage can range from minor to severe.
3. A scapulohumeral luxation is repaired nonsurgically:

a. The wing is taped to the body for a minimum of 3 weeks.
4. If soft tissue trauma is severe, the prognosis for normal wing function is guarded to poor.

C. Elbow Joint Luxation
1. This type of injury is common, and the prognosis for normal function is guarded to poor.
2. Signs include a dropped wing, elbow abducted and extended 90 degrees, crepitus, swelling, and wounds at the elbow.
3. A mild luxation that is stable after manual reduction will respond well if the wing is placed in a figure-eight wrap for 1 week.[8] Physical therapy and limited use are required after 1 week to prevent ankylosis.
4. External fixation has been used with unstable elbow luxations.[8]
   a. Three K-wires are placed in the distal humerus, and three are placed in the proximal ulna.
   b. A type I external skeletal fixation is created by bending wires at right angles and encasing them in methylmethacrylate, forming a crossbar over the dorsal wing.
   c. As an alternative, a type II external fixator can be made by placing one K-wire through the distal humerus and one through the proximal ulna.[8]
   d. The ESF device is removed after 1 to 2 weeks. Physical therapy is then used to help restore the wing to a normal range of motion.[8]

VIII. Skeletal Anatomy of the Pelvic Limb (Fig. 23-11)

A. Pelvic Girdle
1. The pelvic girdle is formed by the ilium, the ischium, and the pubis.[3]
2. The bones are partly fused to each other and to the synsacrum.
3. The two most prominent bones are the ilium (cranially) and the ischium (caudally).
4. The acetabulum and its acetabular foramen are formed between the ilium and the ischium.
5. The pubic bone curves caudally and ventrally from the attachment with the ischium.

B. Femur[1]
1. The femur is a stout bone, the largest diameter bone in the pelvic limb.
2. The femur slopes forward cranially in galliformes but may be fairly straight in raptors that tend to walk more and use their legs for tearing dead prey.
3. The femoral trochanter is found at the proximal end, a prominent location for muscle insertion.
4. A sulcus is found between the lateral and medial condyles. The condyles are directed more in line with the lower leg in birds with a large degree of terrestrial locomotion. The deep grooves with ridges reduce the problem of dislocation.

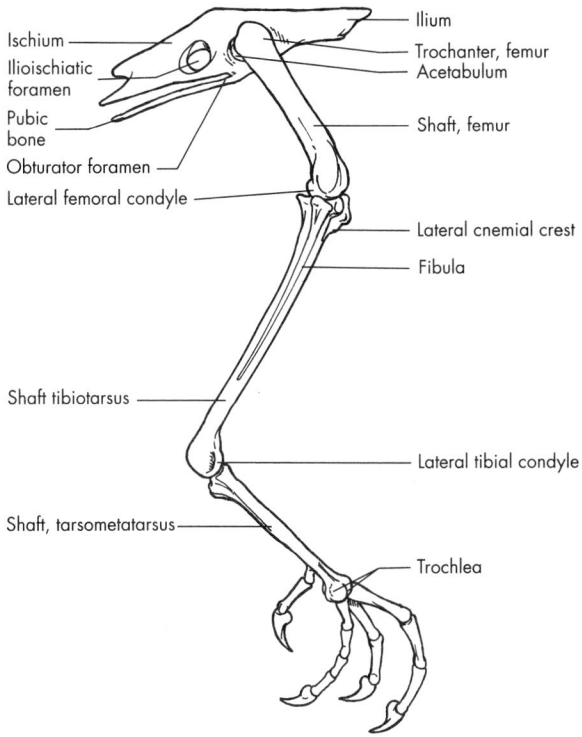

Fig. 23-11 Overview of the skeletal anatomy of the pelvic limb.

C. Patella
1. The patella is present in most birds and is large in waterfowl.[11]
2. The patella is considered a sesamoid bone of the femorotibialis muscle, which acts as a biceps femoris muscle.
3. The patella slides in the sulcus between the femoral condyles.

D. Tibiotarsus[3]
1. The tibiotarsus is formed by the fusion of the proximal tarsal bones with the distal tibia.
2. The proximal end of the tibiotarsus has an articular surface for the femoral condyles.
3. Cranial and lateral cnemial crests serve to increase the surface area for the origin of the extensor muscles of the hock and digits.
4. The cranial cnemial crest is a landmark for bone marrow taps and for the insertion of intraosseous catheters.
5. The tibiotarsus ends in the medial and lateral tibial condyles, which articulate with the tarsometatarsus.

E. Fibula
1. The fibula is much reduced in birds; it usually is a tapering, slender bone that extends only two thirds of the length of the adjacent tibiotarsus.
2. The fibula usually is fused to the tibiotarsus at the proximal end.
3. Because the fibula is small and short, most birds cannot rotate their legs.[11]

F. Tarsometatarsus
1. The tarsometatarsus is formed by the fusion of the distal tarsal and metatarsal bones.
2. The proximal tarsometatarsus has a medial and lateral surface for the articulation with the tibiotarsus.
3. The tarsometatarsus usually is shorter than the tibiotarsus; however, in long-legged wading birds, the two bones are approximately equal in length.[11]
4. The tarsometatarsus ends in three trochleae for articulation with digits II, III, and IV.[3]
5. An interosseous canal is found distally for the tendon of the extensor of digit IV (extensor brevis digit IV). This canal lies between the trochleae for digits III and IV.
6. A fossa is present on the medial side for articulation with the first metatarsal bone.

G. Digits[3]
1. Digit I, often called the hallux, usually has two phalanges, and the first articulates with metatarsal I.
2. Digit II has three phalanges.
3. Digit III has four phalanges.
4. Digit IV has five phalanges.

IX. Muscular Anatomy of the Pelvic Limb, Lateral Thigh, and Leg (Fig. 23-12)

A. Iliotibialis Cranialis Muscle
1. The iliotibialis cranialis muscle is a straplike muscle found along the cranial margin of the thigh; it is comparable to the sartorius muscle in mammals.
2. Origin: Craniodorsal surface of the ilium
3. Insertion: Craniomedial patella and the proximocranial portions of the tibiotarsus
4. Innervation: Femoral nerve
5. The iliotibialis cranialis muscle extends the stifle while flexing the hip (Table 23-2).

B. Iliotibialis Lateralis Muscle
1. Origin: Crest of the ilium

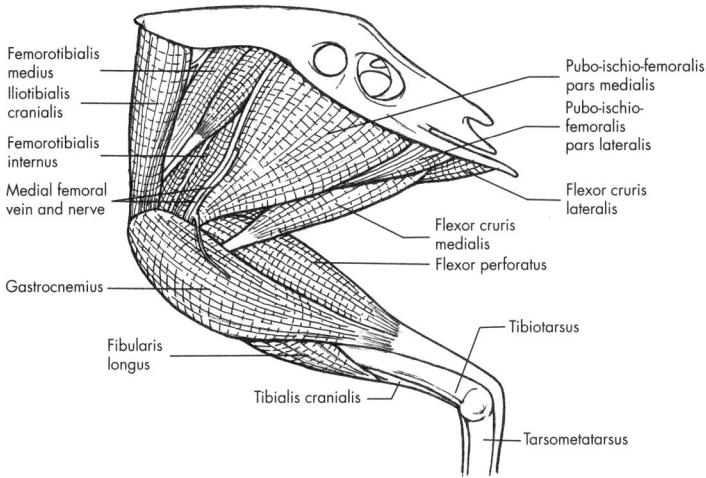

Fig. 23-12 Orientation of the major muscles on the medial pelvic limb.

Table 23-2
Major muscles in the pelvic limb and their associated actions

Muscle	Action
Puboischiofemoralis, iliofemoralis	Hip extension
Iliotibialis cranialis	Hip flexion
Iliotrochantericus caudalis	Medial hip rotation
Ischiofemoralis	Lateral hip rotation
Femorotibialis	Knee extension
Iliofibularis, flexor cruris medialis, flexor cruris lateralis	Knee flexion
Femorotibialis internus	Medial tibiotarsus rotation
Flexor cruris lateralis	Lateral tibiotarsus rotation
Tibialis cranialis	Hock flexion
Gastrocnemius	Hock extension
Extensor digitorum longus	Digit extension
Flexores perforati II, III, and IV	Digit flexion

  2. Insertion: Lateral cnemial crest and by fascial attachments to the iliotibialis cranialis muscle as part of the patellar tendon
  3. Innervation: Femoral nerve
  4. The iliotibialis lateralis muscle functions like the tensor fasciae latae muscle of mammals, which tenses the fascia on the lateral side of the thigh.

5. The iliotibialis lateralis muscle is well developed in galliformes, vultures, and wading birds.[1] It may have as many as three heads and, in these species, underlying muscles. The extent of the iliotibialis lateralis makes it difficult to visualize the femur.
6. In hawks and owls the iliotibialis lateralis muscle is fused to the underlying femorotibialis externus muscle; this muscle is reduced in size in these birds, which allows easier visualization of the femoral shaft for fracture repair.

C. Flexor Cruris Lateralis Muscle (Semitendinosus)
1. Origin: Caudal portion of the ischium
2. Insertion: Medial surface of the gastrocnemius and flexor cruris medialis
3. Innervation: Ischiatic nerve
4. The flexor cruris lateralis muscle, along with the flexor cruris medialis, represents the hamstring muscles of the bird. The flexor cruris lateralis may have an accessory head, except in most swimming birds.
5. The flexor cruris lateralis muscle is absent in hawks, owls, and psittacines.

D. Flexor Cruris Medialis Muscle (Semimembranosus)
1. Origin: By a broad attachment to the ventral portion of the ischium
2. Insertion: On the proximomedial portion of the tibiotarsus in concert with the insertion of the flexor cruris lateralis (when present)
3. Innervation: Ischiatic nerve
4. The flexor cruris medialis muscle extends the hip while flexing the stifle.

E. Iliofibularis Muscle
1. The iliofibularis muscle is more prominent than the iliotibialis lateralis muscle in psittaciformes, accipitridae, and strigiformes.
2. Origin: Iliac crest caudal to acetabulum[1]
3. Insertion: Caudolateral fibula between the heads of the gastrocnemius muscle[1] (its tendon of insertion can be visualized easily because it is surrounded by a retinaculum; the fibular nerve and popliteal vein course through the retinaculum as well)
4. Innervation: Ischiatic nerve
5. The iliofibularis muscle is similar to the biceps femoris muscle of mammals. The muscle flexes the stifle.

F. Femorotibialis Externus Muscle (Portion of the Quadriceps Femoris)
1. The femorotibialis externus muscle often is partly fused to the underlying iliotibialis lateralis and covers the shaft of the femur laterally. It becomes continuous with the medius as part of the quadriceps complex.
2. Origin: Lateral ilium and proximal femoral head

3. Insertion: Cnemial crests of the tibiotarsus
4. Innervation: Femoral nerve
5. The femorotibialis externus muscle extends the stifle.

G. Fibularis Longus Muscle (Peroneus Longus Muscle)
1. The fibularis longus muscle varies; it is absent in owls and ospreys, small in pigeons and parakeets, and large in vultures and fowl. It often is the most cranial muscle of the crus, or leg, but may be deep to the tibialis cranialis in hawks and psittacines.
2. Origin: Lateral side of the tibiotarsus along the cnemial crests
3. Insertions: Two:
   a. On the tibial cartilage to extend and abduct the hock
   b. On the tendon of the flexor perforatus digiti III to flex the third digit
4. Innervation: Fibular nerve

H. Flexor Perforatus Digiti IV Muscle
1. Often divides into three portions to insert on P_1 (two locations) and P_2
2. Insertion: Single tendon of insertion in psittacines; inserts on phalanx I
3. Because the fourth digit in psittacines faces backward, this muscle extends digit IV; in other species, with forward-facing fourth digits, the muscle flexes it.

I. Tibialis Cranialis Muscle
1. Origin: Two heads:
   a. Lateral condyle of the femur
   b. Lateral cnemial crest
2. Insertion: The two heads join into a common tendon that inserts on the dorsal metatarsal groove.
3. Innervation: Fibular or peroneal nerve
4. The tibialis cranialis muscle flexes the intertarsal joint, or hock.
5. This muscle is located superficial to the fibularis longus muscle in psittacines and hawks.

J. Gastrocnemius Muscle
1. The gastrocnemius muscle has two or three heads and is found on the caudal surface of the crus, or leg.
2. Origin: When two heads are present, one arises from the lateral femoral condyle; the other takes origin from the medial femoral condyle. The third head, when present, takes origin from the medial popliteal area.[1]
3. Insertion: Plantar surface of the tarsometatarsus and digital flexor tendons
4. Innervation: Medial and lateral tibial nerves
5. The gastrocnemius muscle is a strong extensor of the hock and flexor of the digits because of its attachments to the tendons of the digital flexors.

X. Muscular Anatomy of the Medial Thigh (see Fig. 23-12)

A. Pubo-ischio-femoralis (Pars Medialis) Muscle
1. This muscle corresponds to the adductor magnus et brevis muscle of mammals. A lateral portion of the avian muscle, the pars lateralis, corresponds to the abductor longus in mammals and can be seen from a deep location, laterally.
2. Origin: Ventral surface of ischium
3. Insertion: Caudal surface of the femur
4. Innervation: Obturator nerve
5. The pubo-ischio-femoralis pars medialis muscle adducts the femur while extending the hip.

B. Ambiens Muscle
1. The ambiens muscle is similar to the pectineus muscle in mammals. The ambiens can be distinguished by its central location on the medial thigh, its fusiform shape, and its long white tendon of insertion.
2. Origin: Medial ilium
3. Insertion: Narrows to a tendon that crosses the patella, inserting on the heads of the long digital flexor muscles
4. Innervation: Femoral nerve
5. The ambiens muscle is considered the perching muscle, because it flexes the toes for grasping when the hock is flexed.
6. The ambiens muscle is absent in finches, canaries, and other passerines, in some psittacines, and in pigeons and doves.

C. Femorotibialis Internus and Femorotibialis Medius Muscles (Portion of the Quadriceps)
1. These two muscles are continuous with each other and form part of the quadriceps femoris group.
2. Origin: From the dorsal femur
3. Insertion: The medius muscle inserts onto the cnemial crest of the tibiotarsus as part of the patellar tendon.
4. Innervation: Femoral nerve
5. The femorotibialis internus and medius muscles extend the stifle.

XI. Fractures of the Pelvic Limb

A. Femoral Fractures
1. General aspects
    a. Femoral fractures generally are not treatable with coaptation, because the medial side is not accessible
2. Surgical approaches
    a. Lateral approach (used most often)[1]
        (1) An incision is made over the shaft of the femur and then through the iliotibialis lateralis muscle (a large muscle in

fowl and vultures but a small one in psittacines, hawks, and owls).
- (2) The iliofibularis and femorotibialis externus muscles are retracted cranially.
- (3) Periosteal elevation of the femorotibialis externus may be needed for better exposure.[1]
- (4) The muscles are always sutured to get good closure and coverage over the bone.[12]

b. Lateral approach to the proximal femur and coxofemoral joint[3]
- (1) This approach is useful for open repair of the proximal femur and removal of a fractured femoral head.
- (2) An incision is made from the dorsolateral crest of the ilium over the femoral trochanter and along the skin over the caudolateral femur.
- (3) The iliotibialis and iliofibularis muscles are separated lengthwise.
- (4) The clinician incises through the musculotendinous insertion of the iliotrochantericus caudalis and iliofemoralis externus.
- (5) The clinician then incises through the clear membrane overlying the femoral neck and the joint capsule overlying the femoral head.
- (6) A fractured femoral head can be repaired (difficult) or removed. Birds as large as Florida sandhill cranes (*Grus canadensis praetensis*) have done well after femoral head ostectomy.

3. Fixation methods for diaphyseal fractures
   a. Normograde intramedullary pinning
   - (1) This technique can be used for closed pinning to avoid damage to the ischiatic nerve[6]
   - (2) Pins are placed in the femur starting at the trochanteric fossa.

   b. Retrograde intramedullary pinning
   - (1) IM pins generally are used only in small birds (<150 g), such as budgerigars and cockatiels.

   c. External skeletal fixation
   - (1) This technique offers an advantage for severely comminuted fractures because the fracture area is not damaged further.
   - (2) Type I external skeletal fixators are most commonly used because of the femur's attachment to the body wall medially.
   - (3) The connecting bar for the external skeletal fixation can be the traditional Kirschner-Ehmer (KE) apparatus of steel clamps and connecting bars.
   - (4) An alternative connecting bar commonly used with birds involves fixing a Penrose drain or flexible tubing to the pins by pushing the tubing through the ends of the pins and then filling the tubing with dental acrylic or horse hoof repair acrylic.
   - (5) Still another method for constructing connecting bars involves cutting long narrow pieces of Hexalite or VTP, softening the material in hot water, and then folding the strap lengthwise over the ends of the pins.

d. An IM-ESF tie-in is the preferred treatment method.
 4. Fractures or luxations of the femoral head
    a. Luxations can be reduced and held in place by running nonabsorbable suture between the greater trochanter and the dorsolateral iliac crest. A second suture goes between the greater trochanter and the cranial acetabular rim.[3]
        (1) The joint capsule is closed with simple interrupted sutures of 5-0 polyglactin suture (Vicryl, Ethicon, Inc., Somerville, NJ). The iliotrochantericus caudalis and the iliofemoralis externus are reattached using a simple continuous pattern. The simple continuous pattern is also used to repair the iliotibialis lateralis and the iliofibularis.
        (2) A spica-type splint is used to support the surgery site for several weeks after surgery.

B. Fractures of the Tibiotarsus
 1. General aspects
    a. These fractures often are oblique, and rotational problems are encountered in repair.
    b. Intramedullary pins have been used but have the disadvantage of damaging either the stifle or the hock when the pin is placed; IM pins also offer little rotational support.
        (1) Coaptation splints (e.g., Schroeder-Thomas type) must be used with intramedullary pinning to achieve rotational stability.
    c. There is good soft tissue coverage of the bone at the proximal end.
 2. Surgical approaches
    a. The medial approach is favored because the large muscle mass laterally over the tibiotarsus makes for a difficult lateral approach[3] (Fig. 23-13).
    b. The skin is incised over the medial shaft of the tibiotarsus, with care taken to avoid the medial metatarsal vein running medially over the hock and then just caudal to the tibiotarsus.
    c. The distal shaft of the tibiotarsus is easily seen after the skin incision is made.
    d. The proximal shaft of the tibiotarsus is exposed by pulling the fibularis longus cranially and the gastrocnemius medial head caudally.
    e. A simple continuous pattern with absorbable suture is used to pull the fibularis longus and gastrocnemius medial head back together. The skin is closed with nonabsorbable suture in a continuous pattern.
 3. Fixation methods
    a. External skeletal fixation
        (1) A type I ESF device works for most of these fractures.
        (2) A type II ESF device with support rods medially and laterally also works well and is needed in larger birds (e.g., >3 kg).[13]
    b. ESF-plastic shuttle pin technique
        (1) A polypropylene plastic rod is inserted into the marrow cavity using shuttle-pin techniques; then the holes for placement of

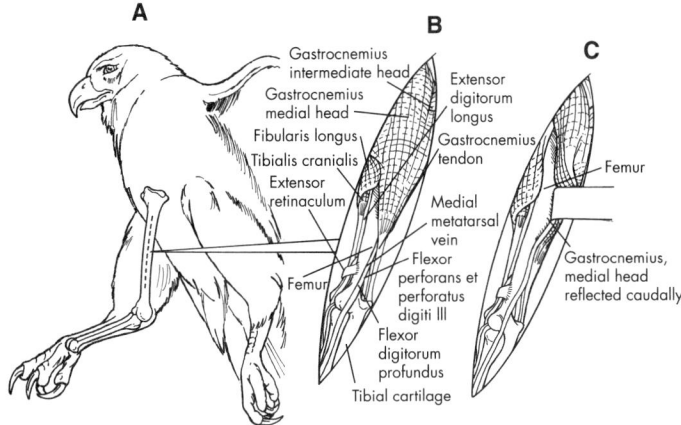

Fig. 23-13 Medial surgical approach to the tibiotarsus. **A,** Dotted line shows the area of the skin incision. **B,** Structures over the medial tibiotarsus as seen through a skin incision. **C,** View after separating the gastrocnemius muscle from the fibularis longus and tibialis cranialis muscles to allow better exposure of the proximal tibiotarsus.

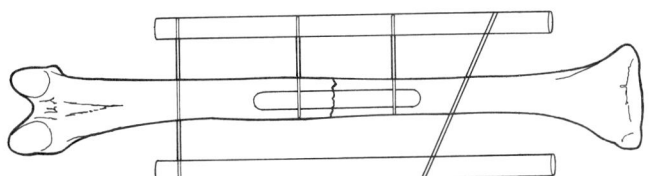

Fig. 23-14 The preferred method of repairing tibiotarsal fractures is placement of two type II ESF pins at the ends and two type I ESF pins near the fracture site.

the type I ESF pins are drilled through the bone and the polypropylene plastic rod (Fig. 23-14).
(2) Two type II pins are placed first, then two type I pins are placed closer to the fracture site (through the support rod is best) (see Fig. 23-14). Positive-profile threaded pins are used.
(3) Stabilizing bars can be fashioned from flexible tubing or Penrose drain tubing filled with acrylic, or from Hexalite.
(4) This technique ensures bone alignment during placement of the ESF pins and is the preferred method for repair of tibiotarsal fractures.

C. Fractures of the Tarsometatarsus and Phalanges
1. General aspects
   a. Very little soft tissue overlies the bones in these locations.

522 *Limb Dysfunction*

   b. Surgical intervention is not needed as often and may compromise the area, leading to nonunion.[15]
   c. Open, comminuted fractures are common.
   d. There are no hollow marrow cavities in these bones; any internal longitudinal pin must be drilled through the bone.[13]
2. Surgical approaches
   a. If open reduction is required, the lateral approach is used.[3]
      (1) The patient is placed in the lateral recumbent position.
      (2) The skin is incised along the lateral shaft of the bone or bones (Fig. 23-15). As the skin is pulled back, the following structures are seen:
         (a) Extensor tendons and blood vessels are cranial to the tarsometatarsus
         (b) Flexor tendons lie caudal to the bone
         (c) At the proximal end is the extensor brevis digiti IV and the abductor digiti IV cranially and caudally.
      (3) The skin is closed with a simple continuous suture pattern.
3. Fixation methods
   a. Fractures respond well to splinting or external skeletal fixation.[13]
      (1) Schroeder-Thomas splints, plantar splints, or tape splints (in small birds) have been used successfully.[6]

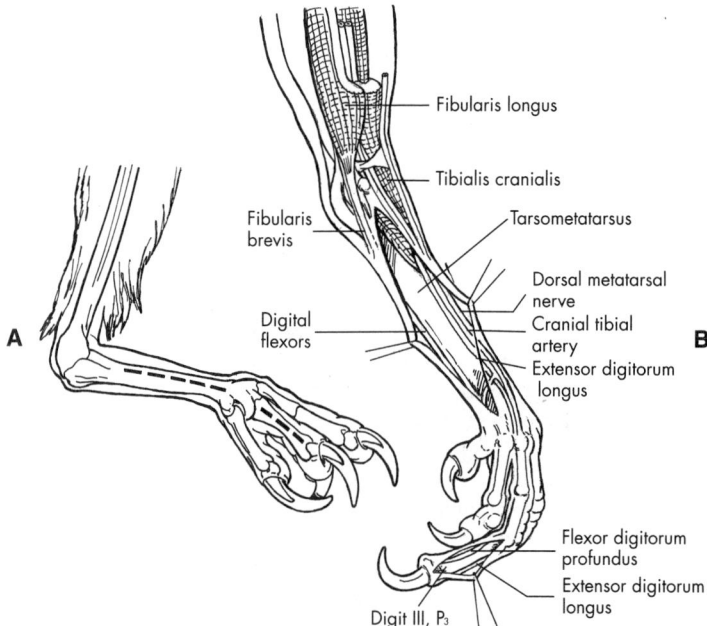

Fig. 23-15 Surgical exposure of the lateral tarsometatarsus and phalanges. **A,** Site of skin incisions (*dotted lines*). **B,** Orientation of the anatomic structures on the lateral tibiotarsus.

b. Type II external skeletal fixation works best; care must be taken with the tendons on the plantar surface and an artery and vein on the cranial surface.[6]
   c. For raptors and some other grasping or perching birds, a ball bandage works well for phalangeal fractures.
 4. Complications
   a. A frequent complication in fractures of the tarsometatarsus is avascular necrosis distal to the fracture site caused by disruption of the blood supply.
      (1) For large wading birds, an artificial leg made from various diameters of polyvinyl chloride pipe can be fitted to the stump, and the bird will do quite well.[14]
      (2) Smaller psittacines and passerines do well in a caged environment after amputation of the necrotic leg.

XII. Pelvic Limb Luxations

A. Coxofemoral Luxation
 1. The coxofemoral joint is a diarthroidal joint with considerable support provided by the collateral and round ligaments. The round and ventral ligaments retain the femoral head in the acetabulum.[16]
 2. Luxations imply severe trauma, usually with all the above ligaments torn. This can occur if the leg is caught and the bird struggles.
 3. Surgical reduction and stabilization are required for repair of acute luxations.
   a. For the approach to the coxofemoral joint and proximal femur, the technique described above (XI, A, 2, b above) is used.
   b. After the luxation has been reduced, place two nonabsorbable or polydioxanone sutures to hold the femur in place.[6,17] The first runs between the greater trochanter and the dorsal iliac crest. The second runs between the greater trochanter and the cranial acetabular rim.[3]
 4. Chronic luxations usually require removal of the femoral head and neck.
   a. The approach is the same as described above (XI, A, 2, b).
   b. The femoral head and neck are removed with rongeurs; the new femoral surface must be smooth.
   c. The femur now will have a tendency to rotate.[6]
      (1) Rotation is prevented by placing support sutures as described above (XII, A, 3, b).
   d. The bird is kept in a spica splint or non-weight-bearing sling for 2 weeks after surgery.

B. Stifle Luxation
 1. Findings include positive drawer sign and also medial or lateral instability.
 2. Transarticular external skeletal fixation is used to maintain the stifle after reduction (Fig. 23-16).

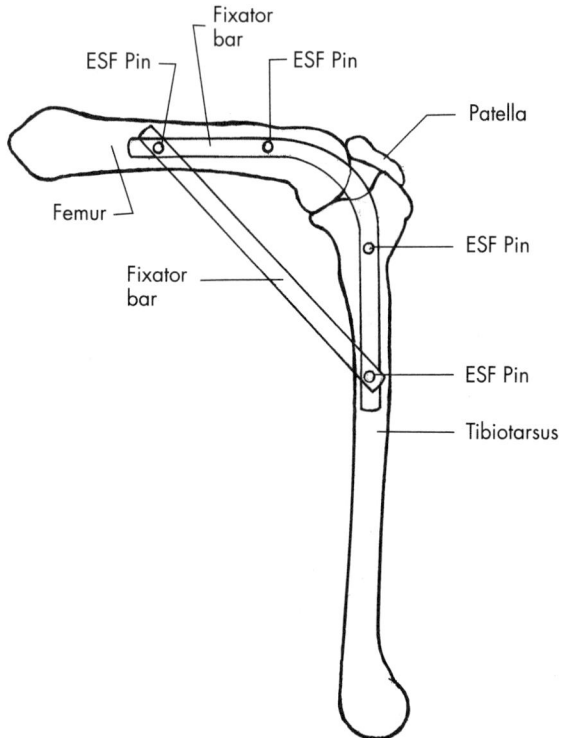

Fig. 23-16 Transarticular type I ESF device used to stabilize a stifle joint after luxation. Two positive-profile ESF pins are placed in both the distal femur and the proximal tibiotarsus on the lateral side of each bone. The fixator bars, which are made from acrylic-filled latex tubing or other suitable material, form a triangle.

    a. At least two pins are placed in both the distal femur and the proximal tibiotarsus.
    b. Type I external skeletal fixation usually is used, with the fixator bars forming a triangle (see Fig. 23-16).
    c. The formation of sufficient scar tissue to stabilize the joint requires 3 to 6 weeks, which is the interval before the pins are removed.
    d. In some larger birds, individual ruptured ligaments in the stifle can be repaired surgically.

XIII. Postoperative Management of Limb Injuries

A. General
1. The patient's general condition should be watched, especially for secondary problems and complications.

2. Cultures should be taken of open wounds and fractures.
3. Serial blood counts should be monitored for increases that might signal infection.

B. Bone Healing
1. The limb is radiographed immediately after surgery, then at 3 weeks after surgery, and then every 2 weeks thereafter until healed.
2. Early in the healing process, osteomyelitis and the formation of periosteal bone cells both appear on radiographs as periosteal proliferation, sclerosis, and increased medullary density.

C. Staged Destabilization
1. Staged destabilization of the fracture site is possible when two forms of stabilization were used.
    a. One apparatus is removed first (as early as 3 weeks after surgery) while the second device is maintained.[6]

D. Activity
1. As with mammals, activity must be restricted after repair of a fracture or luxation.
2. Confinement to a small cage is easily accomplished with most pet birds.

E. Physical Therapy
1. Initially physical therapy (moving each joint that was immobilized) is conducted under general anesthesia.[6]
2. The physical therapy program must be designed according to the type of injury[18]:
    a. Passive range-of-motion exercise is done by holding the limb on either side of the joint and moving it through its full range of motion. This therapy is most useful in the early stages of recovery to avoid soft tissue injury or insult.
    b. Active assisted range-of-motion exercise is useful for preventing changes in joint flexibility. For this technique, the therapist prompts the bird to move its own limbs; for example, the therapist grasps the bird's legs and raises and lowers the bird, causing it to move its wings to maintain balance.
    c. Active range-of-motion exercises are used to build muscular and cardiovascular strength in the mid to late recovery stages. The bird undergoes increasing levels of active exercise.

REFERENCES
1. Orosz SE: Clinical anatomy of the thoracic limb, *Proceedings of the MASAAV Avian Medicine and Surgery Conference,* pp 61-74, 1998.
2. Olsen GH, Carpenter JW, Langenberg JA: Medicine and surgery. In Ellis DH, Gee GF, Mirande CM, editors: *Cranes: their biology, husbandry, and conservation,* Washington, DC, 1996, National Biological Service, pp 137-174.

3. Orosz SE, Ensley PK, Haynes CJ: *Avian surgical anatomy: thoracic and pelvic limbs*, Philadelphia, 1992, Saunders.
4. Fisher HI: Adaptations and comparative anatomy of the locomotor apparatus of New World vultures, *American Midland Naturalist* 35:545-727, 1946.
5. Getty R: *Sisson and Grossman's anatomy of the domestic animals,* vol 2, *Porcine, carnivores, aves,* Philadelphia, 1975, Saunders.
6. Bennett AR: Orthopedic surgery. In Altman RB et al, editors: *Avian medicine and surgery,* Philadelphia, 1997, Saunders, pp 733-766.
7. Howard DJ, Redig PT: Orthopedics of the wing, *Semin Avian Exotic Pet Med* 3(2):51-62, 1994.
8. Martin HD et al: Elbow luxations in raptors: a review of eight cases. In Redig PT et al, editors: *Raptor biomedicine,* Minneapolis, 1993, University of Minnesota Press, pp 199-206.
9. Bennett RA, Kuzma AB: Fracture management in birds, *J Zoo Wildl Med* 23:5-38, 1992.
10. Redig PT: Methods for management of forelimb fractures, *Proceedings of the MASAAV Avian Medicine and Surgery Conference,* pp 86-101, 1997.
11. King AS, McLelland J: *Birds: their structure and function,* London, 1984, Bailliere Tindall.
12. Redig PT, Rousch JC: 1978. Surgical approaches to the long bones of raptors. In Fowler M, editor: *Zoo and wild animal medicine,* Philadelphia, 1978, Saunders, pp 388-401.
13. Hess RE Jr: Management of orthopedic problems of the avian pelvic limb, *Semin Avian Exotic Pet Med* 3(2):63-72, 1994.
14. Olsen GH: Orthopedics in cranes: pediatrics and adults, *Semin Avian Exotic Pet Med* 3(2):73-80, 1994.
15. Redig PT: 1986. A clinical review of orthopedic techniques used in the rehabilitation of raptors. In Fowler M, editor: *Zoo and wild animal medicine,* ed 2, Philadelphia, 1986, Saunders, pp 388-401.
16. MacCoy DM: Excision arthroplasty for management of coxofemoral luxations in pet birds, *J Amer Vet Med Assoc* 194(1):95-97, 1989.
17. Bennett RA et al: Tissue reaction to five suture materials in pigeons (*Columbia livia*), *Proceedings Association of Avian Veterinarians,* pp 212-218, 1992.
18. Martin HD, Ringdahl C, Scherpelz J: Physical therapy for specific injuries in raptors. In Redig PT et al, editors: *Raptor biomedicine,* Minneapolis, 1993, University of Minnesota Press, pp 207-210.

24

Soft Tissue Surgery

Glenn H. Olsen

I. Presurgical Examination

A. Comprehensive Physical Examination
(see Chapter 1)
1. Evaluate patient's nutritional state
   a. Emaciated birds may not be good candidates for anesthesia and surgery because of the stress involved.
   b. Obese birds may have increased respiratory problems resulting from fat deposits reducing air sac spaces.[1]
   c. Birds with calcium deficiencies are susceptible to iatrogenic fractures.[2]
2. Evaluate cardiac and respiratory systems. Stress bird by handling or letting bird fly. Watch for recovery to normal respiratory function (2 to 5 minutes).

B. Hematology and Serum Chemistry Test Results
1. Liver function tests important to surgical patient
   a. Look for elevations in bile acids, aspartate aminotransferase (AST), or serum cholesterol.[3]
2. If possible, correct for or compensate for any clinical pathology values that are beyond the normal range before undertaking surgery.

II. Patient Monitoring During Surgery

A. Body Temperature
1. Body temperature decreases during surgery because of lack of muscle activity and loss to ambient air and to the surface on which the bird is positioned.
2. Monitor body temperature with an electronic thermometer with a probe that allows placement in the cloaca.

3. Conserve body heat by doing the following:
   a. Insulating bird from the stainless steel surgery table by using a circulating warm water pad.
   b. Remove feathers from as small a surgery site as necessary. Use a minimum amount of alcohol in surgical preparation (draws heat when evaporating).
   c. Give warm intravenous, subcutaneous, or intraosseous fluids.

B. Heart Rate and Respiration Monitoring
1. Heart rate and respiration can be monitored with a standard pediatric stethoscope or with an esophageal stethoscope.
2. Heart rate can be mechanically monitored with a cardiac monitor, electrocardiogram tracing, a pulse oximeter, or Doppler sensor apparatus.
3. Respiratory monitors are available that hook between the endotracheal tube and the Ayres T piece on the anesthetic machine.

III. Patient Preparation

A. Patient Handling
1. Minimal
   a. Handle the patient minimally, do all patient preparation after bird is anesthetized.

B. Removing Feathers
1. Remove feathers over the surgery site by plucking parallel to the direction the feathers' follicles are located.
2. Plucking at an angle increases the tendency to cause skin lacerations.
3. Try to avoid removing large tail and flight feathers. If the wing is fractured, it may be better to cut large flight feathers in the area rather than pulling these out, because pulling may cause more damage to the unstable wing.
4. Pluck only the minimal area needed for surgery. Feathers are useful for thermoregulation and are needed by the bird.
5. To keep nearby feathers out of the surgical field, apply sterile water-soluble surgical lubricant to these areas adjacent to the surgery site.
6. Another approach to keeping feathers out of surgical sites on limbs is to wrap the limb (except the plucked surgery site) with a nonadhesive bandage material such as Vetwrap.[1]

C. Preparing Skin
1. Povidone iodine solutions 1% (Betadine) or chlorhexidine diacetate 0.05% (Nolvasan) used to clean skin.
2. Final prep done with alcohol (see caution about radiosurgery below)
3. Avoid using soap solutions, if possible, because the soaps can interfere with oils used by birds to waterproof feathers.

D. Patient Position
   1. Lateral and ventrodorsal positions preferred for most procedures. Some authors believe that placing bird in dorsoventral position is contraindicated, because this position interferes with normal respiratory function.[1]

E. Surgical Drapes
   1. Use transparent lightweight drapes whenever possible. Transparent drapes allow surgeon and assistants to view respiratory movements of the avian patient.
   2. If opaque drape material is used, make sure it is lightweight (such as disposable paper drapes). It may be necessary to tent the drape over the thorax to allow the anesthetist to monitor respiratory movements.

IV. Instrumentation

A. Ophthalmic Instruments or Small, Delicate General Surgery Instruments
   1. Iris forceps or small toothed Brown-Adson forceps are used to grasp tissues.
   2. Eyelid retractors make excellent abdominal retractors.
   3. Ophthalmic needle holders or small 5½ inch Mayo-Haeger needle holders.
   4. Micro-mosquito hemostats
   5. Iris scissors
   6. Small and medium hemoclips for ligation of larger blood vessels that cannot be cauterized (see below) and for providing closure when removing the oviduct.

B. Electrosurgical (Radiosurgical) Instruments
   1. Electrosurgery is the use of electric current to perform surgical manipulations of tissues (primarily cutting or coagulating).
      a. Radiosurgery is the primary form of electrosurgery used in avian practice. Often the two terms are used interchangeably.
      b. Radiosurgery uses a high frequency power oscillator (a radio transmitter) operating at 0.5 to 7.5 MHz (3.8 MHz is ideal).[4]
      c. The radio waves emitted by the tip (handpiece electrode) of the instrument (Fig. 24-1) vaporize intracellular fluid at the point of contact with the tissues, thus separating (cutting) tissues.[5]
      d. In coagulation the high frequency electric current dehydrates cells, coagulating cellular contents, thus sealing small blood vessels.[5]
   2. Four types of current used in electrosurgery[4]
      a. Fully rectified current
         (1) Best for cutting skin and other soft tissues
         (2) Current produces cutting and coagulation in equal parts, to produce excellent cutting and hemostasis

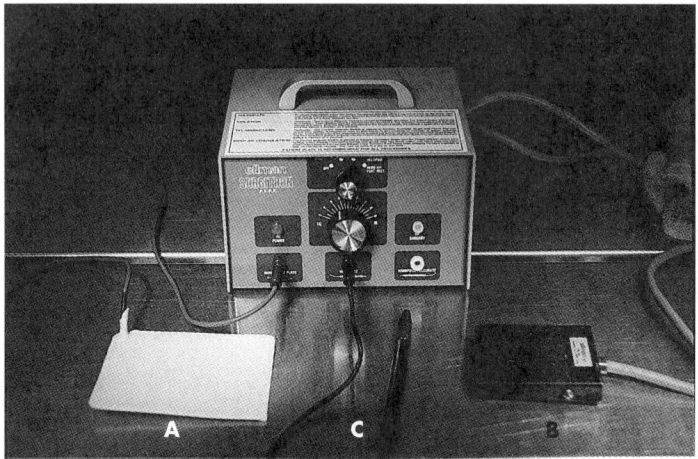

Fig. 24-1 Surgitron unit (Ellman International Mfg., Inc., Hewlett, NY 11557) with ground plate or indifferent electrode (**A**), foot activation switch (**B**), and handpiece or active elctrode (**C**).

    b. Partially rectified current
       (1) Used primarily for hemostasis
       (2) Current produces only 10% cutting while 90% coagulation
    c. Fully rectified filtered current
       (1) Current produces 90% cutting power, but only 10% coagulation
       (2) Most useful in avian surgery to take biopsies without destroying excessive tissues near cutting tip
    d. Fulgurating current
       (1) This current is used to destroy tissues and has little avian surgical application.
3. Cautions in the use of radiosurgery units
    a. Don't use if flammable gases (ether) or liquids (e.g., alcohol) are present. A spark could ignite a fire.
    b. Surgical site must be relatively dry. Remove fluids or blood from the surgical field with a sterile sponge before use.
    c. Certain pacemakers (especially unshielded models and those made before 1973) may be affected by the radiosurgery unit.
4. Proper radiosurgery techniques
    a. Tune the electrode to avoid sparking (which destroys excessive tissues) and drag (which shreds tissues).
    b. Place the ground plate as close to the surgery site as possible and have as much contact with the patient's body as practical.
    c. Use a fine wire electrode for most cutting in avian work.
    d. Depress the foot switch, starting the current, before placing the cutting electrode on the tissues. Do the opposite for coagulation, place the electrode on the tissue, then depress the foot switch.

e. Allow time for electrode and tissues to cool between cuts. Cut for 10 seconds, then allow 10 seconds for cooling before next cut.
f. Bipolar electrodes have both electrodes in the tip and don't require the ground plate. They are especially useful for coagulating small bleeding vessels.

5. Care of electrodes
   a. Clean electrodes by wiping on a wet gauze sponge while the unit is on.
   b. After each surgery, clean the electrode for 3 minutes in an ultrasound cleaner.[4]
   c. Gas or liquid (cold) sterilization is preferred. Autoclaving may damage insulation or materials used in handles, electrodes, or wires.
   d. Always use the instrument properly with the base plate, and ensure it is tuned for best performance.

C. Cotton-Tipped Applicators
1. Sterilized and included in the surgery pack.
2. Can be used to absorb small amounts of blood, or, after moistening, can be used as tissue elevators.

D. Magnification
1. An operating microscope (lens objective 150 mm, binocular objective 12.5 mm) has been recommended by some authors.[1] Disadvantages are expense of equipment, increased time required to perform surgery, and training required to use equipment.
2. Head-mounted ocular lamp with a light source is less expensive and useful without training for most avian procedures. A head mounted ocular loop and light offer more flexibility of movement and light direction than an operating microscope.

E. Suture Materials
1. Recommended sizes for general surgery are 3-0 to 6-0,[5] with ophthalmic sizes (up to 8-0) required for gastrointestinal resections.[6] One author recommends 5-0 suture for birds less than 30 g, 4-0 suture for birds less than 100 g, and 3-0 suture for birds 150 to 1500 g.[6]
   a. Psittacines, despite their powerful prehensile beaks, do not generally pull out sutures. However, they will groom the suture ends and groom feathers over the suture line. They usually do not tolerate bandages as well.[6]
   b. Atraumatic (taper point) needles will work well for suturing muscle, coelomic cavity organs, and skin on psittacines and most other small birds. Cutting-edge needles tend to tear and leave large holes in the avian integument.
2. Recommended types of suture material
   a. Several avian surgeons[1,5,6] prefer polyglactin suture (Vicryl, Ethicon, Inc., Somerville, NJ).

(1) Unlike in mammals where little reaction occurs, in birds polyglactin suture produces an intense inflammatory response.[6]
(2) However, the suture material has excellent tying qualities requiring only three throws for a non-slip square knot and is totally absorbed in 60 days.
(3) Because this suture material is braided multifilament, it is not recommended for infected surgical sites.
  b. Polydioxanone suture (PDS, Ethicon, Inc., Somerville, NJ)
(1) Creates minimal tissue reaction among the absorbable sutures
(2) Is a monofilament product and has the least adherence for bacteria[5]
(3) Requires 4 to 5 throws for a non-slip knot.
(4) Takes 120 days for degradation; sometimes used to attach small radio transmitters to wild birds.[8]
  c. Chromic catgut[1]
(1) Monofilament
(2) Produces marked inflammatory response
(3) Delayed absorption, over 120 days
(4) Poor tying quality, requires at least four throws[5]
  d. Nylon
(1) Nonabsorbable
(2) Monofilament
(3) Requires only three throws to secure knot
  e. Stainless steel wire
(1) Monofilament or multifilament
(2) Difficult to tie but will hold with two throws
(3) Limited use in avian patients
3. Cyanoacrylate tissue adhesives (CAA)
  a. Adhesive converts from liquid to solid in presence of water.
  b. Medical grades of CAA are inert; however, hardware grades contain toxic substances and can cause histologic changes.
  c. CAA will bind tissues together like suture to speed up healing process. However, CAA can form a barrier to wound healing if placed onto the open (debrided) would margins.
  d. Fumes have been noted to produce vomiting in some avian patients.

V. Tissue Healing

A. Stages
1. Stage 1[9]
  a. Hemorrhage controlled by vasoconstriction followed by vasodilation after 30 minutes.
  b. After 120 minutes, heterophils and monocytes begin invading wound margins.
  c. Shift to lymphocytes, monocytes, plasma cells, and macrophages after 12 hours.

d. Phagocytosis of bacteria and necrotic debris occurs.
2. Stage 2
   a. Begins 3 to 4 days after wound occurred.[9]
   b. Fibroblasts appear in late stage 1 and form microfibrils in stage 2.
   c. Microfibrils form fibers.
   d. Capillaries invade wound area, and epithelial cells start forming across wound surface from margins of the wound.[10]
   e. Stage 2 lasts 2 weeks.
3. Stage 3
   a. Fibroblasts decrease while thickened and stronger collagen fibers form.
   b. This stage is variable and can last several weeks to months.[1]

B. **Rate of Wound Healing**
   1. Determined by level of bacterial contamination and amount of tissue damage/necrosis.
   2. Aseptic surgical wounds heal fastest.
   3. Cleanse and debride all traumatic wounds.
   4. Suture wounds, if possible, to reduce healing time.
   5. Cover or protect wound initially if possible. (Tegaderm, 3M Animal Care Products, St. Paul, MN; Dermaheal or DuoDerm, Bristol-Myers Squibb Princeton, NJ)

C. **Suture Patterns**
   1. Keep suture patterns simple and use the least amount of suture material possible.
   2. Simple continuous pattern is quick and useful. One author prefers an interlocking continuous suture pattern for skin sutures.[6]
   3. Use an inverting interrupted suture pattern for hollow organs.
   4. Use an interrupted pattern in the skin if the bird is likely to pick at the area.

D. **Postoperative Care**
   1. Extubation should occur as soon as the swallowing reflex returns. Make sure mouth and trachea remain clear of fluid, mucus, and regurgitated food.
   2. During anesthetic recovery, keep patient in a warm incubator (85°F).
   3. Oxygen can be given in the incubator if the bird is slow in recovering.
   4. Correct for fluid, electrolyte, and glucose imbalances created by stress of anesthesia and surgery.
      a. All patients should be given at least a subcutaneous injection of balanced electrolytes[6] (lactated Ringer's solution).
      b. Bolus intravenous fluids can be given if surgery was prolonged or extensive blood or fluid loss occurred.
      c. Use glucose (dextrose) solutions if presurgical glucose levels were less than 200 mg/dl.
   5. Postoperative pain alleviation (see Chapter 26, p. 570)

VI. Selected Soft Tissue Surgery Procedures

A. Ingluviotomy

1. Common surgery to open crop to remove foreign bodies, food impactions, to pass endoscope into gizzard, or to repair crop rupture from burn, feeding tube puncture, etc. (see Chapter 13)
2. Preoperatively withhold food and water 3 hours
3. Anatomic location: crop is at base of neck on center and right sides.
4. Initial incision of skin parallel with neck; separate any underlying tissues, if any.
5. Enter crop making a small stab incision and extending it (usually make incision parallel with neck).
6. Take any cultures or biopsies immediately if needed before exploring interior of crop.
7. Closure: two layer
   a. First close the crop using a continuous inverting pattern.
   b. Close skin using a continuous suture pattern.
   c. For traumatic crop ruptures, interrupted inverting patterns may be preferred. Use 3-0 to 6-0 Vicryl (Ethicon, Somerville, NJ).[11]
8. Burns sometimes require 3 to 5 days before the full extent of tissue damage is apparent and debridement and suturing is possible. Supportive care by passing a soft feeding tube directly into the proventriculus may be required before surgery.
9. If the healing of the tissues is questionable, use a two-layer closure of the crop. The first layer is an interrupted inverting pattern, the second layer a simple interrupted pattern.[11]
10. Special postoperative care should include soft diet and antibiotics for 7 to 10 days.

B. Exploratory Laparotomy

1. Indications
   a. For egg-bound birds when medical intervention fails to deliver the egg
   b. To repair intussceptions, prolapses, etc.
   c. To retrieve a foreign body from the gastrointestinal tract, especially the ventriculus
   d. To explore or remove abdominal masses palpated or seen on radiographs
   e. In select cases when other noninvasive methods have not yielded a diagnosis
2. Several different approaches are used, depending on the organs to be examined and the size of the patient[2,12-14] (Fig. 24-2).
   a. Ventral midline incision extending from the posterior margin of the sternum to just before the cloacal opening; Useful for surgery in the caudal abdomen and cloacal areas.
   b. Transabdominal incision gives good access to either the anterior or posterior abdomen.

Soft Tissue Surgery **535**

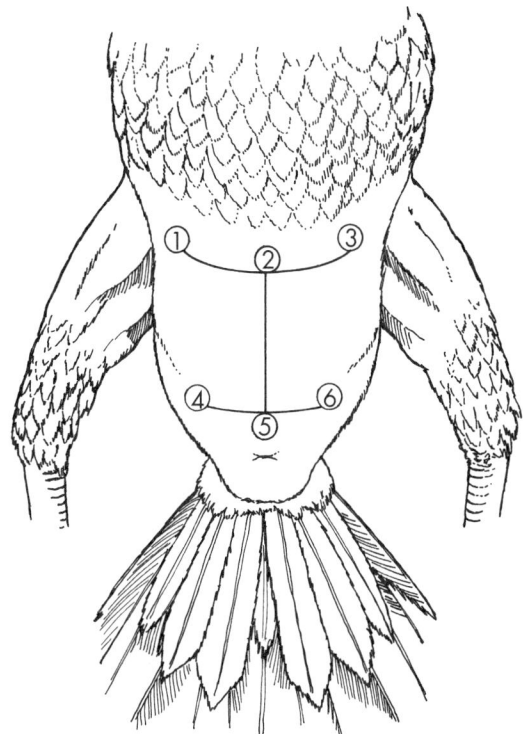

Fig. 24-2 Various approaches to an abdominal incision in the avian patient: standard ventral midline approach (2-5), ventral flap approach (1-2-5-4 or 3-2-5-6), double flap approach (1-2-5-4 plus 3-2 and 5-6, generally not recommended, see text), triangular approach (1-2-5 or 3-2-5, less often 4-5-2 or 6-5-2). Drawing based on Altman.[14]

    c. Partial or full flap incisions have the advantage of giving the greatest abdominal exposure at the expense of potentially weakening the abdominal wall. Exposure is also limited to either the left or right side of the abdomen, as a double full flap incision; even though it offers the best exposure, further weakens the abdominal wall.
3. Making the incision
    a. Tent the skin away from the abdominal musculature to avoid making the initial incision too deep. Make the incision using radiosurgery (above) or scalpel.
    b. Likewise, tent the abdominal musculature when making this incision to avoid damage to underlying organs. Make the incision using radiosurgery or scalpel.
    c. Alternately, a small stab incision can be carefully made into the abdomen, followed by inserting a small groove director in each direction and making the incision over the groove director.

d. Care must be taken in any case where the abdomen is swollen, because the abdominal viscera may be forced against the muscular wall.
  4. Orientation
     a. The right liver lobe is easily seen in some patients.
     b. Also on the right are the duodenal loop and the pancreas.
     c. On the left and anterior is the ventriculus.
     d. The left abdominal air sacs may have to be incised for better exposure. This will complicate gaseous anesthesia, requiring a higher setting or periodically holding the incision closed to prevent the patient from leaving a surgical anesthetic plane.[14]
  5. Closure
     a. A two-layer closure (abdominal muscle/wall and skin) is recommended.[6,14]
     b. Use a continuous suture pattern with absorbable suture material.

C. Proventriculotomy
  1. Indications
     a. Proventricular or ventricular foreign body not retrievable with endoscopy
     b. Food impaction
     c. Removal of heavy metal (e.g., lead, zinc)
     d. Biopsy to confirm psittacine proventricular dilatation syndrome
  2. Surgical approach[6,13,14]
     a. Oblique skin incision from the proximal end of the pubis to the uncinate process of the sixth rib on the left side.
     b. The incision goes through the seventh and eighth ribs. Watch for hemorrhage from intercostal vessels along the cranial border of each rib.
     c. Pull and hold the left leg posteriorly.
     d. Use mini-Balfour ophthalmic retractor to hold the incision open.
     e. Bluntly dissect the two proventricular suspensory ligaments.
     f. Elevate the proventriculus into the incision area. Sometimes it helps to place two stay sutures in the proventriculus, one at the proventricular-ventricular junction and the second as far forward as possible.[6] The external ends of the stay sutures are attached to hemostats.
     g. If the proventriculus is fragile or diseased, care must be taken to avoid damaging it when it is elevated or manipulated.
     h. The incision area is now packed off with gauze sponges soaked in saline solution, and the incision made into the proventriculus.
     i. The proventricular contents should be aspirated using a syringe.
     j. A rigid endoscope is useful for observing in the proventriculus and into the ventriculus.
  3. Closure
     a. Use a two-layer closure in the proventriculus.
        (1) Use 4-0 to 6-0 Vicryl or PDS
        (2) The first suture line should be an inverting continuous pattern.

(3) The second suture line should be a simple, continuous layer.
(4) Some authors[14,15] recommend that the first suture line be simple continuous and the second inverting simple continuous or interrupted.
(5) Pass a soft feeding tube into the proventriculus and fill the proventriculus with warm saline to check for leaks at the suture line.

b. The abdominal muscle wall is closed with absorbable suture in a simple continuous suture pattern with one or two simple interrupted sutures around the ribs to help pull the incision together.
c. The skin is closed with absorbable suture in a simple continuous suture pattern.

4. Postoperative care
   a. Patients are given a broad-spectrum antibiotic for 14 days.[6]
   b. Allow the avian patient to eat small amounts of food as soon as recovery from anesthesia is achieved.
   c. Complications, especially leakage around the sutures in the proventriculus, are most likely 3 to 5 days postoperatively.[14]

D. Cloacapexy

1. Etiology of cloacal prolapses is unknown.
   a. Seen most frequently in cockatoos[14,16]
   b. Chronic enteritis resulting from gram-negative bacteria may be a precursor.[17]
   c. Cloacal attachments may be torn.[16,18]
   d. Severe damage to the cloacal nerve supply is occasionally a problem.[14,16]

2. Vent suturing
   a. Minor prolapses sometimes respond to gently manipulating the prolapsed tissues back into the cloaca and placing a purse-string suture around the cloacal opening.
      (1) 3-0 to 5-0 nylon suture material on a fine needle
      (2) Place a feeding tube in the cloacal to maintain an opening.
      (3) Take small shallow bites of skin around the cloacal opening (usually 5 to 7).
      (4) Tie the knot at the end evenly, and remove the feeding tube. Make sure the bird is able to defecate.
   b. An alternative approach to vent suturing, after returning the prolapsed tissues into the cloaca, is to place two simple interrupted sutures across the opening, about a third of the distance from each side.[19]
   c. Vent suturing can be a temporary measure to allow time to stabilize the patient for cloacapexy.

3. Surgical procedure for cloacapexy
   a. Begin with a midline ventral abdominal incision as described above (laparotomy).
   b. Both caudal abdominal air sacs are incised as the cloacal area is exposed.

c. The cloaca is gently pushed forward with a gloved finger or a sterile cotton-tipped applicator in the orifice.
d. If an adipose pad is attached to the cloaca, it is gently dissected away.
e. Stay sutures of 3-0 Vicryl or PDS are placed through the cloaca penetrating the mucosa. The sutures are placed at 10 and 2 o'clock.
f. The stay sutures are then passed around the posterior rib on each side.
g. The knots are then tied creating enough tension to slightly invert the vent opening.[16]
h. Closure follows that described above (laparotomy).
4. Modifications of the surgical procedure
   a. Stay sutures around pubis[14]
      (1) Start with a transverse abdominal approach about halfway between the vent opening and the caudal border of the sternum.
      (2) After pushing the cloaca forward with the gloved finger or cotton-tipped applicator, place two stay sutures on each side of the cloaca, anchoring these around the pubis bone on each side.
      (3) Incise the cloacal serosa at the anterior aspect of the cloaca.
      (4) Include the incision site in the closure of the abdominal wall.
   b. Lateral body wall technique[20]
      (1) Start with a midline ventral incision.
      (2) Make two 5-mm incisions in the cloacal serosa 10 mm on either side of the midline.
      (3) Make two similar incisions in the peritoneal body wall on either side of the abdominal incision.
      (4) Suture the cloacal serosa to the incision in the body wall on each side using Vicryl or PDS.
      (5) Close the transverse abdominal incision in the manner described above (see exploratory laparotomy above).

E. **Hysterectomy (Salpingohysterectomy)**
1. Indications
   a. Egg binding, especially if uterus is torn, severely prolapsed, or otherwise severely damaged
   b. Chronic egg laying (first try other, less invasive methods of controlling chronic egg laying)
   c. Pathologic conditions in the uterus including severe infections, pyometra-like conditions,[14] and neoplasias
2. Anatomy of the female reproductive tract
   a. Ovary
   b. Infundibulum
   c. Oviduct
   d. Uterus including the shell gland
   e. Vagina/vaginal cloacal orifice
   f. Oviduct has ventral ligament (convoluted, avascular, with a muscular attachment near caudal end) and dorsal ligament (vascular, with branch from the ovarian artery).[21]

3. Surgical approaches
   a. Midline ventral abdominal approach: allows access to both sides of the coelomic cavity but difficult to reach the ovary or oviduct.
   b. Left lateral approach: best exposure of the reproductive tract.
4. Procedure
   a. Locate the oviduct and follow it to the ovary, locating the infundibulum.
   b. The dorsal suspensory ligament is located and ligated (it contains a branch of the ovarian artery).[14]
   c. The ligament and oviduct are cut close to the ovary.
   d. The oviduct/uterus is followed to the vagina.
   e. Ligate the uterus at the vagina, close to the cloaca.
   f. Remove the uterus, cutting the ventral ligament attachments.
5. The ovary is usually not removed.
   a. The ovary is difficult to remove because of its close adherence to the posterior vena cava, aorta, and pelvic nerves.[21]
   b. Ovary usually does not develop and release yolk portion of eggs if the uterus is removed. It is necessary to remove all of the shell gland portion of the uterus to ensure this.
6. Closure
   a. Standard abdominal closure as described above is used (see exploratory laparotomy above).

F. **Sinus Surgery**
   1. Sinus infections often respond to injections or flushes of antibiotics and saline (antibiotic choice based on culture and sensitivity).
      a. Injection site for the infraorbital sinus is ventral to the midpoint of a line between the nares and the medial commissure of the eye. The needle enters the sinus cavity after traveling under the zygomatic arch.
      b. The antibiotic-saline mixture will spread to sinus areas surrounding the eye.
   2. Sinus trephination for the treatment of infections of the supraorbital sinuses
      a. Useful technique for chronic sinusitis or any time purulent material is suspected in the supraorbital sinuses.
      b. Procedure
         (1) Under general anesthesia, remove the feathers from an area over the supraorbital sinuses (dorsal to a line between the medial commissure of the eye and the nares).
         (2) Apply protectant ointment to the eye, then prepare the area with 1% povidone iodine (Betadine).
         (3) Make a skin incision using a radiosurgery unit to control hemorrhage. Expose the underlying bone.
         (4) After the bone is exposed, enter the sinus cavity using a sterile drill bit angling it toward the midline.
         (5) Take swabs for Gram's stain and cultures immediately after entering the sinus space.

(6) Remove any consolidated material, then flush the sinuses with a 1:10 gentamicin-saline solution. The flush solution should drain from the choanal slit.

c. Postsurgical treatment

(1) Flush sinuses twice daily with gentamicin-saline (1:10) or other appropriate antibiotics identified from culture and sensitivity testing. Some authors[16] recommend infusing ophthalmic antibiotic ointment into the sinuses after flushing.

(2) Once the infection is under control and the flushing is stopped, the openings created into the sinus cavities in the trephination process will heal rapidly. Therefore, no specific closure is required.

REFERENCES

1. Altman RB: General surgical considerations. In Altman RB, Clubb SL, Dorrenstein GM, Quesenberry K, editors: *Avian medicine and surgery*, Philadelphia, 1997, WB Saunders, pp 691-703.
2. Bennett RA: Surgical considerations. In Ritchie BW, Harrison GJ, Harrison LR, editors: *Avian medicine: principles and applications*, Lake Worth, Fla, 1994, Wingers, pp 1081-1095.
3. Harrison GJ: Evaluation and support of the surgical patient. In Harrison GH, Harrison LR, editors: *Clinical avian medicine and surgery*, Philadelphia, 1986, WB Saunders, pp 543-549.
4. Altman RB: Electrosurgery. In *Manual of avian laboratory procedures*, Lake Worth, Fla, 1991, Association of Avian Veterinarians Education Office, pp 49-62.
5. Bennet RA: 1993. Instrumentation, preparation, and suture materials for avian surgery, *Semin Avian Exotic Pet Med* 2:62-68, 1993.
6. McCluggage DM: *Surgical principles and common procedures in the avian patient*, Lake Worth, Fla, 1992, Association of Avian Veterinarians Education Office, S4:1-15.
7. Bennett RA, Yeager M, Trapp A, Cambre RC: Tissue reaction to five suture materials in pigeons, *Proc Assoc Avian Vet*, 1992, pp 212-218.
8. Pietz, PJ, Brandt DA, Krapu GL, Buhl, DA: Modified transmitter attachment method for adult ducks, *J Field Ornithol* 66:408-417, 1995.
9. Carlson HC, Allen JR: The acute inflammatory reaction in chicken skin: blood and cellular response, *Avian Dis* 13:817-833, 1969.
10. Degernes LA: Soft tissue wound management in avian patients, *Proc Assoc Avian Vet*, 1992, pp 476-483.
11. Van Sand F: Surgery of the avian gastrointestinal tract, *Semin Avian Exotic Pet Med* 2:91-96, 1993.
12. Harrison GJ: Selected surgical procedures. In Harrison GJ, Harrison LR, editors: *Clinical avian medicine and surgery*, Philadelphia, 1986, WB Saunders, p 831.
13. Jenkins JR: Approaches to the abdominal cavity, *Association of Avian Veterinarians Surgical Symposium*, 1994, pp 20-23.
14. Altman RB: Soft tissue surgical procedures. In Altman RB, Clubb SL, Dorrenstein GM, Quesenberry K, editors: *Avian medicine and surgery*, Philadelphia, 1997, WB Saunders, pp 704-732.
15. McCluggage D: Proventriculotomy: a study of select cases, *Proc Assoc Avian Vet*, 1992, pp 195-198.
16. Rosskopf WJ et al: Pet surgical procedures: an overview. In *Introduction to avian medicine and surgery*, Lake Worth, Fla, 1992, Association of Avian Veterinarians Education Office, S3:1-12.

17. Avgeris S, Rigg D: Cloacapexy in a sulphur-crested cockatoo, *J Am Anim Hosp Assoc* 24:407-410, 1988.
18. Rosskopf WJ, Woerpl RL: Cloacal conditions in pet birds with a cloaca-pexy update, *Proc Assoc Avian Vet,* 1989, pp 156-163.
19. Bennett RA: Soft tissue surgery. In Ritchie BW, Harrison BJ, Harrison LR, editors: *Avian medicine: principles and application,* Lake Worth, Fla, 1994, Wingers, pp 1097-1136.
20. Jenkins JR: Avian soft tissue surgery: part I, *Proc Amer Coll Vet Surg* 1992, pp 631-633.
21. Bennett RA: Techniques in avian thoracoabdominal surgery, *Assoc Avian Vet Ann Meet,* Reno, Nev, 1994, pp 45-57.

25

Necropsy

Robert E. Schmidt

I. General Considerations

A. Organization
1. The examiner must be thorough and consistent.
2. Each necropsy should be conducted in the same manner and order.
3. Each organ or tissue is evaluated and described as the necropsy proceeds.
4. A consistent method ensures that no organ or tissue is overlooked during the procedure.

B. Tissue Preservation
1. All tissues are saved, even if not all will be submitted for histopathologic study.
   a. Changes (or lack of changes) in the organs examined may indicate others that should be studied.
2. The potential for legal action always exists, and a complete examination may be necessary if a lawsuit is contemplated.

C. Samples
1. The samples needed must be taken for microbiologic and toxicologic studies, as well as for other tests.
2. Samples for microbiologic examination should be obtained before manipulation of the carcass.
3. Appropriate samples for toxicologic studies may need to be frozen.

D. Fixatives
1. The proper fixatives must be used.
2. The best all-around fixative is 10% neutral *buffered* formalin.
3. Any necessary fixative can be obtained from a pathology laboratory.

E. Instruments
   1. The examiner must have sufficient instruments for the necropsy.
      a. Minimal equipment would include an autopsy scalpel, scissors, forceps, and equipment to cut bone.

F. Documentation
   1. Changes must be documented, and the examiner should remember this important rule: Describe, do not diagnose.
   2. Parameters should include size, shape, color, consistency, and smell.
   3. The examiner should quantitate when possible.
   4. Redundancy should be avoided (e.g., "brown in color").

G. Possible General Changes
   1. The following general changes may occur in all organs and tissues:
      a. Relative lack or excess of blood
         (1) The amount contributes to the size, color, and consistency of an organ.
      b. Color changes
         (1) These changes may occur before or after death. With practice, the differences become obvious and should be noted.
      c. Consistency
         (1) Consistency can be affected by the postmortem condition and by antemortem changes, including cellular infiltrations and connective tissue proliferation.
      d. Tissue loss
         (1) Tissue loss may result in symmetric or asymmetric changes in organ size and weight. These changes may be indicative of generalized disease or localized problems with blood supply.
      e. Abnormal excess of tissue
         (1) This condition may be due to hypertrophy, hyperplasia, or neoplasia.
         (2) The effect may be symmetric or asymmetric.

II. Systematic Review

A. Integument
   1. Feathers grow in tracts with areas of nonfeathered skin between. The shape and general appearance are similar across species.
   2. It is more difficult to remove feathers from waterfowl than from the usual pet species. With increasing time postmortem, feathers become easier to remove. Ripping of the skin when feathers are removed usually is a sign of advanced autolysis.
   3. Species differences are obvious in the shape of the beak, claws, comb, wattle, and other characteristics.
   4. The examiner should check for signs of trauma, such as bruises and lacerations, as well as for "stress" marks in feathers.
   5. The carcass is examined for lumps and areas of thickening.

a. If a lesion appears to be neoplastic, all internal organs must be checked for metastasis.
   b. Diffuse thickening of the skin may indicate chronic inflammation but also can be seen with cutaneous lymphosarcoma.
6. Poor feathering and loss of feathers have several common causes.
   a. Specific diseases
      (1) Circovirus infection (psittacine beak and feather disease)
      (2) Polyomavirus infection
      (3) Bacterial or mycotic folliculitis
   b. Nonspecific infections
      (1) Organisms that are not specific for avian species but that may cause problems because of immunosuppression or contact with a large number of organisms
   c. Hormonal and metabolic causes
      (1) Endocrinopathic conditions, such as hypothyroidism (confirmatory testing is difficult in avians)
      (2) Nutritional problems, including a deficiency or an excessive amount of fat or of some other dietary component
      (3) Toxins that may lead to feather problems by direct action or as a result of internal organ damage
   d. Allergy
      (1) Food allergies
      (2) External parasites
      (3) Atopy
   e. Behavioral causes
      (1) Must be confirmed or ruled out by an accurate history
   f. Internal disease
      (1) Often is associated with liver or pancreatic problems, but chronic disease in any organ may lead to skin changes

B. Musculoskeletal System
1. The normal color of muscle in most pet species is red-brown. Pale areas and streaks can indicate degeneration or inflammation, or both.
2. The pipping muscle in some hatchlings is markedly hypertrophied; this should not be considered an abnormality.
3. The size and shape of the pectoral muscles are indicators of the bird's general condition and nutritional status.
4. Congenital malformations of the leg or wing probably are due to problems of skeletal development. Beak abnormalities may be due to underlying bone problems.
5. Specific diseases of the musculoskeletal system
   a. Polyomavirus infection
      (1) Polyomavirus infection can cause hemorrhage in skeletal muscle.
   b. *Sarcocystis* infection
      (1) This infection may lead to foci of necrosis and some hemorrhage.

c. Vitamin E deficiency
   (1) Gross changes include pallor and gray-white streaks.
d. Parasitism
   (1) Encysted parasites appear as focal white areas that may be mineralized.
e. Rickets and osteomalacia
   (1) These conditions produce bones that are soft and that bend rather than break; various deformities can be seen grossly.
f. Trauma
   (1) Presenting lesions include hemorrhage, tearing of muscle bundles, and fractures.
6. Neoplasia is unusual in the typical pet species.
   a. The most common neoplasm is lymphosarcoma, which can be multicentric and can infiltrate skeletal muscle from adjacent tissue.

C. Cardiovascular System
1. The pericardial sac should be translucent and have minimal pericardial fluid.
2. The heart usually is roughly conical in shape.
   a. Abnormalities include dilation (flabby heart) and thickening of the myocardium.
   b. Thickening may be due to hypertrophic cardiomyopathy, which may be a gross diagnosis with no specific histologic abnormality.
   c. The left A-V valve is membranous, but the right A-V valve is a muscular flap.
3. Color changes include pallor (degeneration or inflammation), reddening (hemorrhage), and yellowing (often associated with atherosclerosis).
   a. With atherosclerosis, the blood vessels of the heart may be more rigid than normal.
   b. The extent of myocardial involvement in color changes is important; some surface changes may be agonal or may occur after death.
4. The heart should be opened in the path of the blood flow if possible. Complete hemisection often is done in small birds. Major blood vessels should be opened and checked for plaques, either visually or by touch.
5. In cases of sudden death, the heart is one of the important organs and should be examined microscopically.
   a. Specific areas must be examined, and the whole heart should be sent to the pathologist if possible.
6. Fat should not be mistaken for lesions.
7. Specific diseases of the cardiovascular system
   a. Viral infections: Gross lesions include hemorrhage, pale streaks, and foci. Common causes include:
      (1) Polyomavirus
      (2) Herpesvirus (Pacheco's disease)

(3) Proventricular dilation syndrome
   b. Localization of other systemic infection: Adhesions of the pericardium may be present, as well as purulent or fibrinous exudate and pale foci or abscess formation in the myocardium. Common causes include:
      (1) Bacterial infections
      (2) Systemic protozoal infections (e.g., *Sarcocystis* organisms)
   c. Nutritional myopathy: This disorder usually manifests as areas of pale streaking in the myocardium. No exudate should be seen.
   d. Metabolic lesions
      (1) Gout: Gross changes include deposition of white material in or on the pericardium and epicardium. This material, made up of urates, may resemble exudate, and infection must be ruled out.
      (2) Atherosclerosis: This condition manifests grossly as yellow foci or streaks that usually are associated with medium or large intramural arteries.

D. Alimentary System (Including the Liver and Pancreas)
1. The entire alimentary tract must be examined, including the tongue, oral cavity, and cloaca.
2. It is important to remember species variations in the size and shape of the crop and ventriculus, particularly. Some species have cecae.
3. Color changes may be due to postmortem autolysis. Gastrointestinal disease should not be overdiagnosed. The intestinal contents become progressively more bile stained in the distal portion of the tract.
4. Hemorrhage must be differentiated from severe congestion (this may not be possible grossly).
5. All segments should be opened and checked for parasites, unusual contents, and other abnormalities.
6. Specific diseases of the intestinal system
   a. Bacterial, mycotic, and parasitic infections
      (1) Necrosis and exudate are present, and parasite fragments may be found.
   b. Viral enteritis (herpesvirus, reovirus)
      (1) Necrosis and hemorrhage may be seen, although the condition usually does not produce exudate.
   c. Proventricular dilation syndrome
      (1) This is a viral infection that leads to dilation and thinning of the gastrointestinal tract.
      (2) The proventriculus and ventriculus are most affected.
      (3) The small intestine may be affected, but the large intestine rarely shows gross changes.
   d. Neoplasia
      (1) The most common tumor is a proventricular carcinoma.
         (a) The tumor usually is seen at the junction with the ventriculus; it may be flat and difficult to see grossly, and it may have adhered to the peritoneum or liver.

7. Species vary in the size and shape of the liver and gallbladder, and many birds do not have a gallbladder.
8. "Rounded" edges on the liver are not always an indicator of disease.
9. Color changes and consistency are important; autolysis must be differentiated from antemortem lesions.
10. The gross changes caused by a variety of diseases, including viral and bacterial infections and lymphosarcoma, are similar and cannot be distinguished on gross examination. Primary changes include swelling and mottling or discoloration of the liver. Necrotic foci and cellular infiltration may appear similar grossly.
11. Specific diseases of the liver
    a. Chlamydiosis
    b. Bacterial infections
    c. Viral infection (herpesvirus, polyomavirus, and adenovirus)
    d. Metabolic and nutritional disease
    e. Toxic causes
    f. Parasitic infection
    g. Neoplastic conditions
12. The pancreas should fill the duodenal loop; it usually is cream colored or slightly pink. Nodularity or a granular appearance is abnormal.
13. If there is any suspicion of diabetes mellitus, the pancreas must be sampled for histopathologic evidence, because no gross change is seen.
14. Specific diseases of the pancreas
    a. Viral infection (herpesvirus, polyomavirus)
    b. Protozoal infections (in some species)
    c. Acute pancreatic necrosis (most common in Quaker parakeets but also seen in other psittacines; grossly, hemorrhage and discoloration are seen)
    d. Zinc toxicity (may lead to pancreatic degeneration; grossly, discoloration and irregularity of the tissue may be seen)

E. **Hematopoietic and Lymphoid Tissue**
   1. In young birds, hematopoiesis can be seen in the kidney, liver, and spleen, as well as in bone marrow.
      a. The condition may manifest as multifocal to confluent red foci in the liver and kidney.
   2. The primary lymphoid organs are the thymus and the bursa of Fabricius. The examiner should know the location of these organs and be able to sample them even if they are not visible grossly.
      a. In psittacines, thymic tissue is present in the neck from the angle of the mandible to the thoracic inlet.
      b. In most psittacines the bursa of Fabricius is in the dorsal portion of the cloacal wall, slightly to the left of the midline; it is round to oval. In ratites the bursa is composed of papillary structures and is more difficult to see grossly.

3. Long bones can be broken and marrow placed in formalin for histologic examination. Marrow smears also can be made at this time.
4. The spleen is a sensitive indicator of primary and secondary disease in birds. Color changes, enlargement, and atrophy all are seen in various conditions.
5. Specific diseases of hematopoietic and lymphoid tissue
    a. All infectious diseases can affect the spleen as a primary or secondary process. The spleen usually will be enlarged, discolored, and friable.
    b. Viral infections may destroy the bursa or the thymus (or both) in young birds. Histologically, viral inclusion bodies may be present only in the bursa on routine light microscopy.
    c. Lymphoid and myeloid neoplasms usually manifest as enlargement of the affected organ. They also can appear as infiltrative neoplasms in a variety of organs and tissues.
    d. Metabolic diseases ("fatty" spleen) may result in enlargement and discoloration of affected tissue.

F. **Respiratory System**
1. Species differ in the size and shape of the trachea and syrinx.
2. The upper respiratory system (nasal passages and sinuses) must be checked; the pneumatized air spaces in the bones communicate with the respiratory system.
    a. Infections can lead to accumulation of exudate.
    b. Neoplasia may appear as solid, space-occupying masses in the nasal passages or sinuses.
3. An internal examination of the trachea and bronchi should be done.
    a. Plugs can occur because of tumors, inhalation of foreign material, and infection, particularly mycotic infection.
4. The lungs should be removed and all surfaces examined.
5. Careful attention is paid to the air sacs; they should be translucent but opacify with age. They can be examined histologically or with impression smears in some cases.
6. Specific diseases of the respiratory system
    a. Viral, bacterial, mycotic, protozoan, and parasitic infections
        (1) The gross appearance is similar for these infections, and histologic examination and culture usually are necessary for etiologic diagnosis.
    b. Inhalation pneumonia
        (1) Foreign material may be found in the trachea or bronchi
    c. Endogenous lipid pneumonia
    d. Inhalation toxicity
        (1) A classic example is Teflon toxicity, but a variety of substances can cause illness and death in birds. Problems have been reported with aerosol sprays, stoves and heaters, and new furniture and carpets.
        (2) The avian respiratory system is very sensitive to irritation, and birds have problems in environments that do not obviously affect mammals.

(3) The common gross appearance of inhalation toxicity is a red, wet lung.
   e. Hypersensitivity
      (1) This is a common problem in blue and gold macaws or other birds kept in indoor environments. Birds may be allergic to other birds, to bedding material in the cage, and to fungi, bacteria, or endotoxin fragments that are not otherwise pathogenic in the amounts present.
      (2) The gross appearance is similar to that caused by inhalation toxicity, and histologic examination may be necessary to differentiate these problems.

G. Urinary System
1. The kidneys have three main lobes and usually are red-brown or yellow-brown. The cranial lobe of the kidney usually is the largest and often is mistakenly considered abnormal grossly even when no lesion is present and the kidneys are histologically normal.
2. In young birds the kidney often is actively hematopoietic, although this varies with the species. Among the common pet psittacines, the condition is most prominent in African greys.
3. Color and consistency changes can occur antemortem or postmortem.
4. Specific diseases of the urinary system
   a. Viral infections, including polyomavirus, herpesvirus, and adenovirus
   b. Bacterial infections
   c. Renal *Coccidia* organisms (seen in some species, particularly waterfowl; other types of protozoa also may affect the kidney)
   d. Nutritional and metabolic disorders, including nephrotoxins such as excessive vitamin D, calcium and phosphorus imbalances, and dehydration
      (1) The common gross appearance with any of these conditions is paleness of the renal parenchyma and multiple yellow-white foci or streaks.

H. Endocrine System
1. The pituitary gland is located at the base of the brain; it may become neoplastic, particularly in budgerigars.
   a. Distortion or displacement of the brain may result.
2. The thyroid and parathyroid glands are found cranial to the thoracic inlet and are associated with the carotid arteries.
   a. The syringeal muscle is almost the same color as the thyroid glands and often is mistaken for the thyroid. The thyroid and associated parathyroid glands are slightly more cranial and lateral in the neck.
3. The adrenal glands are located at the cranial pole of the kidney. Any enlargement in this area must be carefully localized as to whether it originates in the cranial kidney, the gonad, or the adrenal gland.
4. Specific diseases of the endocrine system
   a. Thyroid dysplasia: The thyroids may be small and discolored.

b. Thyroid hyperplasia: Bilateral and symmetric enlargement of the glands usually is seen.
  c. Parathyroid hyperplasia: The cause may be a nutritional deficiency or the condition may develop secondary to renal disease.
  d. Adrenal adenitis: This condition is seen in birds with proventricular dilation disease; the glands are variably enlarged and discolored.
  e. Adrenal degeneration: This condition is most common in African greys; the affected glands usually are small and pale.
  f. Pancreatitis may lead to necrosis of the islets of Langerhans; the pancreas may be necrotic or hemorrhagic, or both. Birds with diabetes mellitus caused by specific islet cell degeneration usually have no gross lesion.
  g. Neoplasia of any of the endocrine glands usually manifests as asymmetric enlargement.

I. Reproductive System
1. In females the right ovary and oviduct are vestigial.
2. In some species the gonads may be pigmented.
3. In seasonal breeders the testicles may be greatly enlarged.
4. In females the oviduct is flat when inactive but enlarges and becomes tortuous after ovulation. If no ovum is in transit, enlargement and dilation may indicate inflammation.
5. Specific diseases of the reproductive system
   a. Oophoritis: The ovary may be enlarged and discolored, and granulomas can be seen in some infections.
   b. Salpingitis: Exudate may be present in the oviduct, and if the oviduct is perforated there may be associated peritonitis. Common causes include infection and yolk-induced inflammation.
   c. Egg binding or dystocia: These conditions are obvious when an egg or a portion of an egg is present in the oviduct or uterus.
   d. Neoplasia
      (1) Tumors are ovarian or oviductal in origin in females. When they are extensive, it may be difficult to determine the exact origin.
      (2) In males testicular neoplasia must be differentiated from adrenal gland and cranial renal tumors.
      (3) Common testicular tumors include seminomas, Sertoli's cell tumors, and interstitial cell adenomas. Grossly, they cannot be differentiated.

J. Central and Peripheral Nervous Systems
1. In young birds congenital malformations and hydrocephalus may be causes of neurologic signs.
2. Serious disease may be present with little or no gross change.
3. Specific diseases of the central and peripheral nervous systems
   a. Viral encephalitis: Paramyxovirus, polyomavirus, adenovirus, and proventricular dilation disease all can cause clinical signs;

usually no gross lesions are seen. Histologic examination is necessary to confirm and possibly to identify the exact cause.
   b. Bacterial meningoencephalitis: Grossly, exudate may be present in the meninges, particularly on the basilar surface of the brain. Small abscesses can form in the parenchyma.
   c. Toxicity: Usually no gross change is apparent, and there may be minimal histologic lesions.
   d. Neoplasia: Central and peripheral tumors are very uncommon. There may be distortion or obvious tumor formation in the brain, but histologic examination is necessary for exact identification. Pituitary adenomas must be differentiated.
   e. Inflammation of ganglia and peripheral nerves: This condition is seen in proventricular dilation disease. Gross lesions usually involve the gastrointestinal tract, and thinning and dilation of any or all portions of the tract are possible. Histologic examination of tissues is necessary for confirmation.

K. Special Senses
   1. Eye
      a. If the bird is blind or has an obvious ocular lesion, the eye should be removed and examined histologically.
         (1) All extraneous tissue and extraocular muscles should be removed before fixing.
      b. Specific diseases of the eye
         (1) Viral: Poxvirus may cause proliferative lesions on the eyelids. It can also cause acute conjunctivitis without evidence of typical pox lesions on the lids. Viruses that can cause central nervous system lesions potentially can cause retinitis.
         (2) Bacterial: A variety of organisms can cause inflammation of all or part of the eye. Chronic granulomatous inflammation may be associated with generalized mycobacterial infections.
         (3) Mycotic: *Aspergillus* organisms are a common cause of severe keratitis or conjunctivitis.
         (4) Protozoal: Some protozoal infections can cause; the gross lesions are not specific.
         (5) Parasitic: Nematodes may be present in the conjunctival sac or within the eye. Trematodes are a cause of conjunctivitis in waterfowl; they may present as multiple swollen foci in the conjunctiva.
         (6) Neoplasms: Primary tumors of the eye include medulloepithelioma, lymphosarcoma, and melanoma. With the exception of possible pigmentation in melanomas, a specific diagnosis cannot be made grossly. Neoplasms may also be confused with chronic inflammation on gross examination.
   2. Ear
      a. Otitis externa may be grossly obvious, but the inner ear is rarely examined. With vestibular signs, the middle and inner ears should be examined histologically.

b. Specific diseases of the ear
      (1) Viral: Paramyxovirus can cause otitis interna, particularly in finches and *Neophema* sp. Gross lesions are not seen. The head should be fixed in formalin and sent to a pathologist for decalcification and histologic examination of the inner ears.
      (2) Bacterial or fungal: These infections usually are a cause of otitis externa and produce obvious signs of inflammation, including redness and exudation.
      (3) Neoplastic: Carcinoma of the glands of the ear canal occasionally is seen; it usually manifests as a swelling and must be differentiated from chronic infection.

SUGGESTED READING

Latimer KS, Rakich PM: Necropsy examination. In Ritchie BW, Harrison GJ, Harrison LR, editors: *Avian medicine: principles and applications,* Lake Worth, Fla, 1994, Wingers, pp 355-381.

26

Formulary

David J. Rupiper, James W. Carpenter, and Ted Y. Mashima

I. Background

A. Disclaimer
1. Despite intensive efforts to ensure complete accuracy, errors in the original sources or in the preparation of this formulary may have occurred.
2. All users of this reference, therefore, should empirically evaluate all dosages to determine that they are reasonable before use.
3. The authors assume no responsibility for and make no warranty with respect to results obtained from the uses, procedures, or dosages listed, or for any misstatement or error, negligent or otherwise, contained in this formulary.
4. The listing of a drug or commercial product in this book does not indicate approval by the FDA or the manufacturer for use in exotic animals. The following dosages have been specifically selected for companion birds, aviary birds, pigeons, backyard gamebirds, waterfowl, and raptors.
5. Most of the agents and dosages listed here are adapted from Carpenter JW, Mashima TY, Rupiper DJ: *Exotic Animal Formulary,* Manhattan, Kan, 1996, Greystone Publ., 1201 Greystone Place.

B. Abbreviations
d	day
h, hr	hour
IA	intraarticular
ICe	intracoelomic
IM	intramuscularly
IO	intraosseous
IT	intratracheally
IV	intravenously
IU	international units
kg	kilogram
L	liter
LRS	lactated Ringer's solution

mg	milligram
min	minute
mon	month
PD	pharmacokinetic data
PO	orally
prn	as needed
q	every
SC	subcutaneously
wk	week

C. Tables

Table 26-1.	Agents used in emergencies
Table 26-2.	Antimicrobial agents
Box 26-1.	Antimycobacterial agents
Table 26-3.	Antifungal agents
Table 26-4.	Antiviral and immunomodulating agents
Table 26-5.	Antiparasitic agents
Table 26-6.	Chemical restraint/anesthetic/analgesic agents
Table 26-7.	Hormones and steroids
Table 26-8.	Nonsteroidal antiinflammatory agents
Table 26-9.	Nebulizing agents
Table 26-10.	Chelating and related agents
Table 26-11.	Psychotropic agents
Table 26-12.	Nutritional/mineral agents
Table 26-13.	Miscellaneous agents

Table 26-1
Agents used in emergencies

Agent	Dosage	Species/Comments
Atropine	0.5 mg/kg IM, IV, IO, IT[1]	CPR
Calcium gluconate	50–100 mg/kg IV slowly,[2] IM (diluted)[3]	Hypocalcemia
Dexamethasone Na phosphate	2–4 mg/kg IM, IV q12h[1]	Head trauma (until signs abate); shock (one dose); hyperthermia (until stable)
	4 mg/kg IM, SC[4]	Psittacine neonates/shock
Dextrose (50%)	1 ml/kg IV slowly[1]	Hypoglycemia; can dilute with fluids
Diazepam	0.5–1.0 mg/kg IM, IV prn[5]	Sedation or seizures
Doxapram (Dopram-V, Aveco)	20 mg/kg IM, IV, IO[1]	CPR
Epinephrine (1:1,000)	0.5–1.0 ml/kg IM, IV, IO, IT[1]	CPR
Fluids	0.025 × wt (g) = ml fluids IV, IO, SC	Recommended fluids: LRS, Normosol (Abbott), 0.9% NaCl, 2½ % dextrose
Prednisolone Na succinate	10–20 mg/kg IV, IM q15min prn[6]	Head trauma; CPR
Sodium bicarbonate	1–4 meq/kg IV q15-30min to maximum of 4 meq/kg total dose[7]	Severe metabolic acidosis
	5 meq/kg IV, IO once[1]	CPR

Table 26-2
Antimicrobial agents

Agent	Dosage	Species/Comments
Amikacin	—	Maintain hydration during use; frequently used with penicillins or cephalosporins
	10-15 mg/kg IM q12h[8]	Amazon parrots, cockatiels, cockatoos/PD
	10-20 mg/kg IM, IV q8-12h[9]	African grey parrots/PD
	15-20 mg/kg IM q8-12h[10]	Cockatiels/PD
	15 mg/kg IM, SC q12h[11]	Most species
	15–20 mg/kg IM, SC, IV q12h × 5 days maximum[5]	Pigeons
	15–20 mg/kg IM q24h[12]	Raptors
Amoxicillin	100 mg/kg PO q8h[11]	Most species
	200–400 mg/L drinking water[13]	Canaries/aviary use
	2–3 g/gal drinking water[14]	Pigeons
Amoxicillin clavulinate	125 mg/kg PO q12h[15,16]	Most species, including pigeons and raptors/use with allopurinol is contraindicated
Amoxicillin sodium	250 mg/kg IM q12-24h[17]	Pigeons/PD
Amoxicillin trihydrate	100 mg/kg PO q12-24h[17]	Pigeons/PD
Ampicillin	100 mg/kg IM q4h[18]	Most species
	1–2 g/L drinking water[13]	Canaries/aviary use
	155 mg/kg IM q12-24h[19]	Pigeons/PD; amoxicillin sodium preferred over ampicillin for IM use in pigeons
	4 g/gal drinking water[2]	Galliformes/flock use
	15 mg/kg IM q12h[20]	Raptors
Ampicillin sodium	100 mg/kg IM q4h[21]	Amazon parrots/PD
	50 mg/kg IM q6-8h[21]	Amazon parrots/PD; localized infections
	150-200 mg/kg PO q8-12h[21]	Amazon parrots/PD; blue-naped parrots did not obtain therapeutic levels at this dose
Ampicillin trihydrate	120–175 mg/kg PO q12-24h[17]	Pigeons/PD
Azithromycin (Zithromax, Pfizer)	43 mg/kg PO q24h[22]	Most species/antimycobacterial, used with ethambutol and rifabutin
Carbenicillin (Geocillin, Roerig)	100 mg/kg IM q8h[11]	Most species
	100 mg/kg PO q12h[23]	Most species
	100 mg/kg IT q24h[18]	Most species/*Pseudomonas* respiratory infections
	100 mg/kg IM q8-12h[24]	Pigeons
	250 mg/kg IM q12h[25]	Raptors
Cefadroxil	100 mg/kg PO q12h × 7 days[1,14]	Most psittacines, pigeons/14–21 day therapy may be indicated for severe or deep pyodermas; orange-flavored suspensions are preferred
Cefazolin	50–75 mg/kg IM q12h[1]	Most species
	50–100 mg/kg PO, IM q12h[26]	Raptors (owls)
Cefotaxime	75 mg/kg IM q8h[11]	Most species
	100 mg/kg IM q8-12h[14]	Pigeons
	100 mg/kg IM q12h[12]	Raptors
Ceftidime	100 mg/kg IM q8-12h[14]	Pigeons
Ceftiofur (Naxcel, Upjohn)	75–100 mg IM q4-8h[27]	Most species

Continued

Table 26-2
Antimicrobial agents—cont'd

Agent	Dosage	Species/Comments
Ceftriaxone (Rocephin, Roche)	75–100 mg/kg IM q4-8h[27] 100 mg/kg IM[28] q4h	Most species Chickens/PD; nebulization was ineffective; frequency estimated from data
Cephalexin	40–100 mg/kg PO, IM q6h[11]	Most species/see comments for cefadroxil
	100 mg/kg PO q12h[1]	Most species
	100 mg/kg PO q4-6h[2]	Pigeons/PD
	35–50 mg/kg IM q6h[29]	Pigeons/PD
	35–50 mg/kg IM q2-3h[29]	Quail, ducks/PD
Cephalothin	100 mg/kg IM q6h[29]	Most species (pigeons/PD)
	100 mg/kg IM q2-3h[29]	Quail, ducks/PD
Cephradine (Cephradine, BioCraft)	35 mg/kg PO q4-6h[2] 100 mg/kg PO q4-6h[2]	Most species Pigeons
Chloramphenicol palmitate and succinate	—	Public health concern; bacteriostatic activity limits usefulness; palmitate not available in U.S., but can be compounded commercially
	50 mg/kg IM q12h[30]	Budgerigars, chickens, turkeys, ducks, geese/PD
	50–70 mg/kg PO, IM q8h[11]	Most species
	50 mg/kg IM q6h[30]	Macaws, conures/PD
	60–100 mg/kg IM q8h[14]	Pigeons
	250 mg/kg PO q6h[14]	Pigeons
	50 mg/kg PO q6-8h[2,31]	Galliformes
	50 mg/kg IM q24h[30]	Peafowl, eagles/PD
	22 mg/kg IM, IV q3h[32]	Ducks/PD
Chlortetracycline	—	Most species/used to treat chlamydiosis; doxycycline is preferred
	250 mg/16 oz drinking water prepared fresh q8-12h[18]	Most species/limited usefulness
	1,000–2,000 ppm in soft mixed feed[17]	Most psittacines
	0.5% pellets × 30–45 days[6,11,18]	Small psittacines
	1% pellets × 30–45 days[11,33]	Large psittacines
	500 mg/L food × 30–45 days[6]	Nectar feeders
	1,000–1,500 ppm in drinking water[34]	Canaries
	40-50 mg/kg PO q8h (w/grit) or q12h (w/o grit)[17]	Pigeons/PD
	1.0–1.5 g/gal drinking water[14]	Pigeons
Ciprofloxacin	20 mg/kg PO q12h[11]	Most species
	250 mg/L drinking water × 5–10 days[5]	Pigeons
	50 mg/kg PO q12h[35]	Raptors (hawks)/PD
	80 mg/kg PO q 24h[36]	Most species/antimycobacterial, use in combination with other agents
Citric acid	1 g/L drinking water × 2–8 wk[37]	Parakeets, finches/megabacteria; lowers pH; may be an effective control, but rarely a cure
Clarithromycin (Biaxin, Abbot)	85 mg/kg PO q24h[22]	Most species/antimycobacterial; use in combination with ethambutol and rifabutin; allometrically scaled dosage

Table 26-2
Antimicrobial agents—cont'd

Agent	Dosage	Species/Comments
Clindamycin (Antirobe, Upjohn)	100 mg/kg PO q24h x 3-5 days[2]	Pigeons/effective against gram-positive and anaerobic organisms; 150 mg/kg is recommended for osteomyelitis[38]
	200 mg/L drinking water[38]	Pigeons
Doxycycline (Vibramycin, Pfizer)	—	Most species/chlamydiosis, *Mycoplasma*
	25-50 mg/kg PO q24h[39]	Most species
	100 mg capsule/4 oz drinking water[1]	Most species
	1 g/kg soft feed[40]	Large psittacines/10 mg/ml syrup mixed into 29% kidney beans, 29% canned corn, 29% cooked rice, 13% dry oatmeal cereal
	1 g/kg soft feed[34]	Canaries
	7.5 mg/kg (w/o grit), 25 mg/kg (w/ grit) PO q12h[17]	Pigeons/PD
	25 mg/kg PO q12h[12]	Raptors
Doxycycline (Vibravenos, Pfizer)		Not available in U.S.
	60–100 mg/kg IM, SC q5-7d[17,39]	Psittacines, pigeons/PD
	100 mg/kg IM q7d × 4-6 wk[1,11]	Psittacines
	75 mg/kg IM q7d × 4-6 wk[1,11]	Macaws
Enrofloxacin (Baytril, Miles)	10–20 mg/kg PO q24h[17]	Psittacines, pigeons/PD
	100–200 mg/L drinking water[39]	Psittacines, pigeons/PD
	500 ppm in feed[39]	Psittacines/PD
	15–30 mg/kg PO, IM q12h[41]	Most species,[42] (African grey parrots/PD)
	0.19–0.75 mg/ml drinking water[43]	African grey parrots/PD
	5 mg/kg IM, SC q12h[44]	Cockatiels/PD
	10 mg/kg PO q12h[44]	Cockatiels/PD
	5–10 mg/kg IM, SC q24h[17]	Pigeons/PD
	10 mg/kg PO, IM q12h[12]	Raptors
Erythromycin	60 mg/kg PO q12h[45]	Most species
	500 mg/gal drinking water 10 days on, 5 days off, 10 days on[18]	Most species
	125 mg/kg PO q8h[14]	Pigeons
	2-3 g/gal drinking water[14]	Pigeons
Ethambutol (Myambutol, Lederle)	10 mg/kg PO q12h[11]	Most species/antimycobacterial; use in combination with other agents; see Table 26-3 for additional indications
Gentamicin	—	Not generally recommended for administration in pet birds; narrow range between therapeutic and toxic levels[45a] (potentially nephrotoxic); causes polydipsia and polyuria (macaws, cockatoos, hawks)[46,47]; higher concentration detected in renal tissue[48]; patient should be well hydrated; amikacin is a safer alternative
	5–10 mg/kg IM q8-12h[10]	Cockatiels/PD
	5–10 mg/kg IM q4h[49]	Pigeons/PD; *Salmonella*
	5 mg/kg IM q8h[50]	Pheasants/PD
	10 mg/kg IM q6h[50]	Quail/PD
	2.5 mg/kg IM q8h[51]	Raptors/PD
Isoniazid (Isoniazid Tablets, Duramed)	30 mg/kg PO q24h[52,53]	Most species/antimycobacterial; use in combination with other agents; see Table 26-3 for additional indications; *Mycobacterium avium* often develops resistance[2]

Continued

Table 26-2
Antimicrobial agents—cont'd

Agent	Dosage	Species/Comments
Lincomycin	100–200 mg/L drinking water[34]	Canaries
	35–50 mg/pigeon q24h × 7-14 days[54]	Pigeons
	100 mg/kg PO q24h[2]	Raptors
Lincomycin/ spectinomycin (LS-50 Water Soluble, Upjohn)	1/8-1/4 tsp/16 oz drinking water × 10–14 days[18]	Most species
	2 g/gal drinking water through first 5 days of age[55]	Turkeys/PD; *Mycoplasma* airsacculitis
Metronidazole	50 mg/kg PO[1,11] q12-24h × 5–30 days	Most species/effective against most anaerobes; anaerobic and hemorrhagic enteritis; see antiprotozoal dosages
Minocycline (Minocin, Lederle)	15 mg/kg PO q12h[57]	Raptors/moderately effective broad spectrum antibiotic; some anaerobic spectrum; some hematozoa
	0.5% in feed[8,31]	Parakeets/PD
Nitrofurazone (NFZ 9.2, Hess & Clark)	—	May be hepatotoxic; avoid or reduce dosage in hot weather; do not use in finches or pigeons;[5] the authors discourage the use of this agent
	1 tsp/gal drinking water × 7–10 days[23]	Most species
	1/2 tsp/gal drinking water × 7 days[2]	Lories, mynahs/do not put in lory nectar
Oxytetracycline (Liquamycin, LA 200, Pfizer; Terramycin Soluble Powder, Pfizer)	—	IM administration may cause irritation or muscle damage; may be useful for *Chlamydia*
	50–100 mg/kg IM, SC q2-3d[57]	Cockatoos/PD[57]
	58 mg/kg IM q24h[58]	Amazon parrots/PD
	50 mg/kg PO q6-8h[14]	Pigeons
	1,000–1,500 mg/gal drinking water[14]	Pigeons
	5 mg/kg IM, SC q12-24h[59]	Chicken chicks/PD; poor tissue distribution to lung; calcium affects absorption
	43 mg/kg IM q24h[58]	Pheasants/PD
	16 mg/kg IM q24h[58]	Raptors (owls)/PD
Penicillin procaine (Penicillin G Procaine, SmithKline)	100,000 units/kg IM q24-48h[60]	Turkeys/PD; for sulfonamide resistant *Pasteurella multocida*
Piperacillin (Piperacil, Lederle)	150 mg/kg IM q12h[11]	Most species
	100 mg/kg IM q8-12h[24,42,17,12]	Psittacines (PD),[17] pigeons,[24] raptors[12,42]
	200 mg/kg IM q8h[8]	Budgerigars/PD
	0.02 ml (4 mg) in macaw eggs; 0.01 ml (2 mg) in small eggs[61]	Psittacine eggs/200 mg/ml solution; inject into air cell on days 14, 18, and 22 of incubation
Rifabutin (Mycobutin, Pharmacia)	15 mg/kg PO q24h[22]	Most species/antimycobacterial; use in combination with azithromycin or clarithromycin and ethambutol
Rifampin (Rifadin, Marion Merrell Dow)	15 mg/kg PO q12h[11]	Most species/antimycobacterial; use in combination with other agents; see Table 26-3 for additional indications
	45 mg/kg PO q24h[52]	Amazon parrots

Table 26-2
Antimicrobial agents—cont'd

Agent	Dosage	Species/Comments
Spectinomycin (Spectogard, Syntex; Spectam, Sanofi)	200–400 mg/L drinking water[34] 25–35 mg/kg IM q8-12h[24] 165–275 mg/L drinking water[14]	Canaries Pigeons Pigeons
Streptomycin (Streptomycin Sulfate, Roerig)	25–50 mg/kg IM q24h[17]	Chickens/PD; may be nephrotoxic; consider amikacin as alternative; seldom used in birds except in combination with other drugs for antimycobacterial therapy; see Table 26-3 for additional indications
Sulfachlorpyridazine (Vetasulid, Solvay)	1/4-1/2 tsp/L drinking water × 5–10 days[31]	Enteric infections
Tetracycline (Tetracycline Soluble Powder 10 g/6.4 oz, Butler; Panmycin Aquadrops, Upjohn)	200–250 mg/kg gavaged q12-24h[2,18] 200 mg/gal drinking water × 5-10 days[18]	Most species Most species
Tiamulin (Denegard, Fermenta)	25-50 mg/kg PO q24h[17] 30 mg/kg PO q24h × 7 days 60 mg/kg PO q24h × 7 days 400 mg/kg feed × 7 days (1 g powder/ kg feed) 250 mg/L drinking water × 7 days (2.2 g powder/gal) 1 g powder/L water	Most species/approved for use in poultry for *Mycoplasma* Poultry adults Poultry chicks Poultry Poultry Poultry eggs/approved for use in poultry for dipping eggs
Ticarcillin	200 mg/kg IM q8–12h × 7 days[5,11,12]	Most species, including pigeons and raptors
Ticarcillin/clavulinic acid (Timentin, SmithKline)	200 mg/kg IM, IV q12h[1]	Most species
Tobramycin (Tobramycin Sulfate, Injectable, Elkins-Sinn)	5 mg/kg IM q12h[11]	Most species
Trimethoprim (Trimethoprim, Biocraft)	15–20 mg/kg PO q8h[17]	Pigeons/PD
Trimethoprim/ sulfamethoxazole (Bactrim, Roche; Septra, Burroughs Wellcome)	75 mg/kg IM q12h[11] 100 mg/kg PO q12[11] 48 mg/120 ml drinking water[1] 60 mg/kg PO q24h[17] 1,800–3,600 mg/gal drinking water × 7-10 days[24] 48 mg/kg PO, IM q12h[12]	Most species Most species Most species Pigeons/PD Pigeons Raptors
Tylosin (Tylan, Elanco; Tylan Soluble, Elanco)	30 mg/kg IM q12h[11] 1/2 tsp/gal drinking water[1] 250–400 mg/L drinking water[34] 25 mg/kg IM q6h[62] 2 tsp/gal drinking water[24]	Most species Most species Canaries Pigeons, quail/PD Pigeons

Box 26-1

Antimycobacterial agents*

The research concerning mycobacteriosis in humans is extensive. Medications for *Mycobacterium avium* complex under study in humans include macrolides (clarithromycin and azithromycin), aminoglycosides (amikacin, streptomycin, paromomycin, and the new liposomal aminoglycosides), rifamycins (rifabutin and rifapentine), a quinolone (ciprofloxacin), and clofazimine and ethambutol.[63]

Following are four antimycobacterial treatment protocols used in birds:

1. **Ethambutol, isoniazid, and rifampin**[1]
   Ethambutol 200 mg + isoniazid 200 mg + rifampin 300 mg
   Finely crush all together, add to 10 ml of simple syrup (Humco), and administer PO q24h as follows:

Bird weight (g)	Volume (ml)
<100	0.1
100–250	0.2
250–500	0.3
500–1,000	0.4

2. **Ethambutol, isoniazid, and rifampin**[52]
   Ethambutol 30 mg/kg + isoniazid 30 mg/kg + rifampin 45 mg/kg
   Medications are mixed into a dextrose powder, mixed with a small amount of food, and administered PO q24h.

3. **Ethambutol, streptomycin, and rifampin**[11]
   Ethambutol 10 mg/kg PO q12h + streptomycin 30 mg/kg IM q12h + rifampin 15 mg/kg PO q12h

4. **Clarithromycin, ethambutol, and rifabutin**[22]
   - Clarithromycin 85 mg/kg (an allometrically scaled dosage) PO q 24h + ethambutol 30 mg/kg PO q24h + rifabutin 15 mg/kg PO q24h
   - In addition, add fluoroquinolone 15 mg/kg PO q12h or clofazamine 6 mg/kg PO q12h *or* aminoglycoside 15 mg/kg IV, IM q12h
   - All the drugs except for clofazamine can be mixed together in the appropriate ratios in an aqueous base for 7 to 10 days for convenient dosing. Clofazamine can be diluted in vegetable oil.[36]

---

*Individual agents used in the treatment of mycobacteriosis are also listed in Table 26-2.

Table 26-3
Antifungal agents

Agent	Dosage	Species/Comments
Amphotericin B (Fungizone, Squibb)	—	Preferred IV medication for aspergillosis; IT administration for syringeal aspergilloma may cause tracheitis
	1.5 mg/kg IV q8h × 3 days[64,65]	Most species
	1 mg/kg IT q8-12h[2,65]	Psittacines, raptors/aspergillosis; potentially nephrotoxic
	0.05 mg/ml sterile water[64]	Most species/nasal flush
	100 mg/kg gavaged q12h × 10 days[66]	Psittacines, passerines/megabacteriosis
	1 g/L drinking water × 10 days[67]	Psittacines, passerines/megabacteriosis
Clotrimazole (Lotrimin, Schering)	—	Psittacines/used commonly as adjunctive therapy for aspergillosis; administer via air sac, IT, nebulization, or topically
	0.2 ml (2 mg)/kg IT q24h × 5 days[1]	Psittacines/syringeal aspergilloma; apply with catheter directly into syrinx during anesthesia
	Inject 10 mg/kg into air sacs; divide total amount between the four easily accessible air sacs[1]	Psittacines/aspergillosis; dilute in propylene glycol to 2.5 mg/ml; severely toxic and may result in death in African grey parrots and other birds if injected into the muscle or viscera[1]
	10 mg/ml solution[65]	Most species/topical use only; effective against aspergillosis at sites that can be flushed, or in early respiratory infections that can be reached by inhalation or nebulization
	Nebulize 1% solution × 30-60[68] min	Use in combination with amphotericin, flucytosine, and itraconazole
Fluconazole (Diflucan, Roerig)	—	Water soluble at pH 7; safest therapeutic index of the azole antifungals; fungistatic; suspension available
	15 mg/kg PO q12h × 14-60 days or longer[1]	Most species/preferred oral azole antifungal for aspergillosis, mycelial candiasis, and systemic mycoses
	20 mg/kg PO q48h[69]	Cockatoos, African greys, Amazons/PD; candidiasis
	50 mg/L drinking water × 14–60 days[1]	Most species/systemic mycoses; candidiasis
	2–5 mg/kg PO q24h × 7–10 days[70]	Most species/gastrointestinal and systemic candidiasis; CNS and ocular mycoses
Flucytosine (Ancobon, Roche)	—	Use prophylactically in raptors (especially in falcons) and waterfowl to prevent aspergillosis; may be administered in conjunction with amphotericin B; fungistatic
	60 mg/kg PO q12h (birds >500 g) or 150 mg/kg PO q12h (birds <500 g)[68]	Most species, galliformes, swans/syringeal aspergillosis
	100–250 mg/kg PO q12h[4]	Psittacine neonates
	120 mg/kg PO q6h[12]	Raptors/aspergillosis
	50–75 mg/kg PO q8h[56]	Raptors/aspergillosis prophylaxis in conjunction with amphotericin B
	250 mg/kg PO q12h[12]	Raptors/candidiasis

Continued

Table 26-3
Antifungal agents—cont'd

Agent	Dosage	Species/Comments
Itraconazole (Sporanox, Janssen)	—	Most species/systemic mycoses; may be useful in superficial candidiasis or dermatophytosis; may be toxic if used concurrently with clotrimazole; approximately 0.35 mg/granule (approximately 285–290 granules per capsule, but number can vary considerably)[1]; a method of compounding itraconazole with strong acid and orange juice has been reported[70]
	10 mg/kg PO q24h × 14 days with food[70,71]	Psittacines/use in combination with other nonazole antifungals
	2.5–5.0 mg/kg PO q24h[71]	African grey parrots/may be toxic in this species
	6 mg/kg PO q12h[72]	Pigeons/PD; this dosage will achieve fungicidal plasma concentrations, although 26 mg/kg PO q12h is required to reach a therapeutic level of 1 µg/ml in respiratory tissues[72]
	5 mg/kg PO q24h[68]	Galliformes, swans/generalized air sac aspergillosis
	5–10 mg/kg PO q12-24h[12]	Raptors
	5-10 mg/kg PO q12h[2]	Waterfowl/aspergillosis, candidiasis, cryptococcosis
Ketoconazole (Nizoral, Janssen)	—	Most species/systemic mycoses (i.e., aspergillosis) and candidiasis; much less toxic than amphotericin B
	30 mg/kg PO q12h × 7-14 days[73]	Amazon parrots/PD
	10-30 mg/kg PO q12h × 30-60 days[1]	Most species/grind 200 mg tablet and add to 5 ml simple syrup (Humco) (=40 mg/ml); dosages >20 mg/kg may cause regurgitation (if so, discontinue for 1–2 days, then restart and regurgitation usually ceases)
	20 mg/kg PO q12h[4]	Psittacine neonates
	200 mg/L drinking water or 200 mg/kg soft food × 7–14 days[34]	Canaries
	20-40 mg/kg PO q12h × 15–60 days[5]	Pigeons
	30 mg/kg PO q12h[73] × 7–30 days	Pigeons, raptors/prophylactic in raptors for aspergillosis
	15 mg/kg PO q12h[12]	Raptors/candidiasis
	60 mg/kg PO q12h[74]	Raptors (common buzzard/PD; aspergillosis
Nystatin (Mycostatin, Squibb)	—	Most species/oral or gastrointestinal tract candidiasis; medication must contact the organism
	300,000 IU/kg PO q12h × 7-14 days[65]	Most species
	100,000 IU/L drinking water[34]	Canaries
	5,000,000 IU/L drinking water[67]	Goldfinches/megabacteriosis
	100,000 IU/kg PO q12h[12,24]	Pigeons, raptors

Table 26-4
Antiviral and immunomodulating agents

Agent	Dosage	Species/Comments
Acyclovir (Zorivax, Burroughs Wellcome)	80 mg/kg PO q8h × 7 days[75]	Psittacines/PD (Quaker parakeets); psittacine herpes virus prophylaxis or treatment
	1 mg/ml drinking water[76]	Psittacines (Quaker parakeets)/psittacine herpes virus
	330 mg/kg PO q12h × 4-7 days[4]	Psittacine neonates/psittacine herpes virus
	10 mg/kg IM q24h × 14 days starting 3 days postexposure[76]	Chickens/Marek's disease
	333 mg/kg PO q12h × 7–14 days[77]	Raptor neonates/falcon and owl herpes viruses
Amantadine	10 mg/kg PO × 3 days pre- and 18 days postexposure[76]	Turkeys/antiviral for influenza viruses; must be administered before and during virus exposure
	25 mg/kg PO × 10 days after onset of viral infection[76]	Chickens
	0.01% in drinking water[76]	Chickens/can use simultaneously with vaccine
Interferon α2a (Roferon-A Injection, Roche)	—	Glycoprotein; inhibits intracellular virus replication
	60–240 IU/kg IM, SC q12h[78]	Most species/prepare stock by mixing 1 ml with 100 ml sterile water (can freeze as 2 ml vials up to 1 yr); mix 2 ml of stock into 1 L LRS (can refrigerate up to 3 mo)
	300–1,200 IU/kg PO q12h[1]	
	1 IU/ml or 4,000 IU/gal drinking water × 14-28 days[1]	Pigeons/may be useful against circovirus
Levamisole (Levasole, Mallinckrodt)	41 mg/gal drinking water × 3–5 weeks[6]	Most species/immunostimulation
	2 mg/kg SC q14d × 3 treatments[79]	Macaws/immunostimulation
Rimantadine (Flumadine, Forest)	0.01% in drinking water[76]	Chickens/antiviral for influenza viruses; must be used before and during exposure

Table 26-5
Antiparasitic agents

Agent	Dosage	Species/Comments
Albendazole suspension (11.36%) (Valbazen, SmithKline)	50 mg/kg PO × 3–4 days[80]	Doves, rock partridges/capillariasis; may be toxic in some columbiformes at 50–100 mg/kg
Amprolium (Corid, Merck)	—	Coccidiosis; efficacy is reduced by high doses of thiamine
	2 ml of 9.6% solution/gal drinking water × 5 days[81]	Parakeets, finches/9.6% solution
	8 ml of 9.6% solution/gal drinking water[14]	Pigeons/9.6% solution
	1 tsp of 20% powder/gal drinking water × 3–5 days[14]	Pigeons/20% powder
	120 g/ton (0.0175%) feed[82]	Pheasants, poultry/coccidiosis; sarcosporidiosis; interferes with thiamine absorption; approved in pheasants and poultry
	117–234 g/ton (0.0125%–0.025%) feed[44,82]	Poultry/lower dose is prophylactic; higher dose is therapeutic
Carnidazole (Spartrix(tm), Wildlife Pharmaceuticals)	20–30 mg/kg PO once[81]	Most species/trichomoniasis
	5 mg/adult dove or pigeon squab PO[5]	Pigeons, doves
	20 mg/kg PO[14] once	Pigeons
	20 mg/kg q24h PO × 2 days[12]	Raptors
	0.5 mg/finch[83]	Society finches/flagellates, based on 15 g body weight per finch
Chloroquine phosphate (Aralen, Sanofi)	—	Avian malaria (*Plasmodium*); may be used with primaquine for *Haemoproteus, Leucocytozoon*
	25 mg/kg PO at 0 hr, then 15 mg/kg PO at 12, 24, and 48 hr[84,85]	Most species, including raptors/for clinically ill birds; use with 0.75–1.0 mg/kg primaquine at 0 hr
	5 mg/kg PO q24h or in feed[86]	Game birds
	250 mg/4 oz drinking water, grape juice, or orange juice[86]	Game birds/bitter; juice makes drug more palatable
	10 mg/kg PO at 0 hr, then 5 mg/kg at 6, 24, and 48 hr[44]	Raptors/use with 0.3 mg/kg primaquine (at 24 hr following the initial chloroquine dose) q24h × 7 days
Clazuril (Appertex, Janssen)	5–10 mg/kg PO q24h × 3 days, off 2 days, on 3 days[81]	Poultry, pigeons/coccidiostat
	5 mg/kg PO once[14]	Pigeons
Clopidol-25% (Coyden-25, A.L. Laboratories)	113.5–227.0 g/ton (125–250 ppm) (0.0125–0.025%) feed[82]	Game birds, poultry/coccidiosis; *Leucocytozoon; Plasmodium*; approved in poultry
Clorsulon (Curatrem, Merck)	20 mg/kg PO q14d × 3 treatments[2,87]	Psittacines/trematodes
Dimetridazole (Emtryl, Jensen Salisbury)	—	*Trichomonas; Giardia; Spironucleus; Histomonas*; not available in U.S.; low therapeutic index; toxic to robins, lories, some passerines[88]; not recommended for finches; 182 g/6.42 oz powder
	1 tsp/gal drinking water × 5 days[23,81]	Most species
	1/2 tsp/gal drinking water	Lories, mynahs
	100 mg/L drinking water[34]	Canaries/use cautiously
	400 mg/L drinking water × 3 days[89]	Pigeons/PD; bioavailability reduced with food
	1/4–1/2 tsp/gal drinking water × 3–5 days[14]	Pigeons
	0.01875% in feed[82]	Game birds, poultry/not approved for game birds
	3 g powder/L drinking water[82]	Game birds, poultry/not approved for game birds

Table 26-5
Antiparasitic agents—cont'd

Agent	Dosage	Species/Comments
Fenbendazole	—	Most species/safe and efficacious anthelmintic
	25 mg/kg PO, repeat in 14 days[87]	Most species/ascarids
	50 mg/kg PO q24h × 3–5 days[5,87]	Most species, including pigeons/nematodes, including capillariasis; flukes; *Giardia* (efficacy unproven); use the longer duration for capillariasis[81]
	50 mg/kg PO q12h × 5 days[1]	Cockatoos/use with ivermectin at 0.2 mg/kg once; effective as a filarid adulticide
	125 mg/L drinking water × 5 days[81]	Most species/nematodes
	50 mg/L drinking water × 5 days[81]	Finches
	1.5–3.9 mg/kg PO q24h × 3 days[90]	Chickens/PD; *Capillaria*
	54 g/ton of feed × 5–7 days[82]	Game birds/nematodes; capillariasis; gizzard worms; *Gongylonema*; cecal worms; gapeworms; flukes
	10–50 mg/kg PO, repeat in 14 days[12]	Raptors/ascarids, flukes
	5–15 mg/kg PO q24h × 5 days[2]	Anseriformes
Hygromycin B (Hygromix 8, Elanco)	12 g/ton feed × 12 weeks[82]	Game birds/ascarids; cecal worms; some efficacy against capillariasis; approved for chickens
	0.00088%–0.00132% in feed[82]	Game birds/ascarids
	0.0018%–0.0026% in feed × 2 months[82]	Game birds/cecal worms
Ipronidazole (Ipropran, Roche)	—	*Giardia*; *Trichomonas*; *Histomonas*; not available in US; 61 g/2.65 oz
	500 mg/gal drinking water × 7 days[81]	Most species
	960 mg/gal drinking water × 3–7 days[5,23]	Psittacines, pigeons
Ivermectin	—	All species/most nematodes; most ectoparasites (including *Knemidokoptes*); can dilute with water or saline for immediate use; dilute with propylene glycol for extended use; dosages exceeding 0.2 mg/kg are probably unnecessary in birds
	0.2 mg/kg IM, PO once*	Most species including psittacines, parakeets, pigeons, raptors/nematodiasis; acariasis; may be given SC;[81] use in combination with fenbendazole at 50 mg/kg PO q12h × 5 days for microfilaria in cockatoos[1]
	0.8–1.0 mg/L drinking water[34]	Canaries
	0.05 mg/kg topically to eyes[94]	Ocular nematodiasis
Lasalocid (Avatec, Hoffmann-La Roche)	68–113 g/ton feed, continuously[82]	Game birds, chickens/coccidiosis; approved for chukar partridges and chickens
Levamisole (Tramisol, Mallinckrodt)	—	Many species/nematodes; immunostimulant; low therapeutic index (toxic reactions and deaths reported); IM administration may cause severe toxicity
	10–20 mg/kg SC once[81]	Most species/do not use in lories
	1.0–1.5 g (10 ml)/gal drinking water × 1–3 days[14,81]	Most species, including pigeons/13.65% injectable
	20 mg/kg PO once[81]	Psittacines, pigeons, raptors/nematodes
	40 mg/kg PO once[14,81]	Psittacines, pigeons, raptors/*Capillaria*

*References 1, 5, 12, 58, 91, 92, 93.

Continued

Table 26-5
Antiparasitic agents—cont'd

Agent	Dosage	Species/Comments
Levamisole (Tramisol, Mallinckrodt)—cont'd	20-25 mg/kg SC[82]	Game birds/nematodes (i.e., gapeworms, capillarids, gizzard worms, eye worms)
	1-2 g/gal drinking water × 1 day, repeat in 7–14 days[82]	Game birds
	0.03%–0.06% in drinking water[82]	Game birds
	0.04% in feed × 2 days[82]	Game birds
Metronidazole	—	Most species/antiprotozoal, including alimentary tract protozoa (especially flagellates such as *Giardia, Histomonas, Spironucleus,* and *Trichomonas*)
	20–35 mg/kg IM q24h × 2 days[58]	Most species
	50 mg/kg PO q12h × 5 days[1,5]	Most species, pigeons
	30 mg/kg PO q12h × 10 days[95]	Psittacines
	25 mg/kg PO q12h × 2–10 days[4]	Psittacine neonates
	40 mg/kg PO q24h × 7 days[96]	Budgerigars
	100 mg/L drinking water[34]	Canaries
	10–20 mg/kg IM q24h × 2 days[81]	Pigeons
	4 g/gal drinking water[14]	Pigeons
	30 mg/kg PO q12h[100]	Poultry/PD
	1.5 g/gal drinking water × 5-15 days[82]	Game birds
	30 mg/kg PO q12h × 5–7 days[97]	Raptors
Milbemycin oxime (Interceptor, Ciba-Geigy)	2 mg/kg PO, repeat in 4 weeks[98]	Galliformes/ascarids; capillarids; *Heterakis*
Monensin (Coban 45, Elanco)	90 g/ton (94 ppm) of feed[82,99]	Quail/coccidiosis
	90–110 g/ton feed × 8 weeks[82]	Chickens
	54–90 g/ton feed × 10 weeks[82]	Turkeys
Piperazine	—	Most species/ascarids, oxyurids; less efficacious than fenbendazole and ivermectin; seldom used in companion birds
	250 mg/kg PO once[81]	Psittacines, pigeons
	50–100 mg/chicken PO[82]	Chickens/approved in poultry
	100–400 mg/turkey PO[82]	Turkeys/approved in poultry
	0.2%–0.4% in feed[82]	Game birds
	1–2 g/L drinking water × 1–2 days[82]	Game birds
	100 mg/kg PO, repeat in 14 days[12,81]	Raptors
	1 g/L drinking water × 3 days[81]	Raptors
	45–200 mg/kg PO[2]	Anseriformes
Praziquantel (Droncit, Miles)	—	Most species/tapeworms; flukes
	10–20 mg/kg PO, repeat in 10–14 days[101a]	Most species
	5.75 mg (¼ cat tablet)/bird PO, repeat in 14 days[5]	Pigeons
	10 mg/kg PO, 8.5 mg/kg IM, or 11 mg/kg SC once[31]	Chickens
	30 mg/kg PO, IM, repeat in 14 days[12]	Raptors
	10 mg/kg IM q24h × 3 days, then PO × 11 days[101]	Toucans/flukes

Table 26-5
Antiparasitic agents—cont'd

Agent	Dosage	Species/Comments
Primaquine (Primaquine Phosphate, Sanofi)	—	Pigeons, raptors, game birds/hematozoa (i.e., *Plasmodium, Hemoproteus, Leucocytozoon*); use with chloroquine; dosage based on patient receiving a prescribed amount of the active base and not the tablet mg strength
	0.75–1.0 mg/kg PO once[84,86]	Most species, including raptors/use with chloroquine 25 mg/kg at 0 hr then 15 mg/kg at 12, 24, and 48 hr; palliative only
	0.03 mg/kg PO q24h × 3 days[2,18,82]	Game birds
	0.3 mg/kg PO (at 24 hr following the initial chloroquine dose) q24h × 7 days[44]	Raptors/hematozoa; use with chloroquine 10 mg/kg at 0 hr, then 5 mg/kg at 6, 24, and 48 hr
Pyrantel pamoate (Nemex, Strongid, Pfizer)	—	Intestinal nematodes
	7 mg/kg PO, repeat in 14 days[87]	Most species
	20–25 mg/kg PO[24]	Pigeons
Pyrimethamine (Fansidar Tablets, Roche)	—	Toxoplasmosis; atoxoplasmosis; sarcocystosis; may be effective for leucocytozoonosis; supplementing with folic acid or folinic acid may be warranted
	0.5 mg/kg PO q12h × 14–28 days[87]	Most species
	100 mg/kg food[87]	Most species
	0.5–1.0 mg/kg PO q12h × 30 days[102,103]	Eclectus, Amazon parrots/use with 30 mg/kg trimethoprim/sulfadiazine
Quinacrine (Atabrine HCl Tablets, Sanofi)	—	Most species/avian malaria; chloroquine and primaquine are preferred
	5–10 mg/kg PO q24h × 7 days[2]	Most species; generally administered via gavage
	100–300 mg/gal drinking water × 10–21 days[14]	Pigeons
Ronidazole (Ridzol 10%, Merck)	—	*Giardia; Trichomonas; Histomonas; Cochlosoma;* wide margin of safety and greater therapeutic range than other nitroimidazoles; not available in U.S.
	6–10 mg/kg PO q24h × 6–10 days[81]	Most species
	1–2 g 10% powder/L drinking water × 7 days[81]	Cockatiels, pigeons
	400 mg 10% powder/L drinking water × 7 days[34,81]	Passerines, including canaries
	4 g 10% powder/gal drinking water × 3–5 days[14]	Pigeons/treatment of choice for *Trichomonas*
Sulfachlorpyridazine (Vetisulid, Solvay)	—	Coccidiosis
	400–500 mg/L drinking water × 5 days, off 2 days, on 5 days[81]	Most species
	200 mg/16 oz drinking water × 30 days[1]	Cockatiels, budgerigars/mixture is stable for up to 5 days if refrigerated; change daily; mix well
	150–300 mg/L drinking water[34]	Canaries
	1,200 mg/gal drinking water × 7–10 days[14]	Pigeons

Continued

Table 26-5
Antiparasitic agents—cont'd

Agent	Dosage	Species/Comments
Sulfadimethoxine (12.5% solution) (Albon, SmithKline)	25 mg/kg PO q12h × 5 days[1] 250 mg/kg IM q24h × 3 days, off 2 days, on 3 days[104] 1,250–1,500 mg/gal drinking water × 1 day then 750 mg/gal × 4 days[14] 7.5 ml/gal drinking water × 5 days[82] 15 ml/gal drinking water × 6 days[82] 25–50 mg/kg PO q24h × 3 days[12]	Most species/coccidiosis Pigeons/PD; close to toxic limit Pigeons Turkeys/approved for turkeys Chickens/approved for chickens (not laying hens) Raptors
Sulfadimethoxine/ ormetoprim (Rofemaid, Hoffmann-La Roche)	10 ppm in feed[82]	Game birds/coccidiosis; leucocytozoonosis; malaria; sarcosporidiosis
Sulfamethazine (Sulmet, Cyanamid)	— 125 mg/L drinking water × 3 days, off 2 days, on 3 days[81] 75 mg/kg PO q24h × 3 days, off 2 days, on 3 days[81] 1,500 mg/gal drinking water × 1 day, then 750–1,000 mg/gal × 4 days[14]	Coccidiosis Most species Parakeets Pigeons
Sulfaquinoxaline (Sulquin 6-50, Solvay)	— 100 mg/kg PO q24h × 3 days, off 2 days, on 3 days[81] 500 mg/L drinking water × 6 days, off 2 days, on 6 days[5] 400 mg/L drinking water × 6 days, off 2 days, on 6 days 454 g/ton feed, continuously[82] 250 mg/L drinking water × 6 days, off 2 days, on 6 days 227 g/ton feed, continuously[82]	Coccidiosis; approved in poultry; 286.2 mg/ml Lories, pigeons Pigeons/1.8 ml/L provides approximately a 0.05% solution Chickens/1.4 ml/L provides approximately a 0.04% solution Chickens Turkeys/0.9 ml/L provides approximately a 0.025% solution Turkeys
Thiabendazole (TBZ, Omnizole, Thibenzole, Merck)	— 40–100 mg/kg PO q24h × 7 days[81] 100–500 mg/kg PO[81] once 454 g/ton feed × 14 days[82] 100 mg/kg PO once, repeat in 10–14 days[86]	Most species/nematodes; generally less efficacious than fenbendazole Most species Most species Pheasants/approved for gapeworms Raptors
Toltrazuril (Baycox, Bayvet)	— 7 mg/kg PO q24h × 2–3 days[12,105] 75 mg/L drinking water × 2 days/week × 4 week[106] 25 mg/L drinking water × 2 days, repeat in 5 days[107]	Coccidiosis; efficacious for refractory coccidiosis; not available in U.S. Budgerigars, raptors Canaries Geese

Table 26-6
Chemical restraint/anesthetic/analgesic agents*

Agent	Dosage	Species/Comments
Alphaxalone/ alphadalone (Saffan, Glaxcovet Labs)	5–7 mg/kg IV[108]	Pigeons/duration 3–5 min; excellent for short procedures; not available in U.S.
	12–15 mg/kg IM[108]	Pigeons/duration 20–30 min; useful for radiography; not available in U.S.
Atipamezole (Antisedan, Farmos Group)	5 times medetomidine dosage IM, IV[109]	Psittacines, raptors, geese/potent alpha-2 antagonist used to reverse medetomidine; righting reflex regained 3–10 min after administration
Atropine	0.01–0.02 mg/kg IM, SC[2]	Most species/preanesthetic; rarely needed; see Tables 26-1 and 26-14 for additional indications
	0.04–1.0 mg/kg IM, SC[6]	Preanesthetic; rarely needed
Buprenorphine (Buprenex, Reckitt & Colman)	0.01–0.05 mg/kg IM[110]	Most species/analgesia
	6.5 mg/L drinking water[111]	Most species/analgesia
Butorphanol (Torbugesic, Fort Dodge)	3–4 mg/kg IM[112,113]	Analgesia; mild motor deficits observed; minimal cardiovascular and respiratory effects
	1 mg/kg IM[114]	Psittacines/preanesthetic
	0.5–2.0 mg/kg IM[110]	Analgesia
	0.05–0.40 mg/kg IM, SC q6-8h[5]	Pigeons/analgesia; sedation
Diazepam	—	Commonly used alone for sedation, tranquilization, seizures, and appetite stimulation; often used with ketamine; can be used in combination with phenobarbital for seizure control; can be diluted with sterile water if used immediately
	0.6 mg/kg IM[115]	Most species
	1 mg/180 ml drinking water[115]	Most species
	0.5–1.0 mg/kg IM, IV prn[5]	Pigeons/sedation; seizures
	0.25–0.50 mg/kg IM, IV q24h × 2-3 days[116]	Raptors/appetite stimulation
Glycopyrrolate (Robinul-V, Aveco)	0.01 mg/kg IM, IV[117]	Most species/preanesthetic; rarely needed
Isoflurane	4%–5% induction and immediately reduce to 2.0%–2.5% for maintenance[1]	Most species/anesthetic of choice in birds
Ketamine	—	Seldom used alone in most species because of prolonged violent recoveries and poor muscle relaxation; can be used alone for short, non-painful procedures; usually used with other anesthetic agents[78]
	25–50 mg/kg IM[108]	Pigeons/duration 30–60 min; better muscle relaxation if given with diazepam
	15–20 mg/kg IM, IV[117]	Raptors
	15–25 mg/kg IM, IV[117]	Waterfowl
Ketamine (K)/ acepromazine (A)	(K) 25–50 mg/kg/ (A) 0.5–1.0 mg/kg IM[113]	Psittacines/improved muscle relaxation over ketamine alone; thermoregulation may be affected

*For other analgesic recommendations, refer to Tables 26-7 (hormones and steroids) and 26-8 (nonsteroidal antiinflammatory agents).

Continued

Table 26-6
Chemical restraint/anesthetic/analgesic agents—cont'd

Agent	Dosage	Species/Comments
Ketamine (K)/ diazepam (D)	(K) 10–50 mg/kg/ (D) 0.5–2.0 mg/kg IM[113]	Psittacines/improved muscle relaxation over ketamine alone
	(K) 5-25 mg/kg/ (D) 0.5-2.0 mg/kg IV[113]	Psittacines/lower end of range is preferred
	(K) 10–25 mg/kg/ (D) 0.5–1.0 mg/kg IM, IV[5]	Pigeons/low dose administered IV is preferred; useful for oral procedures
Ketamine (K)/ medetomidine (M)	—	Lower (K) doses are needed when used with (M); recovery in 10–40 min; note, medetomidine dosages are in µg/kg
	(K) 3–7 mg/kg/ (M) 75–100 µg/kg IM[109]	Psittacines
	(K) 2–5 mg/kg/ (M) 50–100 µg/kg IV[109]	Psittacines
	(K) 3–5 mg/kg/ (M) 50–100 µg/kg IM[109]	Raptors
	(K) 2-4 mg/kg/ (M) 25–75 µg/kg IV[109]	Raptors
	(K) 5–10 mg/kg/ (M) 100–200 µg/kg IM, IV[109]	Geese
Ketamine (K)/ midazolam (M)	(K) 10–25 mg/kg/ (M) 0.5–1.0 mg/kg IM[113]	Psittacines/combination needs further evaluation
Ketamine (K)/ xylazine (X)	—	(K)/ (X) combinations are seldom used for avian anesthesia due to cardiac depressive effects and harsh recoveries; in waterfowl, allometrically scaled dosages are available[118]
	(K) 10–30 mg/kg/ (X) 2–6 mg/kg IM[113]	Psittacines/birds <250 g require a higher dose than birds >250 g
	(K) 10 mg/kg/ (X) 1 mg/kg IM	Vultures
	(K) 20 mg/kg/ (X) 1 mg/kg IV[118]	Ducks/causes respiratory depression
Medetomidine (Domitor, Farmos Group)	See ketamine/medetomidine	Potent alpha-2 agonist used in combination with ketamine; righting reflex not lost with doses up to 1,000 µg/kg; reversed with atipamazole[89]
Midazolam (Versed, Roche)	—	Short acting benzodiazepine seldom used in birds; needs further evaluation; contraindicated with hepatic disease
	0.8 mg/kg IM, IV[117]	Most species/birds >500 g
	1.5 mg/kg IM, IV[117]	Most species/birds <500 g
	2.0 mg/kg IM[119]	Geese/restraint; sedation 15–20 min
Naloxone (Narcan, DuPont)	2 mg IV[2]	Most species/narcotic antagonist; may be repeated in 14-21 hr[2] to prevent renarcotization
Phenobarbital	1-7 mg/kg PO q8-12h[120]	Most species/mild sedative effect; see Table 26-11 for additional indications
	To effect, not to exceed 5 mg/kg IM, IV[1]	Seizures
Propofol (Rapinovet, Mallinckrodt)	14 mg/kg IV[121]	Pigeons/anesthesia; duration 2–7 min; may cause marked respiratory depression and bradycardia; use is limited in birds[122]
	1.32-1.50 mg/kg IV[122]	Owls/1–6 min anesthesia duration; recovery in 15 min; hypothermia; respiratory depression
	0.95–1.20 mg/kg IV[122]	Buzzards/2–6 min anesthesia duration; recovery in 15 min; respiratory depression

Table 26-6
Chemical restraint/anesthetic/analgesic agents—cont'd

Agent	Dosage	Species/Comments
Sevaflurane (Ultane™, Abbot)	Stepwise induction of up to 7%[123]	Psittacines/anesthesia; similar to isoflurane, may provide better or more rapid recovery[124]
Tiletamine/zolazepam (Telazol, Fort Dodge)	—	Most species/generally not recommended as an anesthetic agent for pet birds[113]
	10–30 mg/kg IM[125]	Most species/restraint; anesthesia; moderate analgesia
	10 mg/kg IM[126]	Owls/restraint
	6.6 mg/kg IM[127]	Swans/anesthesia
Tolazoline (Priscoline, Ciba-Geigy)	15 mg/kg IV[128]	Vultures/xylazine antagonist
Xylazine (Rompun, Miles)	—	Seldom used in pet birds; if used, use in combination with ketamine; may cause cardiac depression; not recommended in debilitated birds; tolazoline and yohimbine are antagonists
Yohimbine (Yobine, Lloyd)	0.125 mg/kg IV[93]	Ratites/xylazine antagonist; cage bird dosage not available

Table 26-7
Hormones and steroids

Agent	Dosage	Species/Comments
Dexamethasone	—	May predispose to aspergillosis and other mycoses
	2-4 mg/kg IM[115]	Most species/shock
	1 mg/kg IM once[23]	Most species/trauma
	3 mg/kg IM, IV[129]	Owls, hawks/PD; antiinflammatory; trauma; shock; enterotoxemia
	0.5 mg/kg IM, IV q24h × 5–7 days[130]	Raptors/head trauma
Dexamethasone Na phosphate	—	Most species/preferred steroidal anti-inflammatory agent for use in birds; suppresses the pituitary-adrenocortical system at doses as low as 50 µg/kg[131]; may predispose birds to aspergillosis and other mycoses
	2–4 mg/kg IM, IV q12h[1]	Head trauma (until signs abate); shock (one dose); hyperthermia (until stable)
	4 mg/kg IM, SC[4]	Psittacine neonates/shock
Estradiol benzoate (in olive oil) (E)/ 1-thyroxine (T)	(E) 1 mg/kg IM/ (T) 1 mg/kg PO q24h × 12 days[132]	Ducks/induces molt; estradiol valerate, 17B-estradiol, goserelin acetate, leuprolide acetate, medroxyprogesterone, and tamoxifen have also been used to induce molt in penguins[132]
Human chorionic gonadotropin (hCG)	500–1,000 IU/kg IM on days 1, 3, 7, 14, 21, then q2-4wk[1]	Feather picking resulting from sexually related disorders; inhibits egg laying; dilute to 1000 IU/ml
Insulin	0.002 IU IM q12-48h[133]	Budgerigars/NPH insulin
	0.01-0.10 IU IM q12-48h[133]	Amazon parrots/NPH insulin
	2 IU/bird, reduce as needed based on glucose curves[134]	Toco toucans/protamine zinc insulin or ultralente (extended) insulin

Continued

Table 26-7
Hormones and steroids—cont'd

Agent	Dosage	Species/Comments
Leuprolide acetate (Lupron, TAP Pharmaceutical)	(number of days for desired effect) × (100 µg/kg) × dosage IM once[135]	GnRH depot drug for preventing ovulation; approximately 100 µg/kg released daily (e.g., in cockatiels, 2800 µg/kg given once will last 28 days)
Levothyroxine (l-thyroxine)	0.02 mg/kg PO q12-24h[133] 0.1 mg/4–12 oz drinking water[133]	Most species/hypothyroidism; treating obesity and lipomas; monitor weight closely; see estradiol benzoate/l-thyroxine for use in molt
Medroxyprogesterone	20 mg/kg IM[1] 0.1% of feed[2]	Most species/reduces aggression; inhibits egg laying; stimulates follicle resorption; some feather picking disorders; may cause polyuria, polydipsia, weight gain, immunosuppression Pigeons/inhibits ovulation
Methylprednisolone acetate	0.5–1.0 mg/kg IM[2]	Most species/may predispose to aspergillosis and other mycoses; low end of dose preferred
Oxytocin	5 IU/kg IM, may repeat q30min[136]	Most species/egg binding; uterovaginal sphincter should be well-dilated before use (see PGE_2);[137] can be used in conjunction with PGE_2[138,139]
PGE_2 (dinoprostone) (Prepidil Gel, Upjohn)	0.2 mg (1 ml)/kg intracloacally on the uterovaginal sphincter[138,139]	Most species/egg binding; relaxes uterovaginal sphincter; can freeze in syringes in 0.1 ml aliquots; may use in conjunction with oxytocin
Prednisolone	0.5–1.0 mg/kg IM, IV once[2] 25 mg/m² PO q24h[140]	Most species/may predispose to aspergillosis and other mycoses Cockatoos/lymphosarcoma
Prednisolone Na succinate	10–20 mg/kg IV, IM q15min prn[6]	Most species/shock
Stanozolol (WinstrolV, Upjohn)	0.5–1.0 mg/kg IM[23] 2 mg/4 oz drinking water[23]	Most species/debilitation, anemia, anorexia, and to increase muscle mass
Tamoxifen (Nolvadex, Zeneca)	40 mg/kg IM[132]	Ducks/induces molt; see also estradiol benzoate
Testosterone	8 mg/kg IM q7d prn[31] Mix 5–10 drops stock/oz drinking water × 5–10 days[1]	Most species, including canaries/anemia; libido; debilitation Canaries/stock solution: 100 mg injectable suspension/30 ml drinking water
Triamcinolone	0.1–0.2 mg/kg IM once[1]	Most species/glucocorticoid; seldom used; may predispose to aspergillosis and other mycoses; may be immunosuppressive

Table 26-8
Nonsteroidal antiinflammatory agents

Agent	Dosage	Species/Comments
Acetylsalicylic acid (aspirin)	5 grains (325 mg)/250 ml drinking water[113]	Most species/musculoskeletal pain and inflammation; antipyretic; not critically evaluated in birds
	10 mg/kg PO q24h × 3 days[1]	
Carprofen	5–10 mg/kg PO[141]	Most species/analgesia
Flunixin meglumine (Banamine, Schering)	—	Preferred antiinflammatory agent used in birds; analgesia; antipyretic; potentially nephrotoxic in cranes and quail[141a]
	1–10 mg/kg IM[113]	Most species
	4 mg/kg IM[114]	Psittacines/dose may be inadequate
Ketoprofen (Ketofen, Fort Dodge)	2 mg/kg IM[142]	Most species/antiinflammatory; analgesia; for musculoskeletal disorders
Phenylbutazone	—	Antiinflammatory; antipyretic; IM, SC administration not recommended
	3.5–7.0 mg/kg q8-12h PO[2]	Psittacines
	20 mg/kg PO q8h[2]	Raptors

Table 26-9
Nebulizing agents

Agent	Dosage	Species/Comments
Acetylcysteine 20% (Mucomyst, Bristol)	200 mg/9 ml sterile water until dissipated[1] 200 mg/8 ml sterile water and 1 ml amikacin or gentamicin until dissipated[1]	Most species/mucolytic; rhinitis; pneumonia; airsacculitis; syringeal aspergilloma; can mix with dexamethasone (4 mg),[1] aminoglycosides, and aminophylline; tracheal irritation has been reported in mammals
Amikacin	50 mg/ 9 ml sterile water until dissipated[1] 50 mg/8 ml sterile water and 1 ml acetylcysteine until dissipated[1]	Most species/rhinitis; pneumonia; airsacculitis; tracheitis
Aminophylline	25 mg/9 ml sterile water or saline × 15 min[1]	Most species/bronchodilator; allergic pulmonary disease; can mix with dexamethasone, aminoglycosides, and acetylcysteine
Amphotericin B (Fungizone, Squibb)	1 mg/ml sterile water × 15 min q12h[18]	Most species/antifungal
Carbenicillin (Geocillin, Roerig)	200 mg/9 ml saline	Psittacines/*Pseudomonas* pneumonia; use in combination with parenteral aminoglycosides
Chloramphenicol	100 mg/9 ml sterile water or saline × 15 min[1]	Most species/aerosolization could represent a public health risk
Clotrimazole (Lotrimin, Schering)	Nebulize 1% solution × 30-60[68,143] min	Use in combination with systemically administered amphotericin, flucytosine, and itraconazole
Doxycycline	200 mg/15ml saline[144]	Psittacines/antibiotic
Enroflaxacin	10 mg/ml saline[144]	Most species/antibiotic
Erythromycin (Erythro, Sanofi)	50 mg/10 ml saline × 15 min q8h[18]	Most species/airsacculitis
Gentamicin	50 mg/9 ml sterile water × 15 min, or with 8 ml water and 1 ml 20% acetylcysteine q8h[1]	Most species/rhinitis; tracheitis; pneumonia; airsacculitis
Oxytetracycline	2 mg/ml × 60 min q4-6h[145]	Parakeets/PD
Spectinomycin	13 mg/ml saline[144]	Most species/antibiotic
Sulfadimethoxine	13 mg/ml saline[144]	Most species/antibiotic
Tylosin (Tylan, Elanco)	1 g powder/50 ml DMSO or distilled water × 1 hour[146]	Pigeons, quail/PD; respiratory tract infections

Table 26-10
Chelating and related agents

Agent	Dosage	Species/Comments
Calcium EDTA	—	Preferred initial chelator for lead and zinc toxicity
	35 mg/kg IM q12h × 5 days, off 3–4 days, repeat prn[147,148]	Most species
	30 mg/kg IM q12h × 3-5 days until asymptomatic[5]	Pigeons
	35 mg/kg IV, IM q8h × 3–4 days, off 2 days, repeat until asymptomatic[12]	Raptors
	25.0–52.2 mg/kg IV q12h[149] until asymptomatic	Geese
Deferoxamine mesylate (Desferal, Ciba-Geigy)	—	Preferred chelator for iron toxicity; hemochromatosis
	20 mg/kg PO initially, then IM q4h until recovery[148]	Most species
	20 mg/kg PO q4h until recovery[148]	Most species
	40 mg/kg IM q24h × 7 days[150]	Mynahs
	100 mg/kg SC q24h × 4 months[151]	Toucans
Dimercaptosuccinic acid (DMSA, Chemet, Bock Pharmacal)	25–35 mg/kg PO q12h × 5 days/wk × 3–5 weeks[148,152]	Preferred oral lead chelator for lead toxicity
Dimercaprol (BAL, Becton Dickinson)	2.5 mg/kg IM q4h × 2 days, then q12h × 10 days or until recovery[148]	Heavy metal toxicity; rarely used
Penicillamine (Cuprimine, Merck)	—	Commercially available oral chelator for lead and zinc toxicity; preferred chelator for copper toxicity
	55 mg/kg PO q12h × 1–2 weeks[147,148]	Most species
	30–55 mg/kg PO q12h[12]	Raptors
Sodium sulfate (GoLytely, Braintree)	0.5–1.0 g/kg PO[148]	Lead toxicity; cathartic; adjunct to chelation; rarely used

Table 26-11
Psychotropic agents

Agent	Dosage	Species/Comments
Amitriptyline (Elavil, Stuart)	1–2 mg/kg PO q12-24h[7]	Most species/feather picking
Carbamazepine (Tegretol, Basel)	3–10 mg/kg PO q24h[88] 20 mg/4 oz drinking water[1]	Most species/authors' preferred psychotropic agent for feather picking; anticonvulsant, antiobsessive, anticompulsive, antimutilating drug; may cause bone marrow depression and hepatotoxicity; combination therapy with chlorpromazine or haloperidol is recommended for initial treatment the first 2 weeks[1]
Chlorpromazine	—	Most species/feather picking; discontinue within 30 days; effectiveness diminishes in 14–30 days when given PO[1]
	Mix 1 ml stock solution/4 oz drinking water (=1 mg/oz), or administer 0.2–1.0 ml stock solution/kg PO q12-24h prn for very mild sedation[1]	Stock solution: crush five 25 mg tablets and mix with 31 ml simple syrup (Humco); start at low dose initially
	0.1–0.2 mg/kg IM once[1]	Cockatoos/use with carbamazepine following removal of E-collar
	0.2 mg/kg IM once prn for sedation[1]	Ringneck parakeets/causes mild sedation and decreases obsessive behaviors
Clomipramine (Anafranil, Basel)	1 mg/kg PO q24h or divided q12h × 6 wk[153]	Most species/feather picking; tricyclic antidepressant; marginal effectiveness; crushed tablet can be suspended in a diluent to form a 1–2 mg/ml solution
Diazepam	1.25–2.50 mg/120 ml drinking water[120]	Most species/feather picking
Diphenhydramine	2–4 mg/kg PO q12h[120]	Most species/feather picking
Doxepin (Sinequan, Roerig)	0.5–1.0 mg/kg PO q12h[120]	Most species/feather picking; tricyclic antidepressant
Fluoxetine (Prozac, Dista)	2 mg/kg PO q12h[1]	Most species/antidepressant; adjunctive treatment for feather picking; lack of scientific information available on its usefulness in birds; needs further investigation
Haloperidol (Haldol, McNeil)	—	Most species/butyrophenone tranquilizer; hyperexcitability and frustration-induced feather picking and behavior disorders

Table 26-11
Psychotropic agents—cont'd

Agent	Dosage	Species/Comments
Haloperidol (Haldol, McNeil)—cont'd	0.15–0.20 mg/kg PO q12h[154]	Amazon parrots, cockatoos/birds <1 kg
	0.10–0.15 mg/kg PO q12h[154]	Amazon parrots, cockatoos/birds >1 kg
	6.4 mg/L drinking water × 7 mo[155]	African grey parrots
	1–2 mg/kg IM q21d[120]	Most species
Human chorionic gonadotropin (hCG)	500–1,000 IU/kg IM on days 1, 3, 7, 14, 21, then q2-4wk[1]	Most species/feather picking resulting from sexually related disorders; inhibits egg laying; dilute to 1000 IU/ml
Hydroxyzine (Atarax, Roerig)	2.0–2.2 mg/kg PO q8h[120,156]	Most species/feather picking
	4 mg/100–120 ml drinking water[120,157]	Most species/feather picking
Medroxyprogesterone	30 mg/kg SC, repeat in 90 days prn[23, 158]	Most species/steroid of choice for sexually related feather picking; seasonal aggression; chronic egg laying; may cause polydipsia, polyphagia, polyuria, weight gain
	20 mg/kg IM, SC, repeat in 30–60 days 1-2×[1]	
Megestrol acetate (Ovaban, Schering)	1.25–2.50 mg/120 ml drinking water × 7–10 days, then 1–2×/wk[120]	Most species/feather picking; seldom used
Naltrexone (Trexan, DuPont)	1.5 mg/kg PO q8-12h × 1-18 months[159]	Most species/opioid agonist; feather picking; contraindicated in patients with liver disease; dissolve tablet in 10 ml sterile water (the preservative does not go into solution); may need to increase dosage 2–6 × to be effective
Nortriptyline (Pamelor, Sandoz)	2 mg/120 ml drinking water[120]	Most species/feather picking; seldom used
Phenobarbital	—	Most species/mild sedative effect; useful alone or in combination with diazepam for seizure control; sedative effects may help with feather picking; see Table 26-6 for additional indications
	1–7 mg/kg PO q8-12h[120]	Most species/feather picking; mild sedative effect
	5 mg/kg/day PO divided q8-12h[1]	Most species/idiopathic epilepsy
	2.0–3.2 mg/kg PO q12h[160]	Amazon parrots/idiopathic epilepsy
	6–10 mg/120 ml drinking water[160]	Amazon parrots/idiopathic epilepsy

Table 26-12
Nutritional/mineral agents

Agent	Dosage	Species/Comments
Calcium glubionate (Neo-calglucon, Sandoz)	— 25 mg/kg PO[161] 150 mg/kg PO q12h[3] 23 mg/kg PO q24h[4] 23 mg/30 ml drinking water[3]	Most species/hypocalcemia Most species Most species Psittacine neonates Most species
Calcium gluconate	50–100 mg/kg IV slowly,[2] IM (diluted)[3,5]	Most species/hypocalcemia
Calcium lactate/calcium glycerophosphate (Calphosan, Glenwood)	5–10 mg/kg IM q7d prn[3,161]	Most species/hypocalcemia
Calcium levulinate	75–100 mg/kg IM, IV[3,23]	Most species/hypocalcemia
Dextrose (50%)	1 ml (500 mg)/kg IV slowly[1] 2 ml (1 g)/kg IV slowly[6]	Most species/hypoglycemia; can dilute with fluids
Iodine (Lugol's iodine)	1 drop/250 ml drinking water daily[3]	Most species/thyroid hyperplasia
Iodine (sodium iodide 20%)	0.3 ml (60 mg)/kg IM[3]	Most species/thyroid hyperplasia
Iron dextran	10 mg/kg IM, repeat in 7–10 days prn[2] 10 mg/kg IM q7d[12]	Most species/iron deficiency anemia; use cautiously in species in which iron storage disease is common (e.g., toucans, mynahs, starlings, birds of paradise, and other passerines) Raptors
Pancreatic enzyme powder	— 1/8 tsp/kg food[31] 1/8 tsp on 2–4 oz lightly oil-coated seed[1] 1/8 tsp/1–4 oz handfeeding formula, prn[162]	Most species/pancreatic insufficiency; maldigestion Most species Most species Psittacine neonates
Vitamin A (Aquasol A Parenteral, Astra)	20,000 IU/kg IM[111] 50,000 USP units/kg IM q7d[4]	Most species/hypovitaminosis A; maximum dose[111] Psittacine neonates
Vitamins A, D_3, E (Vital E- A+D, Schering)	10,000 IU Vit A and 1,000 IU Vit D_3 /300 g q7d[3] IM	Most species/hypovitaminosis A or D_3; hypervitaminosis D may occur with excessive use; product contains alcohol and may sting when administered; a product without alcohol can be compounded commercially
Vitamin B_1 (thiamine)	1–3 mg/kg IM q7d[2]	Most species
Vitamin B_{12} (cyanocobalamin)	250–500 µg/kg IM q7d[7]	Most species/may turn urates pink
Vitamin C	20-50 mg/kg IM q1-7d[12]	Raptors
Vitamin E	15 mg/kg PO once[163]	Raptors/PD; administer without food; serum concentration from PO administration (without food) exceeds concentration from IM administration

Table 26-12
Nutritional/mineral agents—cont'd

Agent	Dosage	Species/Comments
Vitamin E/selenium (E-Se, Schering)	0.05–0.10 mg/kg IM q14d[2] 0.06 mg Se/kg IM q3-14d[3]	Most species/neuromuscular disease (e.g., capture myopathy, white muscle disease, some cardiomyopathies); may be useful in some cockatiels with jaw, eyelid, and tongue paralysis[2]
Vitamin K_1	0.2–2.2 mg/kg IM q4-8h until stable, then q24h × 2 weeks[147,148] 2.5 mg/kg IM q24h until hemostasis, then q7d prn[1]	Most species/rodenticide toxicity Vitamin K responsive disorders (conures); hematochezia (Amazon parrots); coagulopathies (psittacines)

Table 26-13
Miscellaneous agents

Agent	Dosage	Species/Comments
Allopurinol	10 mg/kg PO q4-12h[164]	Most species/gout; prepare 10 mg/ml suspension by crushing a 100 mg tablet and mixing with 10 ml simple syrup (Humco); reduce dose as uricemia decreases
	30 mg/kg PO q12h[115]	Most species
	100 mg/2–4 oz water[1]	Most species
Aluminum hydroxide	30–90 mg/kg PO q12h[1]	Most species/antacid; phosphate binder
Aminopentamide (Centrine, Fort Dodge)	0.11 mg/kg IM, SC q8-12h × 1 day, then q12h × 1 day, then q24h × 1 day[165] 0.05 mg/kg IM, SC q12h × 5 doses maximum[2]	Most species/regurgitation
Aminophylline	4 mg/kg PO q6-12h[1]	Bronchodilation; prepare 10 mg/ml suspension by finely crushing a 100 mg tablet and mixing with 10 ml simple syrup (Humco)
	4 mg/kg IM q12h[1]	
Anticoagulant dextrose (ACD) (Formula A-Baxter)	0.15 ml/1 ml whole blood[110]	Anticoagulant; transfusions
Asparaginase (Elspar, Merck)	400 IU/kg IM q7d[140]	Cockatoos/lymphosarcoma; premedicate with diphenhydramine
Atropine	—	See Tables 26-1 and 26-7 for additional indications
	0.2 mg/kg IM, IV q3-4h[147,166]	Most species/organophosphate toxicity
	0.5 mg/kg IM, IV[1]	Pigeons/organophosphate toxicity
	0.5 mg/kg , 1/2 IV, 1/2 IM[167]	Raptors/organophosphate toxicity
Barium sulfate (Barotrast, Rhone-Poulenc)	25 ml/kg gavaged[1]	Most species/dilute 72% suspension 1:1 with water; dilute 92% suspension 1:2 with water
Bismuth sulfate	1 ml/kg PO[148]	Most species/weak adsorbent and demulcent
Charcoal, activated	2–8 g/kg PO[2,148]	Adsorbs toxins; use 1 g/5-10 ml water
Cimetidine (Tagamet, SmithKline Beecham)	5 mg/kg IM q8-12h[1] 5–10 mg/kg PO q8-12h[1]	Psittacines/proventricular ulcers
Cisapride (Propulsid, Janssen)	1 mg/kg PO q12h[7]	Psittacines/gastrointestinal irritation; gastric stasis

Continued

Table 26-13
Miscellaneous agents—cont'd

Agent	Dosage	Species/Comments
Colchicine (ColBenemid, Merck)	0.04 mg/kg PO q12-24h[168]	Most species/gout; hepatic fibrosis, cirrhosis; tablet also contains probenecid
Cyclophosphamide (Cytoxan, Squibb)	200 mg/m^2 IO q7d[140]	Cockatoos/lymphosarcoma
Digoxin	0.05 mg/kg PO q24h[169]	Quaker parakeets/PD; congestive heart failure; cardiomyopathy
	0.02 mg/kg PO q24h[31,170]	Budgerigars, mynahs, sparrows
Dioctyl Na sulfosuccinate (Diocto, Barre)	1 ml/30 ml drinking water[4]	Psittacine chicks/constipation; use only if chick is drinking
Diphenhydramine	1–4 mg/kg PO q8h[1]	Macaws, Amazon parrots/allergic rhinitis
	2–4 mg/100 ml drinking water[157]	Most species/hypersensitivity
	2 mg/kg IO once[140]	Cockatoos/prior to chemotherapy
Doxapram	5–10 mg/kg IM, IV once[2]	Most species/respiratory depression; see Table 26-1 for additional indications
Doxorubicin	60 mg/m^2 IV diluted with saline q30d[171]	Blue-front Amazon parrots/osteosarcoma; premedicate with diphenhydramine 30 min before; dilute and give IV over 30 min (anesthesia recommended)
	30 mg/m^2 IO q2d[140]	Cockatoos/lymphosarcoma; premedicate with diphenhydramine
Epinephrine (1:1000)	0.5–1.0 ml/kg IM, IV, IO, IT[1]	CPR; cardiac arrest
Furosemide	0.15–2.0 mg/kg IM, SC q12-24h[2]	Most species
	5 mg/120 ml drinking water[1]	Most species; ascites
	2.5–10.0 mg/kg PO q12h × 7-14 days[1]	Cockatiels, budgerigars/ascites
	2.2 mg/kg PO q12h[170]	Mynahs/cardiac disease
	2.2 mg/kg PO, IM, IV q12h prn[5]	Pigeons
	4–6 mg/kg PO, IM[56]	Raptors/pulmonary congestion
Gadopentate dimeglumine (Magnevist, Berlex)	0.25 mmol/kg IV[172]	Contrast agent for magnetic resonance imaging
Glycosaminoglycan (Adequan, Luitpold)	10 mg/kg IM, IA q7d × 3 mo[173,174]	Noninfectious or traumatic joint dysfunction; 250 mg/ml for IA use; 500 mg/ml for IM use; contraindicated in septic arthritis
Hetastarch (Hespan, DuPont)	10-15 ml/kg IV q8h × 1–4 treatments, decrease fluid treatment to 1/3-1/2 normal fluid dose[175]	Most species/chronic hypoproteinemia; an amylopectin plasma expander colloid with a T-1/2 of 25 hr; dextrans are an alternative but have shorter T-1/2s
Hydroxyzine (Atarax, Roerig)	2.0–2.2 mg/kg PO q8h[120,156]	Amazon parrots/pruritus (allergic); feather picking; self mutilation
	4 mg/100–120 ml drinking water[120,157]	Most species/respiratory allergy; feather picking
Iohexol (Omnipaque, Sanofi Wintrop)	25–30 ml/kg gavaged[176]	Cockatoos, Amazon parrots/radiographic gastrointestinal contrast media; may dilute 1:1 with water
Kaolin/pectin (Kaopect, Med-Tech)	2 ml/kg PO q6-8h[115]	Psittacine neonates/intestinal protectant

Table 26-13
Miscellaneous agents—cont'd

Agent	Dosage	Species/Comments
Lactulose (Cephulac, Marion Merrell Dow)	0.3–1.0 ml/kg PO q12h[142]	Most species/hepatic encephalopathy; reduces blood ammonia levels; increases gram-positive bacteria in the gastrointestinal tract
	0.3 ml/kg PO q8-12h[4]	Psittacine neonates
Lorelco (Probacoll, Marion Merrell Dow)	0.25 tsp/day PO × 2-4 mon[7]	Most species/used to control lipemia and suppress lipoma growth
Magnesium hydroxide (Milk of Magnesia, Roxane)	10–12 ml with 1 tsp powdered activated charcoal[148]	Most species/cathartic; adsorbent
Magnesium sulfate	0.5–1.0 g/kg PO[148]	Epsom salts; cathartic; may cause lethargy; combine with dilute peanut butter as a cathartic and bulk diet[148]
Mannitol	0.5 mg/kg q24h IV slowly[2]	Osmotic diuretic; head trauma
Methocarbamol (Robaxin, Fort Dodge)	130 mg/kg IM[7]	Most species/muscle relaxation
Metoclopramide (Reglan, Robins)	0.5 mg/kg PO, IM, IV[2]	Most species/gastrointestinal motility disorders; regurgitation; slow crop motility; the authors recommend a frequency of q12h
Mineral oil	0.3 ml/35 g or 3–5 ml/500 g[148] via gavage	Most species/cathartic
	Mineral oil and peanut butter 1:2[148] via gavage	Most species/cathartic; bulk diet
Peanut butter	Peanut butter and mineral oil 2:1[148] via gavage	Most species/bulk diet; cathartic
	Dilute peanut butter and magnesium sulfate[148] PO	Most species/bulk diet; cathartic
Pralidoxime (2-PAM) (Protopam, Wyeth)	10–100 mg/kg IM q24-48h[147,148]	Organophosphate toxicity; use lower dose in combination with atropine
Propranolol (Inderal, Wyeth-Ayerst)	0.2 mg/kg IM[2] 0.04 mg/kg IV slowly[2]	Most species/cardiac dysrhythmias
Psyllium (Metamucil, Procter & Gamble)	0.5 tsp/60 ml baby food gruel[147,148]	Most species/bulk diet; can try mineral oil as alternative or in addition to psyllium
Sodium bicarbonate	1–4 meq/kg IV q15-30min to maximum of 4 meq/kg total dose[7]	Most species/shock; severe metabolic acidosis; see Table 26-1 for additional indications
	2.5 meq/kg gavaged q12h × 2 doses[1]	Most species neonates/adjunctive treatment for sour crop
Sodium sulfate (GoLytely, Braintree)	0.5–1.0 g/kg PO[148]	Most species/cathartic
Sucralfate (Carafate, Marion Merrell Dow)	25 mg/kg PO q8h[2]	Most species/oral, esophageal, gastric, and duodenal ulcers
Terbutilene (Brethine, Geigy)	0.1 mg/kg PO q12-24h[1]	Macaws, Amazon parrots/bronchodilator; obstructive pulmonary disease; may cause hyperactivity
Vecuronium bromide (Norcuron, Organon)	0.96 mg topically to eye (unilaterally)[177]	Cockatoos/use caution with bilateral application
	2 drops topically to eye q15min × 3 treatments[178]	Raptors/mydriatic agent; 4 mg/ml
Vincristine sulfate (Oncovin, Lilly)	0.75 mg/m^2 IO q7d × 3 weeks[140]	Cockatoos/lymphosarcoma
	0.5 mg/m^2 initial dose, then 0.75 mg/m^2 q7d × 3 weeks[179]	Ducks/antineoplastic; lymphoma; lymphocytic leukemia

REFERENCES

1. Rupiper DJ: Unpublished data.
2. Ritchie BW, Harrison GJ: Formulary. In Ritchie BW, Harrison GJ, Harrison LR, editors: *Avian medicine: principles and application,* Lake Worth, Fla, 1994, Wingers, pp 457-478.
3. Huff DG: Avian fluid therapy and nutritional therapeutics, *Semin Avian Exotic Pet Med* 2:13-16, 1993.
4. Joyner KL: Pediatric therapeutics, *Proc Annu Conf Assoc Avian Vet,* 1991, pp 188-199.
5. Rupiper DJ, Ehrenberg M: Introduction to pigeon practice, *Proc Annu Conf Assoc Avian Vet,* 1994, pp 203-211.
6. Clubb SL: Therapeutics. In Harrison GJ, Harrison LR, editors: *Clinical avian medicine and surgery,* Philadelphia, 1986, WB Saunders, pp 327-355.
7. Tully TN: Formulary. In Altman RB, Clubb SL, Dorrestein GM, Quesenberry K, editors: *Avian medicine and surgery,* Philadelphia, 1997, WB Saunders, pp 671-688.
8. Lumeij JT: Psittacine antimicrobial therapy. In *Antimicrobial therapy in caged birds and exotic pets,* Trenton, NJ, 1995, Veterinary Learning Systems Co, pp 38-48.
9. Gronwall R, Brown MP, Clubb S: Pharmacokinetics of amikacin in African gray parrots, *Am J Vet Res* 50:250-252, 1989.
10. Ramsay EC, Vulliet R: Pharmacokinetic properties of gentamicin and amikacin in the cockatiel, *Avian Dis* 37:628-634, 1993.
11. Bauck L, Hoefer HL: Avian antimicrobial therapy, *Semin Avian Exotic Pet Med* 2:17-22, 1993.
12. Joseph V: Preventive health programs for falconry birds, *Proc Annu Conf Assoc Avian Vet* 1995, pp 171- 178.
13. Haneveld-v Laarhoven MA, Dorrestein GM: IME - Sudden high mortality in canaries, *J Assoc Avian Vet* 4: 82, 1990.
14. Harlin RW: Pigeons, *Vet Clin N Am Small Anim Pract* 24:157-173, 1994.
15. Kasper A: Rehabilitation of California towhees, *Proc Annu Conf Assoc Avian Vet,* 1997, pp 83-90.
16. Moore DM, Rice RL: Exotic animal formulary. In Holt KK et al, editors: *Veterinary values,* ed 5, Lenexa, Kan, 1998, Veterinary Medicine Publishing Group, pp 159-245.
17. Dorrestein GM: Antimicrobial drug use in pet birds. In Prescott JF, Baggot JD, editors: *Antimicrobial therapy in veterinary medicine,* ed 2, Ames, Ia, 1993, Iowa State University Press, pp 491-506.
18. Clubb SL: Birds. In Johnston DE, editor: *The Bristol veterinary handbook of antimicrobial therapy,* ed 2, Trenton, NJ, 1987, Veterinary Learning Systems, pp 188-199.
19. Dorrestein GM et al: Comparative study of ampicillin and amoxicillin after intramuscular, intravenous, intramuscular and oral administration in homing pigeons (*Columba livia*), *Res Vet Sci* 42:343-348, 1987.
20. Frazier DL, Jones MP, Orosz SE: Pharmacokinetic considerations of the renal system in birds: part II. Review of drugs excreted by renal pathways, *J Avian Med Surg* 9:104-121, 1995.
21. Ensley PK, Janssen DL: A preliminary study comparing the pharmacokinetics of ampicillin given orally and intramuscularly to psittacines: Amazon parrots (*Amazona* spp.) and blue-naped parrots (*Tanygnathus lucionensis*), *J Zoo Anim Med* 12:42-47, 1981.
22. Rupley AE: Respiratory bacterial, fungal, and parasitic diseases, *Proc Avian Speciality Advanced Prog/Small Mammal Rep Med Surg* 23-44, 1997.
23. McDonald SE: IME - Summary of medications for use in psittacine birds, *J Assoc Avian Vet* 3:120-127, 1989.
24. Harlin RW: Pigeons, *Proc Annu Conf Assoc Avian Vet,* 1995, pp 361-373.

25. Redig PT: Treatment protocol for bumblefoot types 1 and 2, *AAV Today* 1:207-208, 1987.
26. Porter SL: Vehicular trauma in owls, *Proc Annu Conf Assoc Avian Vet,* 1990, pp 164-170.
27. Flammer K: A review of the pharmacology of antimicrobial drugs in birds, *Proc Avian/Exotic Anim Med Symp,* Davis, Calif, 1994, University of California, pp 65-78.
28. Junge RE et al: Pharmacokinetics of intramuscular and nebulized ceftriaxone in chickens, *J Zoo Wildl Med* 25:224-228, 1994.
29. Bush M et al: Pharmacokinetics of cephalothin and cephalexin in selected avian species, *Am J Vet Res* 42:1014-1017, 1981.
30. Clark CH et al: Plasma concentrations of chloramphenicol in birds, *Am J Vet Res* 43:1949, 1982.
31. Allen DG et al, editors: *Handbook of veterinary drugs,* Philadelphia, 1993, JB Lippincott, pp 573-634.
32. Dein FJ, Monard DF, Kowalczyk DF: Pharmacokinetics of chloramphenicol in Chinese spot-billed ducks, *J Vet Pharmacol Therap* 3:161-168, 1980.
33. Flammer K et al: Blood concentrations of chlortetracycline in macaws fed a medicated pelleted feed, *Avian Dis* 33:199-203, 1989.
34. Dorrestein GM: Infectious diseases and their therapy in Passeriformes. In *Antimicrobial therapy in caged birds and exotic pets,* Trenton, NJ, 1995, Veterinary Learning Systems, pp 11-27.
35. Isaza R et al: Disposition of ciprofloxacin in red-tailed hawks following a single oral dose, *J Zoo Wildl Med* 24:498-502, 1993.
36. VanDerHeyden N: New strategies in the treatment of avian mycobacteriosis, *Semin Avian Exotic Pet Med* 6:25-33, 1997.
37. Gerlach H: Personal communication, *Annu Conf Assoc Avian Vet,* Reno, Calif, 1997.
38. Crosta L, Delli Carri AP: Oral treatment with clindamycin in racing pigeons, *Proc First Conf Euro Comm Assoc Avian Vet,* 1991, pp 293-296.
39. Dorrestein G: Avian chlamydiosis therapy, *Semin Avian Exotic Pet Med* 2:23-29, 1993.
40. Flammer K, Aucoin DP, Whitt D: Preliminary report on the use of doxycycline-medicated feed in psittacine birds, *Proc Annu Conf Assoc Avian Vet,* 1991, pp 1-4.
41. Flammer K, Aucoin DP, Whitt DA: Intramuscular and oral disposition of enrofloxacin in African grey parrots following single and multiple doses, *J Vet Pharmacol Therap* 41:359-366, 1991.
42. Redig PT: Bumblefoot treatment in raptors. In Fowler ME, editor: *Zoo & wild animal medicine: current therapy 3,* Philadelphia, 1993, WB Saunders, pp 181-188.
43. Flammer K et al: Plasma concentrations of enrofloxacin in African grey parrots treated with medicated water, *Avian Dis* 34:1017-1022, 1990.
44. Carpenter JW: Unpublished data.
45. Hoefer HL: Antimicrobials in pet birds. In Kirk RW, editor: *Current veterinary therapy XII - small animal practice,* Philadelphia, 1995, WB Saunders, pp 1278-1283.
45a. Dorrestein GM, Van Gogh H, Rinzema JD: Pharmacokinetic aspects of penicillins, aminoglycosides and chloramphenicol in birds compared to mammals. A review, *Vet Quart* 6:216-224, 1984.
46. Bird JE, Walser MM, Duke GE: Toxicity of gentamicin in red-tailed hawks (*Buteo jamaicensis*), *Am J Vet Res* 44:1289-1293, 1983.
47. Flammer K et al: Adverse effects of gentamicin in scarlet macaws and galahs, *Am J Vet Res* 51:404-407, 1990.
48. Bush M et al: Gentamicin tissue concentration in various avian species following recommended dosage therapy, *Am J Vet Res* 42:2114-2116, 1981.

49. Sabrautzki S: The course of gentamicin concentrations in serum and tissues of pigeons, *Vet Bull* 54:5915, 1983.
50. Custer RS, Bush M, Carpenter JW: Pharmacokinetics of gentamicin in blood plasma of quail, pheasants, and cranes, *Am J Vet Res* 40:892-895, 1979.
51. Bird JE et al: Pharmacokinetics of gentamicin in birds of prey, *Am J Vet Res* 44:1245-1247, 1983.
52. Loudis BG: Soft tissue involvement of avian tuberculosis and attempted treatment: a case study, *Proc Am Assoc Zoo Vet*, 1991, pp 246-247.
53. Rosskopf WJ, Jr, Woerpel RW, Asterino R: Successful treatment of avian tuberculosis in pet psittacines, *Proc Annu Conf Assoc Avian Vet*, 1991, pp 238-251.
54. Marx D: Preventive health care with diagnostics, *AAV Today* 2:92-94, 1988.
55. Hamdy AH, Saif YM, Kasson CW: Efficacy of lincomycin-spectinomycin water medication on *Mycoplasma meleagridis* air sacculitis in commercially raised turkey poults, *Avian Dis* 26:227-233, 1981.
56. Redig P: *Medical management of birds of prey: a collection of notes on selected topics*, St Paul, Minn, 1993, The Raptor Center.
57. Flammer K et al: Potential use of long-acting injectable oxytetracycline for treatment of chlamydiosis in Goffin's cockatoos, *Avian Dis* 34:228-234, 1990.
58. Ritchie BW: Avian therapeutics, *Proc Annu Conf Assoc Avian Vet*, 1990, pp 415-431.
59. Black WD: A study of the pharmacodynamics of oxytetracycline in the chicken, *Poultry Sci* 56:1430-1434, 1977.
60. Hirsch DC et al: Pharmacokinetics of penicillin-G in the turkey, *Am J Vet Res* 39:1219-1221, 1978.
61. McDonald SE: IME - Injecting eggs with antibiotics, *J Assoc Avian Vet* 1:9, 1989.
62. Locke D, Bush M, Carpenter JW: Pharmacokinetics and tissue concentrations of tylosin in selected avian species, *Am J Vet Res* 43:1807-1810, 1982.
63. Killian AD, Drusano GL, Kanyok TP: Pharmacokinetics of drugs used for the therapy of *Mycobacterium avium*-complex infection. In Korvick JA, Benson CA, editors: *Mycobacterium avium- complex infection, progress in research and treatment*, New York, 1996, Marcel Dekker, pp 197-240.
64. Bauck L, Hillyer E, Hoefer H: Rhinitis: case reports, *Proc Annu Conf Assoc Avian Vet* 134-139, 1992.
65. Flammer K: An overview of antifungal therapy in birds, *Proc Annu Conf Assoc Avian Vet* 1993, pp 1-4.
66. Gould WJ: Common digestive tract disorders in pet birds, *Vet Med* (1) 40-52, 1995.
67. Filippich LJ: Megabacteria and proventricular/ventricular disease in psittacines and passerines, *Proc Annu Conf Assoc Avian Vet*, 1990, pp 287-293.
68. Aguilar RF, Redig PT: Diagnosis and treatment of avian aspergillosis. In Kirk RW, editor: *Current veterinary therapy XII - small animal practice*, Philadelphia, 1995, WB Saunders, pp 1294-1299.
69. Flammer K: Fluconazole in psittacine birds, *Proc Annu Conf Assoc Avian Vet*, 1996, pp 203-204.
70. Orosz SE, Frazier DL: Antifungal agents: a review of their pharmacology and therapeutic indications, *J Avian Med Surg* 9:8-18, 1995.
71. Orosz SE, Schroeder EC, Frazier DL: Itraconazole: a new antifungal drug for birds. *Proc Annu Conf Assoc Avian Vet*, 1994, pp 13-19.
72. Lumeij JT, Gorgevska D, Woestenborghs R: Plasma and tissue concentrations of itraconazole in racing pigeons (*Columba livia domestica*), *J Avian Med Surg* 9:32-35, 1995.
73. Ludders JW et al: Effects of ketamine, xylazine, and a combination of ketamine and xylazine in Pekin ducks, *Am J Vet Res* 50:245, 1989.

74. Wagner CH, Hochleitner M, Rausch W-D: Ketoconazole plasma levels in buzzards, *Proc First Conf Euro Comm Assoc Avian Vet,* 1991, 333-340.
75. Norton TM et al: Efficacy of acyclovir against herpesvirus infection in Quaker parakeets, *Am J Vet Res* 52:2007-2009, 1991.
76. Cross G: Antiviral therapy, *Semin Avian Exotic Pet Med* 4:96-102, 1995.
77. Joseph V: Raptor pediatrics, *Semin Avian Exotic Pet Med* 2:142-151, 1993.
78. Sedgwick CJ: Anesthesia of caged birds. In Kirk RW, editor: *Current veterinary therapy vii - small animal practice,* Philadelphia, 1980, WB Saunders, pp 653-656.
79. Breadner S: Chronic *Nocardia* infection in a hyacinth macaw, *Proc Annu Conf Assoc Avian Vet,* 1994, pp 283-285.
80. Stalis IH et al: Possible albendazole toxicity in birds, *Proc Am Assoc Zoo Vet,* 1995, pp 216-217, 1995.
81. Marshall R: Avian anthelmintics and antiprotozoals, *Semin Avian Exotic Pet Med* 2:33-41, 1993.
82. Stadler C, Carpenter JW: Parasites of backyard game birds, *Semin Avian Exotic Pet Med* 5(2):85-96, 1996.
83. Scott JR: Passerine aviary diseases. Diagnosis and treatment, *Proc Annu Conf Assoc Avian Vet,* 1996, pp 39-48.
84. Redig PT, Talbot B, Guarnera T: Avian malaria, *Proc Annu Conf Assoc Avian Vet,* 1993, pp 173-181.
85. Sellers C: Personal communication, 1993.
86. Smith SA: Parasites of birds of prey: their diagnosis and treatment, *Semin Avian Exotic Pet Med* 5(2):97-105, 1996.
87. Clyde VL, Patton S: Diagnosis, treatment and control of common parasites in companion and aviary birds, *Semin Avian Exotic Pet Med* 5(2):75-84, 1996.
88. Ramsay E: Personal communication, 1993.
89. Inghelbrecht S et al: Pharmacokinetics and antitrichomonal efficacy of dimetridazole in homing pigeons (*Columba livia*). In Kösters J, editor: X. Conference of a specialist group on bird diseases, *German Vet Med Soc,* Giessen, 1996, pp 287-290.
90. Taylor SM et al: Efficacy, pharmacokinetics and effect on egg-laying and hatchability of two dose rates of in-feed fenbendazole for the treatment of *Capillaria* species infections in chickens, *Vet Rec* 133:519-521, 1993.
91. Carpenter JW: Infectious and parasitic diseases of cranes. In Fowler ME, editor: *Zoo & wild animal medicine: current therapy 3,* Philadelphia, 1993, WB Saunders, pp 229-237.
92. Hogan HL et al: Efficacy and safety of ivermectin treatment for scaley leg mite infestation in parakeets, *Proc Am Assoc Zoo Vet,* 1984, p 156.
93. Jensen JM, Johnson JH, Weiner ST: *Husbandry & medical management of ostriches, emus & rheas,* College Station, Tex, 1992, Wildlife and Exotic Animal TeleConsultants.
94. Thomas-Baker B, Dew RD, Patton S: Ivermectin treatment of ocular nematodiasis in birds, *J Am Vet Med Assoc* 189:1113, 1986.
95. Murphy J: Psittacine trichomoniasis, *Proc Annu Conf Assoc Avian Vet,* 1992, pp 21-24.
96. Ramsay EC, Drew ML, Johnson B: Trichomoniasis in a flock of budgerigars, *Proc Annu Conf Assoc Avian Vet:* 309-311, 1990.
97. Redig PT: Health management of raptors trained for falconry, *Proc Annu Conf Assoc Avian Vet,* 1992, pp 258-264.
98. Hedberg GE, Bennet RA: Preliminary studies on the use of milbemycin oxime in Galliformes, *Proc Annu Conf Assoc Avian Vet* 261-264, 1994.

99. Carpenter JW, Novilla MN: Safety and efficacy of selected coccidiostats in cranes and implications for other avian species, *Proc Annu Conf Assoc Avian Vet*, 1990, pp 147-149.
100. Cybulski W et al: Disposition of metronidazole in hens (*Gallus gallus*) and quail (*Coturnix coturnix japonica*): pharmacokinetics and whole-body autoradiography, *J Vet Pharmacol Ther* 19:352-358, 1996.
101. Giddings RF: Treatment of flukes in a toucan, *J Am Vet Med Assoc* 193:1555-1556, 1988.
101a. Lung NP, Romagnano A: Current approaches to feather picking. In Kirk RW, editor: *Current veterinary therapy XII - small animal practice*, Philadelphia, 1995, WB Saunders, pp 1303-1307.
102. Page DC et al: Antemortem diagnosis and treatment of sarcocystosis in two species of psittacines, *J Zoo Wildl Med* 23:77-85, 1992.
103. Page DC: A clinical review of apicomplexan parasites in non-domestic birds, *Proc Annu Conf Am Coll Vet Internal Med Forum*, 1995, p 1072.
104. Bruch J von, Aufinger P, Jakoby JR: Pharmacokinetics and efficacy of sulfadimethoxine in healthy and coccidia-infected adult pigeons, *Vet Bull* 56:7830, 1986.
105. Hooimeijer J: Coccidiosis in lorikeets infectious for budgerigar, *Proc Annu Conf Assoc Avian Vet*, 1993, pp 59-61.
106. Cornelissen H: IME - Treatment of common passerine conditions, *J Assoc Avian Vet* 7:103, 1993.
107. Haberkorn A et al: The use of Bay VI 9142 (Baycox), a new coccidiocide, in waterfowl, particularly in the goose, *Proc Conf Avian Diseases*, pp 144-149, 1988.
108. Cooper JE: Anaesthesia of exotic animals, *Anim Technol* 35:13-20, 1984.
109. Jalanka HH: Medetomidine-ketamine and atipamezole: a reversible method for chemical restraint of birds, *Proc First Annu Conf Euro Comm Assoc Avian Vet* 1991, pp 102-104.
110. Jenkins JR: Postoperative care of the avian patient, *Semin Avian Exotic Pet Med* 2:97-102, 1993.
111. Tennant B: *Small animal formulary*, Mid Glamorgan, Wales, UK, 1994, Stephens & George Ltd, pp 23, 26, 168.
112. Bauck L: Analgesics in avian medicine. *Proc Annu Conf Assoc Avian Vet*, 1990, pp 239-244.
113. Wheler C: Avian anesthetics, analgesics, and tranquilizers, *Semin Avian Exotic Pet Med* 2:7-12, 1993.
114. Curro TG: Evaluation of the isoflurane-sparing effects of butorphanol and flunixin in Psittaciformes, *Proc Annu Conf Assoc Avian Vet*, 1994, pp 17-19.
115. Bauck L: *A Practitioner's guide to avian medicine*, Lakewood, Colo, 1993, American Animal Hospital Association, pp 5, 36.
116. Suarez DL: Appetite stimulation in raptors. In Redig PT et al, editors: *Raptor biomedicine*, Minneapolis, Minn, 1993, University of Minnesota Press, pp 225-228.
117. Heard DJ: IME - Overview of avian anesthesia, *AAV Today* 2:92-94, 1988.
118. Kollias GV et al: The use of ketoconazole in birds: preliminary pharmacokinetics and clinical applications, *Proc Annu Conf Assoc Avian Vet* 1986, p 103.
119. Valverde A et al: Determination of a sedative dose and influence of midazolam on cardiopulmonary function in Canada geese, *Am J Vet Res* 51:1071-1074, 1990.
120. Gould WJ: Caring for birds' skin and feathers, *Vet Med*, (1), 53-63, 1995.
121. Fitzgerald G, Cooper JE: Preliminary studies on the use of propofol in the domestic pigeon (*Columba livia*), *Vet Sci* 49:334-338, 1990.
122. Mikaelian I: Intravenously administered propofol for anesthesia of the common buzzard (*Buteo buteo*), the tawny owl (*Strix aluco*), and the barn owl (*Tyto alba*), *Proc First Conf Euro Comm Assoc Avian Vet*, 1991, pp 97-101.

123. Greenacre CB, Quandt JE: Comparison of sevoflurane to isoflurane in psittaciformes, *Proc Annu Conf Assoc Avian Vet,* 1997, pp 123-124.
124. Heard DJ: Avian anesthesia: present and future trends, *Proc Annu Conf Assoc Avian Vet,* 1997, pp 117-122.
125. Hartup BK, Miller EA: *Willowbrook wildlife haven pharmaceutical index,* Glen Ellyn, Ill, 1992, Friends of the Furred and Feathered, p 12.
126. Kreeger TJ et al: Immobilization of raptors with tiletamine and zolazepam (Telazol). In Redig PT et al, editors: *Raptor biomedicine,* Minneapolis, Minn, 1993, University of Minnesota Press, pp 141-144.
127. Schobert E: Telazol use in wild and exotic animals. *Vet Med* (Oct): 1080-1088, 1987.
128. Allen JL, Oosterhuis JE: Effect of tolazoline on xylazine-ketamine-induced anesthesia in turkey vultures, *J Am Vet Med Assoc* 189:1011-1012, 1986.
129. Burns RB, Birrenkott GP: Half-life of dexamethasone and its effect on plasma corticosterone in raptors, *Proc Am Assoc Zoo Vet,* 1988, pp 12-13.
130. Ramsay E: Emergency medicine and critical care of raptors, *Proc Avian/Exotic Anim Med Symp,* Davis, Calif, 1990, University of California, pp 75-79.
131. Westerhof I, Pellicaan CH: Effects of different application routes of glucocorticoids on the pituitary-adrenocortical axis in pigeons (*Columba livia domestica*), *J Avian Med Surg* 9:175-181, 1995.
132. Hines R, Kolattukuty PE, Sharkey P: Pharmacological induction of molt and gonadal involution in birds, *Proc Annu Conf Assoc Avian Vet,* 1993, pp 127-134.
133. Rae M: Endocrine disease in pet birds, *Semin Avian Exotic Pet Med* 4:32-38, 1995.
134. Murphy J: Diabetes in toucans, *Proc Annu Conf Assoc Avian Vet,* 1992, pp 165-170.
135. Millam JR. Leuprolide acetate can reversibly prevent egg laying in cockatiels, *Proc Annu Conf Assoc Avian Vet,* 1993, p 46.
136. Rosskopf WJ, Jr, Woerpel RW: Avian obstetrical medicine, *Proc Annu Conf Assoc Avian Vet,* 1993, pp 323-336.
137. Romagnano A: Avian obstetrics, *Semin Avian Exotic Pet Med* 5:180-188, 1996.
138. Hudleson KS: A review of the mechanisms of avian reproduction and their clinical applications, *Semin Avian Exotic Pet Med* 5:189-198, 1996.
139. Hudleson KS, Hudleson P: A brief review of the female avian reproductive cycle with special emphasis on the role of prostaglandins in clinical application, *J Avian Med Surg* 10:67-74, 1996.
140. France M: Chemotherapy treatment of lymphosarcoma in a Moluccan cockatoo, *Proc Annu Conf Assoc Avian Vet,* 1993, pp 15-19.
141. Johnson-Delaney CA, Harrison LR, editors: *Exotic companion medicine handbook for veterinarians,*Lake Worth, Fla, 1996, Wingers.
141a. Klein PN, Charmatz K, Langenberg J: The effect of flunixin meglumine (Banamine) on the renal function in northern bobwhite (*Colinus virginianus*): an avian model, *Proc Annu Conf Am Assoc Zoo Vet,* 1994, pp 128-131.
142. Malley AD: Practical therapeutics for cage and aviary birds. In Raw ME, Parkinson TJ, editors: *The veterinary annual,* London, 1994, Blackwell Scientific, pp 235-246.
143. Joseph V, Pappagianis D, Reavill DR: Clotrimazole nebulization for the treatment of respiratory aspergillosis, *Proc Annu Conf Assoc Avian Vet,* 1994, pp 301-306.
144. Forbes NA: Respiratory problems. In Beynon PH, Forbes NA, Lawton MPC, editors: *Manual of psittacine birds,* Ames, Iowa, 1996, Iowa State University Press, pp 147.

145. Dyer DC, Van Alstine WG: Antibiotic aerosolization: tissue and plasma oxytetracycline concentrations in parakeets, *Avian Dis* 31:677-679, 1987.
146. Locke D, Bush M: Tylosin aerosol therapy in quail and pigeons, *J Zoo Anim Med* 15:67-72, 1984.
147. LaBonde J: Household poisonings in caged birds. In Kirk RW, editor: *Current veterinary therapy xii - small animal practice,* Philadelphia, 1995, WB Saunders, pp 1299-1303.
148. LaBonde J: Toxicity in pet avian patients, *Semin Avian Exotic Pet Med* 4:23-31, 1995.
149. Murase T et al: Treatment of lead poisoning in wild geese, *J Am Vet Med Assoc* 200:1726-1729, 1992.
150. Norton TM: Bali mynah captive medical management and reintroduction program, *Proc Annu Conf Assoc Avian Vet,* 1995, pp 125-136.
151. Cornelissen H, Ducatelle R, Roels S: Successful treatment of a channel-billed toucan (*Ramphastos vitellinus*) with iron storage disease by chelation therapy: sequential monitoring of the iron content of the liver during the treatment period by quantitative chemical and image analyses, *J Avian Med Surg* 9:131-137, 1995.
152. Degernes LA: Toxicities in waterfowl, *Semin Avian Exotic Pet Med* 4:15-22, 1995.
153. Ramsay EC, Grindlinger H: Use of clomipramine in the treatment of obsessive behavior in psittacine birds, *J Assoc Avian Vet* 8:9, 1994.
154. Lennox AM, VanDerHeyden N: Haloperidol for use in treatment of psittacine self-mutilation and feather plucking, *Proc Annu Conf Assoc Avian Vet,* 1993, pp 119-120.
155. Iglauer F, Rasim R: Treatment of psychogenic feather picking in psittacine birds with a dopamine antagonist, *J Small Anim Pract* 34:564-566, 1993.
156. Krinsley M: IME - Use of DermCaps Liquid and hydroxyzine HCl for the treatment of feather picking, *J Assoc Avian Vet* 7:221, 1993.
157. Fudge AM, Reavill DR, Rosskopf WJ, Jr: Diagnosis and management of avian dyspnea: a review, *Proc Annu Conf Assoc Avian Vet,* 1993, pp 187-195.
158. Carpenter JW: Supportive care procedures in companion bird medicine, *Kansas Vet* (3):10-13, 1991.
159. Turner R: Trexan (naltrexone hydrochloride) use in feather picking in avian species, *Proc Annu Conf Assoc Avian Vet,* 1993, pp 116-118.
160. Quesenberry K: Avian neurologic disorders. In Birchard SJ, Sherding RG, editors: *Saunders manual of small animal practice,* Philadelphia, 1994, WB Saunders, pp 1312-1316.
161. Bauck L: Nutritional problems in pet birds, *Semin Avian Exotic Pet Med* 4:3-8, 1995.
162. Oglesbee BL, McDonald S, Warthen K: Avian digestive system disorders. In Birchard SJ, Sherding RG, editors: *Saunders manual of small animal practice,* Philadelphia, 1994, WB Saunders, pp 1290-1301.
163. Mainka SA et al: Circulating (-tocopherol following intramuscular or oral vitamin E administration in Swainson's hawks (*Buteo swainsonii*), *J Zoo Wildl Med* 25:229-232, 1994.
164. Rupiper DJ: IME - Allopurinol in simple syrup for gout, *J Assoc Avian Vet* 7:219, 1993.
165. Bond MW: IME-Medication for vomiting psittacines, *J Assoc Avian Vet* 7:102, 1993.
166. LaBonde J: Two clinical cases of exposure to household use of organophosphate and carbamate insecticides, *Proc Annu Conf Assoc Avian Vet,* 1992, pp 113-118.
167. Porter SL: Organophosphate/carbamate poisoning in birds of prey, *Proc Am Assoc Zoo Vet/Am Assoc Wildl Vet,* 1992, p 176.

168. Hoefer H. IME - Hepatic fibrosis and colchicine therapy, *J Assoc Avian Vet* 5:193, 1991.
169. Wilson RC et al: Single dose digoxin pharmacokinetics in the Quaker conure (*Myiopsitta monachus*), *J Zoo Wildl Med* 20:432-434, 1989.
170. Rosenthal K, Stamoulis M: Diagnosis of congestive heart failure in an Indian Hill mynah bird (*Gracula religiosa*), *J Assoc Avian Vet* 7:27-30, 1993.
171. Doolen M: Adriamycin chemotherapy in a blue-front Amazon with osteosarcoma, *Proc Annu Conf Assoc Avian Vet,* 1994, pp 89-91.
172. Romagnano A: Magnetic resonance imaging of the avian brain and abdominal cavity. *Proc Annu Conf Assoc Avian Vet,* 1995, pp 307-309.
173. Suedmeyer WK: IME - Use of Adequan in articular diseases of avian species, *J Assoc Avian Vet* 7:105, 1993.
174. Tully TN: A treatment protocol for non-responsive arthritis in companion birds. *Proc Annu Conf Assoc Avian Vet,* 1994, pp 45-49.
175. Stone EG: Preliminary evaluation of hetastarch for the management of hypoproteinemia and hypovolemia, *Proc Annu Conf Assoc Avian Vet,* 1994, pp 197-199.
176. Ernst S et al: Comparison of iohexol and barium sulfate as gastrointestinal contrast media in mid-sized psittacine birds, *J Avian Med Surg* 12:16-20, 1998.
177. Ramer JC et al: Induction of mydriasis in three psittacine species, *Proc Am Assoc Zoo Vet,* 1994, pp 288-289.
178. Mikaelian I, Paillet I, Williams D: Comparative use of various mydriatic drugs in kestrels (*Falco tinnunculus*), *Am J Vet Res* 55:270-272, 1994.
179. Newell SM: Diagnosis and treatment of lymphocytic leukemia and malignant lymphoma in a Pekin duck (*Anas platyrhyncos domesticus*), *J Assoc Avian Vet* 5:83-86, 1991.

27

Blood and Chemistry Tables

Carolyn Cray

I. Blood Values

A. Use of Normal Reference Ranges

1. Several factors can affect the relevance of blood reference ranges. Whenever possible, every attempt must be made to negate these factors or, if that is unsuccessful, to compensate for them appropriately.
   a. Laboratory established ranges: The veterinarian must ensure that the chosen reference service has validated the machinery and techniques for avian use and that laboratory-established reference ranges are available.
   b. Proper sample preparation: When preparing samples, the guidelines supplied by the reference laboratory should be used. More specific guidelines are available.[1,2] In general, the following are recommended:
      (1) Hematology: Whole blood in ethylenediaminetetraacetic acid (EDTA) and smear prepared at the time of sample acquisition
      (2) Chemistry: Lithium heparinized blood, centrifuged with plasma separated.
      (3) Storage: 4° C (39.2° F) until shipment
      (4) Age of sample: 24- to 48-hour delay in analysis acceptable for most determinations
   c. Problems with hemolysis and lipemia: If either is a factor, the chemistry values must be examined accordingly. Several values are still stable despite these changes for some chemistry analyzers.[1] The reference laboratory should be contacted for assistance in general interpretation.

B. **Differences in Analyzers, Techniques, and Species**
   When using normal ranges that were not established by the chosen reference laboratory, the clinician must be aware of the different techniques used.
   1. Hematology: White blood cell (WBC) estimates often are higher than values obtained by the Unopette method. Differentials from automated analyzers usually are different from those obtained through manual determination.
   2. Chemistry: The total protein value will vary, depending on whether a refractometer or a dry or wet chemistry analyzer was used. The albumin value also can vary considerably and is best measured by protein electrophoresis.
   3. Species: For the most part, most chemistry and hematologic determinations are similar for the psittacine species. Reference ranges generally are available for most species and are presented in this chapter (Tables 27-1 and 27-2).[3] If species-specific ranges are not available, general avian ranges should be used.
      a. The clinician should keep in mind that most ranges are based on adult birds, and several determinations (i.e., WBC, total protein, and albumin) generally are different in juvenile birds.[4-6]

Table 27-1
Reference values for selected hematologic determinations

Determination*	African grey	Amazon	Lovebird	Budgerigar	Parakeet	Canary	Cockatiel	Cockatoo	Conure	Macaw	Ecletus	Senegal	Pionus	Quaker	Jardine's	Toucan	Pigeon	Ostrich	Emu
WBC × 10³/μl	5-11	6-11	3-8.5	3-8.5	4.5-9.5	4-9	5-10	5-11	4-11	6-12	4-10	4-11	4-11.5	4-10	4-10	4-11	13-23	10-25	8-25
RBC × 10⁶/μl†	2.4-3.9	2.4-4	2.3-3.9	2.4-4	2.2-3.9	2.5-3.8	2.2-3.9	2.2-4	2.5-4	2.4-4	2.4-3.9	2.4-4	2.4-4	2.3-4	2.4-4	2.5-4.5	3.1-4.5	2.5-4.5	2.4-4.5
HCT (%)‡	38-48	37-50	38-50	38-48	36-48	37-49	36-49	38-48	36-49	35-48	35-47	36-48	35-47	35-46	35-48	35-55	38-50	40-55	40-55
Hb (g/dl)‡	11-16	11-17.5	13-18	12-16	12-16	12-16	11-16	11.5-16	12-16	11-16	11.5-16	11-16	11-16	11-15	11-16	11-16	13-17.5	13-18	13-17
MCV (fl)	90-180	85-200	90-190	90-200	85-195	90-210	90-200	85-200	90-190	90-185	95-220	90-200	85-210	90-200	90-190	90-150	85-200	90-190	90-200
MCH (pg)	28-52	28-55	27-59	25-60	25-60	26-55	28-55	28-60	28-55	27-53	27-55	27-55	26-54	26-55	26-56	30-55	28-55	27-59	25-60
MCHC (g/dl)	23-33	22-32	22-32	23-30	24-31	22-32	22-33	21-34	23-31	23-32	22-33	23-32	24-31	22-32	21-33	25-35	22-33	22-32	23-30
Differential																			
Heterophils (%)	55-75	55-80	50-75	50-75	50-75	50-80	55-80	55-80	55-75	58-78	55-70	55-75	50-75	55-80	55-75	55-65	50-60	55-85	50-75
Eosinophils (%)	0-2	0-1	0-1	0-2	0-2	0-2	0-2	0-2	0-2	0-1	0-1	0-1	0-2	0-1	0-1	0-2	0-3	0-2	0-2
Basophils (%)	0-1	0-1	0-1	0-1	0-1	0-1	0-2	0-1	0-1	0-1	0-2	0-1	0-1	0-2	0-1	0-2	0-3	0-1	0-1
Monocytes (%)	0-3	0-3	0-2	0-2	0-2	0-1	0-2	0-1	0-2	0-3	0-2	0-2	0-2	0-3	0-2	0-2	0-3	0-1	0-1
Lymphocytes (%)	25-45	20-45	25-50	25-45	25-45	20-45	20-45	20-45	25-45	20-45	30-45	25-45	25-45	20-45	25-45	25-35	20-40	10-40	20-40

Reference ranges from the Avian and Wildlife Laboratory, University of Miami, Miami, Fla.
*Red blood cell (RBC) and WBC counts performed by Unopette method using 24-hour-old EDTA blood and smears prepared at sample acquisition.
†Hemoglobin determined by hemoglobinometer.
‡Spun hematocrit.
WBC, White blood cell; *RBC*, red blood cell; *HCT*, hematocrit; *Hb*, hemoglobin; *MCV*, mean corpuscular volume; *MCH*, mean corpuscular hemoglobin; *MCHC*, mean corpuscular hemoglobin concentration.

Table 27-2
Reference values for selected plasma biochemistries

Determination[†]	African grey	Amazon	Lovebird	Budgerigar	Parakeet	Canary	Cockatiel	Cockatoo	Conure	Macaw	Eclectus	Senegal	Pionus	Quaker	Jardine's	Lory	Toucan	Pigeon	Ostrich	Emu
Alkaline phosphatase (U/L)	20-160	15-150	10-90	10-80	20-120	20-135	20-250	15-255	80-250	20-230	150-350	70-300	80-290					140-320	130-230	
ALT (U/L)	5-12	5-11	5-13	5-10	5-12	5-11	5-11	6-12	5-13	5-12	5-11	5-11	5-12					15-25		
AST (U/L)	100-365	130-350	110-345	145-350	145-395	145-345	95-345	145-355	125-345	100-300	120-370	100-350	150-365	150-285	150-275	150-350	120-340	100-350	190-245	75-390
Amylase (U/L)	210-530	205-510					205-490	200-510	100-450	150-550	200-645									
BUN (mg/dl)	3-5.4	3.1-5.3					2.9-5	3.5-1	2.8-5.4	3-5.6	3-5.5	2.9-5.4	3-5.4							
Calcium (mg/dl)	8.5-13	8.5-14	8-14	6.5-11	6.5-13	5.5-13.5	8-13	8-13	7-15	8.5-13	7-13	6.5-13	7-13.5	7-12	7-13	6.5-13	9-15	8-13	13-20	8-13
Cholesterol (mg/dl)	160-425	180-305	95-335	145-275	150-400	150-400	140-360	145-355	120-400	100-390	130-350	130-340	130-295							
Creatinine (mg/dl)	0.1-0.4	0.1-0.4	0.1-0.4	0.1-0.4	0.1-0.4	0.1-0.4	0.1-0.4	0.1-0.4	0.1-0.4	0.1-0.5	0.1-0.4	0.1-0.4	0.1-0.4							
CO₂ total (mmol/L)	13-25	13-26	14-25	14-25	14-24	14-26	13-25	14-25	14-25	14-25	14-24	14-25	14-24							
CPK (U/L)	165-412	55-345	52-245	90-300	50-400	55-350	30-245	95-305	35-355	100-330	220-345	100-330	100-300					220-375	305-875	310-900
GGT (U/L)	1-10	1-12	2.5-18	1-10	1-12	1-14	1-30	1-45	1-15	1-30	1-20	1-15	1-18							
GLDH (U/L)	0-9.9	0-9.9	0-9.9	0-9.9	0-9.9	0-9.9	0-9.9	0-9.9	0-9.9	0-9.9	0-9.9	0-9.9	0-9.9							
Glucose (mg/dl)	190-350	190-345	195-405	190-390	205-345	205-435	200-445	185-355	200-345	145-345	145-245	140-250	125-300	200-350	200-325	200-300	220-350	100-250	150-350	100-310
LDH (U/L)	145-465	155-425	105-355	145-435	145-445	120-450	120-455	220-550	120-390	70-350	200-425	150-350	125-380				200-420	55-155	425-1500	310-1200
Lipase (U/L)	35-325	35-225					30-280	25-275	30-290	30-250	35-275									

Reference ranges from the Avian and Wildlife Laboratory, University of Miami, Miami, Fla.
*Unless otherwise noted, these values were obtained with Kodak Ektachem and Dupont Analyst chemistry analyzers using 24-hour-old lithium heparinized plasma.
†Performed using non-temperature-compensated refractometer.
ALT, Alanine aminotransferase; AST, aspartate aminotransferase; BUN, blood urea nitrogen; CO_2, carbon dioxide; CPK, creatine phosphokinase; GGT, gamma glutamyl transferase; GLDH, glutamic dehydrogenase; LDH, lactate dehydrogenase; T_4, thyroxine; A/G, albumin-globulin (ratio).

Table 27-2
Reference values for selected plasma biochemistries—cont'd

Determination	African grey	Amazon	Lovebird	Budgerigar	Parakeet	Canary	Cockatiel	Cockatoo	Conure	Macaw	Ecletus	Senegal	Pionus	Quaker	Jardine's	Lory	Toucan	Pigeon	Ostrich	Emu
Phosphorus (mg/dl)	3.2-5.4	3.1-5.5	2.8-4.9	3.0-5.2	3.0-5.3	2.9-4.9	3.2-4.8	2.5-5.5	2-10	2-12	2.9-6.5	2.9-9.5	2.9-6.6					2.9-6.5	2.9-6.6	3-6.5
Potassium (mmol/l)	2.9-4.6	3.0-4.5	2.1-4.8	2.2-3.9	2.3-4.2	2.2-4.5	2.4-4.6	2.5-4.5	3-4.5	2-5	3.5-4.3	3-5	3.5-4.6					3.9-4.7	2.2-6.5	2.3-7
Sodium (mmol/l)	157-165	125-155	132-168	139-165	138-166	135-165	130-153	130-155	135-149	140-165	130-145	130-155	145-155					140-150	130-155	135-155
Total bilirubin	0-0.1	0-0.1	0-0.1	0-0.1	0-0.1	0-0.1	0-0.1	0-0.1	0-0.1	0-0.1	0-0.1									
Total protein[†] (g/dl)	3-4.6	3-5	2.8-4.4	2.5-4.5	3-4.4	2.8-4.5	2.4-4.1	3-5	3-4.2	2.1-4.5	2.8-3.8	3.5-4.4	2.2-4	2.8-3.6	2.8-4	2-3.5	3-5	2-5.5	2.5-3.8	3.4-5.5
Triglycerides (mg/dl)	45-145	49-190	45-200	105-265	55-250	60-265	45-200	45-205	50-300	60-135	70-410									
Uric acid (mg/dl)	4.5-9.5	2.3-10	3.5-11	4.5-14	4.5-12	4-12	3.5-10.5	3.5-10.5	2.5-11	2.5-11	2.5-11	2.3-10	3.5-10	3.5-11.5	2.5-12	2.8-11.5	3.8-13	3.5-12	6.5-14.5	5.3-15
Special Biochemistries[‡]																				
Bile acids (μmol/L)	13.90	18-60	13-65	15-70	18-79	23-90	20.85	25-87	15-55	6-35	10-35	20-85	14-60	25-65	25-65	20-65	5-40	8-65	5-50	8-55
T_4 (μg/dl)	0.3-2.1	0.1-1.1	0.2-4.3	0.5-2.1	0.6-2.3	0.7-3.2	0.7-2.4	0.7-4.1	0.5-2	0.5-2.3	0.5-2.5									
Plasma Protein Electrophoresis Values[§]																				
Pre-albumin (g/dl)	0.03-1.35	0.35-1.05	0.6-1.2		0.6-1		0.8-1.6	0.24-1.18	0.18-0.98	0.05-0.7	0.4-1.04	0.19-0.64	0.19-0.93	0.48-1.13	0.18-0.32	0.48-0.76				
Albumin (g/dl)	1.57-3.23	1.9-3.52	2-2.8		1.9-3		0.7-1.8	1.8-3.1	1.9-2.6	1.24-3.11	2.3-2.6	1.45-2.28	2.19-3.29	1.26-2.52	1.85-2.23	1.26-1.96				
Alpha-1 globulins (g/dl)	0.02-0.27	0.05-0.32	0.08-0.21		0.06-0.16		0.05-0.40	0.05-0.18	0.04-0.23	0.04-0.25	0.09-0.33	0.02-0.2	0.1-0.19	0.04-0.25	0.04-0.15	0.04-0.14				
Alpha-2 globulins (g/dl)	0.12-0.31	0.07-0.32	0.08-0.25		0.07-0.2		0.05-0.44	0.04-0.36	0.08-0.26	0.04-0.31	0.11-0.27	0.08-0.16	0.05-0.28	0.05-0.28	0.08-0.15	0.04-0.23				
Beta globulins (g/dl)	0.15-0.56	0.12-0.72	0.19-0.4		0.22-0.45		0.21-0.58	0.22-0.82	0.07-0.47	0.14-0.62	0.17-0.43	0.26-0.58	0.08-0.45	0.20-0.55	0.18-0.38	0.15-0.58				
Gamma globulins (g/dl)	0.11-0.71	0.17-0.76	0.18-0.45		0.19-0.3		0.11-0.43	0.21-0.65	0.12-0.61	0.1-0.62	0.18-0.55	0.14-0.23	0.18-0.4	0.13-0.48	0.12-0.26	0.13-0.29				
A/G ratio	1.6-4.3	1.9-5.9	2.5-4.6		4.0-5.3		1.5-4.3	2.0-4.5	2.2-4.3	1.6-4.3	2.6-4.1	2.2-3.9	3.4-5.5	2.2-3.2	2.9-3.5	2.3-4.0				

[†]Performed by radioimmunoassay.
[§]Performed using Bechman Paragon System and SPEP-II gels.

REFERENCES

1. Kossoff S et al: Standardization of avian diagnostics in hematology and chemistry, *Proc Annu Conf Assoc Avian Vet*:57-63, 1996.
2. Lane RA: Clinical techniques and diagnostics: the quality of results are directly proportional to the quality of the samples sent, *Proc Annu Conf Assoc Avian Vet* :318-322, 1993.
3. Reference ranges from the Avian and Wildlife Laboratory, University of Miami, Miami, Fla.
4. Clubb SL et al: Hematologic and serum biochemical reference intervals in juvenile *Eclectus* parrots *(Eclectus roratus)*, *J Assoc Avian Vet* 4:218-225, 1990.
5. Clubb SL et al: Hematologic and serum biochemical reference intervals in juvenile cockatoos, *J Assoc Avian Vet* 5:16-26, 1991.
6. Clubb SL et al: Hematological and serum biochemical reference intervals in juvenile macaws *(Ara sp)*, *J Assoc Avian Vet* 5:154-162, 1991.

Index

A

ABC; *see* Airway, breathing, circulation
Abdominal air sacs, 50, 456-457
Abdominal distention, 85-94
Abdominal fluid accumulation, 87
Abdominal hernia, 89
Abdominal space, 7
Abdominocentesis, 90-91
Abducent nerve, 156
Abductor alulae muscle, 502
Abnormal droppings, 62-69, 374
Abscess, 114, 277
    lacrimal gland, 284, 287
    lacrimal sac, 277, 287, 288
    orbital, 284
    oropharyngeal, 365-366
    periorbital, 284, 288
    respiratory, 59
    retrobulbar, 287
Accessory sex glands, 328
ACD; *see* Anticoagulant dextrose
Acepromazine, 569
Acetylcysteine (Mucomyst), 574
Acetylsalicylic acid, 572
Achromatosis, 384
Acoustic meatus, external, 452
ACTH; *see* Adrenocorticotropic hormone
Activated charcoal, 229, 579
Active range-of-motion exercises, 525
Activity, postoperative, 525
Acyclovir (Zorivax), 563
Adenocarcinoma, 120
Adenohypophysis, 313-314
Adenosine triphosphate (ATP), 336, 380
Adenovirus, 78
Adequan; *see* Glycosaminoglycan
Adnexa, 264, 268-269
Adrenal adenitis, 550
Adrenal degeneration, 550
Adrenal glands, 341-345
Adrenal neoplasia, 344
α_2-Adrenoceptor agonists, 466
α_2-Adrenoceptor antagonists, 472
Adrenocorticotropic hormone (ACTH), 315
Adversion, 151
Aflatoxin, 253
Afterfeathers, 99
Age, signalment and, 335
Aggression, 188
Air in crop, 222-223
Air capillaries, 50
Air sacculitis, 59, 459, 460
Air sacs, 50, 54, 88, 403-404, 453-457, 459, 460, 461
Airway, breathing, circulation (ABC), 17
ALAD; *see* Aminolevulinic acid dehydratase
Albendazole suspension (Valbazen), 564
Albon; *see* Sulfadimethoxine
Albumen, 14
Aldosterone, 315, 343
Alga, blue-green, neurotoxicoses and, 252
Alimentary system, necropsy and, 546-547

Allantois, 193
Allergens, periorbital swelling and, 289
Allometric scaling of physiologic variables, 479
Allopurinol, 579
Alphachoralose, 242
Alpha-type cells, 314
Alphaxalone/alphadalone (Saffan), 569
Altitude, pet parrots and, 128-129
Altricial species, 213
Aluminum hydroxide, 579
Amantadine, 563
Amazon foot necrosis, 118-119
Amazon hemorrhagic syndrome, 66
Ambiens muscle, 518
Ambient environment, 215
Amikacin, 555, 574
Amino acids, 376, 378, 381
Aminoglycosides, 230-231
Aminolevulinic acid dehydratase (ALAD), 239
Aminopentamide (Centrine), 579
Aminophylline, 574, 579
Amitriptyline (Elavil), 576
Amnion, 193
Amoxicillin, 555
Amoxicillin clavulinate, 555
Amoxicillin sodium, 555
Amoxicillin trihydrate, 555
Amphotericin B (Fungizone), 231, 561, 574
Ampicillin, 555
Ampicillin sodium, 555
Ampicillin trihydrate, 555
Amprolium (Corid), 434, 564
Amyloidosis, 344
Anafranil; *see* Clomipramine
Analgesics, 467, 569-571
Ancobon; *see* Flucytosine
Anemia, 10-11, 23
Anesthesia, 173, 231-232, 392, 393, 464-492, 569-571
Anesthesia machine, 475
Animal-origin products, 377
Anisocoria, 268
Ankyloblepharon, 279

Antebrachium, 499-501
Anterior segment of eye, 265
Anthelmintics, 435-436
Antiacaricides, 436
Antibiotics, 38, 206, 207
Anticoagulant dextrose (ACD), 579
Anticoagulants, 247
Antifungal agents, 61, 561-562
Antimicrobial agents, 37-38, 38, 61, 555-559
Antimycobacterial agents, 560
Antiparasitic agents, 564-568
Antiprotozoal agents, 434-435
Antirobe; *see* Clindamycin, 557
Antisedan; *see* Atipamezole
Antiviral agents, 563
Apicomplexan protozoa, 439-442
Appertex; *see* Clazuril
Appetite, 3
Apteria, 96, 103
Aquasol A Parenteral, 578
Aralen; *see* Chloroquine phosphate
Arginine vasotocin (AVT), 315, 332-333, 341
Arsenic toxicity, 241
Arterial blood pressure, 480
Arthropods, 445-446
Articular gout, 384
Artificial incubation, 195-196, 370
Ascaridiasis, 443
Ascaris, 79, 443
Ascites, 58
Asparaginase (Elspar), 579
Aspartate aminotransferase, 13-14
Aspergillosis, 80, 121, 403-404
Aspergillus, 205
Aspiration, sinus, 53, 54
Aspirin, 572
Assisted hatching, 209-211
Atabrine HCl Tablets; *see* Quinacrine
Atarax; *see* Hydroxyzine
Ataxia of head and limbs, 385-386
Atelectasis, 483
Atherosclerosis, 546
Atipamezole (Antisedan), 569
Atoxoplasma, 440
ATP; *see* Adenosine triphosphate
Atracurium, 478-479

Atretic follicles, 325
Atria, 49
Atropine, 18, 467-468, 554, 569, 579
Attitude, 3, 5, 128-129, 217
Aura, 170
Auscultation, 8, 52, 53, 219, 479-480
Autonomic disorders, seizures and, 180
Avatec; *see* Lasalocid
Avian anesthesia, 173, 231-232, 392, 393, 464-492, 569-571
Avian dermatology, 95-123, 218, 372, 382, 426, 528
Avian endocrinology, 105, 313-358, 549-550
Avian herpesvirus, 86
Avian medicine, 37-40
  abdominal distention, 85-94
  abnormal droppings, 62-69
  anesthesia, 464-492
  behavior problems in pet parrots, 124-147
  bill problems, 359-368
  blood tables, 590-595
  chemistry tables, 590-595
  coelomic distention, 85-94
  dermatology, 95-123
  diagnostic workup plan in, 1-16
  dyspnea and other respiratory signs, 47-61
  embryologic considerations, 189-212
  emergency medical syndromes and, 554
  endocrinology, 313-358
  endoscopic diagnosis, 449-463
  feather picking and, 142
  formulary, 553-589
  imaging interpretation, 391-423
  limb dysfunction, 493-526
  necropsy, 542-552
  neonatal problems, 213-227
  neurologic signs, 148-169
  nutrition, 369-390
  ophthalmic disorders, 264-312
  oropharynx problems, 359-368
  parasitism of caged birds, 424-448

Avian medicine—cont'd
  regurgitation, 70-84
  reproductive disorders, 183-188
  respiratory signs, 47-61
  seizures, 170-182
  soft tissue surgery, 527-541
  straining disorders, 183-188
  supportive care and shock, 17-46
  toxicity and, 82, 228-263
  vomiting and regurgitation, 70-84
Avian nutrition; *see* Nutrition
Avian pox, 58, 292-295, 361
Avian thrombocytes, 12
Avian toxicology, 1, 30, 228-263
  drug, 230-235
  lead, 164, 176-177, 237-239, 416
  metal, 81, 164-165, 235-241
  necropsy and, 205-206, 551
  neuropathic syndrome and, 164-166
  reproductive tract disease and, 336-337
  seizures and, 176-178
  vomiting and regurgitation and, 81-82
  zinc, 164-165, 241
Avian viral serositis, 86
Avicides, 242
Avocado, 251
AVT; *see* Arginine vasotocin
Azithromycin (Zithromax), 555
Azotemia, 43

B

Bacteria, 76, 226
Bacterial meningoencephalitis, 551
Bactericidal antibiotics, 38
Bactericidal antimicrobials, 38
Bacteriostatic antibiotics, 38
Bactrim; *see* Trimethoprim/sulfamethoxazole
BAL; *see* Dimercaprol
Banamine; *see* Flunixin meglumine
Barbiturates, 172-173
Barbs of feather, 101-102
Barbules of feather, 101-102
Barium sulfate (Barotrast), 418, 579
Barotrast; *see* Barium sulfate

Basilic vein, 10, 33-34
Basophils, 12
Baycox; *see* Toltrazuril
Baytril; *see* Enrofloxacin
Beak, 6, 120-122, 221-222, 359-365
"Beak-wrestling," 128-129
Behavior, 3-4
  displacement, 127-128, 187-188
  parrots and, 124-147
  reproductive hormonal pressure and, 187-188
  vomiting and regurgitation and, 82-83
Benzodiazepines, 466-467
Beta lactams, 232
Beta-type cells, 314
Biaxin; *see* Clarithromycin
Bicarbonate therapy, 38-39
Biceps brachii muscle, 498
Bile acids, 13-14
Bile duct obstruction, 15
Bill, 6, 120-122, 221-222, 359-365
Biochemistries, plasma, 593-594
Biopsy, 54, 91, 458-462
Biopsy forceps, 450-451
Bismuth sulfate, 579
Biting, pet parrots and, 124, 133
Blastoderm, 189, 192
Blastodisc, 189, 192
Blastula, 192
Blindness, 305-312
Blood in stool, 69
Blood clotting disorders, 386-387
Blood gas analysis, 483
Blood loss, anemia and, 11, 23
Blood parasites, 13
Blood pressure, arterial, 480
Blood tables, 590-595
Blood transfusions, 36-37, 482
Blood values, normal, 590-591
Blue-green alga, neurotoxicoses and, 252
Body temperature, 45, 484, 527-528
Bond, client-animal, 2-3, 5
Bone healing, 525
Bone marrow, 13
Bone scans, 421
Bone scintigraphy, 421

Botulism, 165, 386
Bowl hygiene, 2
Brachium, 496-498
Brachygnathism, maxillary, 222
Bragnathism, 364-365
Braille sling, 507-508
Brain syndromes, 148, 151-152, 171
Breakout, egg, 200-202
Breathing system, inhalation anesthesia and, 475, 476
Breeders, 330, 380
Brethine; *see* Terbutilene
Brewer's yeast, 377
Bristle feather, 100, 101
Bronchi, 49
Brook patch, 335
Bumblefoot, 118, 385
Buphthalmos, 272-274
Buprenorphine (Buprenex), 569
Burns, 23-24
Butorphanol (Torbugesic), 569
Byproducts, nutrition and, 377

C

CAA; *see* Cyanoacrylate tissue adhesives
Cage, oxygen, 44
Cage hygiene, 1, 5
Cage layer fatigue, 15
Cage problems, 125
Cage territoriality, 125, 134
Caged birds, parasitism of; *see* Parasites
Calcification, 190
Calcitonin cells, 318
Calcium, 14, 39-40, 192, 203, 232, 333-335
Calcium EDTA, 575
Calcium glubionate (Neocalglucon), 40, 578
Calcium gluconate, 40, 554, 578
Calcium lactate/calcium glycerophosphate (Calphosan), 192, 578
Calcium levulinate, 578
Calla lily, 249
Calphosan; *see* Calcium lactate/calcium glycerophosphate
Candida albicans, 79-80, 361

Candidiasis, 79-80, 121
Candling, 199-200
Cannibalism, 108
Capillaria, 78, 367, 443-444
Capillaries, air, 50
Caponization, 329
Capons, 329
Carafate; *see* Sucralfate
Carbamates, 177-178, 243-245
Carbamazepine (Tegretol), 576
Carbenicillin (Geocillin), 555, 574
Carbohydrate metabolism, 345-346
Carbohydrates, 381
Carbon monoxide, necropsy and, 206
Carcinoma, squamous cell, 115
Cardiac failure, 23
Cardiac massage, 18
Cardiogenic shock, 20
Cardiomyopathy, 58
Cardiopulmonary arrest, 23
Cardiopulmonary resuscitation, ABCs of, 17
Cardiovascular emergencies, 20-23
Cardiovascular system, 479-480
    anesthesia and, 481-482
    necropsy and, 545-546
    plant toxicoses and, 250-251
Care, supportive, and shock, 17-46
Carfentanil, 490
Carina, 398
Carnidazole (Spartrix), 434, 564
Carotenes, 103
Carotenoids, 103, 383
Carpometacarpus, 501
Carprofen, 572
Carpus, 501
Castor bean, 249-250
Castration, 329
Casuariiformes, 488-490
Cataracts, 306
Catgut, chromic, 532
Catheters, 44, 451
Caudal thoracic air sac, 50, 455-456
Cefadroxil, 555
Cefazolin, 555
Cefotaxime, 555
Ceftidime, 555

Ceftiofur (Naxcel), 555
Ceftriaxone (Rocephin), 556
Cellulase, 381
Cellulose, 381
Central nervous system (CNS), 148
    necropsy and, 550-551
    nutrition and, 375, 385-386
Central vestibular disease, 158
Centrifugal flotation procedure, 430-431
Centrine; *see* Aminopentamide
Cephalexin, 556
Cephalosporins, 232
Cephalothin, 556
Cephradine, 556
Cephulac; *see* Lactulose
Cere, 6, 50-51
Cerebellar syndrome, 154-155
Cerebellum, 154-155
Cerebral syndrome, 151-152
Cervical air sac, 50
Cervical syndrome, 158-160
Cervicocephalic air sac, 50, 403
Cervicothoracic syndrome, 160
Cestodes, 79, 442
Charcoal, activated, 229, 579
ChE; *see* Cholinesterase
Chelating agents, 575
Chemical restraint agents, 569-571
Chemistry profile, 9, 13-16
Chemistry tables, 590-595
Chlamydiae, 9, 204
Chlamydiosis, 14, 93
Chloramphenicol, 206, 232, 574
Chloramphenicol palmitate, 556
Chloramphenicol succinate, 556
Chlorinated hydrocarbons, 245-246
Chlorine, 81
Chloroquine phosphate (Aralen), 564
Chlorpromazine, 576
Chlorsulon, 435
Chlortetracycline, 556
Choanae, 49, 51, 222
Choanal atresia, 61
Choanal opening, 71
Chocolate, neurotoxicoses and, 252
Cholecalciferol, 247-248

Cholesterol, 15-16
Cholinesterase (ChE), 244
Chorion, 193
Christmas cherry, 251
Chromaffinoma, 344
Chromic catgut, 532
Chronic ulcerative dermatitis (CUD), 116
Cimetadine (Tagamet), 579
Ciprofloxacin, 556
Circovirus, 108, 109
Circulatory shock, 20
Cirrhotic liver, 175-176
Cisapride (Propulsid), 579
Citric acid, 556
Citrinin, 253
Clarithromycin (Biaxin), 556
Clavicle, 494, 503
Clavicular air sac, 50, 453-454
Claws, 117-118
Clazuril (Appertex), 564
Client-animal bond, 2-3, 5, 125, 126, 134-136
Clindamycin (Antirobe), 557
Cloacal papilloma, 185-186
Cloacal papillomatosis, 26
Cloacal prolapse, 26
Cloacal promontory, 328
Cloacapexy, 537-538
Cloacitis, 186
Clomipramine (Anafranil), 576
Clopidol-25% (Coyden-25), 564
Clorsulon (Curatrem), 564
Clotrimazole (Lotrimin), 561, 574
CNS; see Central nervous system
Coaptation splinting, 511
Coban 45; see Monensin
Coccidia, 439-442
Cockatiel paralysis syndrome, 375
Cocoa beans, 252
Coelom, 85
Coelomic distention, 85-94
Coelomic space, 7
Coffee bean, 250, 252
Colchicine (ColBenemid), 580
Colitis, 186
Colloids, 33
Columbiformes, 488

Commands, pet parrots and, 130-131
Commercial hand-feeding formula, 371
Common digital extensor muscle, 499
Complete blood count, 9, 10-13, 53, 89
Complexus muscularis, 113
Computed tomography, 420-421
Conduction, heat loss and, 45
Conformation, 3, 5, 217
Congenital disorders, seizures and, 180
Congenital heart disease, 222
Conjunctival parasites, 283-284, 289
Conjunctivitis, 279-284
Consent, informed, 18
Continuous breeders, 330
Contour feathers, 96-97
Contrast techniques, imaging and, 418-419
Convection, heat loss and, 45
Copper toxicity, 235-236
Coracobrachialis caudalis muscle, 495-496
Coracoid, 495
Coracoid fractures, 502-503
Coracoid luxation, 511
Core body temperature, 45, 484
Corid; see Amprolium
Cornea, diseases of, 295-301
Corticosteroids, 12, 38, 343
Corticosterone, 315, 342-343
Corticotrophin-releasing factor (CRF), 314
Cortisol, 342
Cotton-tipped applicators, 531
Coughing, 4
Courtship behaviors, 82
Coverts, 98
Coxofemoral luxation, 523-524
Coyden-25; see Clopidol-25%
Cranial nerve deficits, 154, 156
Cranial thoracic air sacs, 50, 454-455
Creatine kinase, 13-14
CRF; see Corticotrophin-releasing factor
Crooked toes, 225, 226
Crop, 7, 71-72, 222-223, 374

Crop impaction, 82
Crop stasis, 82-83, 223
Crude oil, 255-256
Cryptophthalmos, 279
Cryptosporidium, 439-440
Crystalloids, 33
CUD; *see* Chronic ulcerative dermatitis
Cultures, 9, 52
Cuprimine; *see* Penicillamine
Curatrem; *see* Clorsulon
Cutaneous glands, 113
Cutaneous papillomas, 290
Cuticle layer, 73
Cutting edge of horny beak, 70
Cyanoacrylate tissue adhesives (CAA), 532
Cyanoacrylics, 363-364
Cyanocobalamin, 578
Cyclophosphamide (Cytoxan), 580
Cyno-Veneer kit, 363-364
Cysts, 105, 338
Cytoxan; *see* Cyclophosphamide

D

Dairy products, 377
Daylight, 330, 351
Death; *see* Necropsy
Deferoxamine mesylate (Desferal), 575
Dehydration and fluid therapy, 31-37
Delta-type cells, 315
Deltoid crest, 504
Deltoideus major muscle, 497
Deltoideus minor muscle, 497
Denegard; *see* Tiamulin
Deoxyribonucleic acid probe, 9
Dermatitis, 284
Dermatology, 95-123, 218, 372, 382, 426, 528
Desferal; *see* Deferoxamine mesylate
Desflurane, 474-475
Destabilization, postoperative, 525
Detergent sedimentation, 432-433
Determinate layers, 330
Dexamethasone, 345, 571
Dexamethasone Na phosphate, 554, 571

Dextrans, 36, 39, 578
Dextrose, 554, 578
Diabetes insipidus, 316
Diabetes mellitus, 346-350
Diagnostic workup plan, 1-16, 47, 48
Diaphyseal fractures, 519-520
Diaphyseal humerus fractures, 504-506
Diarrhea, 25-26, 66, 67
Diazepam, 172, 466-467, 489, 554, 569, 570, 576
Diazepam-ketamine, 471
Diazinon, 206
Diet; *see* Nutrition
Diffenbachia, 249
Diflucan; *see* Fluconazole
Digestive system, 74-75, 405-406
Digital extensor muscle, common, 499
Digitalis glycosides, 232-233
Digitoxin, 232-233
Digits, 501, 514
Digoxin, 232-233, 580
Dimercaprol (BAL), 575
Dimercaptosuccinic acid (DMSA), 575
Dimetridazole (Emtryl), 564
Dimorphism, sexual, 336
Dinoprostone (Prepidil Gel), 94, 192, 572
Dioctyl Na sulfosuccinate (Diocto), 580
Diphenhydramine, 576, 580
Direct flotation procedure, 429-430
Direct smear technique, 427-428
Dirt-floored enclosures, 425
Disequilibrium, 155
Disinfection of eggs, 206-207
Dispharynx, 79
Displacement behaviors, 127-128, 187-188
Distal wing, 501-502
Diuretics, 173
Diverticulum, 48, 54
DMSA; *see* Dimercaptosuccinic acid
Dominance training, 126-127, 129-131, 132, 133-142
Domitor; *see* Medetomidine

Doppler flow apparatus, ultrasonic, 480
Dopram-V; *see* Doxapram
Dorsal antebrachium, 499-500
Dorsal manus, 501-502
Dorsal metencephalon, 154-155
Dorsal wing, 497
Dorsolateral strabismus, 154
Down feathers, 98, 99
Doxapram (Dopram-V), 18, 554, 580
Doxepin (Sinequan), 576
Doxorubicin, 580
Doxycycline (Vibramycin; Vibravenos), 234, 557, 574
Drama reward, parrots and, 128-129
Drapes, surgical, 529
Droncit; *see* Praziquantel
Droppings, abnormal, 4, 62-69, 374
Drug toxicities, 230-235
Drugs; *see* Avian medicine
Dry skin, parasites and, 426
Duck plague virus, 77
Duck viral enteritis, 178
Duck viral hepatitis, 178
Ductus deferens, 328
Dwarfism, 316
Dysmetria, 155
Dyspnea, 19, 47-61
Dystocia, 88, 93-94, 339-341, 386, 550

E

Ear, 7, 218, 222, 551-552
ECG; *see* Electrocardiography
EDTA; *see* Ethylenediaminetetraacetic acid
Egg(s), 190, 191, 196-202, 332-333, 417
  candling of, 199-200
  disinfection of, 206-207
  floating, 200
  nutrition and, 377
  weights of, 198-199, 207
Egg binding, 28, 87-88, 93-94, 183-185, 190-192, 339-341, 386, 550
Egg breakout, 200-202
Egg death, 200-202

Egg hygiene, 196-198, 214-215
Egg laying, excessive, 338-339
Egg membranes, 192-193
Egg necropsy; *see* Necropsy
Egg peritonitis, 28
Egg retention, 87-88
Egg therapeutics, 206-207
Eggshell, 333-335, 386
Eimeria, 439
Elavil; *see* Amitriptyline
Elbow joint luxation, 512
Electrocardiography (ECG), 480
Electrolytes, 42-43
Electrosurgical instruments, 529-531
Elephant ear, 249
ELISA; *see* Enzyme linked immunosorbent assay
Elspar; *see* Asparaginase
Embryology, 189-212
Emergency medical syndromes, 19-30, 554
Emphysema, 61, 88
Emtryl; *see* Dimetridazole
Emu, anesthesia for, 489
Encephalitis, 178, 550-551
Encephalomyelitis, 86, 178-179
Encephalopathy, hepatic, 175-176, 386
Endocrine glands, 80-81
Endocrinology, avian, 105, 313-358, 549-550
Endodocumentation, 462-463
Endoparasites, 427
Endoscopic biopsy, 458-462
Endoscopy, 91, 449-463
Endotracheal tubes, 476
Energy, 41, 42, 380-381
Enophthalmia, 271-272
Enrofloxacin (Baytril), 233, 557, 574
Enteral nutritional support, 41-42
Enteritis, 186
Environment, 1-2, 215
  feather plucking and, 140
  hydrocarbon contamination of, 255-256
  infertility and, 194
Enzyme linked immunosorbent assay (ELISA), 433

Eosinophils, 12
Epidermis, 112-113
Epididymis, 328
Epilepsy, 180-181
Epinephrine, 18, 554, 580
Epiphysis cerebri, 350-351
Epsilon-type cells, 315
Epson salt, 229
Equatorial climate, 329-330
Equine encephalomyelitis, 179
Equipment
    electrosurgical, 529-532
    for endoscopy, 449-451
    for inhalation anesthesia, 475-477
    necropsy and, 543
Ergot, 253
Erythromycin (Erythro), 557, 574
Escherichia coli, 204
E-Se; *see* Vitamin E/selenium
ESF; *see* External skeletal fixation
ESF-IM tie-in, 505-506, 509, 520, 521
Esophageal sac, 71
Esophagus, 71-72, 223
Essential amino acids, 381
Estradiol benzoate, 571
Estrogen, 331
Eta-type cells, 315
Ethambutol (Myambutol), 557
Ethylenediaminetetraacetic acid (EDTA), 590
Etorphine, 490
"Evil eye," 132, 133
Excessive screaming, pet parrots and, 124, 133
Exercises, range-of-motion, 525
Exophthalmus, 274-278
Exotic Animal Formulary, 553
Exploratory laparotomy, 534-536
Expulsion of egg, 332-333
Extensor longus digiti majorus muscle, 500-501
Extensor metacarpal radialis muscle, 499
Extensor metacarpi ulnaris muscle, 499-500
External acoustic meatus, 452
External coaptation splinting, 507-508
External skeletal fixation (ESF), 504-505, 507, 511, 519-520, 524
Extraretinal photoreceptors, 330-331
Extruded diet, 378
Eye, 6-7, 51, 218, 265-266, 268-269
    disorders of, 264-312, 373, 487
    "evil," 132, 133
    necropsy and, 551
Eyelid tears, 279, 289
Eyelids, malformations of, 222, 264, 278-295

F

Facial nerve, 156
Falcon herpesvirus, 77
Falconiformes, anesthesia for, 488
Familial disorders, seizures and, 180
Fansidar Tablets; *see* Pyrimethamine
Fasting, anesthesia and, 465-466
Fatty acids, 381
Fatty infiltration, 92
Fatty liver syndrome, 87
FDIM; *see* Focal directed illumination with magnification
Feather achromatosis, 373
Feather cysts, 105
Feather duster disease, 104
Feather grooming, 139
Feather loss, 103-112, 426
Feather picking, 95, 103-112, 107, 125, 137-142, 188, 383, 426
Feather plucking, 125, 137-141
Feathers, 8, 96-112, 218
    colors of, 103, 383-384
    nutrition and, 372-373, 382-384
    removing, 528
Fecal examination, 9, 66-69, 426-434
Feeding schedule, 371
Feeding utensils, 371
Feet, 117-120
Female infertility, 187
Female reproductive organ enlargement, 86-87
Female reproductive tract, anatomy of, 189-192
Femoral fractures, 518-520
Femoral head fractures, 520
Femorotibialis externus muscle, 517

Femorotibialis internus muscle, 518
Femorotibialis medius muscle, 518
Femur, 512-513
Fenbendazole, 434-435, 565
Fertilization, 189
Fiberoptic endoscopy, 449
Fibrosarcoma, 115
Fibula, 514
Fibularis longus muscle, 517
Figure-eight wrap, 507-508
Filariidea, 445
Filoplumes, 100
Fine needle aspiration (FNA), 90-91
Fistulas, 23-24
Fixatives, necropsy and, 542
Flagellated protozoa, 436-438
Flexible endoscopy, 450
Flexor alulae muscle, 502
Flexor carpi ulnaris muscle, 500
Flexor cruris lateralis muscle, 516
Flexor cruris medialis muscle, 516
Flexor digiti minoris muscle, 502
Flexor digitorum profundus muscle, 500
Flexor digitorum superficialis muscle, 500
Flexor metacarpi caudalis, 501-502
Flexor perforatus digiti IV muscle, 517
Floating eggs, 200
Flock greeting, 137
Flock hierarchy, 129
Flock leader, 127
Flotation solutions, 428-429
Fluconazole (Diflucan), 561
Flucytosine (Ancobon), 561
Fluid, 43, 87, 482, 554
Fluid deficit, 35-36
Fluid therapy, 31-37
Flumadine; *see* Rimantadine
Flunixin meglumine (Banamine), 572
Fluorescein stain, 270
Fluoroscopy, 419
Fluoxetine (Prozac), 576
FNA; *see* Fine needle aspiration
Focal directed illumination with magnification (FDIM), 449

Folic acid, 203
Follicle-stimulating hormone (FSH), 314
Food; *see* Nutrition
Foot holding, 130
Foot pads, 118
Forceps, 450-451
Forearm, 499-501
Foreign bodies, 68, 83, 224
Forgiveness, 133
Formalin ethyl acetate sedimentation, 431-432
Formulary, 553-589
Foster-rearing of neonates, 213, 214
Fovea, 265
Foxglove, 232-233
Fractures, 399
  of bill, 363-364
  of clavicle, 503
  coracoid, 502-503
  diaphyseal, 504-506, 519-520
  femoral, 518-520
  of humerus, 503-507
  of metacarpals, 510-511
  of midshaft, 504-506
  of pelvic limb, 518-523
  of phalanges, 522
  of radius, 507-510
  of scapula, 503
  of tarsometatarsus, 522
  of thoracic limb, 502-511
  of tibiotarsus, 520-521
  of ulna, 507-510
French molt, 109, 110
Fruits, 376
FSH; *see* Follicle-stimulating hormone
Fundus, 306
Fungicides, 242-243
Fungizone; *see* Amphotericin B
Furosemide, 580

G

Gadopentate dimeglumine (Magnevist), 580
Galliformes, anesthesia for, 488
Gamma-type cells, 314
Ganglia, inflammation of, 551

Gaseous anesthesia, 173
Gasoline, 255-256
Gastric ulcers, 83
Gastrocnemius muscle, 517-518
Gastrointestinal blockage, 224
Gastrointestinal emergencies, 23-27
Gastrointestinal stasis, 24
Gastrointestinal system, 218-219, 222-224
    contrast techniques and, 418
    imaging of, 405-406
    nutrition and, 373-374
    parasites and, 426
    plant toxicoses and, 249-250
Gastrulation, 192
Gavage feeding, 24
Gender, signalment and, 336
General anesthesia, 392
Generalized seizures, 170
Gentamicin, 557, 574
Geocillin; *see* Carbenicillin
Georgia Animal Poison Information Center, 230
Geriatric birds, nutrition and, 380
GH; *see* Growth hormone
GHRH; *see* Growth hormone–releasing hormone
Giardia psittaci, 78, 436-438
Girdle
    pelvic, 398-399, 402-403, 512
    thoracic, 398, 401, 493-496
Gizzard, 73
Gizzard malfunction syndrome, 80
Globe, 265, 270-278, 286-287
Glomus, 328
Glossopharyngeal nerve, 156-157
Glucagon, 40, 345-346, 348
Glucocorticoids, 342, 344-345
Glucose monitoring, 485
Glucose tolerance test, 349
Glucosuria, 348
Glycopyrrolate (Robinul-V), 467-468, 569
Glycosaminoglycan (Adequan), 580
Gnathotheca, 120
Goiter, 58, 80-81, 318-319, 386
GoLytely; *see* Sodium sulfate
Gonads, imaging of, 409

Gout, 384, 546
Gram stains, 9, 15, 52
Gram-negative bacteria, 37
Granulomas, 366
Grasping forceps, 451
Growth, poor, parasites and, 426
Growth hormone (GH), 314, 346
Growth hormone–releasing hormone (GHRH), 314

H

Haldol; *see* Haloperidol
Hallux, 514
Haloperidol (Haldol), 142, 576-577
Halothane, 231-232, 473-474
Hand-feeding formulas, 371
Hand-rearing of neonates, 214
Hatching, 207-212, 370
hCG; *see* Human chorionic gonadotropin
Head trauma, 29, 179
Healing, postoperative, 525, 532-533
Heart, 8, 404-405
Heart disease, congenital, 222
Heart rate, 528
Heat, supplemental, 45
Heat loss, 45
Heavy metal toxicity, 81, 164-165, 235-241
Helminths, 442-445
Hematochezia, 25-26
Hematology, 527, 590-595
Hematoma, 113, 289
Hematopoietic tissue, 547-548
Hemochromatosis, 58, 93, 375, 386
Hemocoelom, 58
Hemolysis, 590
Hemoparasites, 442
Hemorrhage, peri-pectin, 306
Hepatic disease, 26-27
Hepatic encephalopathy, 175-176, 386
Hepatic lipidosis, 92, 375
Hepatic system, 374-375
Hepatitis, duck viral, 178
Hepatomegaly, 374, 407, 408, 412
Herbicides, 206, 243
Hernia, abdominal, 89

Herpesvirus, 77, 86, 114, 119, 120, 178
Hetastarch (Hespan), 580
Histomoniasis, 78
History, 1-5, 95-96, 217, 267-270, 464-465
Homemade hand-feeding formula, 371
Hormonal pressure, reproductive, 187-188
Hormonal therapy, 338-341, 571-572
Horner's syndrome, 160, 166
Horny claw, 117
Household compounds, volatile, 81, 254
Human chorionic gonadotropin (hCG), 571, 577
Humerus, 496-498, 503-507
Humidity, artificial incubation and, 195-196, 197
Hunger traces, 343
Husbandry, 194, 215-217, 222-225
Hydration, 31
Hydrocarbons, 245-246, 255-256
Hydrocephalus, 180
Hydroxyzine (Atarax), 577, 580
Hygiene, 1, 2, 5, 196-198, 214-215
Hygromycin B (Hygromix 8), 565
Hyperadrenocorticism, 343-344
Hyperestrogenism, 334-335
Hypermetabolism, 40
Hyperparathyroidism, 322
Hypocalcemia, 14, 15, 174
Hypocalcemic tetany, 385
Hypoglossal nerve, 156-157
Hypoglycemia, 173-174, 386
Hypoparathyroidism, 322-323
Hypopennae, 99
Hypophosphatemia, 44
Hypotension, orthostatic, 480-481
Hypothalamic syndrome, 152-153
Hypothalamic-hypophyseal-adrenal axis, 342
Hypothalamohypophysial complex, 313-317
Hypothalamus, 313-317
Hypothermia, 484
Hypothyroidism, 105, 319-321
Hypovitaminosis A, 106, 116, 290, 291-292, 365-366, 367, 382, 384-385
Hypovitaminosis B_1, 163-164
Hypovitaminosis B_2, 164
Hypovitaminosis B_6, 164
Hypovitaminosis E, 163, 545
Hypovolemic shock, 20
Hypoxia, 163, 174-175
Hysterectomy, 326, 538-539

I

Ictus, 170
Identification of neonates, 216-217
Ileus, 86
Iliofibularis muscle, 516
Iliotibialis cranialis muscle, 514-515
Iliotibialis lateralis muscle, 516
Illinois Animal Poison Information Center, 230
IM pins; *see* Intramedullary pins
IM route; *see* Intramuscular route
Imaging, 53, 391-423
IM-ESF tie-in, 505-506, 509, 520, 521
Immunoassays, 433-434
Immunofluorescence antibody tests, 433
Immunomodulating agents, 563
Imprinting, 339
Incubation, 194-196, 335, 370
Incubation patches, 113
Incubator care, 198
Independence, parrots and, 125, 136-137
Inderal; *see* Propranolol
Indeterminate layers, 330
Induction, inhalation anesthesia and, 477
Infection, 226
  liver dysfunction and, 80
  necropsy and, 203-205
  neurologic syndromes and, 150-151
  neuropathic syndrome and, 166
  polyomavirus, 544
  regurgitation and, 76-80
  Sarcocystis, 544

Infection—cont'd
  seizures and, 178
  vomiting and, 76-80
Infertility, 186-187, 193-194, 386
Infiltrative splanchnic neuropathy, 86
Informed consent, 18
Infraorbital rhinitis, 284
Infraorbital sinus, 48-49, 54, 266-267
Infraorbital sinusitis, 284
Infundibular cleft, 70, 71
Infundibulum, 49, 326
Infusion needle, endoscopy and, 451
Ingluvies, 452-453
Ingluviotomy, 534
Inhalation anesthesia, 473-479
Inhalation toxicity, 548-549
Insect bite hypersensitivity, 106, 107
Insecticides, 81, 205-206, 243-247
Instrumentation; *see* Equipment
Insulin, 346, 348, 349, 571
Insulin-dependent diabetes, 346-350
Integument, 372, 382, 543-544
Interceptor; *see* Milbemycin oxime
Interferon, 563
Internal laying, 189
Interosseous dorsalis muscle, 501
Interosseous ventralis muscle, 502
Interstitial cells, 328
Intestinal peritoneal cavity, 457
Intestinal tract, 74-75
Intoxication; *see* Avian toxicology
Intramedullary (IM) pins, 505, 509-510, 519
Intramuscular (IM) route, 37, 469
Intraocular disease, 301-305
Intraosseous catheterization, 481
Intraosseous (IO) route, 34, 469
Intravenous catheterization, 481
Intravenous (IV) route, 33-34, 469
Intubation, 19, 477-478
IO route; *see* Intraosseous route
Iodine, 578
Iohexol (Omnipaque), 418, 419, 580
Ipronidazole (Ipropran), 565
Iris, 265, 266
Iron dextran, 39, 578
Iron storage disease, 86, 386

Iron toxicity, 236-237
Islets of Langerhans, 345-350
Isoflurane, 451, 474, 489, 569
Isoniazid (Isoniazid Tablets), 557
Isospora, 439
Isotonic crystalloids, 33
Isthmus, 326
Itraconazole (Sporanox), 562
IV route; *see* Intravenous route
Ivermectin, 435, 436, 565

J

Jackson Rees circuit, 476
Jerusalem cherry, 251
Jimsonweed, 251
Jugular vein, 9-10, 33
Junctionopathies, neuromuscular, 162

K

Kaolin/pectin (Kaopect), 580
Kappa-type cells, 315
KE apparatus; *see* Kirschner-Ehmer apparatus
Keel, 398, 493
Keratitis, 295-301
Kerosine, 255-256
Ketamine, 232, 470-471, 489, 569, 570
Ketoconazole (Nizoral), 562
Ketofen; *see* Ketoprofen
Ketonuria, 348
Ketoprofen (Ketofen), 572
Kidneys, 64, 80, 407-408, 458, 459
Kilovolt peak (kVp), 391
Kirschner-Ehmer (KE) apparatus, 519
Knemidokoptic mites, 107, 115, 289, 291, 362, 445-446
Koilin layer, 73
kVp; *see* Kilovolt peak
Kyphosis, 161

L

Laboratory studies, 10, 426-434
Lacrimal gland abscess, 284, 287
Lacrimal sac abscesses, 277, 287, 288

Lactase, 377
Lactate dehydrogenase, 13-14
Lactulose (Cephulac), 581
Laddering, 130, 132, 133
Lafora body neuropathy, 167, 180
Lameness, 385
Laparotomy, exploratory, 534-536
Large intestine, 75
Laryngeal mound, 70
Laryngitis, 58-59
Lasalocid (Avatec), 565
Lateral thigh, 514
Latissimus dorsi muscle, 498
Lavage, tracheal, 53
Lead toxicity, 164, 176-177, 237-239, 416
Left to right lateral (LRL) view, 394
Leg, 514
Legumes, 376
Lens, 266
Leukocytosis, 11-12
Leukopenia, 12
Leukopoiesis, decreased, 12
Leuprolide acetate (Lupron), 572
Levamisole (Levasole; Tramisol), 233, 435, 563, 565-566
Levothyroxine (l-thyroxine), 572
Leydig's cells, 328
LH; *see* Luteinizing hormone
LHRH; *see* Luteinizing hormone–releasing hormone
Lice, 106
Lids, malformations of, 222, 264, 278-295
Light, 330, 351
Lily of the valley, 250
Limb; *see* Pectoral limb; Pelvic limb; Thoracic limb
Limb dysfunction, 493-526
Limberneck, 165, 386
Lincomycin, 558
Lincomycin/spectinomycin (LS-50 Water Soluble), 558
Lipemia, 590
Lipomas, 115
Liposarcoma, 115
Liquamycin; *see* Oxytetracycline
Liquid stool, 66, 67

"Little earthquake," 132
Liver, 13-14, 15, 75, 80, 92-93, 120, 175-176, 407, 408, 458-459, 546-547
Local anesthesia, 468
Locomotion, pineal gland and, 351
Locoweed, 252
Lorelco (Probacoll), 581
Lotrimin; *see* Clotrimazole
Lower motor neuron disease, 162
Lower respiratory tract disease, 48, 57-59, 60-61, 61
LRL view; *see* Left to right lateral view
LS-50 Water Soluble; *see* Lincomycin/spectinomycin
Lugol's iodine, 578
Lumbosacral syndrome, 161-162
Lungs, 49-50, 54, 402, 459-460, 461
Lupron; *see* Leuprolide acetate
Luteinizing hormone (LH), 314, 331
Luteinizing hormone–releasing hormone (LHRH), 314
Luxations, 399
  coxofemoral, 523-524
  elbow joint, 512
  of femoral head, 520
  of pectoral limb, 511-512
  of pelvic limb, 523-525
  scapulohumeral, 512
  shoulder, 512
  stifle, 524-525
Lymphocytosis, 12
Lymphoid tissue, 547-548
Lymphoplasmacytic ganglioneuritis, 86

M

MAC; *see* Minimum anesthetic concentration
Macaw wasting syndrome, 86
Macronutrients, 380-382
MAD; *see* Minimum anesthetic dose
Magnesium hydroxide (Milk of Magnesia), 581
Magnesium sulfate, 229, 581
Magnetic resonance imaging, 421
Magnevist; *see* Gadopentate dimeglumine

Magnification, soft tissue surgery and, 531
Magnum, 326
Maintenance energy (ME), 41
Male infertility, 187
Male reproductive organ enlargement, 87
Malposition, hatching and, 207-212
Mandibular compression, 121
Mandibular diverticulum, 49
Mandibular prognathism, 222
Mannitol, 581
Manufactured diets, 377-378
Manus, 398, 501-502, 510
Marble spleen disease, 78
Masks, inhalation anesthesia and, 476
Masturbation, 187
Mate trauma, 107-108, 120
Maxilla, 121, 221
Maxillary brachygnathism, 222
MDAC; see Multiple dose activated charcoal
ME; see Maintenance energy
Meats, 377
MEC; see Minimum energy cost
Mechanical restraint, 392
Medetomidine (Domitor), 570
Medial thigh, 518
Medications; see Avian medicine
Mediolateral view, 402-403
Medroxyprogesterone, 572, 577
Megabacteria, 76
Megestrol acetate (Ovaban), 577
Melanins, 103
Melanotropic hormone (MSH), 315
Melatonin, 351
Melena, 25-26
Menace response, 268
Meningoencephalitis, 178-179, 551
Mental status, altered, 153
Mercury toxicity, 240
Mesotocin (MT), 315
Metabolic disorders, 80-81, 173-176, 386-387
Metabolic drug scaling, 469-470
Metabolic neuropathies, 386
Metacarpals, fractures of, 510-511

Metal toxicity, 81, 164-165, 235-241
Metamucil; see Psyllium
Metencephalon, 154-155
Methocarbamol (Robaxin), 581
Methoxycyanoacrylate, 363-364
Methoxyflurane, 473
Methylprednisolone acetate, 572
Metoclopramide (Reglan), 581
Metronidazole, 435, 558, 566
MIC, 234
Microbial culture, 269-270
Microbial products, 377
Microbiology, 221
Microchip Transponders, 217
Microfilaria, 115
Microphthalmia, 270-271
Midazolam (Versed), 466-467, 570
Midbrain syndrome, 153-154
Midshaft, fractures of, 504-506
Milbemycin oxime (Interceptor), 566
Milk of Magnesia; see Magnesium hydroxide
Milkweed, 251
Mineral oil, 581
Minerals, 39-40, 203, 376, 377, 379, 578-579
Minimum anesthetic concentration (MAC), 473
Minimum anesthetic dose (MAD), 473
Minimum energy cost (MEC), 469-470
Minocycline (Minocin), 558
Miosis, pupillary, 154
Mites, knemidokoptic, 107, 115, 289, 291, 362, 445-446
Moldy foods or seed, 82
Molt, 336
 French, 109, 110
Monensin (Coban 45), 233, 566
Monocytosis, 12
Morphine, 467
Mortality; see Necropsy
Movement, 3, 5-6
MSH; see Melanotropic hormone
MT; see Mesotocin
Mucomyst; see Acetylcysteine

Mucormycosis, 80
Multifocal syndrome, 148, 151
Multiple dose activated charcoal (MDAC), 229
Muscle relaxants, 478-479
Musculoskeletal system, 219, 220, 224-225
    necropsy and, 544-545
    nutrition and, 373, 384-385
Myambutol; *see* Ethambutol
Mycobacteriosis, 114-115
Mycobutin; *see* Rifabutin
Mycoplasma, 204-205
Mycoses, 79-80
Mycostatin; *see* Nystatin
Mycotoxins, 253-254
Myelencephalon, 155
Myenteric ganglioneuritis, 86

N

Naloxone (Narcan), 570
Naltrexone (Trexan), 577
Narcan; *see* Naloxone
Nares, 6, 48, 51
Nasal cavity, 48
Natal down, 372
Natural incubation, 194-195, 370
Naxcel; *see* Ceftiofur
Nebulization, 44-45, 60, 61
Nebulizing agents, 574
Necropsy, 200-206, 220, 370, 542-552
Needle, infusion, endoscopy and, 451
Negative cultures, 9
Nematodes, 443-445
Nemex; *see* Pyrantel pamoate
Neocalglucon; *see* Calcium glubionate
Neonates, 213-227, 369, 486-487
Neoplasia, 120
    neuropathic syndrome and, 167
    pituitary disorders and, 315
    seizures and, 179-180
    straining and reproductive disorders and, 186
Nephromegaly, 408
Nephron, 65

Nesting birds, 138
Neurohypophysis, 313-314, 315
Neurologic syndromes, 29, 148-169
    nutrition and, 385-386
    parasites and, 426
    vision assessment and, 268
Neuromuscular junctionopathies, 162
Neuronal excitability of brain, 171
Neuropathy, 162-167
    Lafora body, 167, 180
    metabolic, 386
Neurotoxic esterase (NTE), 244
Neurotoxicoses, 252-253
Neutral perch, 129-130
Neutral territory, 129-130
New bird, 4-5
New cage syndrome, 164-165
Newcastle's disease, 178
NFZ 9.2; *see* Nitrofurazone
Nicotine, 206
Nictitans, 264-265
Nightshade, 251
Nitrofurans, 234
Nitrofurazone (NFZ 9.2), 558
Nitrous oxide (N_2O), 475, 489
Nizoral; *see* Ketoconazole
N_2O; *see* Nitrous oxide
Nodules, subcutaneous, parasites and, 426
Nolvadex; *see* Tamoxifen
Noncontinuous breeders, 330
Nonessential amino acids, 381
Nonfeathered skin, masses of, 426
Non-insulin-dependent diabetes, 346-350
Nonisotonic crystalloids, 33
Nonprotein calorie requirement, 42
Nonregenerative anemias, 10-11, 23
Norcuron; *see* Vecuronium bromide
Normograde intramedullary pinning, 519
Nortriptyline (Pamelor), 577
NTE; *see* Neurotoxic esterase
Nuclear scintigraphy, 421-422
Nurse cells, 328
Nursing design, 215
Nutrient pellets, 377

Primary epilepsy, 180-181
Primidone, 172-173
Priscoline; *see* Tolazoline
Probacoll; *see* Lorelco
Procaine, 235
Prodrome, 170
Prognathism, 121, 222
Prolactin, 315
Prolactin-releasing factor, 314
Pronator profundus muscle, 500
Pronator superficialis muscle, 500
Propofol (Rapinovet), 470, 570
Propranolol (Inderal), 581
Propulsid; *see* Cisapride
Propylene glycol, 235
Prostaglandin (PG), 332
Prostaglandin E (PGE), 332, 340-341
Prostaglandin E_2 (PGE_2), 94, 192, 572
Prostaglandin F_2-alpha (PGF_2-alpha), 332, 341
Protein, 42, 381
Protopam; *see* Pralidoxime
Protozoa
  apicomplexan, 439-442
  flagellated, 436-438
Proventer chronic ulcerative dermatitis, 116, 117
Proventricular dilatation disease (PDD), 77, 86, 410, 546
Proventricular hypertrophy, 86
Proventriculotomy, 536-537
Proventriculus, 72-74, 374, 453
Proximal coracoid luxation of pectoral limb, 511
Proximal humerus, 504
Proximal thoracic limb, 496-498
Prozac; *see* Fluoxetine
Pruritus, 426
Pseudolymphoma, 115
Psittaciformes, anesthesia for, 488
Psittacine beak, 362
  and feather disease (PBFD), 13, 77, 108, 109, 110, 111, 121
Psittacine mutilation syndrome, 118
Psittacine proventricular dilation syndrome, 68
Psittacine wasting syndrome, 86
Psychologic disturbances, 140-141
Psychotropic agents, 576-577
Psyllium (Metamucil), 581
Psyllium hydrophilic mucilloid, 229-230
Pterylae, 96
Pterylosis, 96
PTFE; *see* Polytetrafluoroethylene
PTH; *see* Parathyroid hormone
Ptilosis, 96
Puberty, 328
Pubo-ischio-femoralis muscle, 518
Pulmonary air sacs, 50, 403-404
Pulmonary disease, 413
Pulse oximetry, 483
Punishment, 128-129
Pupillary light response (PLR), 154, 268
Pupillary miosis, 154
Pyrantel pamoate (Nemex), 435-436, 567
Pyrethrin powder, 436
Pyrethroids, 246-247
Pyridoxine, 203
Pyridoxine deficiency, 164
Pyrimethamine (Fansidar Tablets), 435, 567

Q

Quadriceps muscle, 517, 518
Quinacrine (Atabrine HCl Tablets), 567

R

Radiation, heat loss and, 45
Radiography, 9, 89, 219-221, 391-394, 409-417, 487
Radiology, 9, 89, 219-221, 391-394, 409-417, 487
Radiosurgical instruments, 529-531
Radius, 499
Range-of-motion exercises, 525
Rapinovet; *see* Propofol
Raptors, anesthesia for, 488
Ratites, anesthesia for, 488-490
Rattlebox, 250
RBC count; *see* Red blood cell count
Rearing methods of neonates, 213-214

Photorefractoriness, 330
Photosensitivity, 330
Phthisis, 271
PHV-1; *see* Pigeon herpesvirus
Physical environment; *see* Environment
Physical examination, 5-13, 18, 217-219
    abdominal distention and, 91-92
    anesthesia and, 464-465
    dehydration and, 31-32
    dermatology and, 95-96
    ophthalmic disorders and, 268
    respiratory system and, 50-52
    soft tissue surgery and, 527
Physical restraint, 392-394, 488-489
Physical therapy, postoperative, 525
Piconavirus, 178
Pigeon herpesvirus (PHV-1), 77, 367
Pigeons, anesthesia for, 488
Pineal gland, 350-351
Pip holes, 210
Piperacillin (Piperacil), 558
Piperazine, 566
Pipping muscle, 113
Pip-to-hatch interval, 209
Pituitary gland, 313-317
Plant toxicoses, 82, 248-254
Plasma, 36
Plasma biochemistries, 593-594
Plasma urea nitrogen, 31
Plasmodium, 79
Pleural cavity, 458
Pleurothotonus, 151
PLR; *see* Pupillary light response
Plucking, feather, 125, 137-141
Plumping, 189
Pneumatized bones, 50, 404
Pneumoconiosis, 461
Pneumonia, 59
PNS; *see* Peripheral nervous system
Pododermatitis, 118, 119-120, 385
Podotheca, 117
Poinsetta, 250
Poison Control Centers, 230
Pokeweed, 250
Polychromasia, 10
Polydioxanone suture, 532
Polydipsia, 65, 374

Polyglactin suture, 531-532
Polyomavirus, 13, 77, 108, 109, 111, 544
Polyostotic hyperostosis, 14, 334
Polytetrafluoroethylene (PTFE), 254
Polyuria, 65, 374
Pontomedullary syndrome, 155-157
Porphyrins, 103
Positioning, patient, 394, 529
Positive cultures, 9
Posterior segment of eye, 265-266, 269
Postictal, 170
Postorbital diverticulum, 48
Postovulatory follicles, 325
Posture of neonate, 217
Postventer chronic ulcerative dermatitis, 116, 117
Potassium, 39
Potatoes, 251
Poultry, anesthesia for, 488
Powder down feathers, 99
Powder-coated seeds, 376
Poximune-C, 365
Poxvirus, 114, 366-367
Pralidoxime (2-PAM; Protopam), 245, 581
Praziquantel (Droncit), 436, 566
Preauditory diverticulum, 48
Precocial species, 213
Prednisolone, 345, 572
Prednisolone Na succinate, 554, 572
Prednisone, 345
Preening, 139
Premedication, anesthesia and, 466-468
Preoperative examination, 527
Preorbital diverticulum, 48, 54
Preovulatory follicles, 325
Preovulatory uterine contractions, 331
Prepatagial chronic ulcerative dermatitis, 116, 117
Prepidil Gel; *see* Dinoprostone
Presurgical examination, 527
Primaquine (Primaquine Phosphate), 435, 567
Primaries, 98
Primary bronchi, 49

Pancreatic neoplasms, 350
Pancreatitis, 550
Panmycin Aquadrops; see Tetracycline
Papilloma, 292
  cloacal, 185-186
  cutaneous, 290
  oropharyngeal, 366
Papillomatosis, 26, 76-77
Papillomavirus, 114
Parabronchi, 49-50
Paradoxic vestibular disease, 158
Paramyxovirus, 178
Paraquat, 206
Parasites, 424-448, 545
  abdominal distention and, 93
  blood, 13
  conjunctival, 283-284, 289
  dermatology and, 106-107
  necropsy and, 205
  oropharynx and, 367
  periorbital swelling and, 285
  regurgitation and, 78-79
  tracheitis and, 58
Parasympatholytics, 467-468
Parathyroid glands, 321-323
Parathyroid hormone (PTH), 321-322
Parathyroid hyperplasia, 550
Parenteral anesthesia, 469-472
Parenteral nutrition, 42-44
Parent-rearing of neonates, 213
Paromomycin sulfate trimethoprim-sulfadiazine, 435
Paromomycin sulfate trimethoprim-sulfamethoxazole, 435
Parrots, 124-147, 488
Pars distalis, 314-315
Pars medialis, 518
Partial seizures, 170-171
Passive range-of-motion exercises, 525
Patella, 513
Patient positioning, 394, 529
Patient preparation, 527-529
PBFD; see Psittacine beak and feather disease
PCV; see Packed cell volume
PDD; see Proventricular dilatation disease

Peanut butter, 581
Pectin, 266, 267, 306
Pectoral limb, 496, 498, 511-512
Pectoral musculature, 7, 495
Pelleted diet, 378
Pelvic girdle and appendages, 398-399, 402-403, 512
Pelvic limb, 8, 512-514, 518-525
Penicillamine (Cuprimine), 575
Penicillin procaine (Penicillin G Procaine), 558
Penicillins, 206, 232
Pentobarbital, 172-173
Perianesthetic monitoring, 479-485
Periorbital abscesses, 284, 288
Periorbital diseases, 278-295
Periorbital hyperplastic lesions, 284, 289-292
Periorbital swelling, 56, 60, 284-289
Peri-pectin hemorrhage, 306
Peripheral nerves, inflammation of, 551
Peripheral nervous system (PNS), 148, 550-551
Peripheral vestibular disease, 157-158
Peritoneal cavities, 457
Peritonitis, egg, 28
Peroneus longus muscle, 517
Pesticides, 81, 165-166, 242-248
Pet parrots, behavior problems in, 124-147
Petroleum products, 205, 255-256
PG; see Prostaglandin
PGE; see Prostaglandin E
PGE_2; see Prostaglandin E_2
PGF_2-alpha; see Prostaglandin F_2-alpha
Phalanges, fractures of, 522-523
Pharynx, 223
Phenobarbital, 172-173, 570, 577
Phenothiazine tranquilizers, 466
Phenylbutazone, 572
Pheochromocytoma, 344
Philodendron, 249
Phobias, 125, 142-143
Photography, still, 462
Photoperiod, 329-330, 331
Photoreceptors, extraretinal, 330-331

Nutrition, 2, 120, 214-215, 216, 369-390, 578-579
   dermatology and, 106, 116
   disorders of, 369-370, 372-375, 382-387
   infertility and, 194
   necropsy and, 203
   neuropathic syndrome and, 163-164
   pet parrots and, 144
   seizures and, 175
   shock and, 40-44
   toxicity and, 82
   vomiting and regurgitation and, 83
Nutritional hyperparathyroidism, 322
Nutritional myopathy, 546
Nylon suture, 532
Nystatin (Mycostatin), 562

O

Oak, 249
Obesity, 87, 384
Obstructive lesions, 83
Oculomotor nerve, 154
Oculonasal discharge, 4
Oil seeds, 376
Old bird, 5
Oleander, 232-233, 250-251
Omnipaque; see Iohexol
Omnizole; see Thiabendazole
Omphalitis, 221
Oncovin; see Vincristine sulfate
One-person bird, 125, 134-136
Oophoritis, 550
Ophthalmic disorders, 264-312, 373, 487
Ophthalmic instruments, 529
Opioid analgesics, 467
Opisthotonus, 153
Opportunistic breeders, 330
Optic disc, 265
Oral cavity, 70-71
Oral route, 35
Orbit, diseases of, 270-278
Orbital abscess, 284
Orbital inflation, 284
Orchidectomy, 329
Organomercurial toxicity, 243
Organophosphates, 81, 177-178, 243-245
Organophosphorus insecticides, 205
Oropharyngeal abscesses, 365-366
Oropharyngeal papillomas, 366
Oropharynx, 6, 51, 365-368, 452, 453
Orthopedics, 493
Orthostatic hypotension, 480-481
Osteodystrophy, 14-15
Osteomalacia, 14-15, 385, 545
Osteomyelosclerosis, 334
Osteoporosis, 15
Ostrich, anesthesia for, 489-490
Otitis externa, 551
Otitis interna, 552
Ovaban; see Megestrol acetate
Ovarian cysts, 338
Ovarian pocket, 189
Ovariectomy, 326
Ovary, 189, 324-325
Overbonding, 125, 134-136
Overdependency, 125, 136-137
Oviduct, 189, 190-192, 325-327
Oviductal prolapse, 27-28
Oviposition, 332-333
Ovulation, 189, 331
Ovulation-oviposition cycle, 333
Owner, 2-3, 5, 125, 126, 134-136
Oxygen cages, 44
Oxygen therapy, 44
Oxytetracycline (Liquamycin; Terramycin Soluble Powder), 234, 558, 574
Oxytocin, 341, 572

P

Pacheco's disease, 13, 77, 86
Packed cell volume (PCV), 10, 11, 32
Paired choanae, 49
Palpation, 6-8, 89
2-PAM; see Pralidoxime
Pamelor; see Nortriptyline
Pancreas, 75, 81, 345-350, 546-547
Pancreatic enzyme powder, 578

Stifle luxation, 524-525
Stifle problems, 224-225
Still photography, 462
Stomatitis, 367-368
Stool, 9, 66-69, 426-434
Strabismus, dorsolateral, 154
Straining and reproductive disorders, 183-188
Straw feathering, 105
Streptococci, 204
Streptomycin (Streptomycin Sulfate), 559
Stress, 140
Stress bars, 372
Stress marks, 343
Strigiformes, anesthesia for, 488
Struthioniformes, anesthesia for, 488-490
Strychnine, 248
Stunting, parasites and, 426
Subcutaneous emphysema, 61
Subcutaneous nodules, parasites and, 426
Subcutaneous (SC) route, 34-35
Suborbital diverticulum, 48
Sucralfate (Carafate), 581
Sulfa drugs, 206
Sulfachlorpyridazine (Vetasulid), 559, 567
Sulfadimethoxine (Albon), 568, 574
Sulfadimethoxine/ormetoprim (Rofemaid), 568
Sulfamethazine (Sulmet), 568
Sulfaquinoxaline (Sulquin 6-50), 568
Sulfonamides, 234
Sulmet; *see* Sulfamethazine
Sulquin 6-50; *see* Sulfaquinoxaline
Supinator muscle, 499
Supplemental heat, 45
Supportive care and shock, 17-46
Supracoracoideus muscle, 495
Surgical drapes, 529
Surgical sexing, 488
Surgitron unit, 530
Survival skills, 141
Sustentacular cells, 328
Suture materials, 531-532
Swallowing, 71
Swelling, periorbital, 56, 60, 284-289
Syngamus trachea, 444-445
Syrinx, 49

T

T_3; *see* Triiodothyronine
T_4; *see* Thyroxine
Table foods, 377
Tagamet; *see* Cimetadine
Tail bobbing, 8, 52
Tamoxifen (Nolvadex), 572
Tarsometatarsus, 514, 522-523
Tarsus pad, 114
TBZ; *see* Thiabendazole
Tea, 252
Teflon toxicity, 548-549
Tegretol; *see* Carbamazepine
Telazol; *see* Tiletamine/zolazepam
Temperate climate, 329-330
Temperature, 45, 195-196, 484, 527-528
Tenesmus, 183-186
Tensor propatagialis muscle, 496-497
Terbutilene (Brethine), 581
Terminal oviposition, 333
Terramycin Soluble Powder; *see* Oxytetracycline
Territoriality, cage, 125, 134
Tertiary bronchi, 49-50
Testis, 327-335
Testosterone, 314, 329, 572
Tetany, hypocalcemic, 385
Tetracycline (Panmycin Aquadrops; Tetracycline Soluble Powder), 206, 234, 559
Thermal burns, 23-24
Thiabendazole (Omnizole; Thibenzole; TBZ), 568
Thiamine, 578
  deficiency of, 163-164
Thibenzole; *see* Thiabendazole
Thigh, 514, 518
Thirst, 3
Thoracic air sac, 454, 456, 461
Thoracic girdle and appendages, 398, 401, 493-496
Thoracic limb, 8, 496-498, 502-511
Thoracolumbar syndrome, 161

Thrombocytes, avian, 12
Thrombocytopenias, 12
Through-the-lens (TTL) metering system, 462
Thyroglobulin, 318
Thyroid dysplasia, 549
Thyroid glands, 80-81, 317-321
Thyroid hyperplasia, 80-81, 550
Thyroid-stimulating hormone (TSH), 314, 315, 320
Thyrotropin-releasing hormone (TRH), 314, 320
Thyroxine (T_4), 105, 314
l-Thyroxine; *see* Levothyroxine
Tiamulin (Denegard), 559
Tibialis cranialis muscle, 517
Tibiotarsus, 513-514, 520-521
Ticarcillin, 559
Ticarcillin/clavulinic acid (Timetin), 559
Tiletamine/zolazepam (Telazol), 571
Timetin; *see* Ticarcillin/clavulinic acid
Tissue healing, 532-533
Tobacco, 252-253
Tobramycin (Tobramycin Sulfate, Injectable), 559
Tocopherol, 163
Toe constrictions, 225
Toenail clip, 10
Toes, 225, 226, 373
Tolazoline (Priscoline), 571
Toltrazuril (Baycox), 568
Tomia, 70
Tongue, 70
Torbugesic; *see* Butorphanol
Total body water, 31
Total protein level, 31
Toxicity; *see* Avian toxicology
Toxoplasma gondii, 434, 441
Toys, 139-140
Trace minerals, 376, 377, 379
Trachea, 7, 49, 52, 60-61, 399
Tracheal lavage, 53
Tracheal mites, 445
Tracheal mucosa, 453
Tracheitis, 58
Training, dominance, 126-127, 129-131, 132, 133-142

Training perch, moving, 130-131
Tramisol; *see* Levamisole
Tranquilizers, phenothiazine, 466
Transarticular external skeletal fixators, 507, 524
Transfusions, blood, 36-37, 482
Transient diabetes mellitus, 347
Trauma, 30
   to beak, 120
   head, 29
   mate, 107-108, 120
   necropsy and, 545
   neuropathic syndrome and, 166-167
   seizures and, 179
Trematodes, 79, 442
Trephination, sinus, 539-540
Trexan; *see* Naltrexone
TRH; *see* Thyrotropin-releasing hormone
Triage, 17-18
Triamcinolone, 572
Triceps brachii muscle, 498
Trichomonas gallinae, 78, 367, 438
Trichomoniasis, 78
Trigeminal nerve, 156
Triiodothyronine (T_3), 314
Trimethoprim, 559
Trimethoprim-sulfadiazine, 435
Trimethoprim/sulfamethoxazole (Bactrim; Septra), 435, 559
Trimethoprim-sulfonamides, 234
Triosseal canal, 398
Trochlear nerve, 154
TSH; *see* Thyroid-stimulating hormone
TTL metering system; *see* Through-the-lens metering system
Turgescence, 31
Tylosin (Tylan; Tylan Soluble), 559, 574

U

Ulcerative lesions, 116
Ulcers, gastric, 83
Ulna, 499
Ulnaris lateralis, 499-500
Ulnometacarpalis dorsalis muscle, 501-502

Ultane; *see* Sevaflurane
Ultimobranchial glands, 323
Ultrasonic Doppler flow apparatus, 480
Ultrasonography, 89-90, 420
Upper airway obstruction, 487-488
Upper respiratory tract disease, 48, 55-56, 59-60
Urates, abnormal, 65-66, 68-69, 374
Uremic encephalopathy, 176
Uric acid, 15
Urinary system, 549
Urine, red, 374
Urogenital emergencies, 27-28
Urography, 419
Uropygial gland, 113, 384-385, 395
Uterine contractions, preovulatory, 331
Uterus, 189, 326
Uveitis, 301-305

V

Vagina, 326
Vaginal sphincter, 189
Vagus nerve, 156-157
Valbazen; *see* Albendazole suspension
Valgus deformity, 225
Vapors, 254-256
Vascular access, 43
Vasoactive intestinal peptide (VIP), 335
Vasogenic shock, 20
VD view; *see* Ventrodorsal view
Vecuronium bromide (Norcuron), 581
Vegetables, nutrition and, 376
Velogenic viscerotropic Newcastle's disease, 178
Venous collections sites, 9-10
Vent, 8
Vent suturing, 537
Ventilation, 483
Ventral antebrachium, 500-501
Ventral hepatic peritoneal cavity, 457
Ventral manus, 502
Ventral metencephalon, 155
Ventral wing, 497
Ventriculus, 72-74, 374, 453, 460-462
Ventrodorsal (VD) view, 393, 394, 398, 400-401, 410, 412, 413, 414-415, 416, 417
Versed; *see* Midazolam
Vestibular syndrome, 157-158
Vestibulocochlear nerve, 156
Vetasulid; *see* Sulfachlorpyridazine
Veterinarian, anesthesia administration by, 464
Vibramycin; *see* Doxycycline
Vibravenos; *see* Doxycycline
Videotaping, 463
Vincristine sulfate (Oncovin), 581
Violence, 132
VIP; *see* Vasoactive intestinal peptide
Viral encephalitis, 550-551
Viruses, 12, 58, 76-78, 204
Vision; *see* Eye
Vital E- A+D, 578
Vitamin(s), 39-40, 43, 116, 376, 377, 378
Vitamin A, 39, 203, 234-235, 365, 578
  deficiency of, 106, 116, 290, 291-292, 365-366, 367, 382, 384-385
Vitamin B_1, 578
  deficiency of, 163-164
Vitamin B_2, 203
  deficiency of, 164, 370
Vitamin B_6, 203
  deficiency of, 164
Vitamin B_{12}, 578
Vitamin B complex, 39
Vitamin C, 578
Vitamin D_3, 39, 203, 235, 578
Vitamin E, 39, 578
  deficiency of, 163, 545
Vitamin E/selenium (E-Se), 579
Vitamin K_1, 39, 579
Vitamin toxicities, 230-235
Vitteline membrane, 193
Voice, 4
Voice box, 49
Volatile household compounds, 254
Vomiting and regurgitation, 25, 70-83

W

Wallerian "dying back" degeneration of axons, 244
Water, 2, 31, 214-215, 371, 382
Water deprivation test, 316
WBC count; *see* White blood cell count
Weaning, 216, 370, 371
Weight, egg, 198-199, 207
Wet bulb readings, 196, 197
White blood cell (WBC) count, 10, 11-12, 591
White uric acid, 4
Wild birds, 425
Wild pea, 250
Wing, 97, 497, 501-502
WinstrolV; *see* Stanozolol
Workup plan, diagnostic, 1-16, 47, 48
Wound healing, 533

X

Xanthoma, 115
Xanthophyll, 103, 383
Xylazine (Rompun), 232, 466, 489, 570, 571
Xylazine-ketamine, 471-472

Y

Yew, 251
Yohimbine (Yobine), 571
Yolk, 189, 190, 325
Yolk sac, 211-212, 221

Z

Zearalenone, 253-254
Zinc phosphide, 248
Zinc sulfate, 428
Zinc toxicity, 164-165, 241
Zithromax; *see* Azithromycin
Zolazepam, 466-467
Zolazepam-tiletamine, 472, 489
Zorivax; *see* Acyclovir

polydipsia — extreme drinking
neoplasia — ~~tumor~~ abnormal cell growth
iatrogenic — med./vet caused it
melena — dark, tarry stools — blood in stool altered by intestinal juices
tenesmus — painful anal spasm, straining, little stool
paresis — partial paralysis
tachypnea — rapid breathing
hypo (or per) kalemic — potassium in the blood
Leukopenia — lower than normal leukocytes
hematochezia — blood in stool
parenteral — by means other than GI
Idiopathic — unknown cause, disease of

618 *Index*

Schiff-Sherrington posture, 161
Schiötz tonometer, 270
Schirmer tear test, 270
Scintigraphy, nuclear, 421-422
Scissors, endoscopy and, 451
Scissors beak, 221, 364
Scoliosis, 220
Screaming, 124, 133, 188
Scurvy, 387
Seasonal breeders, 330
Secondaries, 98
Secondary bronchi, 49
Secondary nutritional hyperparathyroidism, 322
Seeds
    moldy, 82
    nutrition and, 375-377
    in stool, 68
Seizures, 29, 170-182, 385
Selenium toxicity, 203, 240-241
Self-mutilation, 107, 112, 383, 426
Semimembranosus muscle, 516
Seminal sac, 328
Seminiferous tubules, 327-328
Semiplume feathers, 98
Semitendinosus muscle, 516
Senses, necropsy and, 551-552
Septic shock, 20
Septicemia, 12
Septra; *see* Trimethoprim/sulfamethoxazole
Serologic tests, 433-434
Sertoli's cells, 328
Sevaflurane (Ultane), 474-475, 571
Sex glands, accessory, 328
Sexual dimorphism, 336
Sexual immaturity, 143
Shaft of feather, 100
Sheather's sugar solution, 428
Shell gland, 189, 326
Shock, supportive care and, 17-46
Shoulder luxations, 512
Shouldering, 130
Signalment, 183, 335-336
Silver nitrate, 363
Sinequan; *see* Doxepin
Sinus, infraorbital, 54, 266-267
Sinus aspiration, 53, 54

Sinus surgery, 539-540
Sinus trephination, 539-540
Sinusitis, 56, 277-278, 284, 288
Skeletal system, 394-399
Skin, 95-123, 218, 372, 382, 426, 528
Skin fold elasticity, 31
Skull, 394-395
Smadels' disease, 367
Small intestine, 74
Sneezing, 4
Socialization, 141
Sodium bicarbonate, 554, 581
Sodium iodide 20%, 578
Sodium nitrate, 428-429
Sodium sulfate (GoLytely), 229, 575, 581
Soft tissue surgery, 527-541
Solvents, 254-256
Somatostatin, 314, 349-350
Somatotrophic hormone (STH), 314
Spartrix; *see* Carnidazole
Specimen collection, 458-462
Spectinomycin (Spectogard), 559, 574
Spinal accessory nerve, 156-157
Spinal cord syndromes, 148, 158-160
Spine, 395, 397, 398
Spiropter bacteria, 79
Spiruroidea, 444
Splay leg, 224
Splayed toes, 373
Spleen, 406-407
Splenomegaly, 407, 411
Splinting, coaptation, 511
Sporanox; *see* Itraconazole
Spraddle leg, 224
Sprains, 399
Spurs, 117
Squamous cell carcinoma, 115
Staged destabilization, 525
Stainless steel wire suture, 532
Stanozolol (WinstrolV), 572
Staphylococci, 203-204
Starvation, 16
Sternostoma tracheacolum, 445
Sternum, 493-494
Steroids, 173, 571-572
STH; *see* Somatotrophic hormone

Recovery from anesthesia, 485-486
Rectrices, 98
Red blood cell (RBC) count, 10-11
Regenerative anemias, 11, 23
Reglan; see Metoclopramide
Regurgitation, 4, 25, 70-83, 187, 223-224
Remiges, 98
Renal disease, 27
Renal failure, 27
Renal mineralization, 416
Renal system, 334, 374
Renal transport, 334
Renomegaly, 414-415
Reprimand, 132
Reproduction, 81, 330, 351
Reproductive disorders, 183-188, 550
Reproductive hormones, 187-188
Respiration rate, 482-483
Respiratory diagnostic workup plan, 47
Respiratory disease, 47-61
   anesthesia and, 487-488
   dyspnea and, 19, 47-61
   emergencies of, 19
Respiratory monitoring, 482-484
Respiratory recovery time, 8
Respiratory system, 8
   disease of; see Respiratory disease
   imaging of, 399-404
   necropsy and, 548-549
   ophthalmic disorders and, 287
   oxygen therapy and, 44
   parasites and, 426
   periorbital swelling and, 287
   soft tissue surgery and, 528
Resting heart rate, 479
Restraint, 41
   chemical, 569-571
   physical, 392-394, 488-489
Resuscitation, 485
Retained egg, 417
Retina, 265
Retinal detachment, 306
Retrobulbar abscess, 287
Retrobulbar hematomas, 289

Retrograde intramedullary pinning, 519
Rewards, 128-129
Rheiformes, anesthesia for, 488-490
Rhinitis, 56, 59-60, 284
Rhinorrhea, chronic, 361
Rhinotheca, 120
Riboflavin, 203
   deficiency of, 164, 370
Ribs, 494
Rickets, 14, 545
Ridzol 10%; see Ronidazole
Rifabutin (Mycobutin), 558
Rifampin (Rifadin), 558
Rigid endoscopy, 449, 450
Rima glottis, 49
Rimantadine (Flumadine), 563
Robaxin; see Methocarbamol
Robinul-V; see Glycopyrrolate
Rocephin; see Ceftriaxone
Rodenticides, 247-248
Rofemaid; see Sulfadimethoxine/ormetoprim
Roferon-A Injection, 563
Rompun; see Xylazine
Ronidazole (Ridzol 10%), 567
Rostral diverticulum, 48
Rostrum, 6, 120-122, 221-222, 359-365
Runners, 109

S

Sacculitis, 403-404, 459, 460
Saffan; see Alphaxalone/alphadalone
Salinomycin, 233
Salivary glands, 71
Salmonellae, 204
Salpingectomy, 326
Salpingitis, 550
Salpingohysterectomy, 538-539
Sandostatin, 349
Sarcocystis infection, 441-442, 544
SC route; see Subcutaneous route
Scales, 117
Scapula, 494, 503
Scapulohumeral luxations, 512
Scavenge system, inhalation anesthesia and, 476-477